T0317758

A FIRST COURSE IN MATHEMATICAL LOGIC AND SET THEORY

A FIRST COURSE IN MATHEMATICAL LOGIC AND SET THEORY

Michael L. O'Leary
College of DuPage

For general information on our other products and services or for technical support, please contact our Customer Care
Department within the United States at (800) 762-2974, outside the United States at (317) 572-3993 or fax (317) 572-
4002.

Wiley also publishes its books in a variety of electronic formats. Some content that appears in print
may not be available in electronic formats. For more information about Wiley products, visit our web site at www.wiley.com.

Library of Congress Cataloging-in-Publication Data applied for.

ISBN: 9780470905883

For my parents

CONTENTS

PREFACE

This book is inspired by *The Structure of Proof: With Logic and Set Theory* published by Prentice Hall in 2002. My motivation for that text was to use symbolic logic as a means by which to learn how to write proofs. The purpose of this book is to present mathematical logic and set theory to prepare the reader for more advanced courses that deal with these subjects either directly or indirectly. It does this by starting with propositional logic and first-order logic with sections dedicated to the connection of logic to proof-writing. Building on this, set theory is developed using first-order formulas. Set operations, subsets, equality, and families of sets are covered followed by relations and functions. The axioms of set theory are introduced next, and then sets of numbers are constructed. Finite numbers, such as the natural numbers and the integers, are defined first. All of these numbers are actually sets constructed so that they resemble the numbers that are their namesakes. Then, the infinite ordinal and cardinal numbers appear. The last chapter of the book is an introduction to model theory, which includes applications to abstract algebra and the proofs of the completeness and compactness theorems. The text concludes with a note on Gödel's incompleteness theorems.

MICHAEL L. O'LEARY

Glen Ellyn, Illinois
July 2015

ACKNOWLEDGMENTS

Thanks are due to Susanne Steitz-Filler, Senior Editor at Wiley, for her support of this project. Thanks are also due to Sari Friedman and Katrina Maceda, both at Wiley, for their work in producing this book. Lastly, I wish to thank the anonymous reviewer whose comments proved beneficial.

On a personal note, I would like to express my gratitude to my parents for their continued caring and support; to my brother and his wife, who will make sure my niece learns her math; to my dissertation advisor, Paul Eklof, who taught me both set theory and model theory; to Robert Meyer, who introduced me to symbol logic; to David Elfman, who taught me about logic through programming on an Apple II; and to my wife, Barb, whose love and patience supported me as I finished this book.

SYMBOLS

Symbol	Page(s)	Symbol	Page(s)
:=	7	=	68, 226
¬	7, 11	A	68
∧	7, 11	ST	68
∨	7, 11	NT	69
→	7, 12	AR	69
↔	7, 13	AR'	69
v	10	GR	69
⊨	21, 336, 346, 352	RI	69
⊭	23, 338	OF	69
⇒	24	TERMS(A)	70
⊢	26	S	70
⇔	34	L(S)	73
⊢$_*$	40	$\dfrac{t}{x}$	75
⊬	52	$p(x)$	80
Con	52, 395	$\hat{a}$	88
VAR	67	|	98
∀	68, 72	$|c|$	116, 307
∃	68, 72		

A First Course in Mathematical Logic and Set Theory, First Edition. Michael L. O'Leary.
© 2016 John Wiley & Sons, Inc. Published 2016 by John Wiley & Sons, Inc.

Symbol	Page(s)	Symbol	Page(s)
CH	313	$\mathfrak{A} \to \mathfrak{B}$	375
GCH	314	$\ker(\psi)$	379
$\beth$	316	$\leq$	389
cf	328	$\underset{\sim}{\prec}$	393
$\mathfrak{A}$	333	$\mathfrak{A}_{\sim}$	404
$(A, \mathfrak{a})$	333	$\mathrm{Th}(\mathfrak{A})$	408
I_x^a	336	S	410
Sat_S	340	P	410
$\mathrm{M}_n(\mathbb{R})$	345	$\mathfrak{P}$	410
$\mathrm{M}_n^*(\mathbb{R})$	345	PA	411
$\mathrm{GL}(n, \mathbb{R})$	345	$\mathfrak{P}'$	411
Sat	352	V_α	417
$\mathfrak{M}_n(\mathbb{R})$	355	V	419
$\langle a \rangle$	363	$\mathrm{TC}(A)$	419
$\bigcup_{\gamma \in \kappa} \mathfrak{A}_\gamma$	372	$\mathfrak{B}_\alpha$	420
$\varphi : \mathfrak{A} \to \mathfrak{B}$	375		

CHAPTER 1

PROPOSITIONAL LOGIC

1.1 SYMBOLIC LOGIC

Let us define **mathematics** as the study of number and space. Although representations can be found in the physical world, the subject of mathematics is not physical. Instead, mathematical objects are abstract, such as equations in algebra or points and lines in geometry. They are found only as ideas in minds. These ideas sometimes lead to the discovery of other ideas that do not manifest themselves in the physical world as when studying various magnitudes of infinity, while others lead to the creation of tangible objects, such as bridges or computers.

Let us define **logic** as the study of arguments. In other words, logic attempts to codify what counts as legitimate means by which to draw conclusions from given information. There are many variations of logic, but they all can be classified into one of two types. There is **inductive logic** in which if the argument is good, the conclusion will probably follow from the hypotheses. This is because inductive logic rests on evidence and observation, so there can never be complete certainty whether the conclusions reached do indeed describe the universe. An example of an inductive argument is:

A First Course in Mathematical Logic and Set Theory, First Edition. Michael L. O'Leary.

> A red sky in the morning means that a storm is coming.
> We see a red sky this morning.
> Therefore, there will be a storm today.

Whether this is a trust-worthy argument or not rests on the strength of the predictive abilities of a red sky, and we know about that by past observations. Thus, the argument is inductive. The other type is **deductive logic**. Here the methods yield conclusions with complete certainty, provided, of course, that no errors in reasoning were made. An example of a deductive argument is:

> All geometers are mathematicians.
> Euclid is a geometer.
> Therefore, Euclid is a mathematician.

Whether Euclid refers to the author of the *Elements* or is Mr. Euclid from down the street is irrelevant. The argument works because the third sentence must follow from the first two.

As anyone who has solved an equation or written a proof can attest, deductive logic is the realm of the mathematician. This is not to say that there are not other aspects to the discovery of mathematical results, such as drawing conclusions from diagrams or patterns, using computational software, or simply making a lucky guess, but it is to say that to accept a mathematical statement requires the production of a deductive proof of that statement. For example, in elementary algebra, we know that given

$$2x - 5 = 11,$$

we can conclude

$$2x = 6$$

and then

$$x = 3.$$

As each of the steps is legal, it is certain that the conclusion of $x = 3$ follows. In geometry, we can write a two-column proof that shows that

$$\angle B \cong \angle D$$

is guaranteed to follow from

> *ABCD is a parallelogram.*

The study of these types of arguments, those that are deductive and mathematical in content, is called **mathematical logic**.

Propositions

To study arguments, one must first study sentences because they are the main parts of arguments. However, not just any type of sentence will do. Consider

> *all squares are rectangles.*

The purpose of this sentence is to affirm that things called squares also belong to the category of things called rectangles. In this case, the assertion made by the sentence is correct. Also, consider,

circles are not round.

This sentence denies that things called circles have the property of being round. This denial is incorrect. If a sentence asserts or denies accurately, the sentence is **true**, but if it asserts or denies inaccurately, the sentence is **false**. These are the only **truth values** that a sentence can have, and if a sentence has one, it does not have the other. As arguments intend to draw true conclusions from presumably true given sentences, we limit the sentences that we study to only those with a truth value. This leads us to our first definition.

■ **DEFINITION 1.1.1**

A sentence that is either true or false is called a **proposition**.

Not all sentences are propositions, however. Questions, exclamations, commands, or self-contradictory sentences like the following examples can neither be asserted nor be denied.

- *Is mathematics logic?*

- *Hey there!*

- *Do not panic.*

- *This sentence is false.*

Sometimes it is unclear whether a sentence identifies a proposition. This can be due to factors such as imprecision or poor sentence structure. Another example is the sentence

it is a triangle.

Is this true or false? It is impossible to know because, unlike the other words of the sentence, the meaning of the word *it* is not determined. In this sentence, the word *it* is acting like a variable as in $x + 2 = 5$. As the value of x is undetermined, the sentence $x + 2 = 5$ is neither true nor false. However, if x represents a particular value, we could make a determination. For example, if $x = 3$, the sentence is true, and if $x = 10$, the sentence is false. Likewise, if *it* refers to a particular object, then *it is a triangle* would identify a proposition.

There are two types of propositions. An **atom** is a proposition that is not comprised of other propositions. Examples include

the angle sum of a triangle equals two right angles

and

some quadratic equations have real solutions.

A proposition that is not an atom but is constructed using other propositions is called a **compound proposition**. There are five types.

- A **negation** of a given proposition is a proposition that denies the truth of the given proposition. For example, the negation of $3 + 8 = 5$ is $3 + 8 \neq 5$. In this case, we say that $3 + 8 = 5$ has been **negated**. Negating the proposition *the sine function is periodic* yields *the sine function is not periodic.*

- A **conjunction** is a proposition formed by combining two propositions (called **conjuncts**) with the word *and*. For example,

    the base angles of an isosceles triangle are congruent,
    and a square has no right angles

    is a conjunction with *the base angles of an isosceles triangle are congruent* and *a square has no right angles* as conjuncts.

- A **disjunction** is a proposition formed by combining two propositions (called **disjuncts**) with the word *or*. The sentence

    the base angles of an isosceles triangle are congruent,
    or a square has no right angles

    is a disjunction.

- An **implication** is a proposition that claims a given proposition (called the **antecedent**) entails another proposition (called the **consequent**). Implications are also known as **conditional propositions**. For example,

    if rectangles have four sides, then squares have for sides   (1.1)

    is a conditional proposition. Its antecedent is *rectangles have four sides*, and its consequent is *squares have four sides*. This implication can also be written as

    rectangles have four sides implies that squares have four sides,

    squares have four sides if rectangles have four sides,

    rectangles have four sides only if squares have four sides,

    and

    if rectangles have four sides, squares have four sides.

    A conditional proposition can also be written using the words *sufficient* and *necessary*. The word *sufficient* means "adequate" or "enough," and *necessary* means "needed" or "required." Thus, the sentence *rectangles having four sides is sufficient for squares to have four sides*

translates (1.1). In other words, the fact that rectangles have four sides is enough for us to know that squares have four sides. Likewise,

squares having four sides is necessary for rectangles to have four sides

is another translation of the implication because it means that squares must have four sides because rectangle have four sides. Summing up, the antecedent is sufficient for the consequent, and the consequent is necessary for the antecedent.

- A **biconditional proposition** is the conjunction of two implications formed by exchanging their antecedents and consequents. For example,

if rectangles have four sides, then squares have four sides,
and if squares have four sides, then rectangles have four sides.

To remove the redundancy in this sentence, notice that the first conditional can be written as

rectangles have four sides only if squares have four sides

and the second conditional can be written as

rectangles have four sides if squares have four sides,

resulting in the biconditional being written as

rectangles have four sides if and only if squares have four sides

or the equivalent

rectangles having four sides is necessary and sufficient
for squares to have four sides.

Propositional Forms

As a typical human language has many ways to express the same thought, it is beneficial to study propositions by translating them into a notation that has a very limited collection of symbols yet is still able to express the basic logic of the propositions. Once this is done, rules that determine the truth values of propositions using the new notation can be developed. Any such system designed to concisely study human reasoning is called a **symbolic logic**. Mathematical logic is an example of symbolic logic.

Let p be a finite sequence of characters from a given collection of symbols. Call the collection an **alphabet**. Call p a **string** over the alphabet. The alphabet chosen so that p can represent a mathematical proposition is called the **proposition alphabet** and consists of the following symbols.

- **Propositional variables**: Uppercase English letters, $P, Q, R, \ldots$, or uppercase English letters with subscripts, $P_n, Q_n, R_n, \ldots$, where $n = 0, 1, 2, \ldots$

- **Connectives**: $\neg, \wedge, \vee, \rightarrow, \leftrightarrow$

- **Grouping symbols**: (,), [,].

The sequences $P \vee Q$ and $P_1 Q_1 \wedge \leftrightarrow ((($ and the **empty string**, a string with no characters, are examples of strings over this alphabet, but only certain strings will be chosen for our study. A string is selected because it is able to represent a proposition. These strings will be determined by a method called a **grammar**. The grammar chosen for our present purposes is given in the next definition. It is given **recursively**. That is, the definition is first given for at least one special case, and then the definition is given for other cases in terms of itself.

■ **DEFINITION 1.1.2**

A **propositional form** is a nonempty string over the proposition alphabet such that

- every propositional variable is a propositional form.

- $\neg p$ is a propositional form if p is a propositional form.

- $(p \wedge q)$, $(p \vee q)$, $(p \rightarrow q)$, and $(p \leftrightarrow q)$ are propositional forms if p and q are propositional forms.

We follow the convention that parentheses can be replaced with brackets and outermost parenthesis or brackets can be omitted. As with propositions, a propositional form that consists only of a propositional variable is an **atom**. Otherwise, it is **compound**.

The strings P, Q_1, $\neg P$, $(P_1 \vee P_2) \wedge P_3$, and $(P \rightarrow Q) \wedge (R \leftrightarrow \neg P)$ are examples of propositional forms. To prove that the last string is a propositional form, proceed using Definition 1.1.2 by noting that $(P \rightarrow Q) \wedge (R \leftrightarrow \neg P)$ is the result of combining $P \rightarrow Q$ and $R \leftrightarrow \neg P$ with $\wedge$. The propositional form $P \rightarrow Q$ is from P and Q combined with $\rightarrow$, and $R \leftrightarrow \neg P$ is from R and $\neg P$ combined with $\leftrightarrow$. These and $\neg P$ are propositional forms because P, Q, and R are propositional variables. This derivation yields the following **parsing tree**:

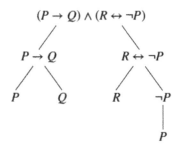

The parsing tree yields the **formation sequence** of the propositional form:

$$P, Q, R, \neg P, P \to Q, R \leftrightarrow \neg P, (P \to Q) \wedge (R \leftrightarrow \neg P).$$

The sequence is formed by listing each distinct term of the tree starting at the bottom row and moving upwards.

■ **EXAMPLE 1.1.3**

Make the following **assignments**:

$$p := R \leftrightarrow (P \wedge Q),$$
$$q := (R \leftrightarrow P) \wedge Q.$$

The symbol $:=$ indicates that an assignment has been made. It means that the propositional form on the right has been assigned to the lowercase letter on the left. Using these designations, we can write new propositional forms using p and q. The propositional form $p \wedge q$ is

$$[R \leftrightarrow (P \wedge Q)] \wedge [(R \leftrightarrow P) \wedge Q]$$

with the formation sequence,

$$P, Q, R, P \wedge Q, R \leftrightarrow P,$$
$$R \leftrightarrow (P \wedge Q), (R \leftrightarrow P) \wedge Q, [R \leftrightarrow (P \wedge Q)] \wedge [(R \leftrightarrow P) \wedge Q],$$

and $\neg q \to p$ is

$$\neg[(R \leftrightarrow P) \wedge Q] \to [R \leftrightarrow (P \wedge Q)]$$

with the formation sequence

$$P, Q, R, R \leftrightarrow P, P \wedge Q, (R \leftrightarrow P) \wedge Q, R \leftrightarrow (P \wedge Q),$$
$$\neg[(R \leftrightarrow P) \wedge Q], \neg[(R \leftrightarrow P) \wedge Q] \to [R \leftrightarrow (P \wedge Q)].$$

Interpreting Propositional Forms

Notice that determining whether a string is a propositional form is independent of the meaning that we give the symbols. However, as we do want these symbols to convey meaning, we assume that the propositional variables represent atoms and set this interpretation on the connectives:

$\neg$	not
$\wedge$	and
$\vee$	or
$\to$	implies
$\leftrightarrow$	if and only if

Because of this interpretation, name the compound propositional forms as follows:

$$\neg p \quad \text{negation}$$
$$p \wedge q \quad \text{conjunction}$$
$$p \vee q \quad \text{disjunction}$$
$$p \rightarrow q \quad \text{implication}$$
$$p \leftrightarrow q \quad \text{biconditional}$$

■ EXAMPLE 1.1.4

To see how this works, assign some propositions to some propositional variables:

$P :=$ *The sine function is not one-to-one.*

$Q :=$ *The square root function is one-to-one.*

$R :=$ *The absolute value function is not onto.*

The following symbols represent the indicated propositions:

- $\neg R$

  The absolute value function is onto.

- $\neg P \vee \neg Q$

  The sine function is one-to-one, or the square root function is not one-to-one.

- $Q \rightarrow R$

  If the square root function is one-to-one, the absolute function is not onto.

- $R \leftrightarrow P$

  The absolute value function is not onto if and only if the sine function is not one-to-one.

- $P \wedge Q$

  The sine function is not one-to-one, and the square root function is one-to-one.

- $\neg P \wedge Q$

  The sine function is one-to-one, and the square root function is one-to-one.

- $\neg(P \wedge Q)$

  It is not the case that the sine function is not one-to-one and the square root function is one-to-one.

The proposition

> *the absolute value function is not onto if and only if*
> *both the sine function is not one-to-one and the square root function is one-to-one*

is translated as $R \leftrightarrow (P \wedge Q)$ and

> *the absolute value function is not onto if and only if the sine function is not*
> *one-to-one, and the square root function is one-to-one*

is translated as $(R \leftrightarrow P) \wedge Q$. If the parenthesis are removed, the resulting string is $R \leftrightarrow P \wedge Q$. It is simpler, but it is not clear how it should be interpreted. To eliminate its ambiguity, we introduce an **order of connectives** as in algebra. In this way, certain strings without parentheses can be read as propositional forms.

■ **DEFINITION 1.1.5 [Order of Connectives]**

To interpret a propositional form, read from left to right and use the following precedence:

- propositional forms within parentheses or brackets (innermost first),

- negations,

- conjunctions,

- disjunctions,

- conditionals,

- biconditionals.

■ **EXAMPLE 1.1.6**

To write the propositional form $\neg P \vee Q \wedge R$ with parentheses, we begin by interpreting $\neg P$. According to the order of operations, the conjunction is next, so we evaluate $Q \wedge R$. This is followed by the disjunction, and we have the propositional form $\neg P \vee (Q \wedge R)$.

■ **EXAMPLE 1.1.7**

To interpret $P \wedge Q \vee R$ correctly, use the order of operations. We discover that it has the same meaning as $(P \wedge Q) \vee R$, but how is this distinguished from $P \wedge (Q \vee R)$ in English? Parentheses are not appropriate because they are not used as grouping symbols in sentences. Instead, use *either... or*. Then, using the assignments from Example 1.1.4, $(P \wedge Q) \vee R$ can be translated as

> *either the sine function is not one-to-one*
> *and the square root function is one-to-one,*
> *or the absolute value function is not onto.*

Notice that *either... or* works as a set of parentheses. We can use this to translate $P \wedge (Q \vee R)$:

> *the sine function is not one-to-one,*
> *and either the square root function is one-to-one*
> *or the absolute value function is not onto.*

Be careful to note that the *either-or* phrasing is logically inclusive. For instance, some colleges require their students to take either logic or mathematics. This choice is meant to be exclusive in the sense that only one is needed for graduation. However, it is not logically exclusive. A student can take logic to satisfy the requirement yet still take a math class.

■ **EXAMPLE 1.1.8**

Let us interpret $\neg(P \wedge Q)$. We can try translating this as *not P and Q*, but this represents $\neg P \wedge Q$ according to the order of operations. To handle a propositional form such as $\neg(P \wedge Q)$, use a phrase like *it is not the case* or *it is false* and the word *both*. Therefore, $\neg(P \wedge Q)$ becomes

it is not the case that both P and Q

or

it is false that both P and Q.

For instance, make the assignments.

$P :=$ *quadratic equations have at most two real solutions,*

$Q :=$ *the discriminant can be negative.*

Then,

*quadratic equations do not have at most two real solutions,
and the discriminant can be negative*

is a translation of $\neg P \wedge Q$. On the other hand, $\neg(P \wedge Q)$ can be

*it is not the case that both quadratic equations have at most two real solutions
and the discriminant can be negative.*

To interpret $\neg P \wedge \neg Q$, use *neither-nor*:

*neither do quadratic equations have at most two real solutions,
nor can the discriminant be negative.*

Valuations and Truth Tables

Propositions have truth values, but propositional forms do not. This is because every propositional form represents any one of infinitely many propositions. However, once a propositional form is identified with a proposition, there should be a process by which the truth value of the proposition is associated with the propositional form. This is done with a rule v called a **valuation**. The input of v is a propositional form, and its output is T or F. Suppose that P is a propositional variable. If P has been assigned a proposition,

$$v(P) = \begin{cases} T & \text{if } P \text{ is true,} \\ F & \text{if } P \text{ is false.} \end{cases}$$

For example, if $P := 2 + 3 = 5$, then $v(P) = T$, and if $P := 2 + 3 = 7$, then $v(P) = F$. If P has not been assigned a proposition, then $v(P)$ can be defined arbitrarily as either T or F.

The valuation of a compound propositional form is defined using **truth tables**. Let p and q be given propositional forms. Along the top row write p and, if needed, q. Draw a vertical line. To its right identify the desired propositional form consisting of p, possibly q, and a single connective. In the body of the table, on the left of the vertical line are all combinations of T and F for p and possibly q. On the right are the results of applying the connective. Each connective will have its own truth table, and we want to define these tables so that they match our understanding of the meaning of each connective.

Since the truth value of the negation of a given proposition is the opposite of that proposition's truth value,

p	$\neg p$
T	F
F	T

This means that $v(\neg p) = F$ if $v(p) = T$ and $v(\neg p) = T$ if $v(p) = F$.

The conjunction,

$$3 + 6 = 9, \text{ and all even integers are divisible by two,}$$

is true, but

$$\text{all integers are rational, and 4 is odd}$$

is false because the second conjunct is false. The disjunction

$$3 + 7 = 9, \text{ or all even integers are divisible by three,}$$

is false since both disjuncts are false. On the other hand,

$$3 + 7 = 9, \text{ or circles are round}$$

is true. This illustrates that

- a conjunction is true when both of its conjuncts are true, and false otherwise, and

- a disjunction is true when at least one disjunct is true, and false otherwise.

We use these principles to define the truth tables for $p \wedge q$ and $p \vee q$:

p	q	$p \wedge q$
T	T	T
T	F	F
F	T	F
F	F	F

p	q	$p \vee q$
T	T	T
T	F	T
F	T	T
F	F	F

We must remember that only one disjunct needs to be true for the entire disjunction to be true. For this reason, the logical disjunction is sometimes called an **inclusive or**. The propositional form for the **exclusive or** is

$$(p \vee q) \wedge \neg(p \wedge q).$$

There are many ways to understand an implication. Sometimes it represents causation as in

if I score at least 70 on the exam, I will earn a passing grade.

Other times it indicates what would have been the case if some past event had gone differently as in

if I had not slept late, I would not have missed the meeting.

Study of such conditional propositions is a very involved subject, one that need not concern us here because in mathematics a simpler understanding of the implication is enough. Suppose that P and Q are assigned propositions so that $P \to Q$ is a true implication. In mathematics, this means that it is not the case that P is true but Q is false. This understanding of the conditional is known as **material implication**. For example,

if rectangles have four sides, then squares have four sides,

if rectangles have three sides, then squares have four sides,

and

if rectangles have three sides, then squares have three sides

are all true, but

if rectangles have four sides, then squares have three sides

is false. Generalizing, in mathematics, $p \to q$ means $\neg(p \wedge \neg q)$, which has the following truth table:

p	q	$\neg q$	$p \wedge \neg q$	$\neg(p \wedge \neg q)$
T	T	F	F	T
T	F	T	T	F
F	T	F	F	T
F	F	T	F	T

The truth table for $p \to q$ is then defined as follows:

p	q	$p \to q$
T	T	T
T	F	F
F	T	T
F	F	T

The truth table for $p \leftrightarrow q$ is simpler because we understand $p \leftrightarrow q$ to mean

$$(p \to q) \wedge (q \to p).$$

The truth table for this propositional form requires five columns:

p	q	$p \to q$	$q \to p$	$(p \to q) \wedge (q \to p)$
T	T	T	T	T
T	F	F	T	F
F	T	T	F	F
F	F	T	T	T

Therefore, define the truth table of $p \leftrightarrow q$ as:

p	q	$p \leftrightarrow q$
T	T	T
T	F	F
F	T	F
F	F	T

This understanding of the biconditional is known as **material equivalence**.

Using these truth tables, the valuation of an arbitrary propositional form can be defined.

■ DEFINITION 1.1.9

Let p and q be propositional forms.

- $v(\neg p) = \begin{cases} T & \text{if } v(p) = F, \\ F & \text{if } v(p) = T. \end{cases}$

- $v(p \wedge q) = \begin{cases} T & \text{if } v(p) = T \text{ and } v(q) = T, \\ F & \text{otherwise.} \end{cases}$

- $v(p \vee q) = \begin{cases} F & \text{if } v(p) = F \text{ and } v(q) = F, \\ T & \text{otherwise.} \end{cases}$

- $v(p \to q) = \begin{cases} F & \text{if } v(p) = T \text{ and } v(q) = F, \\ T & \text{otherwise.} \end{cases}$

- $v(p \leftrightarrow q) = \begin{cases} T & \text{if } v(p) = v(q), \\ F & \text{otherwise.} \end{cases}$

■ EXAMPLE 1.1.10

Consider the propositional form $(P \leftrightarrow Q) \vee (R \to P)$ where

$$v(P) = F, \, v(Q) = T, \text{ and } v(R) = F.$$

Then, $v(P \leftrightarrow Q) = F$ because $v(P) \neq v(Q)$, and $v(R \to P) = T$ because $v(R) = F$. Therefore, because $v(R \to P) = T$,

$$v([P \leftrightarrow Q] \vee [R \to P]) = T,$$

We now generalize the definition of a truth table to create truth tables for more complicated propositional forms and then use the tables to find the valuation of a propositional form given the valuations of its proposition variables.

■ **EXAMPLE 1.1.11**

To write the truth table of $P \to Q \land \neg P$, identify the column headings by drawing the parsing tree for this form:

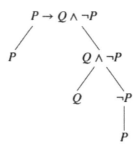

Reading from the bottom, we see that a formation sequence for the propositional form is

$$P, Q, \neg P, Q \land \neg P, P \to Q \land \neg P.$$

Hence, the truth table for this form is

P	Q	$\neg P$	$Q \land \neg P$	$P \to Q \land \neg P$
T	T	F	F	F
T	F	F	F	F
F	T	T	T	T
F	F	T	F	T

So, if $v(P) = T$ and $v(Q) = F$,

$$v(P \to Q \land \neg P) = F.$$

That is, any proposition represented by $P \to Q \land \neg P$ is false when the proposition assigned to P is true and the proposition assigned to Q is false.

The propositional form in the next example has three propositional variables. To make clear the truth value pattern that is to the left of the vertical line, note that if there are n variables, the number of rows is twice the number of rows for $n - 1$ variables. To see this, start with one propositional variable. Such a truth table has only two rows. Add a variable, we obtain four rows. The pattern is obtained by writing the one variable case twice. For the first time, it has a T written in front of each row. The second copy has an F in front of each row. To obtain the pattern for three variables, copy the two-variable pattern twice as in Figure 1.1. To generalize, if there are n variables, there will be 2^n rows.

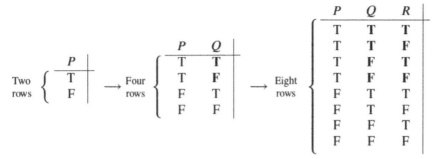

Figure 1.1   Valuation patterns.

■ **EXAMPLE 1.1.12**

Use a truth table to find the truth value of

> *if the derivative of the sine function is the cosine function*
> *and the second derivative of the sine function is the sine function,*
> *then the third derivative of the sine function is the cosine function.*

Define:

> $P :=$ *the derivative of the sine function is the cosine function,*
>
> $Q :=$ *the second derivative of the sine function is the sine function,*
>
> $R :=$ *the third derivative of the sine function is the cosine function.*

So P represents a true proposition, but Q and R represent false propositions. The proposition is represented by

$$P \wedge Q \rightarrow R$$

with truth table:

P	Q	R	$P \wedge Q$	$P \wedge Q \rightarrow R$
T	T	T	T	T
T	T	F	T	F
T	F	T	F	T
T	F	F	F	T
F	T	T	F	T
F	T	F	F	T
F	F	T	F	T
F	F	F	F	T

Notice that we could have determined the truth value by simply writing one line from the truth table:

P	Q	R	$P \wedge Q$	$P \wedge Q \rightarrow R$
T	F	F	F	T

We see that $v(P \wedge Q \rightarrow R) = T$ when $v(P) = T$, $v(Q) = F$, and $v(R) = F$. Therefore, the proposition is true.

■ EXAMPLE 1.1.13

Both $P \vee \neg P$ and $P \rightarrow P$ share an important property. Their columns in their truth tables are all T. For example, the truth table of $P \vee \neg P$ is:

P	$\neg P$	$P \vee \neg P$
T	F	T
F	T	T

Therefore, $v(P \vee \neg P)$ always equals T, no matter the choice of v.

However, the columns for $P \wedge \neg P$ and $P \leftrightarrow \neg P$ are all F. To check the first one, examine its truth table:

P	$\neg P$	$P \wedge \neg P$
T	F	F
F	T	F

This means that $v(P \wedge \neg P)$ is always F for every valuation v.

Based on the last example, we make the next definition.

■ DEFINITION 1.1.14

A propositional form p is a **tautology** if $v(p)$ always equals T for every valuation v, and p is a **contradiction** if $v(p)$ always equals F for every v. A propositional form that is neither a tautology nor a contradiction is called a **contingency**.

Exercises

1. Identify each sentence as either a proposition or not a proposition. Explain.
    (a) *Trisect the angle.*
    (b) *Some exponential functions are increasing.*
    (c) *All exponential functions are increasing.*
    (d) $3 + 8 = 18$
    (e) $3 + x = 18$
    (f) *Yea, logic!*
    (g) *A triangle is a three-sided polygon.*
    (h) *The function is differentiable.*
    (i) *This proposition is true.*
    (j) *This proposition is not true.*

2. Identify the antecedent and the consequent for the given implications.
    (a) *If the triangle has two congruent sides, it is isosceles.*
    (b) *The polynomial has at most two roots if it is a quadratic.*
    (c) *The data is widely spread only if the standard deviation is large.*
    (d) *The function being constant implies that its derivative is zero.*
    (e) *The system of equations is consistent is necessary for it to have a solution.*

(f) *A function is even is sufficient for its square to be even.*

3. Give the truth value of each proposition.
   (a) *A system of equations always has a solution, or a quadratic equation always has a real solution.*
   (b) *It is false that every polynomial function in one variable is differentiable.*
   (c) *Vertical lines have no slope, and lines through the origin have a positive y-intercept.*
   (d) *Every integer is even, or every even natural number is an integer.*
   (e) *If every parabola intersects the x-axis, then an ellipse has only one vertex.*
   (f) *The sine function is periodic if and only if every exponential function is always nonnegative.*
   (g) *It is not the case that $2 + 4 \neq 6$.*
   (h) *The distance between two points is always positive if every line segment is horizontal.*
   (i) *The derivative of a constant function is zero is necessary for the product rule to be true.*
   (j) *The derivative of the sine function being cosine is sufficient for the derivative of the cosine function being sine.*
   (k) *Any real number is negative or positive, but not both.*

4. For each sentence, fill in the blank using as many of the words *and, or, if,* and *if and only if* as possible to make the proposition true.
   (a) *Triangles have three sides* _____ $3 + 5 = 6$.
   (b) $3 + 5 = 6$ _____ *triangles have three sides.*
   (c) *Ten is the largest integer* _____ *zero is the smallest integer.*
   (d) *The derivative of a constant function is zero* _____ *tangent lines for increasing functions have positive slope.*

5. Extend Figure 1.1 by writing the typical pattern of Ts and Fs for the truth table of a propositional form with four propositional variables and then with five propositional variables.

6. Use a parsing tree to show that the given string is a propositional form.
   (a) $P \wedge Q \vee R$
   (b) $Q \leftrightarrow R \vee \neg Q$
   (c) $P \rightarrow Q \rightarrow R \rightarrow S$
   (d) $\neg P \wedge Q \vee (P \rightarrow Q) \wedge \neg S$
   (e) $(P \wedge Q \rightarrow Q) \wedge P \rightarrow Q$
   (f) $\neg\neg P \vee P \wedge S \rightarrow Q \vee [R \rightarrow \neg P \rightarrow \neg(Q \vee R)]$

7. Define:

$$P := \text{the angle sum of a triangle is } 180,$$
$$Q := 3 + 7 = 10,$$
$$R := \text{the sine function is continuous.}$$

Translate the given propositional forms into English.

    (a) $P \lor Q$

    (b) $P \land Q$

    (c) $P \land \neg Q$

    (d) $Q \lor \neg R$

    (e) $Q \leftrightarrow \neg R$

    (f) $R \to Q$

    (g) $P \lor R \to \neg Q$

    (h) $Q \leftrightarrow R \land \neg Q$

    (i) $\neg(P \land Q)$

    (j) $\neg(P \lor Q)$

    (k) $P \lor Q \land R$

    (l) $(P \lor Q) \land R$

8. Write the following sentences as propositional forms using the variables P, Q, and R as defined in Exercise 7.

    (a) *The sine function is continuous, and $3 + 7 = 10$.*

    (b) *The angle sum of a triangle is 180, or the angle sum of a triangle is 180.*

    (c) *If $3 + 7 = 10$, then the sine function is not continuous.*

    (d) *The angle sum of a triangle is 180 if and only if the sine function is continuous.*

    (e) *The sine function is continuous if and only if $3 + 7 = 10$ implies that the angle sum of a triangle is not 180.*

    (f) *It is not the case that $3 + 7 \neq 10$.*

9. Let $v(P) = T$, $v(Q) = T$, $v(R) = F$, and $v(S) = F$. Find the given valuations. (See Exercise 6.)

    (a) $P \land Q \lor R$

    (b) $Q \leftrightarrow R \lor \neg Q$

    (c) $P \to Q \to R \to S$

    (d) $\neg P \land Q \lor (P \to Q) \land \neg S$

    (e) $(P \land Q \to Q) \land P \to Q$

    (f) $\neg\neg P \lor P \land S \to Q \lor [R \to \neg P \to \neg(Q \lor R)]$

10. Write the truth table for each of the given propositional forms.

    (a) $\neg P \to P$

    (b) $P \to \neg Q$

    (c) $(P \lor Q) \land \neg(P \land Q)$

    (d) $(P \to Q) \lor (Q \leftrightarrow P)$

    (e) $P \land (Q \lor R)$

    (f) $P \lor Q \to R$

    (g) $P \to Q \land \neg(R \lor P)$

    (h) $P \to Q \leftrightarrow R \to S$

    (i) $P \lor (\neg Q \leftrightarrow R) \land Q$

(j) $(\neg P \vee Q) \wedge ([P \rightarrow Q] \vee \neg S)$

11. Check the truth value of these propositions using truth tables as in Example 1.1.12.
    (a) *If $2 + 3 = 7$, then $5 - 9 \neq 0$.*
    (b) *If a square is round implies that some functions have a derivative at $x = 2$, then every function has a derivative at $x = 2$.*
    (c) *Either four is odd or two is even implies that three is even.*
    (d) *Every even integer is divisible by 4 if and only if either 7 divides 21 or 9 divides 12.*
    (e) *The graph of the tangent function has asymptotes, and if sine is an increasing function, then cosine is a decreasing function.*

12. If possible, find propositional forms p and q such that
    (a) $p \wedge q$ is a tautology.
    (b) $p \vee q$ is a contradiction.
    (c) $\neg p$ is a tautology.
    (d) $p \rightarrow q$ is a contradiction.
    (e) $p \leftrightarrow q$ is a tautology.
    (f) $p \leftrightarrow q$ is a contradiction.

1.2 INFERENCE

Now that we have a collection of propositional forms and a means by which to interpret them as either true or false, we want to define a system that expands these ideas to include methods by which we can prove certain propositional forms from given propositional forms. What we will define is familiar because it is similar to what Euclid did with his geometry. Take, for example, the familiar result,

> *opposite angles in a parallelogram are congruent.*

In other words,

> *if ABCD is a parallelogram, then $\angle B \cong \angle D$,*

which translates to:

> Given: *ABCD* is a parallelogram,
> Prove: $\angle B \cong \angle D$.

To demonstrate this, we draw a diagram,

and then write a proof:

1.	$ABCD$ is a parallelogram	Given
2.	Join $\overline{AC}$	Postulate
3.	$\angle DAC \cong \angle BCA$	Alternate interior angles
4.	$\angle ACD \cong \angle CAB$	Alternate interior angles
5.	$\overline{AC} \cong \overline{AC}$	Reflexive
6.	$\triangle ACD \cong \triangle CAB$	ASA
7.	$\angle B \cong \angle D$	Corresponding parts

Euclid's geometry consists of geometric propositions that are established by proofs like the above. These proofs rely on rules of logic, previously proved propositions (**lemmas**, **theorems**, and **corollaries**), and propositions that are assumed to be true (the **postulates**). Using this system of thought, we can show which geometric propositions follow from the postulates and conclude which propositions are true, whatever it means for a geometric proposition to be true. Euclidean geometry serves as a model for the following modern definition.

■ **DEFINITION 1.2.1**

A **logical system** consists of the following:

- An alphabet

- A grammar

- Propositional forms that require no proof

- Rules that determine truth

- Rules that are used to write proofs.

Although Euclid did not provide an alphabet or a grammar specifically for his geometry, his system did include the last three aspects of a logical system. In this chapter we develop the logical system known as **propositional logic**. Its alphabet, grammar, and rules that determine truth were defined in Section 1.1. The remainder of this chapter is spent establishing the other two components.

Consider the following collection of propositions:

> *If squares are rectangles, then squares are quadrilaterals.*
> *Squares are rectangles.*
> *Therefore, squares are quadrilaterals.*

This is an example of a **deduction**, a collection of propositions of which one is supposed to follow necessarily from the others. In this particular case,

> *if squares are rectangles, then squares are quadrilaterals*

and

$$squares\ are\ rectangles$$

are the premises, and

$$squares\ are\ quadrilaterals$$

is the conclusion. We recognize that in this case, the conclusion does follow from the premises because whenever the premises are true, the conclusion must also be true. When this is the case, the deduction is **semantically valid**, else it is **semantically invalid**.

Notice that not only do we see that the deduction works because of the meaning of the propositions, but we also see that it is valid based on the forms of the sentences. In other words, we also recognize this deduction as valid:

> *If Hausdorff spaces are preregular, their points can be separated.*
> *Hausdorff spaces are preregular.*
> *Therefore, their points can be separated.*

Although we might not know the terms *Hausdorff space*, *preregular*, and *separated*, we recognize the deduction as valid because it is of the same pattern as the first deduction:

$$
\begin{array}{c}
p \to q \\
\underline{\quad p \quad} \\
\therefore\ q
\end{array}
\tag{1.2}
$$

When the deduction is found to work based on its form, the deduction is **syntactically valid**, else it is **syntactically invalid**.

We study both types of validity by examining general patterns of deductions and choosing rules that determine which forms correspond to deductions that are valid semantically and which forms correspond to deductions that are valid syntactically.

Semantics

The study of meaning is called **semantics**. We began this study when we wrote truth tables. These are characterized as semantic because the truth value of a proposition is based on its meaning. Our goal is to use truth tables to determine when an **argument form**, an example being (1.2), corresponds to a deduction that is semantically valid. We begin with a definition.

■ **DEFINITION 1.2.2**

Let $p_0, p_1, \ldots, p_{n-1}$ and q be propositional forms.

- If q is a tautology, write $\vDash q$.

- Define $p_0, p_1, \ldots, p_{n-1}$ to **logically imply** q if

$$\vDash p_0 \land p_1 \land \cdots \land p_{n-1} \to q.$$

When $p_0, p_1, \ldots, p_{n-1}$ logically imply q, write

$$p_0, p_1, \ldots, p_{n-1} \vDash q.$$

and say that q is a **consequence** of $p_0, p_1, \ldots, p_{n-1}$. Call the propositional forms $p_0, p_1, \ldots, p_{n-1}$ the **premises** of the implication and q the **conclusion**.

Notice that if $p_0, p_1, \ldots, p_{n-1} \vDash q$, then for any valuation v, whenever $v(p_i) = T$ for all $i = 0, 1, \ldots, n - 1$, it must be the case that $v(q) = T$. Moreover, any deduction with premises represented by $p_0, p_1, \ldots, p_{n-1}$ and conclusion by q is semantically valid if $p_0, p_1, \ldots, p_{n-1} \vDash q$.

■ **EXAMPLE 1.2.3**

Because of Example 1.1.13, both $\vDash P \to P$ and $\vDash P \vee \neg P$.

■ **EXAMPLE 1.2.4**

Prove: $P \to Q, P \vDash Q$

To accomplish this, show that the propositional form

$$(P \to Q) \wedge P \to Q,$$

with antecedent equal to the conjunction of the premises and consequent consisting of the conclusion is a tautology.

P	Q	$P \to Q$	$(P \to Q) \wedge P$	$(P \to Q) \wedge P \to Q$
T	T	T	T	T
T	F	F	F	T
F	T	T	F	T
F	F	T	F	T

Therefore, $\vDash (P \to Q) \wedge P \to Q$, so

$$P \to Q, P \vDash Q,$$

and any deduction based on this form is semantically valid.

■ **EXAMPLE 1.2.5**

Prove: $P \vee Q \to Q, P \vDash Q$

P	Q	$P \vee Q$	$P \vee Q \to Q$	$(P \vee Q \to Q) \wedge P$	$(P \vee Q \to Q) \wedge P \to Q$
T	T	T	T	T	T
T	F	T	F	F	T
F	T	T	T	F	T
F	F	F	T	F	T

If it is possible for $v(p_i) = T$ for $i = 0, 1, \ldots, n - 1$ yet $v(q) = F$, the propositional form

$$p_0 \wedge p_1 \wedge \cdots \wedge p_{n-1} \to q$$

is not a tautology, so q is not a consequence of $p_0, p_1, \ldots, p_{n-1}$. If this is the case, write

$$p_0, p_1, \ldots, p_{n-1} \nvDash q.$$

■ **EXAMPLE 1.2.6**

Prove: $P \wedge Q \to Q, P \nvDash Q$

P	Q	$P \wedge Q$	$P \wedge Q \to Q$	$(P \wedge Q \to Q) \wedge P$	$(P \wedge Q \to Q) \wedge P \to Q$
T	T	T	T	T	T
T	F	F	T	T	F
F	T	F	T	F	T
F	F	F	T	F	T

Notice that F appears for $(P \wedge Q \to Q) \wedge P \to Q$ on a line when

$$v(P \wedge Q \to Q) = v(P) = \text{T}$$

yet $v(Q) = \text{F}$. Because of this, we can shorten the procedure for showing that a propositional form is not a consequence of other propositional forms.

■ **EXAMPLE 1.2.7**

Prove: $P \wedge Q \to R \nvDash P \to R$

P	Q	R	$P \wedge Q \to R$	$P \to R$
T	F	F	T	F

Observe that this shows that the valuation of $P \wedge Q \to R$ can be T at the same time that the valuation of $P \to R$ is F.

Syntactics

Although we will return to semantics, it is important to note that using truth tables to check for logical implication has its limitations. If the argument form involves many propositional forms or if the propositional forms are complicated, the truth table used to show or disprove the logical implication can become unwieldy. Another issue is that in practice, truth tables are not the method of choice when determining whether a conclusion follows from the premises. What is typically done is to follow Euclid's example (page 20), basing the conclusions on the **syntax** of the argument form, namely, based only on its pattern and structure.

Let us return to the deduction on page 20 and work with it differently. Start with the two propositions,

if squares are rectangles, then squares are quadrilaterals

and

squares are rectangles.

Because of the combined structure of the two sentences, we know that we can **write** *squares are quadrilaterals.*

This act of writing (on paper or a blackboard or in the mind) means that we have the proposition and that it follows from the first two. Similarly, if we start with

squares are triangles, or squares are rectangles

and

squares are not triangles,

we can write

squares are rectangles.

Determining a method that will model this reasoning requires us to find rules by which propositional forms can be written from other propositional forms. Since every logical system requires a starting point, the first step in this process is to choose which propositional forms can be written without any prior justification. Each such propositional form is called an **axiom**. Playing the same role as that of a postulate in Euclidean geometry, an axiom can be considered as a rule of the game. Certain propositional forms lend themselves as good candidates for axioms because they are regarded as obvious. That is, they are **self-evident**. Other propositional forms are good candidates to be axioms, not because they are necessarily self-evident, but they are helpful. In either case, the number of axioms should be as few as possible so as to minimize the number of assumptions. For propositional logic, we choose only three. They were first found in work of Gottlob Frege (1879) and later in that of Jan Łukasiewicz (1930).

■ **AXIOMS 1.2.8 [Frege–Łukasiewicz]**

Let p, q, and r be propositional forms.

- **[FL1]** $p \to (q \to p)$
- **[FL2]** $p \to (q \to r) \to (p \to q \to [p \to r])$
- **[FL3]** $\neg p \to \neg q \to (q \to p)$.

The next step in defining propositional logic is to state when it is legal to write a propositional form from given propositional forms.

■ **DEFINITION 1.2.9**

The propositional forms $p_0, p_1, \ldots, p_{n-1}$ **infer** q if q can be written whenever $p_0, p_1, \ldots, p_{n-1}$ are written. Denote this by

$$p_0, p_1, \ldots, p_{n-1} \Rightarrow q.$$

This is known as an **inference**.

To make rigorous which propositional forms can be inferred from given forms, we establish some rules. These are chosen because they model basic reasoning. They are also not proved, so they serve as postulates for our logic.

■ **INFERENCE RULES 1.2.10**

Let p, q, r, and s be propositional forms.

- *Modus Ponens* [MP]
  $p \rightarrow q, p \Rightarrow q$

- *Modus Tolens* [MT]
  $p \rightarrow q, \neg q \Rightarrow \neg p$

- **Constructive Dilemma [CD]**
  $(p \rightarrow q) \wedge (r \rightarrow s), p \vee r \Rightarrow q \vee s$

- **Destructive Dilemma [DD]**
  $(p \rightarrow q) \wedge (r \rightarrow s), \neg q \vee \neg s \Rightarrow \neg p \vee \neg r$

- **Disjunctive Syllogism [DS]**
  $p \vee q, \neg p \Rightarrow q$

- **Hypothetical Syllogism [HS]**
  $p \rightarrow q, q \rightarrow r \Rightarrow p \rightarrow r$

- **Conjunction [Conj]**
  $p, q \Rightarrow p \wedge q$

- **Simplification [Simp]**
  $p \wedge q \Rightarrow p$

- **Addition [Add]**
  $p \Rightarrow p \vee q.$

To use Inference Rules 1.2.10, match the form exactly. For example, even though $P \rightarrow R$ appears to follow from $(P \wedge Q) \rightarrow R$ as an application of simplification, it does not. The problem is that simplification can only be applied to propositional forms with the $p \wedge q$ pattern, but $(P \wedge Q) \rightarrow R$ is of the form $p \rightarrow q$. With this detail in mind, we make some inferences.

■ **EXAMPLE 1.2.11**

Each inference is justified by the indicated rule.

- *Modus ponens*
  $P \wedge Q \rightarrow \neg R, P \wedge Q \Rightarrow \neg R$

- Addition
  $P \Rightarrow P \vee Q \wedge R$

- *Modus tolens*
  $\neg \neg P, Q \vee R \rightarrow \neg P \Rightarrow \neg(Q \vee R).$

■ EXAMPLE 1.2.12

Since it is possible that some propositional forms are not needed for the inference, we also have the following:

- *Modus ponens*
  $$P \wedge Q \to \neg R, P \vee Q, P \wedge Q, Q \leftrightarrow S \Rightarrow \neg R$$

- Addition
  $$P, R, S \to T \Rightarrow P \vee Q \wedge R$$

- *Modus tolens*
  $$\neg S, \neg\neg P, P \wedge T, Q \vee R \to \neg P \Rightarrow \neg(Q \vee R).$$

Inference is a powerful tool, but it can only be used to check simple deductions. Sometimes multiple inferences are needed to move from a collection of premises to a conclusion. For example, if we write

$$p \vee q, \neg p, q \to r,$$

based on the first two propositional forms, we can write

$$q$$

by DS, and then based on this propositional form and the third of the given propositional forms, we can write

$$r$$

by MP. This is a simple example of the next definition.

■ DEFINITION 1.2.13

- A **formal proof** of the propositional form q (the **conclusion**) from the propositional forms $p_0, p_1, \ldots, p_{n-1}$ (the **premises**) is a sequence of propositional forms,

$$p_0, p_1, \ldots, p_{n-1}, q_0, q_1, \ldots, q_{m-1},$$

such that $q_{m-1} = q$, and for all $i = 0, 1, \ldots, m-1$, either q_i is an axiom,

$$\text{if } i = 0, \text{ then } p_0, p_1, \ldots, p_{n-1} \Rightarrow q_i, \text{ or}$$

$$\text{if } i > 0, \text{ then } p_0, p_1, \ldots, p_{n-1}, q_0, q_1, \ldots, q_{i-1} \Rightarrow q_i.$$

If there exists a formal proof of q from $p_0, p_1, \ldots, p_{n-1}$, then q is **proved** or **deduced** from $p_0, p_1, \ldots, p_{n-1}$ and we write

$$p_0, p_1, \ldots, p_{n-1} \vdash q.$$

- If there are no premises, a **formal proof** of q is a sequence,

$$q_0, q_1, \ldots, q_{m-1},$$

such that q_0 is an axiom, $q_{m-1} = q$, and for all $i > 0$, either q_i is an axiom or

$$q_0, q_1, \ldots, q_{i-1} \Rightarrow q_i.$$

In this case, write $\vdash q$ and call q a **theorem**.

Observe that any deduction with premises represented by $p_0, p_1, \ldots, p_{n-1}$ and conclusion by q is syntactically valid if $p_0, p_1, \ldots, p_{n-1} \vdash q$.

We should note that although $\Rightarrow$ and $\vdash$ have different meanings as syntactic symbols, they are equivalent. If $p \Rightarrow q$, then $p \vdash q$ using the proof p, q. Conversely, suppose $p \vdash q$. This means that there exists a proof

$$p, q_0, q_1, \ldots, q_{n-1}, q,$$

so every time we write down p, we can also write down q. That is, $p \Rightarrow q$. We summarize this as follows.

■ **THEOREM 1.2.14**

For all propositional forms p and q, $p \Rightarrow q$ if and only if $p \vdash q$.

We use a particular style to write formal proofs. They will be in two-column format with each line being numbered. In the first column will be the sequence of propositional forms that make up the proof. In the second column will be the reasons that allowed us to include each form. The only reasons that we will use are

- *Given* (for premises),

- FL1, FL2, or FL3 (for an axiom),

- An inference rule.

An inference rule is cited by giving the line numbers used as the premises followed by the abbreviation for the rule. Thus, the following proves $P \vee Q \rightarrow Q \wedge R, P \vdash Q$:

1.	$P \vee Q \rightarrow Q \wedge R$	Given
2.	P	Given
3.	$P \vee Q$	2 Add
4.	$Q \wedge R$	1, 3 MP
5.	Q	4 Simp

Despite the style, we should remember that a proof is a sequence of propositional forms that satisfy Definition 1.2.13. In this case, the sequence is

$$P \vee Q \rightarrow Q \wedge R, P, P \vee Q, Q \wedge R, Q.$$

The first two examples involve proofs that use the axioms.

■ **EXAMPLE 1.2.15**

Prove: $\vdash P \to Q \to (P \to P)$

1.	$P \to (Q \to P)$	FL1
2.	$P \to (Q \to P) \to (P \to Q \to [P \to P])$	FL2
3.	$P \to Q \to (P \to P)$	MP

This proves that $P \to Q \to (P \to P)$ is a theorem. Also, by adding $P \to Q$ as a given and an application of MP at the end, we can prove

$$P \to Q \vdash P \to P.$$

This result should not be surprising since $P \to P$ is a tautology. We would expect any premise to be able to prove it.

■ **EXAMPLE 1.2.16**

Prove: $\neg(Q \to P), \neg P \vdash \neg Q$

1.	$\neg(Q \to P)$	Given
2.	$\neg P$	Given
3.	$\neg P \to \neg Q \to (Q \to P)$	FL3
4.	$\neg P \to \neg Q$	1, 3 MT
5.	$\neg Q$	2, 4 MP

The next three examples do not use an axiom in their proofs.

■ **EXAMPLE 1.2.17**

Prove: $P \to Q, Q \to R, S \vee \neg R, \neg S \vdash \neg P$

1.	$P \to Q$	Given
2.	$Q \to R$	Given
3.	$S \vee \neg R$	Given
4.	$\neg S$	Given
5.	$P \to R$	1, 2 HS
6.	$\neg R$	3, 4 DS
7.	$\neg P$	5, 6 MT

■ **EXAMPLE 1.2.18**

Prove: $P \to Q, P \to Q \to (T \to S), P \vee T, \neg Q \vdash S$

1.	$P \to Q$	Given
2.	$P \to Q \to (T \to S)$	Given
3.	$P \vee T$	Given

4.	$\neg Q$	Given
5.	$T \to S$	1, 2 MP
6.	$(P \to Q) \wedge (T \to S)$	1, 5 Conj
7.	$Q \vee S$	3, 6 CD
8.	S	4, 7 DS

■ EXAMPLE 1.2.19

Prove: $P \to Q, Q \to R, \neg R \vdash \neg Q \vee \neg P$

1.	$P \to Q$	Given
2.	$Q \to R$	Given
3.	$\neg R$	Given
4.	$(Q \to R) \wedge (P \to Q)$	1, 2 Conj
5.	$\neg R \vee \neg Q$	3 Add
6.	$\neg Q \vee \neg P$	4, 5 DD

Exercises

1. Show using truth tables.
   (a) $\neg P \vee Q, \neg Q \models \neg P$
   (b) $\neg(P \wedge Q), P \models \neg Q$
   (c) $P \to Q, P \models Q \vee R$
   (d) $P \to Q, Q \to R, P \models R$
   (e) $P \vee Q \wedge R, \neg P \models R$

2. Show the following using truth tables.
   (a) $\neg(P \wedge Q) \nvDash \neg P$
   (b) $P \to Q \vee R, P \nvDash Q$
   (c) $P \wedge Q \to R \nvDash Q \to R$
   (d) $(P \to Q) \vee (R \to S), P \vee R \nvDash Q \vee S$
   (e) $\neg(P \wedge Q) \vee R, P \wedge Q \vee S \nvDash R \wedge S$
   (f) $P \vee R, Q \vee S, R \leftrightarrow S \nvDash R \wedge S$

3. Identify the rule from Inference Rules 1.2.10.
   (a) $P \to Q \to P, P \to Q \Rightarrow P$
   (b) $P, Q \vee R \Rightarrow P \wedge (Q \vee R)$
   (c) $P \Rightarrow P \vee (R \leftrightarrow \neg P \wedge \neg[Q \to S])$
   (d) $P, P \to (Q \leftrightarrow S) \Rightarrow Q \leftrightarrow S$
   (e) $P \vee Q \vee Q, (P \vee Q \to Q) \wedge (Q \to S \wedge T) \Rightarrow Q \vee S \wedge T$
   (f) $P \vee (Q \vee S), \neg P \Rightarrow Q \vee S$
   (g) $P \to \neg Q, \neg\neg Q \Rightarrow \neg P$
   (h) $(P \to Q) \wedge (Q \to R), \neg Q \vee \neg R \Rightarrow \neg P \vee \neg Q$
   (i) $(P \to Q) \wedge (Q \to R) \Rightarrow P \to Q$

4. Arrange each collection of propositional forms into a proof for the given deductions and supply the appropriate reasons.

(a) $P \to Q, R \to S, P \vdash Q \lor S$
- P
- $Q \lor S$
- $R \to S$
- $(P \to Q) \land (R \to S)$
- $P \lor R$
- $P \to Q$

(b) $P \to Q, Q \to R, P \vdash R \lor Q$
- P
- $P \to R$
- $P \to Q$
- R
- $Q \to R$
- $R \lor Q$

(c) $(P \to Q) \lor (Q \to R), \neg(P \to Q), \neg R, Q \lor S \vdash S$
- $\neg(P \to Q)$
- S
- $(P \to Q) \lor (Q \to R)$
- $Q \to R$
- $\neg R$
- $\neg Q$
- $Q \lor S$

(d) $(P \lor Q) \land R, Q \lor S \to T, \neg P \vdash \neg P \land T$
- $P \lor Q$
- Q
- $Q \lor S$
- T
- $(P \lor Q) \land R$
- $Q \lor S \to T$
- $\neg P$
- $\neg P \land T$

5. Prove using Axioms 1.2.8.

(a) $\vdash P \to P$

(b) $\vdash \neg\neg P \to P$

(c) $\vdash P \to (P \to [Q \to P])$

(d) $\neg P \vdash P \to Q$

(e) $P \to Q, Q \to R \vdash P \to R$ (Do not use HS.)

(f) $P \to Q, \neg Q \vdash \neg P$ (Do not use MT.)

6. Prove. Axioms 1.2.8 are not required.
    (a) $P \to Q, P \vee (R \to S), \neg Q \vdash R \to S$
    (b) $P \to Q, Q \to R, \neg R \vdash \neg P$
    (c) $P \to Q, R \to S, \neg Q \vee \neg S \vdash \neg P \vee \neg R$
    (d) $[P \to (Q \to R)] \wedge [Q \to (R \to P)], P \vee Q, \neg (Q \to R), \neg P \vdash \neg R$
    (e) $P \to Q \to (R \to S), S \to T, P \to Q, R \vdash T$
    (f) $P \to Q \wedge R, Q \vee S \to T \wedge U, P \vdash T$
    (g) $P \to Q \wedge R, \neg (Q \wedge R), Q \wedge R \vee (\neg P \to S) \vdash S$
    (h) $P \vee Q \to \neg R \wedge \neg S, Q \to R, P \vdash \neg Q$
    (i) $N \to P, P \vee Q \vee R \to S \vee T, S \vee T \to T, N \vdash T$
    (j) $P \to Q, Q \to R, R \to S, S \to T, P \vee R, \neg R \vdash T$
    (k) $P \vee Q \to R \vee S, (R \to T) \wedge (S \to U), P, \neg T \vdash U$
    (l) $P \to Q, Q \to R, R \to S, (P \vee Q) \wedge (R \vee S) \vdash Q \vee S$
    (m) $P \vee \neg Q \vee R \to (S \to P), P \vee \neg Q \to (P \to R), P \vdash S \to R$

1.3 REPLACEMENT

There are times when writing a formal proof that we want to substitute one propositional
form for another. This happens when two propositional forms have the same valuations.
It also happens when a particular sentence pattern should be able to replace another
sentence pattern. We give rules in this section that codify both ideas.

Semantics

Consider the propositional form $\neg (P \vee Q)$. Its valuation equals T when it is not the
case that $v(P) = T$ or $v(Q) = T$ (Definition 1.1.9). This implies that $v(P) = F$ and
$v(Q) = F$, so $v(\neg P \wedge \neg Q) = T$. Conversely, the valuation of $\neg P \wedge \neg Q$ is T implies that
the valuation of $\neg (P \vee Q)$ is T for similar reasons. Since no additional premises were
assumed in this discussion, we conclude that

$$v(\neg [P \vee Q]) = v(\neg P \wedge \neg Q), \tag{1.3}$$

and this means that
$$\neg (P \vee Q) \leftrightarrow \neg P \wedge \neg Q$$

is a tautology. There is a name for this.

■ **DEFINITION 1.3.1**

Two propositional forms p and q are **logically equivalent** if $\vDash p \leftrightarrow q$.

Observe that Definition 1.3.1 implies the following result.

■ **THEOREM 1.3.2**

All tautologies are logically equivalent, and all contradictions are logically equivalent.

Because of (1.3), we can use a truth table to prove logical equivalence.

■ EXAMPLE 1.3.3

Prove: $\models \neg(P \vee Q) \leftrightarrow \neg P \wedge \neg Q$.

P	Q	$P \vee Q$	$\neg(P \vee Q)$	$\neg P$	$\neg Q$	$\neg P \wedge \neg Q$
T	T	T	F	F	F	F
T	F	T	F	F	T	F
F	T	T	F	T	F	F
F	F	F	T	T	T	T

■ EXAMPLE 1.3.4

Because $\models P \rightarrow Q \leftrightarrow \neg P \vee Q$, we can replace $P \rightarrow Q$ with $\neg P \vee Q$, and vice versa, at any time. When this is done, the resulting propositional form is logically equivalent to the original. For example,

$$\models Q \wedge (P \rightarrow Q) \leftrightarrow Q \wedge (\neg P \vee Q).$$

To see this, examine the truth table

P	Q	$P \rightarrow Q$	$Q \wedge (P \rightarrow Q)$	$\neg P$	$\neg P \vee Q$	$Q \wedge (\neg P \vee Q)$
T	T	T	T	F	T	T
T	F	F	F	F	F	F
F	T	T	T	T	T	T
F	F	T	F	T	T	F

■ EXAMPLE 1.3.5

Consider $R \wedge (P \rightarrow Q)$. Since $\models P \rightarrow Q \leftrightarrow \neg Q \rightarrow \neg P$ [Exercise 2(f)], we can replace $P \rightarrow Q$ with $\neg Q \rightarrow \neg P$ giving

$$\models R \wedge (P \rightarrow Q) \leftrightarrow R \wedge (\neg Q \rightarrow \neg P).$$

When studying an implication, we sometimes need to investigate the different ways that its antecedent and consequent relate to each other.

■ DEFINITION 1.3.6

The **converse** of a given implication is the conditional proposition formed by exchanging the antecedent and consequent of the implication (Figure 1.2).

■ DEFINITION 1.3.7

The **contrapositive** of a given implication is the conditional proposition formed by exchanging the antecedent and consequent of the implication and then replacing them with their negations (Figure 1.3).

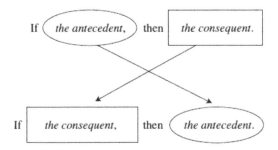

Figure 1.2   Writing the converse.

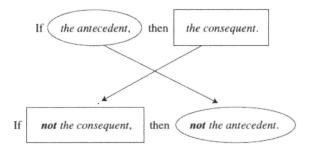

Figure 1.3   Writing the contrapositive.

For example, the converse of

> *if rectangles have four sides, squares have for sides*

is

> *if squares have four sides, rectangles have four sides,*

and its contrapositive is

> *if squares do not have four sides, rectangles do not have four sides.*

Notice that a biconditional proposition is simply the conjunction of a conditional with its converse.

■ EXAMPLE 1.3.8

The propositional form $P \to Q$ has $Q \to P$ as its converse and $\neg Q \to \neg P$ as its contrapositive. The first and fourth columns on the right of the next truth table show $\vDash P \to Q \leftrightarrow \neg Q \to \neg P$, while $\nvDash P \to Q \leftrightarrow Q \to P$ is shown by the first and last columns.

P	Q	$P \to Q$	$\neg Q$	$\neg P$	$\neg Q \to \neg P$	$Q \to P$
T	T	T	F	F	T	T
T	F	F	T	F	F	T
F	T	T	F	T	T	F
F	F	T	T	T	T	T

Syntactics

If we limit our proofs to Inference Rules 1.2.10, we quickly realize that there will be little of interest that we can prove. We would have no reason on which to base such clear inferences as

$$P \vdash Q \vee P$$

or

$$P \vee Q, \neg Q \vdash P.$$

To fix this, we expand our collection of inference rules with a new type.

Suppose that we know that the form $p \wedge q$ can replace $q \wedge p$ at any time and vice versa. For example, in the propositional form

$$P \wedge Q \to R, \tag{1.4}$$

$P \wedge Q$ can be replaced with $Q \wedge P$ so that we can write the new form

$$Q \wedge P \to R. \tag{1.5}$$

This type of rule is called a **replacement rule** and is written using the $\Leftrightarrow$ symbol. For example, the replacement rule that allowed us to write (1.5) from (1.4) is

$$p \wedge q \Leftrightarrow q \wedge p.$$

Similarly, when the replacement rule

$$\neg(p \wedge q) \Leftrightarrow \neg p \vee \neg q$$

is applied to $P \vee (\neg Q \vee \neg R)$, the result is $P \vee \neg(Q \wedge R)$. We state without proof the standard replacement rules.

■ REPLACEMENT RULES 1.3.9

Let p, q, and r be propositional forms.

- **Associative Laws [Assoc]**
  $p \wedge q \wedge r \Leftrightarrow p \wedge (q \wedge r)$
  $p \vee q \vee r \Leftrightarrow p \vee (q \vee r)$

- **Commutative Laws [Com]**
  $p \wedge q \Leftrightarrow q \wedge p$
  $p \vee q \Leftrightarrow q \vee p$

- **Distributive Laws [Distr]**
  $p \wedge (q \vee r) \Leftrightarrow p \wedge q \vee p \wedge r$
  $p \vee q \wedge r \Leftrightarrow (p \vee q) \wedge (p \vee r)$

- **Contrapositive Law [Contra]**
  $p \to q \Leftrightarrow \neg q \to \neg p$

- **Double Negation [DN]**

  $p \Leftrightarrow \neg\neg p$

- **De Morgan's Laws [DeM]**

  $\neg(p \wedge q) \Leftrightarrow \neg p \vee \neg q$

  $\neg(p \vee q) \Leftrightarrow \neg p \wedge \neg q$

- **Idempotency [Idem]**

  $p \wedge p \Leftrightarrow p$

  $p \vee p \Leftrightarrow p$

- **Material Equivalence [Equiv]**

  $p \leftrightarrow q \Leftrightarrow (p \rightarrow q) \wedge (q \rightarrow p)$

  $p \leftrightarrow q \Leftrightarrow p \wedge q \vee \neg p \wedge \neg q$

- **Material Implication [Impl]**

  $p \rightarrow q \Leftrightarrow \neg p \vee q$

- **Exportation [Exp]**

  $p \wedge q \rightarrow r \Leftrightarrow p \rightarrow (q \rightarrow r).$

A replacement rule is used in a formal proof by appealing to the next inference rule.

■ **INFERENCE RULE 1.3.10**

For all propositional forms p and q, if p if obtained from q using a replacement rule, $p \Rightarrow q$ and $q \Rightarrow p$.

As with Inference Rules 1.2.10, the replacement rule used when applying Inference Rule 1.3.10 must be used exactly as stated. This includes times when it seems unnecessary because it appears obvious. For example, $P \vee Q$ does not follow directly from $\neg P \rightarrow Q$ using Impl. Instead, include a use of DN to give the correct sequence,

$$\neg P \rightarrow Q, \neg\neg P \vee Q, P \vee Q.$$

Similarly, the inference rule Add does not allow for $Q \vee P$ to be derived from P. Instead, we derive $P \vee Q$. Follow this by Com to conclude $Q \vee P$.

When writing formal proofs that appeal to Inference Rule 1.3.10, do not cite that particular rule but reference the replacement rule's abbreviation using Replacement Rule 1.3.9 and the line which serves as a premise to the replacement. We use this practice in the following examples.

■ **EXAMPLE 1.3.11**

Although it is common practice to move parentheses freely when solving equations, inferences such as

$$P \wedge Q \wedge (R \wedge S) \vdash P \wedge (Q \wedge R) \wedge S$$

must be carefully demonstrated. Fortunately, this example only requires two applications of the associative law. Using the boxes as a guide, notice that

$$\boxed{P \wedge Q} \wedge (\,\boxed{R} \wedge \boxed{S}\,)$$

is of the same form as the right-hand side of the associative law. Hence, we can remove the parentheses to obtain

$$\boxed{P \wedge Q} \wedge \boxed{R} \wedge \boxed{S}.$$

Next, view the propositional form as

$$\boxed{P \wedge Q \wedge R} \wedge \boxed{S}.$$

One more application within the first box yields the result,

$$\boxed{P \wedge (Q \wedge R)} \wedge \boxed{S}.$$

We may, therefore, write a sequence of inferences by Inference Rule 1.3.10,

$$P \wedge Q \wedge (R \wedge S) \Rightarrow (P \wedge Q \wedge R) \wedge S \Rightarrow [P \wedge (Q \wedge R)] \wedge S,$$

and we have a proof of $[P \wedge (Q \wedge R)] \wedge S$ from $P \wedge Q \wedge (R \wedge S)$:

1.	$P \wedge Q \wedge (R \wedge S)$	Given
2.	$P \wedge Q \wedge R \wedge S$	1 Assoc
3.	$P \wedge (Q \wedge R) \wedge S$	2 Assoc

■ EXAMPLE 1.3.12

Prove: $R \wedge S \vdash \neg R \to P$

1.	$R \wedge S$	Given
2.	R	1 Simp
3.	$R \vee P$	2 Add
4.	$\neg\neg R \vee P$	3 DN
5.	$\neg R \to P$	4 Impl

■ EXAMPLE 1.3.13

Prove: $P \to Q, R \to Q \vdash P \vee R \to Q$

1.	$P \to Q$	Given
2.	$R \to Q$	Given
3.	$(P \to Q) \wedge (R \to Q)$	1, 2 Conj
4.	$(\neg P \vee Q) \wedge (\neg R \vee Q)$	3 Impl
5.	$(Q \vee \neg P) \wedge (Q \vee \neg R)$	4 Com

6.	$Q \vee \neg P \wedge \neg R$	5 Dist
7.	$\neg P \wedge \neg R \vee Q$	6 Com
8.	$\neg (P \vee R) \vee Q$	7 DeM
9.	$P \vee R \rightarrow Q$	8 Impl

■ **EXAMPLE 1.3.14**

Prove: $P \wedge Q \vee R \wedge S \vdash (P \vee S) \wedge (Q \vee R)$

1.	$P \wedge Q \vee R \wedge S$	Given
2.	$(P \wedge Q \vee R) \wedge (P \wedge Q \vee S)$	1 Dist
3.	$(R \vee P \wedge Q) \wedge (S \vee P \wedge Q)$	2 Com
4.	$(R \vee P) \wedge (R \vee Q) \wedge [(S \vee P) \wedge (S \vee Q)]$	3 Dist
5.	$(R \vee Q) \wedge (R \vee P) \wedge [(S \vee P) \wedge (S \vee Q)]$	4 Com
6.	$(R \vee Q) \wedge ([R \vee P] \wedge [(S \vee P) \wedge (S \vee Q)])$	5 Assoc
7.	$R \vee Q$	6 Simp
8.	$(S \vee P) \wedge (S \vee Q) \wedge [(R \vee Q) \wedge (R \vee P)]$	5 Com
9.	$(S \vee P) \wedge (S \vee Q)$	8 Simp
10.	$S \vee P$	9 Simp
11.	$(S \vee P) \wedge (R \vee Q)$	7, 10 Conj
12.	$(P \vee S) \wedge (Q \vee R)$	11 Com

■ **EXAMPLE 1.3.15**

Both $\vdash P \rightarrow P$ and $\vdash P \vee \neg P$. Here is the proof for the second theorem:

1.	$P \rightarrow (P \rightarrow P)$	FL1
2.	$\neg P \vee (\neg P \vee P)$	1 Impl
3.	$(\neg P \vee \neg P) \vee P$	2 Assoc
4.	$\neg P \vee P$	3 Idem
5.	$P \vee \neg P$	4 Com

This proof can be generalized to any propositional form p, so that we also have $\vdash p \rightarrow p$ and $\vdash p \vee \neg p$.

The theorem $p \vee \neg p$ is known as the **law of the excluded middle**, and the theorem $\neg (p \wedge \neg p)$ is the **law of noncontradiction**. Notice that by De Morgan's law with Com and DN, we have for all propositional forms p

$$\vdash (p \vee \neg p) \leftrightarrow \neg (p \wedge \neg p).$$

Exercises

1. The **inverse** of a given implication is the contrapositive of the implication's converse. Write the converse, contrapositive, and inverse for each conditional proposition in Exercise 1.1.2.

2. Prove using truth tables.

   (a) $\vDash P \lor \neg P \leftrightarrow P \to P$

   (b) $\vDash P \lor \neg P \leftrightarrow P \lor Q \lor \neg(P \land Q)$

   (c) $\vDash P \land Q \leftrightarrow (P \leftrightarrow Q) \land (P \lor Q)$

   (d) $\vDash R \land (P \to Q) \leftrightarrow R \land (\neg Q \to \neg P)$

   (e) $\vDash P \land Q \to R \leftrightarrow P \land \neg R \to \neg Q$

   (f) $\vDash P \to Q \leftrightarrow \neg Q \to \neg P$

   (g) $\vDash P \to Q \land R \leftrightarrow (P \to Q) \land (P \to R)$

   (h) $\vDash P \to Q \to (S \to R) \leftrightarrow (P \to Q) \land S \to R$

3. A propositional form is in **disjunctive normal form** if it is a disjunction of conjunctions. For example, the propositional form $P \land (\neg Q \lor R)$ is logically equivalent to

$$(P \land Q \land R) \lor (P \land \neg Q \land \neg R) \lor (\neg P \land Q \land \neg R),$$

which is in disjunctive normal form. Find propositional forms in disjunctive normal form that are logically equivalent to each of the following.

   (a) $P \lor Q \land (P \lor \neg R)$

   (b) $(P \lor Q) \land (\neg P \lor \neg Q)$

   (c) $(P \land \neg Q \lor R) \land (Q \land R \lor P \land \neg R)$

   (d) $P \lor (\neg Q \lor [P \land \neg R \lor P \land \neg Q])$

4. Identify the rule from Replacement Rule 1.3.9.

   (a) $(P \to Q) \lor (Q \to R) \lor S \Leftrightarrow (P \to Q) \lor ([Q \to R] \lor S)$

   (b) $\neg\neg P \leftrightarrow Q \land R \Leftrightarrow P \leftrightarrow Q \land R$

   (c) $P \lor Q \lor R \Leftrightarrow Q \lor P \lor R$

   (d) $\neg P \land \neg(Q \lor R) \Leftrightarrow \neg(P \lor [Q \lor R])$

   (e) $\neg(P \lor Q) \lor R \Leftrightarrow P \lor Q \to R$

   (f) $P \leftrightarrow Q \to R \Leftrightarrow P \land (Q \to R) \lor \neg P \land \neg(Q \to R)$

   (g) $P \lor Q \leftrightarrow Q \land Q \Leftrightarrow P \lor Q \leftrightarrow Q$

   (h) $P \land Q \land R \Leftrightarrow R \land (P \land Q)$

   (i) $(P \lor Q) \land (Q \lor R) \Leftrightarrow (P \lor Q) \land Q \lor (P \lor Q) \land R$

   (j) $(P \land [R \to Q]) \to S \Leftrightarrow P \to (R \to Q \to S)$

5. For each given propositional form p, find another propositional form q such that $p \Leftrightarrow q$ using Replacement Rules 1.3.9.

   (a) $\neg\neg P$

   (b) $P \lor Q$

   (c) $P \to Q$

   (d) $\neg(P \land Q)$

   (e) $(P \leftrightarrow Q) \land (\neg P \leftrightarrow Q)$

   (f) $(P \to Q) \lor (Q \to S)$

   (g) $(P \to Q) \lor P$

   (h) $\neg(P \to Q)$

   (i) $(P \to Q) \land (Q \leftrightarrow R)$

(j) $(P \lor \neg Q) \leftrightarrow T \land Q$
(k) $P \lor (\neg Q \leftrightarrow T) \land Q$

6. Arrange each collection of propositional forms into a proof for the given deduction and supply the appropriate reasons.

(a) $P \lor Q \to R \vdash (P \to R) \land (Q \to R)$
- $R \lor \neg P \land \neg Q$
- $(\neg P \lor R) \land (\neg Q \lor R)$
- $\neg (P \lor Q) \lor R$
- $(P \to R) \land (Q \to R)$
- $(R \lor \neg P) \land (R \lor \neg Q)$
- $\neg P \land \neg Q \lor R$
- $P \lor Q \to R$

(b) $\neg (P \land Q) \to R \lor S, \neg P, \neg S \vdash R$
- $\neg (P \land Q) \to R \lor S$
- $\neg S$
- $S \lor R$
- R
- $\neg (P \land Q)$
- $\neg P$
- $R \lor S$
- $\neg P \lor \neg Q$

(c) $P \to (Q \to R), \neg P \to S, \neg Q \to T, R \to \neg R \vdash \neg T \to S$
- $S \lor T$
- $\neg R \lor \neg R$
- $R \to \neg R$
- $\neg R$
- $\neg (P \land Q)$
- $T \lor S$
- $\neg \neg T \lor S$
- $\neg P \lor \neg Q$
- $P \land Q \to R$
- $P \to (Q \to R)$
- $\neg P \to S$
- $\neg Q \to T$
- $\neg T \to S$
- $(\neg P \to S) \land (\neg Q \to T)$

7. Prove.

(a) $\neg P \vdash P \to Q$
(b) $P \vdash \neg Q \to P$
(c) $\neg Q \lor (\neg R \lor \neg P) \vdash P \to \neg (Q \land R)$
(d) $P \to Q \vdash P \land R \to Q$

(e) $P \rightarrow Q \wedge R \vdash P \rightarrow Q$

(f) $P \vee Q \rightarrow R \vdash \neg R \rightarrow \neg Q$

(g) $P \rightarrow (Q \rightarrow R) \vdash Q \wedge \neg R \rightarrow \neg P$

(h) $P \rightarrow (Q \rightarrow R) \vdash Q \rightarrow (P \rightarrow R)$

(i) $P \wedge Q \vee R \wedge S \vdash \neg S \rightarrow P \wedge Q$

(j) $Q \rightarrow R \vdash P \rightarrow (Q \rightarrow R)$

(k) $P \rightarrow \neg(Q \rightarrow R) \vdash P \rightarrow \neg R$

(l) $P \vee (Q \vee R \vee S) \vdash (P \vee Q) \vee (R \vee S)$

(m) $Q \vee P \rightarrow R \wedge S \vdash Q \rightarrow R$

(n) $P \leftrightarrow Q \wedge R \vdash P \rightarrow Q$

(o) $P \leftrightarrow Q \vee R \vdash Q \rightarrow P$

(p) $P \vee Q \vee R \rightarrow S \vdash Q \rightarrow S$

(q) $(P \vee Q) \wedge (R \vee S) \vdash P \wedge R \vee P \wedge S \vee (Q \wedge R \vee Q \wedge S)$

(r) $P \wedge (Q \vee R) \rightarrow Q \wedge R \vdash P \rightarrow (Q \rightarrow R)$

(s) $P \leftrightarrow Q, \neg P \vdash \neg Q$

(t) $P \rightarrow Q \rightarrow R, \neg R \vdash \neg Q$

(u) $P \rightarrow (Q \rightarrow R), R \rightarrow S \wedge T \vdash P \rightarrow (Q \rightarrow T)$

(v) $P \rightarrow (Q \rightarrow R), R \rightarrow S \vee T \vdash (P \rightarrow S) \vee (Q \rightarrow T)$

(w) $P \rightarrow Q, P \rightarrow R \vdash P \rightarrow Q \wedge R$

(x) $P \vee Q \rightarrow R \wedge S, \neg P \rightarrow (T \rightarrow \neg T), \neg R \vdash \neg T$

(y) $(P \rightarrow Q) \wedge (R \rightarrow S), P \vee R, (P \rightarrow \neg S) \wedge (R \rightarrow \neg Q) \vdash Q \leftrightarrow \neg S$

(z) $P \wedge (Q \wedge R), P \wedge R \rightarrow S \vee (T \vee M), \neg S \wedge \neg T \vdash M$

1.4   PROOF METHODS

The methods of Sections 1.2 and 1.3 provide a good start for writing formal proofs. However, in practice we rarely limit ourselves to these rules. We often use inference rules that give a straightforward way to prove conditional propositions and allow us to prove a proposition when it is easier to disprove its negation. In both the cases, the new inference rules will be justified using the rules we already know.

Deduction Theorem

Because of Axioms 1.2.8, not all of the inference rules are needed to write the proofs found in Sections 1.2 and 1.3. This motivates the next definition.

■ DEFINITION 1.4.1

Let p and q be propositional forms. The notation

$$p \vdash_* q$$

means that there exists a formal proof of q from p using only Axioms 1.2.8, MP, and Inference Rule 1.3.10, and the notation

$$\vdash_* q$$

means that there exists a formal proof of q from Axioms 1.2.8 using only MP and Inference Rule 1.3.10

For example, $P, Q, \neg P \vee (Q \rightarrow R) \vdash_* R$ because

1.	P	Given
2.	Q	Given
3.	$\neg P \vee (Q \rightarrow R)$	Given
4.	$P \rightarrow (Q \rightarrow R)$	3 Impl
5.	$Q \rightarrow R$	1, 4 MP
6.	R	2, 5 MP

is a formal proof using MP and Impl as the only inference rules.

We now observe that the propositional forms that can be proved using the full collection of rules from Sections 1.2 and 1.3 are exactly the propositional forms that can be proved when all rules are deleted from Inference Rules 1.2.10 except for MP.

■ THEOREM 1.4.2

For all propositional forms p and q, $p \vdash_* q$ if and only if $p \vdash q$.

PROOF

Trivially, $p \vdash_* q$ implies $p \vdash q$, so suppose that $p \vdash q$. We show that the remaining parts of Inference Rules 1.2.10 are equivalent to using only Axioms 1.2.8, MP, and Replacement Rules 1.3.9. We show three examples and leave the proofs of the remaining inference rules to Exercise 6. The proof

1.	$p \rightarrow q$	Given
2.	$\neg q$	Given
3.	$\neg\neg p \rightarrow \neg\neg q$	DN
4.	$\neg\neg p \rightarrow \neg\neg q \rightarrow (\neg q \rightarrow \neg p)$	FL3
5.	$\neg q \rightarrow \neg p$	3, 4 MP
6.	$\neg p$	2, 5 MP

shows that

$$p \rightarrow q, \neg q \vdash_* \neg p.$$

Thus, we do not need MT. The proof

1.	p	Given
2.	$p \rightarrow (\neg q \rightarrow p)$	FL1
3.	$\neg q \rightarrow p$	1, 2 MP
4.	$\neg\neg q \vee p$	3 Impl
5.	$q \vee p$	4 DN
6.	$p \vee q$	5 Com

shows that

$$p \vdash_* p \vee q,$$

so we do not need Add. This implies that we can use Add to demonstrate that

$$p \to q, q \to r \vdash_* p \to r.$$

The proof is as follows:

1.	$p \to q$	Given
2.	$q \to r$	Given
3.	$\neg q \lor r$	2 Impl
4.	$\neg q \lor r \lor \neg p$	3 Add
5.	$\neg p \lor (\neg q \lor r)$	4 Com
6.	$p \to (q \to r)$	5 Impl
7.	$p \to (q \to r) \to (p \to q \to [p \to r])$	FL2
8.	$p \to q \to (p \to r)$	6, 7 MP
9.	$p \to r$	1, 8 MP

This implies that we do not need HS. ■

At this point, there is an obvious question: If only MP is needed from Inference Rules 1.2.10, why were the other rules included? The answer is because the other inference rules are examples of common reasoning and excluding them would introduce unnecessary complications to the formal proofs. Try reproving some of the deductions of Section 1.2 with only MP and the axioms to confirm this.

When a formal proof in Section 1.3 involved proving an implication, the replacement rule Impl would often appear in the proof. However, as we know from geometry, this is not the typical strategy used to prove an implication. What is usually done is that the antecedent is assumed and then the consequent is shown to follow. That this procedure justifies the given conditional is the next theorem. Its proof requires a lemma.

■ **LEMMA 1.4.3**

Let p and q be propositional forms. If $\vdash_* q$, then $\vdash_* p \to q$.

PROOF

Let $\vdash_* q$. By FL1, we have that $\vdash_* q \to (p \to q)$, so $\vdash_* p \to q$ follows by MP. ■

■ **THEOREM 1.4.4 [Deduction]**

For all propositional forms p and q, $p \vdash q$ if and only if $\vdash p \to q$.

PROOF

Let p and q be propositional forms. Assume that $\vdash p \to q$, so there exists propositional forms $r_0, r_1, \ldots, r_{n-1}$ such that

$$r_0, r_1, \ldots, r_{n-1}, p \to q$$

is a proof, where r_0 is an axiom, r_i is an axiom or $r_0, r_1, \ldots, r_{i-1} \Rightarrow r_i$ for $i > 0$, and $p \to q$ follows from $r_0, r_1, \ldots, r_{n-1}$. Then,

$$p, r_0, r_1, \ldots, r_{n-1}, p \to q, q$$

is also a proof, where the last inference is due to MP. Therefore, $p \vdash q$.

By Theorem 1.4.2, to prove the converse, we only need to prove that

$$\text{if } p \vdash_* q, \text{ then } \vdash_* p \to q.$$

Assume $p \vdash_* q$. First note that if q is an axiom, then $\vdash_* q$, so $\vdash_* p \to q$ by Lemma 1.4.3. Therefore, assume that q is not an axiom. We begin by checking four cases.

- Suppose that the proof has only one propositional form. In this case, we have that $p = q$, so the inference is of the form $p \vdash_* p$. By FL1,

$$\vdash_* p \to (p \to p).$$

Because

$$\begin{aligned} p \to (p \to p) &\Rightarrow \neg p \lor (\neg p \lor p) \\ &\Rightarrow (\neg p \lor \neg p) \lor p \\ &\Rightarrow \neg p \lor p \\ &\Rightarrow p \to p, \end{aligned}$$

we conclude that $\vdash_* p \to p$.

- Next, suppose the proof has two propositional forms and cannot be reduced to the first case. This implies that $p \vdash_* q$ by a single application of a replacement rule. Thus,

$$p \to (\neg q \to p) \vdash_* p \to (\neg q \to q)$$

by a single application of the same replacement rule. Therefore,

$$\begin{aligned} p \to (\neg q \to p) &\Rightarrow p \to (\neg q \to q) \\ &\Rightarrow p \to (\neg\neg q \lor q) \\ &\Rightarrow p \to (q \lor q) \\ &\Rightarrow p \to q. \end{aligned}$$

This implies that $\vdash_* p \to q$.

- We now consider the case when the proof of $p \vdash_* q$ has three propositional forms and q follows by a rule of replacement. Let p, r, q be the proof. This implies that $p \vdash_* r$, which implies

$$\vdash_* p \to r \tag{1.6}$$

because either r is an axiom and Lemma 1.4.3 applies or the previous two cases apply. If q follows from p by a rule of replacement, then $\vdash p \to q$ by the previous case, so assume that q follows from r by a rule of replacement. Thus, $\vdash_* r \to q$, which implies by Lemma 1.4.3 that

$$\vdash_* p \to (r \to q). \tag{1.7}$$

By FL2,
$$\vdash_* p \to (r \to q) \to (p \to r \to [p \to q]). \tag{1.8}$$

Therefore, by (1.7) and (1.8) with MP,

$$\vdash_* p \to r \to (p \to q), \tag{1.9}$$

and by (1.6) and (1.9) with MP,

$$\vdash_* p \to q.$$

- Again, let the proof have three propositional forms and write it as p, s, q. Suppose that the inference that leads to q is MP. This means that r and $r \to q$ are in the proof. Because either $p = r$ and $p \vdash_* r \to q$ or $p = r \to q$ and $p \vdash_* r$,

$$\vdash_* p \to r$$

and

$$\vdash_* p \to (r \to q).$$

Thus, as in the previous case, using (1.8), we obtain $\vdash_* p \to q$.

These four cases exhaust the ways by which q can be proved from p with a proof with at most three propositional forms. Therefore, since these cases can be generalized to proofs of arbitrary length (Exercise 7), we conclude that $p \vdash q$ implies $\vdash p \to q$. ■

The deduction theorem (1.4.4) yields the next result. Its proof is left to Exercise 8.

■ COROLLARY 1.4.5

For all propositional forms $p_0, p_1, \ldots, p_{n-1}, q, r$,

$$p_0, p_1, \ldots, p_{n-1}, q \vdash r$$
$$\text{if and only if}$$
$$p_0, p_1, \ldots, p_{n-1} \vdash q \to r.$$

Direct Proof

Most propositions that mathematicians prove are implications. For example,

if a function is differentiable at a point, it is continuous at that same point.

As we know, this means that whenever the function f is differentiable at $x = a$, it must also be the case that f is continuous at $x = a$. Proofs of conditionals like this are typically very difficult if we are only allowed to use Inference Rules 1.2.10 and Replacement Rules 1.3.9. Fortunately, in practice another inference rule is used. To prove the differentiability result, what is usually done is that f is assumed to be differential at $x = a$ and then a series of steps that lead to the conclusion that f is continuous at $x = a$ are followed. We copy this strategy in our formal proofs using the next rule. Sometimes known as **conditional proof**, this inference rule follows by Corollary 1.4.5 and Theorem 1.2.14.

■ **INFERENCE RULE 1.4.6 [Direct Proof (DP)]**

For propositional forms $p_0, p_1, \ldots, p_{n-1}, q, r$,

$$\text{if } p_0, p_1, \ldots, p_{n-1}, q \vdash r, \text{ then } p_0, p_1, \ldots, p_{n-1} \Rightarrow q \to r.$$

PROOF

Suppose $p_0, p_1, \ldots, p_{n-1}, q \vdash r$. Then, by Corollary 1.4.5,

$$p_0, p_1, \ldots, p_{n-1} \vdash q \to r.$$

Therefore, by Simp and Com,

$$p_0 \wedge p_1 \wedge \cdots \wedge p_{n-1} \vdash q \to r,$$

so by Theorem 1.2.14,

$$p_0 \wedge p_1 \wedge \cdots \wedge p_{n-1} \Rightarrow q \to r.$$

Finally, we have by Conj and Theorem 1.2.14 that

$$p_0, p_1, \ldots, p_{n-1} \Rightarrow q \to r. \ ■$$

To see how this works, let us use direct proof to prove

$$P \vee Q \to (R \wedge S) \vdash P \to R.$$

To do this, we first prove

$$P \vee Q \to (R \wedge S), P \vdash R.$$

Here is the proof:

1.	$P \vee Q \to (R \wedge S)$	Given
2.	P	Given
3.	$P \vee Q$	2 Add
4.	$R \wedge S$	1, 3 MP
5.	R	4 Simp

Therefore, by Inference Rule 1.4.6,

$$P \vee Q \rightarrow (R \wedge S) \Rightarrow P \rightarrow R.$$

A proof of the original deduction can now be written as

1.	$P \vee Q \rightarrow (R \wedge S)$	Given
2.	$P \rightarrow R$	1 DP

However, instead of writing the first proof off to the side, it is typically incorporated into the proof as follows:

1.	$P \vee Q \rightarrow (R \wedge S)$	Given
2.	$\rightarrow P$	Assumption
3.	$P \vee Q$	2 Add
4.	$R \wedge S$	1, 3 MP
5.	R	4 Simp
6.	$P \rightarrow R$	2–5 DP

The proof that P infers R is a **subproof** of the main proof. To separate the propositional forms of the subproof from the rest of the proof, they are indented with a vertical line. The line begins with the assumption of P in line 2 as an additional premise. Hence, its reason is **Assumption**. This assumption can only be used in the subproof. Consider it a local hypothesis. It is only used to prove $P \rightarrow R$. If we were allowed to use it in other places of the proof, we would be proving a theorem that had different premises than those that were given. Similarly, all lines within the subproof cannot be referenced from the outside. We use the indentation to isolate the assumption and the propositional forms that follow from it. When we arrive at R, we know that we have proved $P \rightarrow R$. The next line is this propositional form. It is entered into the proof with the reason DP. The lines that are referenced are the lines of the subproof.

■ **EXAMPLE 1.4.7**

Prove: $P \rightarrow \neg Q, \neg R \vee S \vdash R \vee Q \rightarrow (P \rightarrow S)$

1.	$P \rightarrow \neg Q$	Given
2.	$\neg R \vee S$	Given
3.	$\rightarrow R \vee Q$	Assumption
4.	$\rightarrow P$	Assumption
5.	$\neg Q$	1, 4 MP
6.	$Q \vee R$	3 Com
7.	R	5, 6 DS
8.	$\neg \neg R$	7 DN
9.	S	2, 8 DS
10.	$P \rightarrow S$	4–9 DP
11.	$(R \vee Q) \rightarrow (P \rightarrow S)$	3–10 DP

■ **EXAMPLE 1.4.8**

Prove: $P \wedge Q \to R \to S, \neg Q \vee R \vdash S$

1.	$P \wedge Q \to R \to S$	Given
2.	$\neg Q \vee R$	Given
3.	$\quad P \wedge Q$	Assumption
4.	$\quad Q \wedge P$	3 Com
5.	$\quad Q$	4 Simp
6.	$\quad \neg\neg Q$	5 DN
7.	$\quad R$	2, 6 DS
8.	$P \wedge Q \to R$	3–7 DP
9.	S	1, 8 MP

■ **EXAMPLE 1.4.9**

Prove: $\vdash P \vee \neg P$

1.	$\quad P$	Assumption
2.	$P \to P$	1 DP
3.	$\neg P \vee P$	2 Impl
4.	$P \vee \neg P$	3 Com

Note that lines 1–2 prove $\vdash P \to P$.

Indirect Proof

When direct proof is either too difficult or not appropriate, there is another common approach to writing formal proofs. Sometimes going by the name of **proof by contradiction** or *reductio ad absurdum*, this inference rule can also be used to prove propositional forms that are not implications.

■ **INFERENCE RULE 1.4.10 [Indirect Proof (IP)]**

For all propositional forms p and q,

$$\neg q \to (p \wedge \neg p) \Rightarrow q.$$

PROOF

Notice that instead of repeating the argument from Example 1.4.9 in this proof, the example is simply cited as the reason on line 2.

1.	$\neg q \to (p \wedge \neg p)$	Given
2.	$p \to p$	Example 1.4.9
3.	$\neg(p \wedge \neg p) \to \neg\neg q$	1 Contra
4.	$\neg(p \wedge \neg p) \to q$	3 DN

5.	$\neg p \lor \neg\neg p \to q$	4 DeM
6.	$\neg p \lor p \to q$	5 DN
7.	$p \to p \to q$	6 Impl
8.	q	2, 7 MP

The rule follows from Theorem 1.2.14. ∎

To use indirect proof, assume each premise and assume the negation of the conclusion. Then, proceed with the proof until a contradiction is reached. (In Inference Rule 1.4.10, the contradiction is represented by $p \land \neg p$.) At this point, deduce the original conclusion.

■ EXAMPLE 1.4.11

Prove: $P \lor Q \to R, R \lor S \to \neg P \land T \vdash \neg P$.

1.	$P \lor Q \to R$	Given
2.	$R \lor S \to \neg P \land T$	Given
3.	$\neg\neg P$	Assumption
4.	P	3 DN
5.	$P \lor Q$	4 Add
6.	R	1, 5 MP
7.	$R \lor S$	6 Add
8.	$\neg P \land T$	2, 7 MP
9.	$\neg P$	8 Simp
10.	$P \land \neg P$	4, 9 Conj
11.	$\neg P$	3–10 IP

Since IP involves proving an implication, the formal proof takes the same form as a proof involving DP.

Indirect proof can also be nested within another indirect subproof. As with direct proof, we cannot appeal to lines within a subproof from outside of it.

■ EXAMPLE 1.4.12

Prove: $P \to Q \land R, Q \to S, \neg P \to S \vdash S$.

1.	$P \to Q \land R$	Given
2.	$Q \to S$	Given
3.	$\neg P \to S$	Given
4.	$\neg S$	Assumption
5.	$\neg Q$	2, 4 MT
6.	P	Assumption
7.	$Q \land R$	1, 6 MP
8.	Q	7 Simp
9.	$Q \land \neg Q$	5, 8 Conj

10.	$\neg P$	6-9 IP
11.	S	3, 10 MP
12.	$S \wedge \neg S$	4, 11 Conj
13.	S	4–12 IP

Notice that line 11 was not the end of the proof since it was within the first subproof. It followed under the added hypothesis of $\neg S$.

■ EXAMPLE 1.4.13

Prove: $P \rightarrow R \vdash P \wedge Q \rightarrow R \vee S$.

1.	$P \rightarrow R$	Given
2.	$P \wedge Q$	Assumption
3.	$\neg R$	Assumption
4.	$\neg P$	1, 3 MT
5.	P	2 Simp
6.	$P \wedge \neg P$	4, 5 Conj
7.	R	3-6 IP
8.	$R \vee S$	7 Add
9.	$P \wedge Q \rightarrow R \vee S$	2–8 DP

Exercises

1. Find all mistakes in the given proofs.

    (a) "$P \vee Q \rightarrow \neg R, R \rightarrow \neg Q \rightarrow S \vee Q \vdash S$"

Attempted Proof

1.	$P \vee Q \rightarrow \neg R$	Given
2.	$R \rightarrow \neg Q \rightarrow S \vee Q$	Given
3.	R	Assumption
4.	$\neg \neg R$	Assumption
5.	$\neg (P \vee Q)$	1, 4 MT
6.	$\neg P \wedge \neg Q$	5 DeM
7.	$\neg Q$	6 Simp
8.	$R \rightarrow \neg Q$	3–7 DP
9.	$S \vee Q$	2, 8 MP
10.	$Q \vee S$	9 Com
11.	S	7, 10 DS

    (b) "$\neg P \vee Q \vdash P \rightarrow Q \rightarrow R$"

Attempted Proof

1.	$\neg P \vee Q$	Given
2.	P	Assumption

3.	$\neg\neg P$	2 DN
4.	Q	1, 3 DS
5.	$P \to Q$	2–4 DP
6.	R	MP
7.	$P \to Q \to R$	2–6 DP

(c) "$\neg R \wedge S, \neg P \vee Q \to R \vdash \neg P \vee Q \to Q$"

Attempted Proof

1.	$\neg R \wedge S$	Given
2.	$\neg P \vee Q \to R$	Given
3.	$\neg P$	Assumption
4.	$\neg P \vee Q$	Assumption
5.	R	2, 3 MP
6.	$\neg R$	1 Simp
7.	$R \wedge \neg R$	5, 6 Conj
8.	P	4–7 IP
9.	$\neg\neg P$	8 DN
10.	Q	4, 9 DS
11.	$\neg P \vee Q \to Q$	3–10 DP

2. Prove using direct proof.
   (a) $P \to Q \wedge R \vdash P \to Q$
   (b) $P \vee Q \to R \vdash P \to R$
   (c) $P \vee (Q \vee R) \to S \vdash Q \to S$
   (d) $P \to Q, R \to Q \vdash P \vee R \to Q$
   (e) $P \to Q, P \to R \vdash P \to Q \wedge R$
   (f) $P \to (Q \to R) \vdash Q \wedge \neg R \to \neg P$
   (g) $R \to \neg S \vdash P \wedge Q \to (R \to \neg S)$
   (h) $P \to (Q \to R) \vdash Q \to (P \to R)$
   (i) $P \to (Q \to R), Q \to (R \to S) \vdash P \to (Q \to S)$
   (j) $P \to (Q \to R), R \to S \wedge T \vdash P \to (Q \to T)$
   (k) $P \leftrightarrow Q \wedge R \vdash P \to Q$
   (l) $P \wedge Q \vee R \to Q \wedge R \vdash P \to (Q \to R)$

3. Prove using indirect proof.
   (a) $P \to Q, Q \to R, \neg R \vdash \neg P$
   (b) $P \vee Q \wedge R, P \to S, Q \to S \vdash S$
   (c) $P \vee Q \wedge \neg R, P \to S, Q \to R \vdash S$
   (d) $P \leftrightarrow Q, \neg P \vdash \neg Q$
   (e) $P \vee Q \vee R \to Q \wedge R \vdash \neg P \vee Q \wedge R$
   (f) $P \to Q, Q \to R, S \to T, P \vee R, \neg R \vdash T$
   (g) $P \to \neg Q, R \to \neg S, T \to Q, U \to S, P \vee R \vdash \neg T \vee \neg U$

(h) $P \to Q \wedge R, Q \vee S \to T \wedge U, P \vdash T$

4. Prove using both direct and indirect proof.

    (a) $P \to Q, P \vee (R \to S), \neg Q \vdash R \to S$

    (b) $P \to \neg(Q \to \neg R) \vdash P \to R$

    (c) $P \to Q \vdash P \wedge R \to Q$

    (d) $P \Leftrightarrow Q \vee R \vdash Q \to P$

5. Prove by using direct proof to prove the contrapositive.

    (a) $P \to Q, R \to S, S \to T, \neg Q \vdash \neg T \to \neg(P \vee R)$

    (b) $P \wedge Q \vee R \wedge S \vdash \neg S \to P \wedge Q$

    (c) $P \vee Q \to \neg R, S \to R \vdash P \to \neg S$

    (d) $\neg P \to \neg Q, (\neg R \vee S) \wedge (R \vee Q) \vdash \neg S \to P$

6. Prove to complete the proof of Theorem 1.4.2.

    (a) $p \vee q, \neg p \vdash_* q$

    (b) $p, q \vdash_* p \wedge q$

    (c) $(p \to q) \wedge (r \to s), p \vee r \vdash_* q \vee s$

    (d) $(p \to q) \wedge (r \to s), \neg q \vee \neg s \vdash_* \neg p \vee \neg r$

    (e) $p \wedge q \vdash_* p$

7. Given there is a proof of q from p with four propositional forms, prove $\vdash p \to q$. Generalize the proof for n propositional forms.

8. Prove Corollary 1.4.5.

9. Can MP be replaced with another inference rule in Definition 1.4.1 and still have Theorem 1.4.2 hold true? If so, find the inference rules.

10. Can any of the replacement rules be removed from Definition 1.4.1 and still have Theorem 1.4.2 hold true? If so, how many can be removed and which ones?

1.5   THE THREE PROPERTIES

We finish our introduction to propositional logic by showing that this logical system has three important properties. These are properties that are shared with Euclid's geometry, but they are not common to all logical systems.

Consistency

Since we can need to consider infinitely many propositional forms, we now write our lists of propositional forms as

$$p_0, p_1, p_2, \ldots,$$

allowing this sequence to be finite or infinite. Since proofs are finite, the notation

$$p_0, p_1, p_2, \ldots \vdash q$$

means that there exists a subsequence $i_0, i_1, \ldots, i_{n-1}$ of $0, 1, 2, \ldots$ such that

$$p_{i_0}, p_{i_1}, \ldots, p_{i_{n-1}} \vdash q.$$

The notation

$$p_0, p_1, p_2, \ldots \nvdash q$$

means that no such subsequence exists.

■ **DEFINITION 1.5.1**

- The propositional forms $p_0, p_1, p_2, \ldots$ are **consistent** if for every propositional form q,

$$p_0, p_1, p_2, \ldots \nvdash q \wedge \neg q,$$

and we write $\mathrm{Con}(p_0, p_1, p_2, \ldots)$. Otherwise, $p_0, p_1, p_2, \ldots$ is **inconsistent**.

- A logical system is **consistent** if no contradiction is a theorem.

We have two goals. The first is to show that propositional logic is consistent. The second is to discover properties of sequences of consistent propositional forms that will aid in proving other properties of propositional logic. The next theorem is important to meet both of these goals. The equivalence of the first two parts is known as the **compactness theorem**

■ **THEOREM 1.5.2**

If $p_0, p_1, p_2, \ldots$ are propositional forms, the following are equivalent in propositional logic.

- $\mathrm{Con}(p_0, p_1, p_2, \ldots)$.

- Every finite subsequence of $p_0, p_1, p_2, \ldots$ is consistent.

- There exists a propositional form p such that $p_0, p_1, p_2, \ldots \nvdash p$.

PROOF

We have three implications to prove.

- Suppose there is a finite subsequence $p_{i_0}, p_{i_1}, \ldots, p_{i_{n-1}}$ that proves $q \wedge \neg q$ for some propositional form q. This implies that there is a formal proof of $q \wedge \neg q$ from $p_0, p_1, p_2, \ldots$, therefore, not $\mathrm{Con}(p_0, p_1, p_2, \ldots)$.

- Assume that $p_0, p_1, p_2, \ldots$ proves every propositional form. In particular, if we take a propositional form q, we have that $p_0, p_1, p_2, \ldots \vdash q \wedge \neg q$. This means that there is a finite subsequence $p_{i_0}, p_{i_1}, \ldots, p_{i_{n-1}}$ that proves $q \wedge \neg q$.

- Lastly, assume that there exists a propositional form q such that

$$p_0, p_1, p_2, \ldots \vdash q \wedge \neg q.$$

This means that there exist subscripts $i_0, i_1, \ldots, i_{n-1}$ and propositional forms $r_0, r_1, \ldots, r_{m-1}$ such that

$$p_{i_0}, p_{i_1}, \ldots, p_{i_{n-1}}, r_0, r_1, \ldots, r_{m-1}, q \wedge \neg q$$

is a proof. Take any propositional form p. Then,

$$p_{i_0}, p_{i_1}, \ldots, p_{i_{n-1}}, r_0, r_1, \ldots, r_{m-1}, \neg p, q \wedge \neg q, p$$

is a proof of p by IP. Therefore, $p_0, p_1, p_2, \ldots \vdash p$. ■

A sequence of propositional forms, such as $P \rightarrow Q, P, Q$, although consistent, has the property that there are propositional forms that can be added to the sequence so that the resulting list remains consistent. When the sequence can no longer take new forms and remain consistent, we have arrived at a sequence that satisfies the next definition.

■ DEFINITION 1.5.3

A sequence of propositional forms $p_0, p_1, p_2, \ldots$ is **maximally consistent** whenever $\text{Con}(p_0, p_1, p_2, \ldots)$ and for all propositional forms p, $\text{Con}(p, p_0, p_1, p_2, \ldots)$ implies that $p = p_i$ for some i.

It is a convenient result of propositional logic that every consistent sequence of propositional forms can be extended to a maximally consistent sequence. This is possible because all possible propositional forms can be put into a list. Following Definition 1.1.2, we first list the propositional variables:

$$
\begin{array}{cccccccc}
A, & B, & C, & \ldots & X, & Y, & Z, \\
A_0, & A_1, & A_2, & \ldots \\
B_0, & B_1, & B_2, & \ldots \\
\vdots & \vdots & \vdots \\
Z_0, & Z_1, & Z_2, & \ldots
\end{array}
$$

Then, we list all propositional forms with only one propositional variable:

$$
\begin{array}{cccccccc}
\neg A, & \neg B, & \neg C, & \ldots & \neg X, & \neg Y, & \neg Z, \\
\neg A_0, & \neg A_1, & \neg A_2, & \ldots \\
\neg B_0, & \neg B_1, & \neg B_2, & \ldots \\
\vdots & \vdots & \vdots \\
\neg Z_0, & \neg Z_1, & \neg Z_2, & \ldots
\end{array}
$$

Next, we list all propositional forms with exactly two propositional variables starting by writing A on the right:

$$
\begin{array}{cccccccc}
A \vee A, & B \vee A, & C \vee A, & \ldots & X \vee A, & Y \vee A, & Z \vee A, \\
A_0 \vee A, & A_1 \vee A, & A_2 \vee A, & \ldots \\
B_0 \vee A, & B_1 \vee A, & B_2 \vee A, & \ldots \\
\vdots & \vdots & \vdots \\
Z_0 \vee A, & Z_1 \vee A, & Z_2 \vee A, & \ldots
\end{array}
$$

$$A \wedge A, \quad B \wedge A, \quad C \wedge A, \quad \ldots \quad X \wedge A, \quad Y \wedge A, \quad Z \wedge A,$$
$$A_0 \wedge A, \quad A_1 \wedge A, \quad A_2 \wedge A, \quad \ldots$$
$$B_0 \wedge A, \quad B_1 \wedge A, \quad B_2 \wedge A, \quad \ldots$$
$$\vdots \qquad\quad \vdots \qquad\quad \vdots$$
$$Z_0 \wedge A, \quad Z_1 \wedge A, \quad Z_2 \wedge A, \quad \ldots$$

$$A \to A, \quad B \to A, \quad C \to A, \quad \ldots \quad X \to A, \quad Y \to A, \quad Z \to A,$$
$$A_0 \to A, \quad A_1 \to A, \quad A_2 \to A, \quad \ldots$$
$$B_0 \to A, \quad B_1 \to A, \quad B_2 \to A, \quad \ldots$$
$$\vdots \qquad\quad \vdots \qquad\quad \vdots$$
$$Z_0 \to A, \quad Z_1 \to A, \quad Z_2 \to A, \quad \ldots$$

$$A \leftrightarrow A, \quad B \leftrightarrow A, \quad C \leftrightarrow A, \quad \ldots \quad X \leftrightarrow A, \quad Y \leftrightarrow A, \quad Z \leftrightarrow A,$$
$$A_0 \leftrightarrow A, \quad A_1 \leftrightarrow A, \quad A_2 \leftrightarrow A, \quad \ldots$$
$$B_0 \leftrightarrow A, \quad B_1 \leftrightarrow A, \quad B_2 \leftrightarrow A, \quad \ldots$$
$$\vdots \qquad\quad \vdots \qquad\quad \vdots$$
$$Z_0 \leftrightarrow A, \quad Z_1 \leftrightarrow A, \quad Z_2 \leftrightarrow A, \quad \ldots$$

After following the same pattern by attaching A on the left, we continue by writing $\neg A$ on the right, and then on the left, and then we adjoin B and $\neg B$, etc., and then use 3 propositional variables, and then four, etc. Following a careful path through this infinite list, we arrive at a sequence

$$q_0, q_1, q_2, \ldots$$

of all propositional forms. We are now ready for the theorem.

■ **THEOREM 1.5.4**

A consistent sequence of propositional forms is a subsequence of a maximally consistent sequence of propositional forms.

PROOF

Let $p_0, p_1, p_2, \ldots$ be consistent and $q_0, q_1, q_2, \ldots$ be a sequence of all propositional forms. Define the sequence r_i as follows:

- Let $r_0 = p_0$ and

$$r_1 = \begin{cases} q_0 & \text{if } \mathrm{Con}(q_0, p_0, p_1, p_2, \ldots), \\ p_0 & \text{otherwise,} \end{cases}$$

  so the sequence at this stage is p_0, q_0 or p_0, p_0. Both of these are consistent.

- Let $r_2 = p_1$ and

$$r_3 = \begin{cases} q_1 & \text{if } \mathrm{Con}(q_1, r_0, r_1, r_2, p_0, p_1, p_2, \ldots), \\ p_1 & \text{otherwise.} \end{cases}$$

At this stage, the sequence is still consistent and of the form r_0, r_1, p_1, q_1 or r_0, r_1, p_1, p_1. The first sequence is consistent by the definition of r_3, and the second sequence is consistent because $\text{Con}(r_0, r_1, p_1)$.

- Generalizing, let $r_{2k} = p_k$ and

$$r_{2k+1} = \begin{cases} q_k & \text{if } \text{Con}(q_k, r_0, r_1, \ldots, r_{2k}, p_0, p_1, p_2, \ldots), \\ p_k & \text{otherwise}, \end{cases}$$

resulting in a consistent sequence of the form

$$r_0, r_1, r_2, r_3, \ldots, p_k, q_k$$

or

$$r_0, r_1, r_2, r_3, \ldots, p_k, p_k.$$

Since $p_0, p_1, p_3, \ldots$ is a subsequence of $r_0, r_1, r_2, \ldots$, it only remains to show that the new sequence is maximally consistent.

- Let s be a propositional form such that $r_0, r_1, r_2, \ldots \vdash s \wedge \neg s$. This implies that there exists a sequence i_j such that $i_0 < i_1 < \cdots < i_k$ and

$$r_{i_0}, r_{i_1}, \ldots, r_{i_k} \vdash s \wedge \neg s,$$

but by Theorem 1.5.2, this is impossible because $\text{Con}(r_0, r_1, \ldots, r_{i_k})$.

- Suppose that s is a propositional form so that $\text{Con}(s, r_0, r_1, r_2, \ldots)$. Write $s = q_i$ for some i. Therefore,

$$\text{Con}(q_i, r_0, r_1, \ldots, r_{2i}, p_0, p_1, p_2, \ldots),$$

which means that s is a term of the sequence r_i because q_i was added at step $2i + 1$. ∎

Soundness

We have defined two separate tracks in propositional logic. One track is used to assign T or F to a propositional form, and thus it can be used to determine the truth value of a proposition. The other track focused on developing methods by which one propositional form can be proved from other propositional forms. These methods are used to write proofs in various fields of mathematics. The question arises whether these two tracks have been defined in such a way that they get along with each other. In other words, we want the propositional forms that we prove always to be assigned T, and we want the propositional forms that we always assign T to be provable. This means that we want semantic methods to yield syntactic results and syntactic methods to yield semantic results.

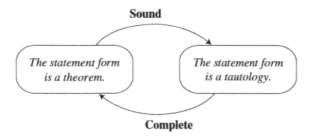

Figure 1.4   Sound and complete logics.

■ DEFINITION 1.5.5

- A logic is **sound** if every theorem is a tautology.

- A logic is **complete** if every tautology is a theorem.

There is no guarantee that the construction of the two tracks for a logic will have these two properties (Figure 1.4), but it does in the case of propositional logic.

To prove that propositional logic is sound, we need three lemmas. The proof of the first is left to Exercise 1.

■ LEMMA 1.5.6

The propositional forms of Axioms 1.2.8 are tautologies.

■ LEMMA 1.5.7

Let p, q, and r be propositional forms.

- If $p \Rightarrow r$, then $p \rightarrow r$ is a tautology.

- If $p, q \Rightarrow r$, then $p \wedge q \rightarrow r$ is a tautology.

PROOF

This is simply a matter of checking Inference Rules 1.2.10 and 1.3.10.

For example, to check the theorem for De Morgan's law, we must show that

$$\neg(p \wedge q) \rightarrow \neg p \vee \neg q$$

and

$$\neg p \vee \neg q \rightarrow \neg(p \wedge q)$$

are tautologies. To do this, examine the truth table

p	q	$p \wedge q$	$\neg(p \wedge q)$	$\neg p$	$\neg q$	$\neg p \vee \neg q$	$\neg(p \wedge q) \rightarrow \neg p \vee \neg q$	$\neg p \vee \neg q \rightarrow \neg(p \wedge q)$
T	T	T	F	F	F	F	T	T
T	F	F	T	F	T	T	T	T
F	T	F	T	T	F	T	T	T
F	F	F	T	T	T	T	T	T

We also have to show that

$$\neg(p \vee q) \rightarrow \neg p \wedge \neg q$$

and

$$\neg p \wedge \neg q \rightarrow \neg(p \vee q)$$

are tautologies.

As another example, to check that the disjunctive syllogism leads to an implication that is a tautology, examining the truth table:

p	q	$p \vee q$	$\neg p$	$(p \vee q) \wedge \neg p$	$(p \vee q) \wedge \neg p \rightarrow q$
T	T	T	F	F	T
T	F	T	F	F	T
F	T	T	T	T	T
F	F	F	T	F	T

The other rules are checked similarly (Exercise 2). ∎

■ LEMMA 1.5.8

If $p \rightarrow q$ and p are tautologies, then q is a tautology.

PROOF

This is done by examining the truth table of $p \rightarrow q$ (page 12) where we see that $v(q)$ is constant and equal to T because $v(p)$ and $v(p \rightarrow q)$ are constant and both equal to T. ∎

We now prove the first important property of propositional logic.

■ THEOREM 1.5.9 [Soundness]

Every theorem of propositional logic is a tautology.

PROOF

Let p be a theorem. Let $p_0, p_1, \dots, p_{n-1}$ be propositional forms such that

$$p_0, p_1, \dots, p_{n-1}$$

is a proof for p such that p_1 is an axiom, p_i with $i > 0$ is an axiom or follows by a rule of inference, and $p_{n-1} = p$ (Definition 1.2.13). We now examine the propositional forms of the proof.

- By Lemma 1.5.6, p_0 is a tautology.

- The propositional form p_1 is a tautology for one of two reasons. If p_1 is an axiom, it is a tautology (Lemma 1.5.6). If it follows from p_0 because $p_0 \Rightarrow p_1$, then $p_0 \rightarrow p_1$ is a tautology (Lemma 1.5.7), so p_1 is a tautology by Lemma 1.5.8.

- If p_2 follows from p_0 or p_1, reason as in the previous case. Suppose $p_0, p_1 \Rightarrow p_2$. Then, by Lemma 1.5.7, $(p_0 \wedge p_1) \rightarrow p_2$ is a tautology. Because $p_0, p_1 \Rightarrow p_0 \wedge p_1$ by Conj, $p_0 \wedge p_1$ is a tautology. Thus, p_2 is a tautology by Lemma 1.5.8.

Since every p_i with $i > 0$ is an axiom, follows from some p_j with $j < i$, or follows from some p_j, p_k with $j, k < i$, continuing in this manner, we find after finitely many steps that p is a tautology. ∎

■ COROLLARY 1.5.10

For all propositional forms $p_0, p_1, \dots, p_{n-1}, q$,

$$\text{if } p_0, p_1, \dots, p_{n-1} \vdash q, \text{ then } p_0, p_1, \dots, p_{n-1} \vDash q.$$

The Law of Noncontradiction being a theorem of propositional logic (page 37) suggests that we have the following result, which is the second important property of propositional logic.

■ COROLLARY 1.5.11

Propositional logic is consistent.

PROOF

Let p be a propositional form. Suppose that $p \wedge \neg p$ is a theorem. This implies that it is a tautology by the soundness theorem (1.5.9), but $p \wedge \neg p$ is a contradiction. ∎

Completeness

We use the consistency of propositional logic to prove that propositional logic is complete. For this we need a few lemmas.

■ LEMMA 1.5.12

If not $\text{Con}(\neg q, p_0, p_1, p_2, \dots)$, then $p_0, p_1, p_2, \dots \vdash q$.

PROOF

If not $\text{Con}(p_0, p_1, p_2, \dots)$, then $p_0, p_1, p_2, \dots \vdash q$ by Theorem 1.5.2, so suppose $\text{Con}(p_0, p_1, p_2, \dots)$. Assume that there exists a propositional form r such that

$$\neg q, p_0, p_1, p_2, \dots \vdash r \wedge \neg r.$$

This implies that there exists a formal proof

$$p_{i_0}, p_{i_1}, \dots, p_{i_{n-1}}, \neg q, s_0, s_1, \dots, s_{m-1}, r \wedge \neg r, \tag{1.10}$$

where $\neg q$ is in the proof because $\text{Con}(p_0, p_1, p_2, \dots)$. Then,

$$p_{i_0}, p_{i_1}, \dots, p_{i_{n-1}}, q$$

is a proof, where q follows by IP with (1.10) as the subproof. Therefore,

$$p_0, p_1, p_2, \ldots \vdash q. \; \blacksquare$$

■ LEMMA 1.5.13

If $p_0, p_1, p_2, \ldots$ are maximally consistent, then for every propositional form q, either $q = p_i$ or $\neg q = p_i$ for some i.

PROOF

Since $p_0, p_1, p_2, \ldots$ are consistent, both q and $\neg q$ cannot be terms of the sequence. Suppose that it is $\neg q$ that is not in the list. By the definition of maximal consistency, we conclude that $\neg q, p_0, p_1, p_2, \ldots$ are not consistent. Therefore, by Lemma 1.5.12, we conclude that $p_0, p_1, p_2, \ldots \vdash q$, and since the sequence is maximally consistent, $q = p_i$ for some i. ■

To prove the next lemma, we use a technique called **induction on propositional forms**. It states that a property will hold true for all propositional forms if two conditions are met:

- The property holds for all propositional variables.

- If the property holds for p and q, the property holds for $\neg p$, $p \wedge q$, $p \vee q$, $p \rightarrow q$, and $p \leftrightarrow q$.

In proving the second condition, we first assume that

$$\text{the property holds for the propositional forms } p \text{ and } q. \qquad (1.11)$$

This assumption (1.11) is known as an **induction hypothesis**. Because

$$p \vee q \Leftrightarrow \neg p \rightarrow q,$$

$$p \wedge q \Leftrightarrow \neg(p \rightarrow \neg q),$$

and

$$p \leftrightarrow q \Leftrightarrow (p \rightarrow \neg q) \rightarrow \neg(\neg p \rightarrow q),$$

we need only to show that the induction hypothesis implies that the property holds for $\neg p$ and $p \rightarrow q$.

■ LEMMA 1.5.14

If $\text{Con}(p_0, p_1, p_2, \ldots)$, there exists a valuation v such that

$$v(p) = T \text{ if and only if } p = p_i$$

for some $i = 0, 1, 2, \ldots$.

PROOF

Since $\text{Con}(p_0, p_1, p_2, \dots)$, we know that

$$p_0, p_1, p_2, \cdots \tag{1.12}$$

can be extended to a maximally consistent sequence of propositional forms by Theorem 1.5.4. If we find the desired valuation for the extended sequence, that valuation will also work for the original sequence, so assume that (1.12) is maximally consistent. Let $X_0, X_1, X_2, \dots$ represent all of the possible propositional variables (page 53). Define

$$v(X_j) = \begin{cases} T & \text{if } X_j \text{ is a propositional variable of } p_i \text{ for some } i, \\ F & \text{otherwise.} \end{cases}$$

We prove that this is the desired valuation by induction on propositional forms. We first claim that for all j,

$$v(X_j) = T \text{ if and only if } X_j = p_i \text{ for some } i.$$

To prove this, first note that if X_j is a term of (1.12), then $v(X_j) = T$ by definition of v. To show the converse, suppose that $v(X_j) = T$ but X_j is not a term of (1.12). By Lemma 1.5.13, there exists i such that $\neg X_j = p_i$. This implies that $v(\neg X_j) = T$, and then $v(X_j) = F$ (Definition 1.1.9), a contradiction.

Now assume that

$$v(q) = T \text{ if and only if } q = p_i \text{ for some } i$$

and

$$v(r) = T \text{ if and only if } r = p_i \text{ for some } i.$$

We first prove that

$$v(\neg q) = T \text{ if and only if } \neg q = p_i \text{ for some } i.$$

- Suppose that $v(\neg q) = T$. Then, $v(q) = F$, and by induction, q is not in (1.12). Since $p_0, p_1, p_2, \dots$ is maximally consistent, $\neg q$ is in the list (Lemma 1.5.13).

- Conversely, let $\neg q = p_i$ for some i. By consistency, q is not in the sequence. Therefore, by the induction hypothesis, $v(q) = F$, which implies that $v(\neg q) = T$.

We next prove that

$$v(q \to r) = T \text{ if and only if } q \to r = p_i \text{ for some } i.$$

- Assume that $v(q \to r) = T$. We have two cases to check. First, let $v(q) = v(r) = T$, so both q and r are in (1.12) by the induction hypothesis. Suppose $q \to r$ is not a term of the sequence. This implies that its negation $\neg(q \rightarrow r)$ is a term of the sequence by Lemma 1.5.13. Therefore, $q \wedge \neg r$ is in the sequence by maximal consistency, so $\neg r$ is also in the sequence, a contradiction. Second, let $v(q) = F$. By induction, q is not a term of the sequence, which implies that $\neg q$ is a term. Hence, $\neg q \vee r$ is in the sequence, which implies that $q \rightarrow r$ is also in the sequence.

- To prove the converse, suppose that $q \rightarrow r = p_i$ for some i. Assume $v(q \rightarrow r) = F$. This means that $v(q) = T$ and $v(r) = F$. Therefore, by induction, q is a term of the sequence but r is not. This implies that $\neg r$ is in the sequence, so $\neg q$ is in the sequence by MT. This contradicts the consistency of (1.12). ∎

■ THEOREM 1.5.15 [Completeness]

Every tautology of propositional logic is a theorem.

PROOF

Let p be a propositional form such that $FL1, FL2, FL3 \nvdash p$. By Lemma 1.5.12, $Con(FL1, FL2, FL3, \neg p)$. Therefore, by Lemma 1.5.14, there exists a valuation such that $v(p) = F$, which implies that $\nvDash p$. ∎

■ COROLLARY 1.5.16

If $p_0, p_1, p_2, \ldots \vDash q$, then $p_0, p_1, p_2 \vdash q$ for all propositional forms $p_0, p_1, p_2, \ldots$.

We conclude by Theorems 1.5.9 and 1.5.15 and their corollaries that the notions of semantically valid and syntactically valid coincide for deductions in propositional logic.

Exercises

1. Prove Lemma 1.5.6.

2. Provide the remaining parts of the proof of Lemma 1.5.7.

3. Let $p, p_0, p_1, p_2, \ldots$ be propositional forms. Prove the following.
   (a) If $p_0, p_1, p_2, \cdots \nvdash p$, then $Con(\neg p, p_0, p_1, p_2, \ldots)$.
   (b) If $Con(p_0, p_1, p_2, \ldots)$ and $p_0, p_1, p_2, \cdots \vdash p$, then $Con(p, p_0, p_1, p_2, \ldots)$.
   (c) If $Con(p_0, p_1, p_2, \ldots)$, then $Con(p, p_0, p_1, p_2, \ldots)$ or $Con(\neg p, p_0, p_1, p_2, \ldots)$.

4. Let $p_0, p_1, p_2, \ldots$ be a maximally consistent sequence of propositional forms. Let p and q be propositional forms. Prove the following.
   (a) If $p_0, p_1, p_2, \cdots \vdash p$, then $p = p_k$ for some k.
   (b) $p \wedge q = p_k$ for some k if and only if $p = p_i$ and $q = p_j$ for some i and j.
   (c) If $(p \rightarrow q) = p_i$ and $p = p_j$ for some i and j, then $q = p_k$ for some k.

5. Use truth tables to prove the following. Explain why this is a legitimate technique.
   (a) $\neg Q \vee (\neg R \vee \neg P) \vdash P \rightarrow \neg(Q \wedge R)$
   (b) $P \vee Q \rightarrow R \vdash \neg R \rightarrow \neg Q$

(c) $P \to \neg(Q \to R) \vdash P \to \neg R$

(d) $P \leftrightarrow Q \vee R \vdash Q \to P$

(e) $P \vee Q \to R \wedge S, \neg P \to (T \to \neg T), \neg R \vdash \neg T$

6. Write a formal proof to show the following. Explain why this is a legitimate technique.

(a) $\neg P \vee Q, \neg Q \vDash \neg P$

(b) $\neg(P \wedge Q), P \vDash \neg Q$

(c) $P \to Q, P \vDash Q \vee R$

(d) $P \to Q, Q \to R, P \vDash R$

(e) $P \vee Q \wedge R, \neg P \vDash R$

7. Write a formal proof to show the following.

(a) $\neg(P \wedge Q) \nvDash \neg P$

(b) $(P \to Q) \vee (R \to S), P \vee R \nvDash Q \vee S$

(c) $P \vee R, Q \vee S, R \leftrightarrow S \nvDash R \wedge S$

8. Write a formal proof to demonstrate the following.

(a) $p \vee \neg p$ is a tautology.

(b) $p \wedge \neg p$ is a contradiction.

9. Modify propositional logic by removing all replacement rules (1.3.9). Is the resulting logic consistent? sound? complete?

10. Modify propositional logic by removing all inference rules (1.2.10) except for Inference Rule 1.3.10. Is the resulting logic consistent? sound? complete?

CHAPTER 2

FIRST-ORDER LOGIC

2.1 LANGUAGES

We developed propositional logic to model basic proof and truth. We did so by using propositional forms to represent sentences that were either true or false. We saw that propositional logic is consistent, sound, and complete. However, the sentences of mathematics involve ideas that cannot be fully represented in propositional logic. These sentences are able to characterize objects, such as numbers or geometric figures, by describing properties of the objects, such as being even or being a rectangle, and the relationships between them, such as equality or congruence. Since propositional logic is not well suited to handle these ideas, we extend propositional logic to a richer system.

Predicates

Consider the sentence *it is a real number*. This sentence has no truth value because the meaning of the word *it* is undetermined. As noted on page 3, the word *it* is like a gap in the sentence. It is as if the sentence was written as

A First Course in Mathematical Logic and Set Theory, First Edition. Michael L. O'Leary.
© 2016 John Wiley & Sons, Inc. Published 2016 by John Wiley & Sons, Inc.

$\boxed{}$ *is a real number.*

However, that gap can be filled. Let us make some **substitutions** for *it*:

5 *is a real number.*
$\pi/7$ *is a real number.*
Fido is a real number.
My niece's toy is a real number.

With each replacement, the word that is undetermined is given meaning, and then the sentence has a truth value. In the examples, the first two propositions are true, and the last two are false.

Notice that $\boxed{}$ changes, whereas *is a real number* remains fixed. This is because these two parts of the sentence have different purposes. The first is a reference to an object, and the second is a property of the object. What we did in the examples was to choose a property and then test whether various objects have that property:

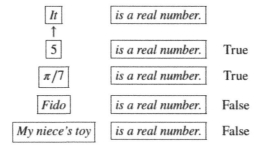

Depending on the result, these two parts are put together to form a sentence that either accurately or inaccurately affirms that a particular property is an attribute of an object. For example, the first sentence states that *being a real number* is a characteristic of 5, which is correct, and the last sentence states that *being a real number* is a characteristic of *my niece's toy*, which is not correct. We have terminology for all of this.

- The subject of a sentence is the expression that refers to an object.

- The predicate of a sentence is the expression that ascribes a property to the object identified by the subject.

Thus, 5, $\pi/7$, *Fido*, and *my niece's toy* are subjects that are substituted for *it*, and *is a real number* is a predicate.

Substituting for the subject but not substituting for the predicate is a restriction that we make. We limit the extension of propositional logic this way because it is to be part of mathematical logic and this is what we do in mathematics. For example, take the sentence,

$$x + 2 = 7.$$

On its own it has no truth value, but when we substitute $x = 5$ or $x = 10$, the sentence becomes a proposition. In this sense, $x + 2 = 7$ is like *it is a real number*. Both the sentences have a gap that is assigned a meaning giving the sentence a truth value.

However, the mathematical sentence is different from the English sentence in that it is unclear as to what part is the subject. Is it x or $x + 2$? For our purposes, the answer is irrelevant. This is because in mathematical logic, the subject is replaced with a variable or, sometimes, with multiple variables. This change leads to a modification of what a predicate is.

■ **DEFINITION 2.1.1**

A **predicate** is an expression that ascribes a property to the objects identified by the variables of the sentence.

Therefore, the sentence $x + 2 = 7$ is a predicate. It describes a characteristic of x. When expressions are substituted for x, the resulting sentence will be either a proposition that affirms that the value added to 2 equals 5 or another predicate. If $x = 5$ is substituted, the result is the true proposition $5 + 2 = 7$. That is, $x + 2 = 7$ is **satisfied** by 5. If $x = 10$ and the substitution is made, the resulting proposition $10 + 2 = 7$ is false. In mathematics, it is also common to substitute with undetermined values. For example, if the substitution is $x = y$, the result is the predicate $y + 2 = 7$, and if the substitution is $x = \sin \theta$, the result is the predicate $\sin \theta + 2 = 7$. Substituting $x = y + 2z^2$ yields $y + 2z^2 + 2 = 7$, a predicate with multiple variables. If we substitute $x = y^2 - 7y$, the result is $y^2 - 7y + 2 = 7$, which is a predicate with multiple **occurrences** of the same variable.

Assume that x represents a real number and consider the sentence

$$\textit{there exists } x \textit{ such that } x + 2 = 7. \tag{2.1}$$

Although there is a variable with 2 occurrences, the sentence is a proposition, so in propositional logic, it is an atom and would be represented by P. This does not tell us much about the sentence, so we instead break the sentence into two parts:

Quantifier		Predicate
$\downarrow$		$\downarrow$
there exists x	such that	$x + 2 = 7$

A **quantifier** indicates how many objects satisfy the sentence. In this case, the quantifier is the **existential quantifier**, making the sentence an **existential proposition**. Such a proposition claims that there is at least one object that satisfies the predicate. In particular, although other sentences mean the same as (2.1) assuming that x represents a real number, such as

$$\textit{there exists a real number } x \textit{ such that } x + 2 = 7,$$

$$x + 2 = 7 \textit{ for some real number } x,$$

and

$$\textit{some real number } x \textit{ satisfies } x + 2 = 7,$$

they all claim along with (2.1) that there is a real number x such that $x + 2 = 7$. This is true because 5 satisfies $x + 2 = 7$ (Figure 2.1).

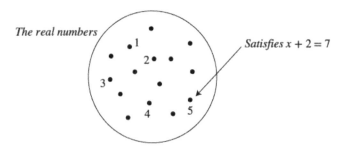

There is at least one object that satisfies $x + 2 = 7$. Therefore,
there exists a real number x such that $x + 2 = 7$ is true.

Figure 2.1

Again, assume that x is a real number. The sentence

$$\text{for all } x, \; x + 5 = 5 + x \tag{2.2}$$

claims that $x + 5 = 5 + x$ is true for every real number x. To see this, break (2.2) like (2.1):

Quantifier		Predicate
↓		↓
for all x	,	$x + 5 = 5 + x$

This quantifier is called the **universal quantifier**. It states that the predicate is satisfied by each and every object. Including propositions that have the same meaning as (2.2), such as

for all real numbers x, $x + 5 = 5 + x$,

and

$x + 5 = 5 + x$ *for every real number x,*

(2.2) is true because each substitution of a real number for x satisfies the predicate (Figure 2.2). These are examples of **universal propositions**,.

A proposition can have multiple quantifiers. Take the equation $y = 2x^2 + 1$. Before we learned the various techniques that make graphing this equation simple, we graphed it by writing a table with one column holding the x values and another holding the y values. We then chose numbers to substitute for x and calculated the corresponding y. Although we did not explicitly write it this way, we learned that

$$\begin{array}{c} \textit{for every real number x,} \\ \textit{there exists a real number y such that } y = 2x^2 + 1. \end{array} \tag{2.3}$$

This is a universal proposition, and its predicate is

there exists a real number y such that $y = 2x^2 + 1$.

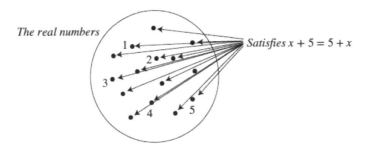

Every object satisfies $x + 5 = 5 + x$. Therefore,
for all real numbers x, x + 5 = 5 + x is true.

Figure 2.2

We conclude that (2.3) is true because whenever x is replaced with any real number, there is a real number y that satisfies $y = 2x^2 + 1$.

The symbolic logic that we define in this chapter is intended to model propositions that have a predicate and possibly a quantifier. As with propositional logic, an alphabet and a grammar will be chosen that enable us to write down the appropriate symbols to represent these sentences. Unlike propositional logic where we worked on both its syntax and semantics at the same time, this logic starts with a study only of its syntax.

Alphabets

No matter what mathematical subject we study, whether it is algebra, number theory, or something else, we can write our conclusions as propositions. These sentences usually involve mathematical symbols particular to the subject being studied. For example, both (2.1) and (2.2) are algebraic propositions. We know this because we recognize the symbols from algebra class. For this reason, the symbolic logic that we define will consist of two types of symbols.

■ **DEFINITION 2.1.2**

Logic symbols consist of the following:

- **Variable symbols**: Sometimes simply referred to as **variables**, these symbols serve as placeholders. On their own, they are without meaning but can be replaced with symbols that do have meaning. A common example are the variable symbols in an algebraic equation. Variable symbols can be lowercase English letters x, y, and z, or lowercase English letters with subscripts, x_n, y_n, and z_n. Depending on the context, variable symbols are sometimes expanded to include uppercase English letters, with or without subscripts, as well as Greek and Hebrew letters. Denote the collection of variable symbols by VAR.

- Connectives: $\neg, \wedge, \vee, \rightarrow, \leftrightarrow$

- **Quantifier symbols**: $\forall$, $\exists$

- **Equality symbol**: $=$

- Grouping symbols: (,), [,], {, }

Theory symbols consist of the following:

- **Constant symbols**: These are used to represent important objects in the subject that do not change. Common constant symbols are 0 and e.

- ***N*-ary function symbols**: The term *n-ary* refers to the number of arguments. For example, these symbols can represent **unary** functions that take one argument such as cosine or **binary** functions such as multiplication that take two arguments.

- ***N*-ary relation symbols**: These symbols are used to represent relations. For example, $<$ represents the binary relation of *less than* and R can represent the unary relation *is an even number*.

It is not necessary to choose any theory symbols. However, if there are any, theory symbols must be chosen so that they are not connectives, quantifier symbols, the equality symbol, or grouping symbols, and the selection of theory symbols has precedence over the selection of variable symbols. This means that these two collections must have no common symbols. Moreover, although this is not the case in general, we assume that the logic symbols that are not variable symbols will be the same for all applications. On the other hand, theory symbols (if there are any) vary depending on the current subject of study. A collection of all logic symbols and any theory symbols is called a **first-order alphabet** and is denoted by A.

The term **theory** refers to a collection of propositions all surrounding a particular subject. Since different theories have different notation (think about how algebraic notation differs from geometric notation), alphabets change depending on the subject matter. Let us then consider the alphabets for a number of theories that will be introduced later in the text. The foundational theory is the first example.

■ **EXAMPLE 2.1.3**

Set theory is the study of collections of objects. The $\in$ is the only relation symbol, and it is binary. It has no other theory symbols. The theory symbols of set theory are denoted by ST and are summarized in the following table.

ST	Constant symbols	Function symbols	Relation symbol
			$\in$

■ **EXAMPLE 2.1.4**

Number theory is the study of the natural numbers. The symbols + and · represent regular addition and multiplication, respectively. As such they are binary function symbols. These and its other theory symbols are indicated in the following table and are denoted by NT.

NT	Constant symbols	Function symbols	Relation symbols
	0 1	+ ·	

Another approach to number theory is called **Peano arithmetic** (Peano 1889). It is the study of the natural numbers using Peano's axioms. It has a constant symbol 0 and a unary function symbol S. Denote these symbols by AR.

AR	Constant symbol	Function symbol	Relation symbols
	0	S	

The Peano arithmetic symbols are sometimes extended to include symbols for the operations of addition and multiplication and the less-than relation. Denote this extended collection by AR'.

AR'	Constant symbol	Function symbols	Relation symbol
	0	S + ·	<

■ **EXAMPLE 2.1.5**

Group theory is the study of groups. A group is a set with an operation that satisfies certain properties. Typically, the operation is denoted by the binary function symbol ∘. There is also a constant represented by e. **Ring theory** is the study of rings. A ring is a set with two operations that satisfy certain properties. The operations are usually denoted by the binary function symbols ⊕ and ⊗. The constant symbol is ○. These collections of theory symbols are denoted by GR and RI, respectively, and are summarized in the following tables.

GR	Constant symbol	Function symbol	Relation symbols
	e	∘	

RI	Constant symbols	Function symbols	Relation symbol
	○	⊕ ⊗	

Field theory is the study of fields. A field is a type of ring with extra properties, so the theory symbols RI can be used to write about fields. However, if an order is defined on a field, a binary relation symbol is needed, and the result is the theory of **ordered fields**. Denote these symbols by OF.

OF	Constant symbol	Function symbols	Relation symbol
	○	⊕ ⊗	<

Notice that the collections of theory symbols in the previous examples had at most two constant symbols. This is typical since subjects of study usually have only a few objects that require special recognition. However, there will be times when some extra constants are needed to reference objects that may or may not be named by the constant theory symbols. To handle these situations, we expand the given theory symbols by adding new constant symbols.

■ **DEFINITION 2.1.6**

Let A be a first-order alphabet with theory symbols S. When constant symbols not in S are combined with the symbols of S, the resulting collection of theory symbols is denoted by $\overline{S}$. The number of new constant symbols varies depending on need.

For example, suppose that we are working in a situation where we need four constant symbols in addition to the ones in OF. Denote these new symbols by c_1, c_2, c_3, and c_4. Then, $\overline{OF}$ consists of these four constant symbols and the symbols from OF.

Terms

For a string to represent a proposition or a predicate from a particular theory, each nonlogic symbol of the string must be a theory symbol of that subject. For example,

$$\forall x (\in x A)$$

and

$$\lor \exists \in ab_4 \{\} \land x \neg y$$

are strings for set theory. However, some strings have a reasonable chance of being given meaning, others do not. As with propositional logic, we need a grammar that will determine which strings are legal for the logic. Because a predicate might have variables, the types of representations that we want to make are more complicated than those of propositional logic. Hence, the grammar also will be more complicated. We begin with the next inductive definition (compare page 59).

■ **DEFINITION 2.1.7**

Let A be a first-order alphabet with theory symbols S. An **S-term** is a string over A such that

- a variable symbol from A is an S-term,

- a constant symbol from S is an S-term,

- $f t_0 t_1 \cdots t_{n-1}$ is an S-term if $t_0, t_1, \ldots, t_{n-1}$ are S-terms and f is a function symbol from S.

Denote the collection of strings over A that are S-terms by TERMS(A).

The string $f t_0 t_1 \cdots t_{n-1}$ is often written as $f(t_0, t_1, \ldots, t_{n-1})$ because it is common to write functions using this notation. We must remember, however, that this notation can always be replaced with the notation of Definition 2.1.7. Furthermore, when writing about a general S-term where it is not important to mention S, we often simply write using the word *term* without the S. We will follow this convention when writing about similar definitions that require the theory symbols S.

■ **EXAMPLE 2.1.8**

Here are examples of terms for each of the indicated theories. Assume that x, y and z_1 are variable symbols.

- 0 [Peano arithmetic]

- $+xy$ [number theory]

- $o0z_1$ [group theory].

The string $+xy$ is typically written as $x+y$, and the string $o0a_1$ is typically written as $0 \circ a_1$. If NT is expanded to $\overline{\text{NT}}$ by adding the constants c and d, then $+cd$ is an $\overline{\text{NT}}$-term.

As suggested by Example 2.1.8, the purpose of a term is to represent an object of study. A variable symbol represents an undetermined object. A constant symbol represents an object that does not change, such as a number. A function symbol is used to represent a particular object given another object. For example, the NT-term $+x2$ represents the number that is obtained when x is added to 2.

Formulas

As propositional forms are used to symbolize propositions, the next definition is the grammar used to represent propositional forms and predicates. The definition is given inductively and resembles the parentheses-less **prefix notation** invented by Łukasiewicz (1951) in the early 1920s.

■ **DEFINITION 2.1.9**

Let S be the theory symbols from a first-order alphabet A. An **S-formula** is a string over A such that

- $t_0 = t_1$ is an S-formula if t_0 and t_1 are S-terms.

- $R t_0 t_1 \cdots t_{n-1}$ is an S-formula if R is an n-ary relation symbol from S and $t_0, t_1, \ldots, t_{n-1}$ are S-terms.

- $\neg p$ is an S-formula if p is an S-formula.

- $p \wedge q$, $p \vee q$, $p \rightarrow q$, and $p \leftrightarrow q$ are S-formulas if p and q are S-formulas.

- $\forall x p$ and $\exists x p$ are S-formulas if p is an S-formula and x is a variable symbol from A.

The string $Rt_0 t_1 \cdots t_{n-1}$ is often represented as $R(t_0, t_1, \ldots, t_{n-1})$. A formula of the form $\forall x p$ is called a **universal formula**, and a formula of the form $\exists x p$ is an **existential formula**. Parentheses can be used around an S-formula for readability, especially when quantifier symbols are involved. For example, $\forall x \exists y p$ is the same formula as $\forall x (\exists y p)$ and $\forall x \exists y (p)$.

■ **EXAMPLE 2.1.10**

Let x and y be variable symbols. Let c be a constant symbol; f, g, and h be unary function symbols; and R be a binary relation symbol from S. The following are S-formulas.

- $x = c$

- $Rcfy$

- $\neg(y = gc)$

- $Rxfx \rightarrow Rfxx$

- $\forall x \neg(fx = fc)$

- $\exists x \forall y (Rfxgy \wedge Rfxhy)$.

In practice, $Rcfy$ is usually written as $R(c, f(y))$, $\neg(y = gc)$ as $y \neq g(c)$, $Rxfx$ as $R(x, f(x))$, $Rfxx$ as $R(f(x), x)$, and $Rfxgy$ as $R(f(x), g(y))$.

■ **EXAMPLE 2.1.11**

These are some ST-formulas with their standard translations.

- $\neg \in x\{\ \}$
  x is not an element of $\{\ \}$.

- $\forall x (\in x A \rightarrow \in x B)$
  For all x, if $x \in A$, then $x \in B$.

- $\neg \exists x \forall y (\in y x)$
  It is not the case that there exists x such that for all y, $y \in x$.

- $x = y \vee \in y x \vee \in x y$
  $x = y$ or $y \in x$ or $x \in y$.

■ **EXAMPLE 2.1.12**

Here are some NT-formulas with their corresponding predicates.

- $\forall x \forall y \forall z [++xyz = +x+yz]$
  $(x + y) + z = x + (y + z)$ for all x, y, and z.

- $\neg(x = 0) \rightarrow \forall y \exists z (z = \cdot xy)$
  If $x \neq 0$, then for all y, there exists z such that $z = xy$.

- $\forall x(\cdot 1x = x)$
  For all x, $1x = x$.

We now name the system just developed.

■ DEFINITION 2.1.13

An alphabet combined with a grammar is called a **language**. The language given by Definitions 2.1.2 and 2.1.9 is known as a **first-order** language. A formula of a first-order language is a **first-order formula**, and all of the formulas of the first-order language with theory symbols S is denoted by L(S).

The first-order language developed to represent predicates, either with quantifiers or without them, is summarized in Figure 2.3. An alphabet that has both logic and theory symbols is chosen. Using a grammar, terms are defined, and then by extending the grammar, formulas are defined. A natural question to ask regarding this system is what makes it *first-order?* Look at the last part of Definition 2.1.9. It gives the rule that allows the addition of a quantifier symbol in a formula. Only $\forall x$ or $\exists x$ are permitted, where x is a variable symbol representing an object of study. Thus, only propositions that begin as *for all x ...* or *there exists x such that ...* can be represented as a first-order formula. To quantify over function and relation symbols, we need to define a **second-order formula**. Augment the alphabet A with function and relation symbols, which creates what is know as a **second-order alphabet**, and then add

$$\forall f p \text{ and } \exists f p \text{ are S-formulas if } p \text{ is an S-formula}$$
$$\text{and } f \text{ is a function symbol from A}$$

and

$$\forall R p \text{ and } \exists R p \text{ are S-formulas if } p \text{ is an S-formula}$$
$$\text{and } R \text{ is a relation symbol from A}$$

to Definition 2.1.9. For example, if the first-order formula

$$\forall x(\in xA \rightarrow \in xB)$$

is intended to be true for any A and B, it can be written as the second-order formula

$$\forall A \forall B \forall x(Ax \rightarrow Bx),$$

where Ax represents the relation $x \in A$ and Bx represents the relation $x \in B$.

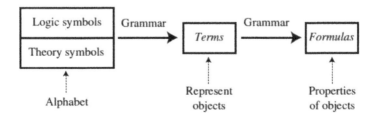

Figure 2.3   A first-order language.

Exercises

1. Determine whether the given strings are GR-terms. If a string is not a GR-term, find all issues that prevent it from being one.

    (a) $2e$

    (b) xyo

    (c) o

    (d) $<xe$

    (e) $ooxye$

2. Write the GR-terms from Exercise 1 in their usual form (Example 2.1.10).

3. Extend ST to ST′ by adding the constant symbol $\varnothing$ and the unary function symbol P. Determine whether the given strings are ST′-terms. If a string is not a ST′-term, find all issues that prevent it from being one.

    (a) $\in x\varnothing$

    (b) $\varnothing$

    (c) x

    (d) $P\in xy$

    (e) Px

4. Determine whether the given strings are OF-formulas. If a given string is not a OF-formula, find all issues that prevent it from being one.

    (a) $< \oplus xy\otimes xy$

    (b) $\oplus x = \otimes y$

    (c) $\forall x \exists y (+xy = 0)$

    (d) $\forall x \forall y (\otimes xy = \oplus yx) \to <xy$

    (e) $\neg(<\otimes xy \wedge \oplus xy) \leftrightarrow \forall u \forall v(<\otimes xy\oplus uv)$

5. Write the RI-formulas from Exercise 4 in their usual form.

6. Extend ST to ST″ by adding the constant symbol $\varnothing$ and the binary relation symbol R. Determine whether the given strings are ST″-formulas. If a given string is not a ST″-formula, find all issues that prevent it from being one.

    (a) $Rx\varnothing \vee \in xy$

    (b) $Rx\varnothing \to \in +xyz$

    (c) $\forall x \exists \varnothing (\neg \in x\varnothing)$

    (d) $\forall x \vee \forall y \vee \forall z(\in \varnothing xyz) \vee \in \varnothing\varnothing$

    (e) $[\exists x(Rx\varnothing) \to \exists u \exists v \exists w(\neg \in uv \leftrightarrow \neg \in uw)] \vee \varnothing = x_1$

7. Write the ST″-formulas from Exercise 6 in their usual form.

8. Translate the given sentence to an S-formula for the given theory symbol.

    (a) For all x, $S(x) \neq 0$. (S = AR)

    (b) For every number x, there is a number y such that $x \circ y = e$. (S = GR)

    (c) If $x < y$ and $y < z$, then $x < z$. (S = OF)

    (d) It is false that $x \in y$, $y \in z$, and $z \in x$ for all x, y, and z. (S = ST)

(e) For every u and v, there exists w such that if $u = v$, then $u + w = v + w$.
($S = RI$ and $+$ should be translated as $\oplus$ in the RI-formula.)

9. Extend NT to NT$'$ by adding the numerals $2, 3, \ldots, 9$. Answer the following questions.

(a) Is $+34 = 7$ a NT$'$-term? Explain.
(b) Is $\cdot 4{+}39$ a NT$'$-term? Explain.
(c) Is $\exists x(+\cdot 4x8 = +3x)$ a NT$'$-formula? If it is, find x.
(d) If possible, give an example of a NT-formula that is not a NT$'$-formula.
(e) If possible, give an example of a NT$'$-formula that is not a NT-formula.

2.2 SUBSTITUTION

As noted in the beginning of Section 2.1, there are times when a substitution will be made for a predicate's variable. For example, in algebra, if $f(x) = 3x^2 + x + 1$, then $f(y) = 3y^2 + y + 1$, $f(2) = 3(2)^2 + 2 + 1$, and $f(\sin x) = 3\sin^2 x + \sin x + 1$. That is, we can substitute with variables, constants, and the results from functions. Therefore, to represent this in formulas, we can replace a variable with a term.

Terms

We begin by defining what it means to substitute for a variable in a term. We use $\Leftrightarrow$ because one string can be replaced with the other string.

■ **DEFINITION 2.2.1 [Substitution in Terms]**

Let S be theory symbols from a first-order alphabet A. Let x be a variable symbol from A and t be an S-term. The notation

$$\frac{t}{x}$$

means that x is replaced with t at every appropriate occurrence of x. For terms, this means the following.

- If y is a variable symbol from A,

$$y\frac{t}{x} \Leftrightarrow \begin{cases} t & \text{if } x = y, \\ y & \text{if } x \neq y. \end{cases}$$

- If c is a constant symbol from S,

$$c\frac{t}{x} \Leftrightarrow c.$$

- If f is an n-ary function symbol from S and $s_0, s_1, \ldots, s_{n-1}$ are S-terms,

$$(f s_0 s_1 \cdots s_{n-1})\frac{t}{x} \Leftrightarrow f\left(s_0\frac{t}{x}s_1\frac{t}{x}\cdots s_{n-1}\frac{t}{x}\right).$$

Observe that when a substitution of a term is made into a term, the result is another term.

■ **EXAMPLE 2.2.2**

Let x, y, and z be distinct variable symbols; c and d be constant symbols; f be a binary function symbol; and g and h be unary function symbols. This means that $fcggd$ is typically written as $f(c, g(g(d)))$.

- $x\dfrac{y}{x} \Leftrightarrow y$

- $x\dfrac{c}{x} \Leftrightarrow c$

- $y\dfrac{fxz}{x} \Leftrightarrow y$

- $c\dfrac{x}{x} \Leftrightarrow c$

- $d\dfrac{c}{x} \Leftrightarrow d$

- $c\dfrac{gx}{x} \Leftrightarrow c$

- $(fxy)\dfrac{c}{y} \Leftrightarrow fxc$

- $\left((fyz)\dfrac{hc}{y}\right)\dfrac{gy}{x} \Leftrightarrow fhcz$

- $\left(\left[\left(z\dfrac{c}{x}\right)\dfrac{d}{y}\right]\dfrac{c}{z}\right)\dfrac{d}{x} \Leftrightarrow c$

- $\left(\left[\left(fxy\dfrac{c}{x}\right)\dfrac{gx}{y}\right]\dfrac{x}{z}\right)\dfrac{gd}{x} \Leftrightarrow fcggd.$

Free Variables

Substitution for a variable in a formula is a bit more involved. This is because of the influence of any possible quantifiers. For example, take the formula

$$x = y \lor x = fy. \tag{2.4}$$

By Definition 2.2.1, we know that we can substitute constants c for x and d for y in the terms x, y, and fy. We expect that we should also be able to make this substitution into the formula resulting in

$$c = d \lor c = fd.$$

However, in the formula

$$\forall x \exists y(x = y \lor x = fy), \tag{2.5}$$

the situation is different because of the quantifiers. Consider the **corresponding occurrences** in (2.4) and its quantified (2.5):

$$x = y \lor x = f y$$

$$\forall x \exists y (x = y \lor x = f y)$$

Even though each occurrence of x and y in (2.4) can receive a substitution, each corresponding occurrence in (2.5) cannot because of the quantifiers.

■ DEFINITION 2.2.3

Let S be theory symbols. Let $t_0, t_1, \ldots, t_{n-1}$ be S-terms, R be an n-ary relation symbol from S, and p and q be S-formulas. An occurrence of a variable in a formula is **free** or not free only according to the following rules:

- A variable occurrence in $t_0 = t_1$ and $R t_0 t_1 \cdots t_{n-1}$ is free.

- A variable occurrence in $\neg p$ is free if and only if the corresponding occurrence in p is free.

- A variable occurrence in $p \land q$, $p \lor q$, $p \to q$, and $p \leftrightarrow q$ is free if and only if the corresponding occurrence in p or q is free.

- Any occurrence of x in $\forall x p$ and $\exists x p$ is not free.

- Any occurrence of $y \neq x$ is free in $\forall x p$ and $\exists x p$ if and only if the corresponding occurrence of y is free in p.

If an occurrence of a variable is not free, it is **bound**. All free occurrences of x in p are within the **scope** of the universal quantifier in $\forall x p$ and the existential quantifier in $\exists x p$.

■ EXAMPLE 2.2.4

Let f be a unary function symbol and R be a 3-ary relation symbol. In the formula

$$\forall x \exists y (x = y \lor f x = y \to R x y z),$$

all occurrences of x and y are bound, but the occurrence of z is free. In the formula

$$\exists y (x = y \lor f x = y \to R x y z),$$

all occurrences of x and z are free, but the occurrences of y are bound. In

$$x = y \lor f x = y \to R x y z,$$

all occurrences are free because all occurrences are free in $x = y$, $f x = y$, and $R x y z$.

We need to know whether a formula has a free occurrence of a variable, so we make the next definition.

■ **DEFINITION 2.2.5**

A **free variable** of the S-formula p is a variable that has a free occurrence in p.

■ **EXAMPLE 2.2.6**

Using the f and R from Example 2.2.4, both x and y are free variables of the formula

$$Rxyc \to \exists x(fy = c),$$

even though x has both free and bound occurrences.

Formulas

We can now define what it means to make a substitution into a formula.

■ **DEFINITION 2.2.7 [Substitution in Formulas]**

Let S be theory symbols from a first-order alphabet A. Let x and y be variable symbols of A and R be an n-ary relation symbol from S. Suppose that p and q are S-formulas and $t, t_0, t_1, \ldots, t_{n-1}$ are S-terms.

- $(t_0 = t_1)\dfrac{t}{x} \Leftrightarrow t_0\dfrac{t}{x} = t_1\dfrac{t}{x}$

- $(Rt_0t_1\cdots t_{n-1})\dfrac{t}{x} \Leftrightarrow Rt_0\dfrac{t}{x}t_1\dfrac{t}{x}\cdots t_{n-1}\dfrac{t}{x}$

- $(\neg p)\dfrac{t}{x} \Leftrightarrow \neg\left(p\dfrac{t}{x}\right)$

- $(p \wedge q)\dfrac{t}{x} \Leftrightarrow p\dfrac{t}{x} \wedge q\dfrac{t}{x}$

- $(p \vee q)\dfrac{t}{x} \Leftrightarrow p\dfrac{t}{x} \vee q\dfrac{t}{x}$

- $(p \to q)\dfrac{t}{x} \Leftrightarrow p\dfrac{t}{x} \to q\dfrac{t}{x}$

- $(p \leftrightarrow q)\dfrac{t}{x} \Leftrightarrow p\dfrac{t}{x} \leftrightarrow q\dfrac{t}{x}$

- $(\forall y p)\dfrac{t}{x} \Leftrightarrow \begin{cases} \forall y\left(p\dfrac{t}{x}\right) & \text{if } x \neq y \text{ and } y \text{ is not a symbol of } t \\ \forall y p & \text{otherwise} \end{cases}$

- $(\exists y p)\dfrac{t}{x} \Leftrightarrow \begin{cases} \exists y\left(p\dfrac{t}{x}\right) & \text{if } x \neq y \text{ and } y \text{ is not a symbol of } t \\ \exists y p & \text{otherwise.} \end{cases}$

The condition on the term t in the last two parts of Definition 2.2.7 is important. Consider the RI-formula

$$p := \exists y(x \oplus y = \bigcirc).$$

The usual interpretation of p is that given x, there exists an number y such that $x+y = 0$. Let f be a unary function symbol and z be a variable symbol. Since y is not a symbol of z,

$$p\frac{z}{x} \Leftrightarrow \exists y(z \oplus y = \bigcirc),$$

which has the same standard interpretation as p. Since y is not a symbol of fz, we can substitute to find that

$$p\frac{fz}{x} \Leftrightarrow \exists y(fz \oplus y = \bigcirc).$$

This is a reasonable substitution because it states that for the number given by fz, there is a number y that when added to fz gives 0. This is very similar to the standard interpretation of p. Both of these substitutions work because the number of free occurrences is unchanged by the substitution. However, if we allow the term to include y among its symbols, the substitution $p\frac{y}{x}$ would yield

$$\exists y(y \oplus y = \bigcirc). \tag{2.6}$$

The typical interpretation of (2.6) is that there exists a number y such that $y + y = 0$. This is a reasonable proposition, but not in the spirit of the original formula p. The change of interpretation is due to the change in the number of free occurrences. The formula p has one free occurrence, while (2.6) has none. Therefore, when making a substitution, the number of free occurrences should not change, and for this reason,

$$\exists y(x \oplus y = \bigcirc)\frac{y}{x} \Leftrightarrow \exists y(x \oplus y = \bigcirc).$$

■ **EXAMPLE 2.2.8**

Let p be the NT formula

$$\forall x_0 \forall x_1(x_0 = x_1 \rightarrow x_0 + x_2 = x_1 + x_2).$$

Notice that x_2 is a free variable of p. However, all occurrences of x_0 and x_1 are bound. Therefore,

$$p\frac{0}{x_0} \Leftrightarrow \forall x_0 \forall x_1(x_0 = x_1 \rightarrow x_0 + x_2 = x_1 + x_2),$$

and then letting y be a variable symbol,

$$\left(p\frac{0}{x_0}\right)\frac{y}{x_1} \Leftrightarrow \forall x_0 \forall x_1(x_0 = x_1 \rightarrow x_0 + x_2 = x_1 + x_2),$$

and finally,

$$\left[\left(p\frac{0}{x_0}\right)\frac{y}{x_1}\right]\frac{1}{x_2} \Leftrightarrow \forall x_0 \forall x_1(x_0 = x_1 \rightarrow x_0 + 1 = x_1 + 1).$$

This last formula has no free variables. A standard interpretation of this formula is for every integer x_0 and x_1, if $x_0 = x_1$, then $x_0 + 1 = x_1 + 1$.

Furthermore,

$$p\frac{x_1}{x_2} \Leftrightarrow [\forall x_0 \forall x_1 (x_0 = x_1 \rightarrow x_0 + x_2 = x_1 + x_2)]\frac{x_1}{x_2}$$

$$\Leftrightarrow \forall x_0 [\forall x_1 (x_0 = x_1 \rightarrow x_0 + x_2 = x_1 + x_2)]\frac{x_1}{x_2}$$

$$\Leftrightarrow \forall x_0 \forall x_1 (x_0 = x_1 \rightarrow x_0 + x_2 = x_1 + x_2).$$

Let p represent the NT-formula $+y2 = 7$. Observe that p has y as a free variable. To emphasize this, instead of writing

$$p := +y2 = 7,$$

we often denote the formula by $p(y)$ and write

$$p(y) := +y2 = 7.$$

Although x is not a variable in the equation, we can also write

$$q(x, y) := +y2 = 7.$$

For example, if we wanted to interpret the formula as an equation with one variable, we would use $p(y)$. If we wanted to view it as the horizontal line $y = 5$, we would use $q(x, y)$.

■ **DEFINITION 2.2.9**

Let A be a first-order alphabet with theory symbols S. Let p be an S-formula. If p has no free variables or the free variables of p are among the distinct variables $x_0, x_1, \ldots, x_{n-1}$ from A, define

$$p(x_0, x_1, \ldots, x_{n-1}) \Leftrightarrow p.$$

The notation of Definition 2.2.9 can also be used to represent substitutions. Consider the formula $p := x + y = 0$. Observe that

$$\left(p\frac{x}{y}\right)\frac{y}{x} \Leftrightarrow y + y = 0$$

and

$$\left(p\frac{y}{x}\right)\frac{x}{y} \Leftrightarrow x + x = 0.$$

However, suppose that we want to substitute into p so that the result is $y + x = 0$. To accomplish this, we need two new and distinct variable symbols, u and v. Then,

$$\left(p\frac{u}{x}\right)\frac{v}{y} \Leftrightarrow u + v = 0, \tag{2.7}$$

and then,

$$\left(p\frac{y}{u}\right)\frac{x}{v} \Leftrightarrow y + x = 0. \tag{2.8}$$

Therefore,

$$\left(\left[\left(p\frac{u}{x}\right)\frac{v}{y}\right]\frac{y}{u}\right)\frac{x}{v} \Leftrightarrow y + x = 0.$$

This works because in (2.7), each variable symbol is replaced by a new symbol in such a way that the resulting formula has the same meaning as the original. In this way, when the original variable symbols are brought back in (2.8), all of the substitutions are made into distinct variables so that there are no conflicts and the switch can be made. This process can be generalized to any number of variable symbols in any order.

■ **DEFINITION 2.2.10**

Let $p(x_0, x_1, \ldots, x_{n-1})$ be an S-formula and $u_0, u_1, \ldots, u_{n-1}$ be distinct variable symbols not among $x_0, x_1, \ldots, x_{n-1}$. Define

$$\overline{p}(u_0, u_1, \ldots, u_{n-1}) \Leftrightarrow \left(\cdots\left[\left(p\frac{u_0}{x_0}\right)\frac{u_1}{x_1}\right]\cdots\right)\frac{u_{n-1}}{x_{n-1}}.$$

Then, for all S-terms $t_0, t_1, \ldots, t_{n-1}$, define

$$p(t_0, t_1, \ldots, t_{n-1}) \Leftrightarrow \left(\cdots\left[\left(\overline{p}\frac{t_0}{u_0}\right)\frac{t_1}{u_1}\right]\cdots\right)\frac{t_{n-1}}{u_{n-1}}.$$

This is called a **simultaneous substitution** and is equivalent to replacing all free occurrences of $x_0, x_1, \ldots, x_{n-1}$ in p with $t_0, t_1, \ldots, t_{n-1}$, respectively.

Observe that if p does not have free variables, then $p(t_0, t_1, \ldots, t_{n-1}) \Leftrightarrow p$.

■ **EXAMPLE 2.2.11**

To illustrate Definition 2.2.10, let RI′ be the theory symbols of RI combined with constant symbols 1 and 2. Let p be the RI′-formula

$$x \otimes (y \oplus z) = x \otimes y \oplus x \otimes z.$$

Since x, y, and z are free variables, represent p by $p(x, y, z, w)$. Then,

$$p(1, x, y, 2) \Leftrightarrow \left(\left[\left[\left(\overline{p}\frac{1}{u_1} \right) \frac{x}{u_2} \right] \frac{y}{u_3} \right) \frac{2}{u_4} \right.$$

$$\Leftrightarrow \left(\left[\left[\left([u_1 \otimes (u_2 \oplus u_3) = u_1 \otimes u_2 \oplus u_1 \otimes u_3] \frac{1}{u_1} \right) \frac{x}{u_2} \right] \frac{y}{u_3} \right) \frac{2}{w} \right.$$

$$\Leftrightarrow \left(\left[\left[(1 \otimes (u_2 \oplus u_3) = 1 \otimes u_2 \oplus 1 \otimes u_3) \frac{x}{u_2} \right] \frac{y}{u_3} \right) \frac{2}{w} \right.$$

$$\Leftrightarrow \left([1 \otimes (x \oplus u_3) = 1 \otimes x \oplus 1 \otimes u_3] \frac{y}{u_3} \right) \frac{2}{w}$$

$$\Leftrightarrow (1 \otimes (x \oplus y) = 1 \otimes x \oplus 1 \otimes y) \frac{2}{w}$$

$$\Leftrightarrow 1 \otimes (x \oplus y) = 1 \otimes x \oplus 1 \otimes y.$$

■ **EXAMPLE 2.2.12**

Let S have constant symbols 5 and 9. Define $p(x, y)$ to be the S-formula

$$\forall x \exists z \, [q(x, y) \wedge r(z)] \vee \exists y \, [r(y) \to s(x)] \,.$$

The first occurrence of y is free, and since $\forall x$ applies only to variables within the brackets on the left, the last occurrence of x is also free. Therefore, $p(x, y)$ is the disjunction of two formulas that we call $u(y)$ and $v(x)$:

$$\overbrace{\forall x \exists z \, [q(x, y) \wedge r(z)]}^{u(y)} \vee \overbrace{\exists y \, [r(y) \to s(x)]}^{v(x)},$$

from which we derive

$$p(9, 5) \Leftrightarrow u(5) \vee v(9) \Leftrightarrow \forall x \exists z [q(x, 5) \wedge r(z)] \vee \exists y [r(y) \to s(9)]$$

As in Example 2.2.11, finding $p(9, 5)$ is equivalent to simply replacing the free occurrences of x with 9 and the occurrences of y with 5.

■ **EXAMPLE 2.2.13**

Consider the NT-formula,

$$\forall x \forall y \forall z \, [++xyz = +x+yz] , \tag{2.9}$$

which is often written as

$$\forall x \forall y \forall z \, [(x + y) + z = x + (y + z)] . \tag{2.10}$$

The formula within the scope of the first quantifier symbol in (2.10) is

$$p(x) := \forall y \forall z \, [(x + y) + z = x + (y + z)] . \tag{2.11}$$

Notice that the occurrences of x are free in (2.11), but the occurrences of y and z are bound. For example, we can make the substitution

$$p(2) \Leftrightarrow \forall y \forall z [(2 + y) + z = 2 + (y + z)].$$

Now, letting

$$q(x, y) := \forall z [(x + y) + z = x + (y + z)],$$

we have that $p(x) \Leftrightarrow \forall y q(x, y)$. We can also define

$$r(x, y, z) := (x + y) + z = x + (y + z)$$

so that $q(x, y) \Leftrightarrow \forall z r(x, y, z)$. What we have done is to break apart an NT-formula that contains multiple quantifiers into a sequence of formulas:

$$\overbrace{\forall x \forall y \forall z \underbrace{[(x + y) + z = x + (y + z)]}_{r(x, y, z)}}^{p(x)}.$$

$$\underbrace{}_{q(x, y)}$$

We conclude that (2.10) can be written as

$$\forall x \, p(x),$$

$$\forall x \forall y \, q(x, y),$$

or

$$\forall x \forall y \forall z \, r(x, y, z).$$

Observe that (2.9) has no free variables. It is a type of formula of particular importance because it represents a proposition. In other words, (2.9) is a propositional form.

■ **DEFINITION 2.2.14**

An S-formula with no free variables is called an **S-sentence**.

Exercises

1. Given a term, make the indicated substitution.

   (a) $\left(4\dfrac{5}{x}\right)\dfrac{6}{y}$

   (b) $\left[\left(x\dfrac{a}{y}\right)\dfrac{f(x)}{x}\right]\dfrac{b}{x}$

   (c) $\left[(x + y)\dfrac{a}{x}\right]\dfrac{g(8)}{y}$

   (d) $\left[\left(\left[(x + y + z)\dfrac{r}{x}\right]\dfrac{b}{y}\right)\dfrac{b}{z}\right]\dfrac{5}{x}$

2. Given a formula, make the indicated substitution.

(a) $\left[(x < 2)\dfrac{y}{x}\right]\dfrac{4}{x}$

(b) $\left(\left[(x < 5 \rightarrow x + 4 < 9)\dfrac{y}{x}\right]\dfrac{x}{y}\right)\dfrac{3}{x}$

(c) $\left(\left[\left(\left[(x = 5 \wedge x = y)\dfrac{5}{x}\right]\dfrac{u}{y}\right)\dfrac{v}{x}\right]\dfrac{6}{u}\right)\dfrac{b}{v}$

(d) $\left([\exists x(x - 4 = y) \vee \forall y(y + z = x + 3)]\dfrac{x}{y}\right)\dfrac{x}{z}$

3. Identify all free occurrences of x in the given formulas.

(a) $x + 4 < 10$

(b) $\exists x \forall y(x + y = 0)$

(c) $\forall x \forall y(x + y = y + x) \vee \forall z(x + y = z - 3)$

(d) $\forall x[\forall z \forall y(x + y = 2 \cdot z) \leftrightarrow x + x = 2 \cdot x]$

4. Identify all bound occurrences of x in the given formulas.

(a) $(\forall x)[p(x) \rightarrow (\exists y)q(y)]$

(b) $(\exists x)p(x, y) \rightarrow (\forall y)q(x)$

(c) $(\exists x)(x > 4) \wedge x < 10$

(d) $[(\forall x)(x + 3 = 1) \wedge x = 9] \vee (\exists y)(x < 0)$

5. Make the simultaneous substitution $p(a)$ for each of the given formulas.

(a) $p(x) := 2x + a = 9$

(b) $p(x) := \forall x(x + y = y + x)$

(c) $p(x) := \exists x q(x) \rightarrow \forall y r(x, y)$

(d) $p(x) := x + 4 = 0 \wedge \exists x(y + z = x) \vee \exists z(x + z = 3)$

6. Make the simultaneous substitution $q(1, 2, 3)$ for each of the given formulas.

(a) $q(x, y, z) := u + v + w$

(b) $q(x, y, z) := r(x, y) \rightarrow (\forall x)[p(x) \wedge (\exists y)t(y, z)]$.

(c) $q(x, y, z) := \exists x(x + y = z) \wedge \forall y(x + y + z) \vee \forall x \forall y(x + y + z)$

(d) $q(x, y, z) := \exists x[p(x) \wedge \forall y(p(x) \wedge p(y) \rightarrow x = y \vee y = z)]$

7. Identify the formula to the right of each quantifiers in the given formulas.

(a) $(\forall x)[p(x) \rightarrow (\exists y)q(y)]$

(b) $(\exists x)p(x, y) \rightarrow (\forall y)q(y)$

(c) $(\forall x)(\exists y)(\exists z)p(x, y, z) \wedge r(w)$

(d) $p(x) \wedge (\exists x)(\forall y)q(x, y) \vee (\forall z)r(z)$

8. Which of the following are sentences.

(a) $\forall x \forall y[q(x) \vee r(y)] \Rightarrow \forall y[q(a) \vee r(y)]$

(b) $q(2) \wedge t(3) \Rightarrow \exists y[q(2) \wedge t(y)]$

(c) $\exists w[\exists x[p(x) \vee \exists z q(z) \leftrightarrow \exists y[p(x) \wedge q(y)] \rightarrow \forall x r(x)] \rightarrow \forall y r(x)]$

(d) $\forall x \forall y(p(x) \rightarrow [q(y) \wedge r(z)]) \rightarrow \exists x \neg q(x)$

2.3 SYNTACTICS

Since formulas without free variables are propositional forms, we can write proofs involving them using the rules from Sections 1.2 through 1.4. However, since the inference rules did not involve quantification, we need new rules to deal with universal and existential formulas. We need rules covering not only negations but also rules that enable the removal (**instantiation**) and adjoining (**generalization**) of quantifiers. We add these rules to Definition 1.2.13 to obtain a stronger notion of proof. Furthermore, since every sentence is a propositional form, every reference to a propositional form in Definition 1.2.13 can be considered to be a reference to a sentence. This allows us to write formal proofs using first-order languages.

Quantifier Negation

Consider the proposition

all rectangles are squares.

This sentence is false because there is a rectangle in which one side is twice the length of the adjacent side, so

not all rectangles are squares.

That is,

some rectangle is not a square

is true. Generalizing, we conclude that

the negation of *all P are Q* is *some P are not Q*.

This should be translated as an inference rule for formulas, so we assume

$$\neg \forall x p \Rightarrow \exists x \neg p. \tag{2.12}$$

Now consider the proposition

some rectangles are round.

This is false because there are no round rectangles, so

all rectangles are not round

is true. Generalizing, we conclude that

the negation of *some P are Q* is *all P are not Q*.

Again, this should be translated as an inference rule for formulas, so we assume

$$\neg \exists x p \Rightarrow \forall x \neg p. \tag{2.13}$$

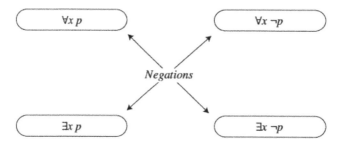

Figure 2.4   The modern square of opposition.

Furthermore, by DN and (2.13),

$$\exists x\neg p \Rightarrow \neg\neg\exists x\neg p \Rightarrow \neg\forall x\neg\neg p \Rightarrow \neg\forall x p,$$

and by DN with (2.12),

$$\forall x\neg p \Rightarrow \neg\neg\forall x\neg p \Rightarrow \neg\exists x\neg\neg p \Rightarrow \neg\exists x p.$$

We summarize assumptions (2.12) and (2.13) and the two conclusions in the following replacement rule.

■ REPLACEMENT RULES 2.3.1 [Quantifier Negation (QN)]

Let S be theory symbols and p be an S-formula.

- $\neg\forall x p \Leftrightarrow \exists x\neg p$

- $\neg\exists x p \Leftrightarrow \forall x\neg p.$

QN is illustrated with the **modern square of opposition** (Figure 2.4). Negations of quantified formulas are found at opposite corners. A version of the Square is found in Aristotle's *De Interpretatione*, dating around 350 BC (Aristotle 1984).

Whenever we negate formulas of the form $\forall x p$ or $\exists x p$, to make it easier to read, the final form should not have a negation immediately to the left of any quantifier, and using the replacement rules, the negation should be as far into the formula p as possible. We say that such negations are in **positive form**.

■ EXAMPLE 2.3.2

Find the negation of $\forall x(p \wedge q)$ and put the final answer in positive form.

$$\neg\forall x(p \wedge q) \Leftrightarrow \exists x\neg(p \wedge q) \Leftrightarrow \exists x(\neg p \vee \neg q).$$

The next example will use De Morgan's law as the last one did. It also needs material implication and double negation.

◼ EXAMPLE 2.3.3

Find the negation of $\forall x \exists y[p(x) \rightarrow q(y)]$ and put it into positive form.

$$\neg\forall x \exists y[p(x) \rightarrow q(y)] \Leftrightarrow \exists x\neg\exists y[p(x) \rightarrow q(y)]$$
$$\Leftrightarrow \exists x\forall y\neg[p(x) \rightarrow q(y)]$$
$$\Leftrightarrow \exists x\forall y\neg[\neg p(x) \vee q(y)]$$
$$\Leftrightarrow \exists x\forall y[p(x) \wedge \neg q(y)].$$

Proofs with Universal Formulas

Consider the sentence *all multiples of 4 are even*. This implies, for instance, that 8, 100, and -16 are even. To generalize this reasoning to formulas means that whenever we have $\forall x p(x)$, we also have $p(a)$, where a might be either a constant symbol such as 8, 100, and -16 or a randomly chosen constant symbol. This gives the first inference rule that involves a quantifier. We do not prove it, but take it to be an axiom. Observe the use of Definition 2.1.6

◼ INFERENCE RULE 2.3.4 [Universal Instantiation (UI)]

If $p(x)$ is an S-formula, then for every constant symbol a from $\overline{S}$,

$$\forall x p(x) \Rightarrow p(a),$$

We make two observations about UI. First, since the resulting formula is to be part of a proof, the substitution must yield a sentence, so a must be a constant symbol. Second, we use the notation $p(x)$ to represent the formula instead of p because $\forall x p(x)$ will be part of a proof, which means, again, that it must be a sentence. Writing $p(x)$ limits the formula to have only x as a possible free variable, so $\forall x p(x)$ is a sentence. If the formula p had free variables other than x, then $\forall x p$ would not be a sentence and not suitable for a proof.

◼ EXAMPLE 2.3.5

Let p, q, and r be S-formulas and a and b be constant symbols from $\overline{S}$. The following are legitimate uses of UI. Notice that each of the inferences results in an $\overline{S}$-formula.

- $\forall x [p(x) \rightarrow q(x)] \Rightarrow p(a) \rightarrow q(a)$

- $\forall x [p(x) \vee \forall y q(y)] \Rightarrow p(a) \vee \forall y q(y)$

- $\forall x\forall y [q(x) \vee r(y)] \Rightarrow \forall y [q(a) \vee r(y)]$

- $\forall y [q(a) \vee r(y)] \Rightarrow q(a) \wedge r(a)$

- $\forall y [q(a) \vee r(y)] \Rightarrow q(a) \wedge r(b).$

Before we can write formal proofs, we need a rule that will attach a universal quantifier. It will be different from universal instantiation, for it requires a criterion on the constant.

■ DEFINITION 2.3.6

Let a be a constant symbol introduced into a formal proof by a substitution. If the first occurrence of a is in a sentence that follows by UI, then a is **arbitrary**. Otherwise, a is **particular** and can be denoted by $\hat{a}$ to serve as a reminder that the symbol is not arbitrary.

The idea behind Definition 2.3.6 is that if a constant symbol a is arbitrary, it represents a randomly selected object, but if a is particular, then a represents an object with at least one known property. This property can be identified by a formula $p(x)$ so that we have $p(a)$.

■ EXAMPLE 2.3.7

The constant symbol a in the following is not arbitrary because its first occurrence of a in line 1 is not the result of UI.

$$
\begin{array}{llll}
1. & p(a) & & \text{Given} \\
2. & p(a) \vee q(a) & & \text{Add}
\end{array}
$$

Therefore, these two lines should be written using $\hat{a}$ instead of a.

$$
\begin{array}{llll}
1. & p(\hat{a}) & & \text{Given} \\
2. & p(\hat{a}) \vee q(\hat{a}) & & \text{Add}
\end{array}
$$

However, a in the following is arbitrary because its first occurrence is in line 2, and line 2 follows from line 1 by UI.

$$
\begin{array}{llll}
1. & \forall x\, p(x) & & \text{Given} \\
2. & p(a) & & \text{1 UI} \\
3. & p(a) \vee q(a) & & \text{Add}
\end{array}
$$

We can now introduce the rule of inference that allows us to attach universal quantifiers to formulas. We state it as an axiom.

■ INFERENCE RULE 2.3.8 [Universal Generalization (UG)]

If $p(x)$ is an S-formula with no particular constant symbols and a is an arbitrary constant symbol from $\overline{S}$,

$$p(a) \Rightarrow \forall x p(x).$$

Consider the following argument:

> *All squares are rectangles.*
> *All rectangles are quadrilaterals.*
> *Therefore, all squares are quadrilaterals.*

Representing the premises by $\forall x[s(x) \to r(x)]$ and $\forall x[r(x) \to q(x)]$, a formal proof of this includes UG:

1.	$\forall x[s(x) \to r(x)]$	Given
2.	$\forall x[r(x) \to q(x)]$	Given
3.	$s(a) \to r(a)$	1 UI
4.	$r(a) \to q(a)$	2 UI
5.	$s(a) \to q(a)$	3, 4 HS
6.	$\forall x[s(x) \to q(x)]$	5 UG

Since a was introduced in line 3 by UI, it is an arbitrary constant symbol. In addition, $s(a) \to q(a)$ contains no particular constant symbols that appeared in the proof by substitution. Hence, the application of UG in line 6 is legal.

■ **EXAMPLE 2.3.9**

These are illegal uses of universal generalization.

- Let $p(x)$ be an S-formula with c a constant symbol.

1.	$p(c)$	Given
2.	$\forall x p(x)$	1 UG [error]

   The constant symbol c in line 1 is particular, even without being written as $\hat{c}$. It was not introduced to the proof by UI. Therefore, universal generalization does not apply.

- Suppose that a is an arbitrary constant symbol.

1.	$a + \hat{b} = 0$	
2.	$\forall x(x + \hat{b} = 0)$	1 UG [error]

   The restriction against $p(x)$ containing particular symbols prevents the errant conclusion in line 2.

- The following is an attempt to prove $\forall x \forall y(x + y = 2 \cdot x)$ from the formula $\forall x(x + x = 2 \cdot x)$.

1.	$\forall x(x + x = 2 \cdot x)$	Given
2.	$a + a = 2 \cdot a$	1 UI
3.	$\forall y(a + y = 2 \cdot a)$	2 UG [error]
4.	$\forall x \forall y(x + y = 2 \cdot x)$	3 UG

   Although the constant symbol a in line 2 is arbitrary, the proof is not valid. The reason is that an illegal substitution was made in line 3. To see this, let

$$p(x) := x + x = 2 \cdot x.$$

Applying universal generalization gives

$$\forall y(y + y = 2 \cdot y)$$

because $p(y) \Leftrightarrow y + y = 2 \cdot y$, but this is not what was written in line 3.

Now for some formal proofs.

■ **EXAMPLE 2.3.10**

Prove: $\forall x \forall y p(x, y) \vdash \forall y \forall x p(x, y)$

1.	$\forall x \forall y p(x, y)$	Given
2.	$\forall y p(a, y)$	1 UI
3.	$p(a, b)$	2 UI
4.	$\forall x p(x, b)$	3 UG
5.	$\forall y \forall x p(x, y)$	4 UG

Since both a and b first appear because of UI, they are arbitrary and universal generalization can be applied to both constant symbols.

■ **EXAMPLE 2.3.11**

Prove: $\forall x\, [p(x) \to q(x)]\, , \forall x \neg\, [q(x) \lor r(x)] \vdash \forall x \neg p(x)$

1.	$\forall x\, [p(x) \to q(x)]$	Given
2.	$\forall x \neg\, [q(x) \lor r(x)]$	Given
3.	$p(a) \to q(a)$	1 UI
4.	$\neg\, [q(a) \lor r(a)]$	2 UI
5.	$\neg q(a) \land \neg r(a)$	4 DeM
6.	$\neg q(a)$	5 Simp
7.	$\neg p(a)$	3, 6 MT
8.	$\forall x \neg p(x)$	7 UG

Notice that the a in line 3 was introduced because of UI. Hence, a is arbitrary throughout the proof.

Proofs with Existential Formulas

Since $4 + 5 = 9$, we conclude that there exists x such that $x + 5 = 9$. Since we can construct an isosceles triangle, we conclude that isosceles triangles exist. This motivates the next inference rule.

■ **THEOREM 2.3.12 [Existential Generalization (EG)]**

If $p(x)$ is an S-formula and a is a constant symbol from $\overline{S}$,

$$p(a) \Rightarrow \exists x p(x).$$

PROOF

Assume $p(a)$. Suppose that $\forall x \neg p(x)$. By UI we have that $\neg p(a)$. Therefore, $\neg\forall x\neg p(x)$ by IP, from which $\exists x p(x)$ follows by QN. ■

■ **EXAMPLE 2.3.13**

Each of the following is a valid use of existential generalization.

- $p(a) \wedge \neg r(a) \Rightarrow \exists x [p(x) \wedge \neg r(x)]$

- $q(a) \wedge t(b) \Rightarrow \exists y [q(a) \wedge t(y)]$.

Before we write some proofs, here is the inference rule that allows us to detach existential quantifiers.

■ **THEOREM 2.3.14 [Existential Instantiation (EI)]**

If $p(x)$ is an S-formula,
$$\exists x p(x) \Rightarrow p(\hat{a}),$$
where a is a constant symbol from $\overline{S}$ that has no occurrence in the formal proof prior to $p(\hat{a})$.

PROOF

Assume that $\exists x p(x)$ does not infer $p(b)$ for any constant symbol b. Suppose $\exists x p(x)$. Combined with the assumption, this implies that $\neg p(b)$ for every constant symbol b by the law of the excluded middle (page 37) and DS. That is, $\neg p(a)$ for an arbitrary constant symbol a. Therefore, $\forall x\neg p(x)$ by UG, so $\neg\exists x p(x)$ by QN, which is a contradiction. Therefore, $\exists x p(x) \Rightarrow p(a)$ for some constant symbol a, and a is particular by Definition 2.3.6. ■

The constant symbol $\hat{a}$ obtained by EI is called a **witness** of $\exists x p(x)$.

The reason that the constant symbol a must have no prior occurrence in a proof when applying EI is because a used symbol already represents some object, so if a had appeared in an earlier line, we would have no reason to assume that we could write $p(a)$ from $\exists x p(x)$. For example, given

$$\exists x(x + 4 = 5) \tag{2.14}$$

and

$$\exists x(x + 2 = 13), \tag{2.15}$$

we write $a + 4 = 5$ for some constant symbol a by EI and (2.14). By EI and (2.15), we conclude that $b + 2 = 13$ for some constant symbol b, but inferring $a + 2 = 13$ is invalid because $a \neq b$.

■ EXAMPLE 2.3.15

The following are legal uses of existential instantiation assuming that a and b have no prior occurrences.

- $\exists x \, [p(x) \wedge q(x)] \Rightarrow p(\hat{a}) \wedge q(\hat{a})$
- $\exists y \, [r(a, y, c) \rightarrow r(a, y, c)] \Rightarrow r(a, \hat{b}, c) \rightarrow r(a, \hat{b}, c)$
- $\exists x \forall y \exists z q(x, y, z) \Rightarrow \forall y \exists z q(\hat{a}, y, z)$.

■ EXAMPLE 2.3.16

Existential Instantiation cannot be used to justify either of the following.

- $\exists z \, [p(z) \vee q(z)]$ does not imply $p(\hat{b}) \vee q(z)$ by EI because the substitution was not made correctly. The result should have been $p(\hat{b}) \vee q(\hat{b})$.

- $\exists x \exists y p(a, x, y)$ does not imply $\exists y p(a, a, y)$ by EI because a has a prior occurrence. Also, notice that the hat notation was not used.

■ EXAMPLE 2.3.17

Assume that x is a real number and consider the following.

1.	$\forall x \exists y (x + y = 0)$	Given
2.	$\exists y (a + y = 0)$	1 UI
3.	$a + \hat{b} = 0$	2 EG
4.	$\forall x (x + \hat{b} = 0)$	3 UG [error]
5.	$\exists y \forall x (x + y = 0)$	4 EG

The conclusion in line 5 is incorrect. The problem lies in line 4 where UG was applied despite the particular constant symbol in line 3 (compare Example 2.3.9). This example makes clear why there is a restriction on particular elements in UG (Inference Rule 2.3.8). Since b represents a particular real number, line 3 cannot be used to conclude that all real numbers plus that particular b equals 0. We know that b is the witness to line 2, but that is all we know about it. To correct the argument, we can essentially reverse the steps to arrive back at the premise.

1.	$\forall x \exists y (x + y = 0)$	Given
2.	$\exists y (a + y = 0)$	1 UI
3.	$a + \hat{b} = 0$	2 EG
4.	$\exists y (x + y = 0)$	3 EG
5.	$\forall x \exists y (x + y = 0)$	4 UG

Notice that there is no particular constant symbol in line 4, so line 5 does legally follow by UG.

Here are some formal proofs that use existential instantiation and generalization.

■ EXAMPLE 2.3.18

Prove: $\exists x \, [p(x) \wedge q(x)] , \forall x \, [p(x) \rightarrow r(x)] \vdash \exists x r(x)$

1.	$\exists x\,[p(x) \wedge q(x)]$	Given
2.	$\forall x\,[p(x) \rightarrow r(x)]$	Given
3.	$p(\hat{c}) \wedge q(\hat{c})$	1 EI
4.	$p(\hat{c}) \rightarrow r(\hat{c})$	2 UI
5.	$p(\hat{c})$	3 Simp
6.	$r(\hat{c})$	4, 5 MP
7.	$\exists x r(x)$	6 EG

■ EXAMPLE 2.3.19

Prove: $\forall x \exists y\,[q(x) \wedge t(y)] \vdash \forall x\,[q(x) \wedge (\exists y)t(y)]$

1.	$\forall x \exists y\,[q(x) \wedge t(y)]$	Given
2.	$\exists y\,[q(a) \wedge t(y)]$	1 UI
3.	$q(a) \wedge t(\hat{c})$	2 EI
4.	$t(\hat{c}) \wedge q(a)$	3 Com
5.	$t(\hat{c})$	4 Simp
6.	$\exists y t(y)$	5 EG
7.	$q(a)$	3 Simp
8.	$q(a) \wedge (\exists y)t(y)$	6, 7 Conj
9.	$\forall x\,[q(x) \wedge (\exists y)t(y)]$	8 UG

■ EXAMPLE 2.3.20

Prove : $p(a) \rightarrow \exists x\,[q(x) \wedge r(x)]\,, p(a) \vdash \exists x r(x)$

1.	$p(\hat{a}) \rightarrow \exists x\,[q(x) \wedge r(x)]$	Given
2.	$p(\hat{a})$	Given
3.	$\exists x\,[q(x) \wedge r(x)]$	1, 2 MP
4.	$q(\hat{b}) \wedge r(\hat{b})$	3 EI
5.	$r(\hat{b}) \wedge q(\hat{b})$	4 Com
6.	$r(\hat{b})$	5 Simp
7.	$\exists x r(x)$	6 EG

Exercises

1. Use QN and other replacement rules to determine whether the following are legal replacements.

(a) $\neg \forall x p(x) \Leftrightarrow \forall x \neg p(x)$

(b) $\neg \exists x p(x) \Leftrightarrow \neg \forall x p(x)$

(c) $\forall x \neg p(x) \Leftrightarrow \exists x p(x)$

(d) $\exists x\,[p(x) \rightarrow q(x)] \Leftrightarrow \neg \forall x\,[\neg p(x) \rightarrow \neg q(x)]$

(e) $\neg \forall x \exists y p(x, y) \Leftrightarrow \exists x \forall y \neg p(x, y)$

(f) $\neg \forall x \exists y p(x, y) \Leftrightarrow \exists y \forall x p(x, y)$

2. Negate and put into positive form.

   (a) $\exists x \, [q(x) \rightarrow r(x)]$

   (b) $\forall x \exists y \, [p(x) \wedge q(y)]$

   (c) $\exists x \exists y \, [p(x) \vee q(x, y)]$

   (d) $\forall x \forall y \, [p(x) \vee (\exists z)q(y, z)]$

   (e) $\exists x \neg r(x) \vee \forall x \, [q(x) \leftrightarrow \neg p(x)]$

   (f) $\forall x \forall y \exists z (p(x) \rightarrow [q(y) \wedge r(z)]) \rightarrow \exists x \neg q(x)$

3. Determine whether each pair of propositions are negations. If they are not, write the negation of both.

   (a) *Every real number has a square root.*
       Every real number does not have a square root.

   (b) *Every multiple of four is a multiple of two.*
       Some multiples of two are multiples of four.

   (c) *For all x, if x is odd, then x^2 is odd.*
       There exists x such that if x is odd, then x^2 is even.

   (d) *There exists an integer x such that $x + 1 = 10$.*
       For all integers x, $x + 1 \neq 10$.

4. Write the negation of the following propositions in positive form and in English.

   (a) *For all x, there exists y such that $y/x = 9$.*

   (b) *There exists x so that $xy = 1$ for all real numbers y.*

   (c) *Every multiple of ten is a multiple of five.*

   (d) *No interval contains a rational number.*

   (e) *There is an interval that contains a rational number.*

5. Let f be a function and c be a real number in the open interval I. Then, f is **continuous** at c if for every $\epsilon > 0$, there exists $\delta > 0$ such that for all x in I, if $0 < |x - c| < \delta$, then $|f(x) - f(c)| < \epsilon$.

   (a) Write what it means for f to be not continuous at c.

   (b) The function f is continuous on an open interval if it is continuous at every point of the interval. Write what it means for f to be not continuous on an open interval.

6. Let f be a function and c a real number in the open interval I. Then, f is **uniformly continuous** on I means that for every $\epsilon > 0$, there exists $\delta > 0$ such that for all c and x in I, if $0 < |x - c| < \delta$, then $|f(x) - f(c)| < \epsilon$.

   (a) Write what it means for f to be not uniformly continuous on I.

   (b) How does f being continuous on I differ from f being uniformly continuous on I?

7. Prove using QN.

   (a) $\exists x p(x) \vdash \forall x \neg p(x) \rightarrow \forall x q(x)$

   (b) $\forall x [p(x) \rightarrow q(x)], \forall x [q(x) \rightarrow r(x)], \neg \forall x r(x) \vdash \exists x \neg p(x)$

   (c) $\forall x p(x) \rightarrow \forall y [q(y) \rightarrow r(y)], \exists x [q(x) \wedge \neg r(x)] \vdash \exists x \neg p(x)$

   (d) $\exists x p(x) \rightarrow \exists y q(y), \forall x \neg q(x) \vdash \forall x \neg p(x)$

8. Find all errors in the given proofs.

    (a) "$\exists x\,[p(x) \lor q(x)]\,, \exists x\neg q(x) \vdash \exists x p(x)$"

Attempted Proof

1.	$\exists x\,[p(x) \lor q(x)]$	Given
2.	$\exists x\neg q(x)$	Given
3.	$p(c) \lor q(c)$	1 EI
4.	$\neg q(c)$	2 EI
5.	$\neg p(c)$	3, 4 DS
6.	$\exists x\neg p(x)$	5 EG

    (b) "$\forall x p(x) \vdash \exists x \forall y\,[p(x) \lor q(y)]$"

Attempted Proof

1.	$\forall x p(x)$	Given
2.	$p(\hat{c})$	1 UI
3.	$p(\hat{c}) \lor q(a)$	2 Add
4.	$\forall y\,[p(\hat{c}) \lor q(y)]$	3 UG
5.	$\exists y \forall x\,[p(x) \lor q(y)]$	4 EG

    (c) "$\exists x p(x), \exists x q(x) \vdash \forall x\,[p(x) \land q(x)]$"

Attempted Proof

1.	$\exists x p(x)$	Given
2.	$\exists x q(x)$	Given
3.	$p(\hat{c})$	1 EI
4.	$q(\hat{c})$	2 EI
5.	$p(\hat{c}) \land q(c)$	3, 4 Conj
6.	$\forall x\,[p(x) \land q(x)]$	5 UG

9. Prove.

    (a) $\forall x p(x) \vdash \forall x\,[p(x) \lor q(x)]$

    (b) $\forall x p(x), \forall x\,[q(x) \to \neg p(x)] \vdash \forall x\neg q(x)$

    (c) $\forall x\,[p(x) \to q(x)]\,, \forall x p(x) \vdash \forall x q(x)$

    (d) $\forall x\,[p(x) \lor q(x)]\,, \forall x\neg q(x) \vdash \forall x p(x)$

    (e) $\exists x p(x) \vdash \exists x\,[p(x) \lor q(x)]$

    (f) $\exists x \exists y p(x, y) \vdash \exists y \exists x p(x, y)$

    (g) $\forall x \forall y p(x, y) \vdash \forall y \forall x p(x, y)$

    (h) $\forall x\neg p(x), \exists x\,[q(x) \to p(x)] \vdash \exists x\neg q(x)$

    (i) $\exists x p(x), \forall x\,[p(x) \to q(x)] \vdash \exists x q(x)$

    (j) $\forall x\,[p(x) \to q(x)]\,, \forall x\,[r(x) \to s(x)]\,, \exists x\,[p(x) \lor r(x)] \vdash \exists x\,[q(x) \lor s(x)]$

    (k) $\exists x\,[p(x) \land \neg r(x)]\,, \forall x\,[q(x) \to r(x)] \vdash \exists x\neg q(x)$

    (l) $\forall x p(x), \forall x([p(x) \lor q(x)] \to [r(x) \land s(x)]) \vdash \exists x s(x)$

    (m) $\exists x p(x), \exists x p(x) \to \forall x \exists y\,[p(x) \to q(y)] \vdash \forall x q(x) \lor \exists x r(x)$

   (n) $\exists x p(x), \exists x q(x), \exists x \exists y [p(x) \wedge q(y)] \rightarrow \forall x r(x) \vdash \forall x r(x)$

10. Prove the following:

   (a) $p(x) \wedge \exists y q(y) \Leftrightarrow \exists y [p(x) \wedge q(y)]$
   (b) $p(x) \vee \exists y q(y) \Leftrightarrow \exists y [p(x) \vee q(y)]$
   (c) $p(x) \wedge \forall y q(y) \Leftrightarrow \forall y [p(x) \wedge q(y)]$
   (d) $p(x) \vee \forall y q(y) \Leftrightarrow \forall y [p(x) \vee q(y)]$
   (e) $\exists x p(x) \rightarrow q(y) \Leftrightarrow \forall x [p(x) \rightarrow p(y)]$
   (f) $\forall x p(x) \rightarrow q(y) \Leftrightarrow \exists x [p(x) \rightarrow p(y)]$
   (g) $p(x) \rightarrow \exists y q(q) \Leftrightarrow \exists y [p(x) \rightarrow p(y)]$
   (h) $p(x) \rightarrow \forall y q(q) \Leftrightarrow \forall y [p(x) \rightarrow p(y)]$

11. An S-formula is in **prenex normal form** if it is of the form

$$Q_0 x_0 Q_1 x_1 \cdots Q_{n-1} x_{n-1} q,$$

where $Q_0, Q_1, \ldots, Q_{n-1}$ are quantifier symbols and q is a S-formula. Every S-formula can be replaced with an S-formula in prenex normal form, although variables might need to be renamed. Use Exercise 10 and QN to put the given formulas in prenex normal form.

   (a) $[p(x) \vee q(x)] \rightarrow \exists y [q(y) \rightarrow r(y)]$
   (b) $\neg \forall x [p(x) \rightarrow \neg \exists y q(y)]$
   (c) $\exists x p(x) \rightarrow \forall x \exists y [p(x) \rightarrow q(y)]$
   (d) $\forall x [p(x) \rightarrow q(y)] \wedge \neg \exists y \forall z [r(y) \rightarrow s(z)]$

2.4   PROOF METHODS

The purpose of the propositional logic of Chapter 1 is to model the basic reasoning that one does in mathematics. Rules that determine truth (semantics) and establish valid forms of proof (syntactics) were developed. The logic developed in this chapter is an extension of propositional logic. It is called **first-order logic**. As with propositional logic, first-order logic provides a working model of a type of deductive reasoning that happens when studying mathematics, but with a greater emphasis on what a particular proposition is communicating about its subject. Geometry can serve as an example. To solve geometric problems and prove geometric propositions means to work with the axioms and theorems of geometry. What steps are legal when doing this are dictated by the choice of the logic. Because the subjects of geometric propositions are mathematical objects, such as points, lines, and planes, first-order logic is a good choice. However, sometimes other logical systems are used. An example of such an alternative is **second-order logic**, which allows quantification over function and relation symbols (page 73) in addition to quantification over variable symbols. Whichever logic is chosen, that logic provides the general rules of reasoning for the mathematical theory. Since it has its own axioms and theorems, a logic itself is a theory, but because it is intended to provide rules for other theories, it is sometimes called a **metatheory** (Figure 2.5).

   Although all mathematicians use logic, they usually do not use symbolic logic. Instead, their proofs are written using sentences in English or some other human language

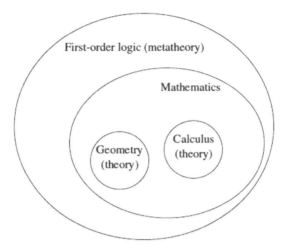

Figure 2.5   First-order logic is a metatheory of mathematical theories.

and usually do not provide all of the details. Call these **paragraph proofs**. Their intention is to lead the reader from the premises to the conclusion, making the result convincing. In many instances, a proof could be translated into the first-order logic, but this is not needed to meet the need of the mathematician. However, that it could be translated means that we can use first-order logic to help us write paragraph proofs, and in this section, we make that connection.

Universal Proofs

Our first paragraph proofs will be for propositions with universal quantifiers. To prove $\forall x p(x)$ from a given set of premises, we show that every object satisfies $p(x)$ assuming those premises. Since the proofs are mathematical, we can restrict the objects to a particular **universe**. A proposition of the form $\forall x p(x)$ is then interpreted to mean that $p(x)$ holds *for all x from a given universe*. This restriction is reasonable because we are studying mathematical things, not airplanes or puppies. To indicate that we have made a restriction to a universe, we randomly select an object of the universe by writing an **introduction**. This is a proposition that declares the type of object represented by the variable. The following are examples of introductions:

> *Let a be a real number.*
> *Take a to be an integer.*
> *Suppose a is a positive integer.*

From here we show that $p(a)$ is true. This process is exemplified by the next diagram:

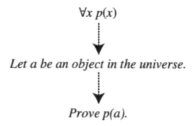

$$\forall x \, p(x)$$

Let a be an object in the universe.

Prove $p(a)$.

These types of proofs are called **universal proofs**.

■ EXAMPLE 2.4.1

To prove that for all real numbers x,

$$(x - 1)^3 = x^3 - 3x^2 + 3x - 1,$$

we introduce a real number and then check the equation.

PROOF

Let a be a real number. Then,

$$\begin{aligned}
(a - 1)^3 &= (a - 1)(a - 1)^2 \\
&= (a - 1)(a^2 - 2a + 1) \\
&= a^3 - 3a^2 + 3a - 1. \quad \blacksquare
\end{aligned}$$

For our next example, we need some terminology.

■ DEFINITION 2.4.2

For all integers a and b, a **divides** b (written as $a \mid b$) if $a \neq 0$ and there exists an integer k such that $b = ak$.

Therefore, 6 divides 18, but 8 does not divide 18. In this case, write $6 \mid 18$ and $8 \nmid 18$. If $a \mid b$, we can also write that b is **divisible** by a, a is a **divisor** or a **factor** of b, or b is a **multiple** of a. For this reason, to translate a predicate like *4 divides a*, write

$$a = 4k \text{ for some integer } k.$$

A common usage of divisibility is to check whether an integer is divisible by 2 or not.

■ DEFINITION 2.4.3

An integer n is **even** if $2 \mid n$, and n is **odd** if $2 \mid (n - 1)$.

We are now ready for the example.

■ EXAMPLE 2.4.4

Let us prove the proposition

the square of every even integer is even.

This can be rewritten using a variable:

for all even integers n, n^2 is even.

The proof goes like this.

PROOF

Let n be an even integer. This means that there exists an integer k such that $n = 2k$, so we can calculate:

$$n^2 = (2k)^2 = 4k^2 = 2(2k^2).$$

Since $2k^2$ is an integer, n^2 is even. ∎

Notice how the definition was used in the proof. After the even number was introduced, a proposition that translated the introduction into a form that was easier to use was written. This was done using Definitions 2.4.2 and 2.4.3.

Existential Proofs

Suppose that we want to write a paragraph proof for $\exists x p(x)$. This means that we must show that there exists at least one object of the universe that satisfies the formula $p(x)$. It will be our job to find that object. To do this directly, we pick an object that we think will satisfy $p(x)$. This object is called a **candidate**. We then check that it does satisfy $p(x)$. This type of a proof is called a **direct existential proof**, and its structure is illustrated as follows:

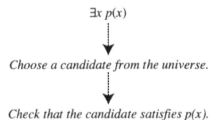

$\exists x \, p(x)$

Choose a candidate from the universe.

Check that the candidate satisfies p(x).

■ **EXAMPLE 2.4.5**

To prove that there exists an integer x such that $x^2 + 2x - 3 = 0$, we find an integer that satisfies the equation. A basic factorization yields

$$(x + 3)(x - 1) = 0.$$

Since $x = -3$ or $x = 1$ will work, we choose (arbitrarily) $x = 1$. Therefore,

$$(1)^2 + 2(1) - 3 = 0,$$

proving the existence of the desired integer.

Suppose that we want to prove that there is a function f such that the derivative of f is $2x$. After a quick mental calculation, we choose $f(x) = x^2$ as a candidate and check to find that

$$\frac{d}{dx}x^2 = 2x. \tag{2.16}$$

Notice that d/dx is a function that has functions as its inputs and outputs. Let d represent this function. That is, d is a function symbol such that

$$d(f)(x) = \frac{d}{dx}f(x).$$

A formula that represents the proposition that was just proved is

$$\exists f(d(f)(x) = 2x). \tag{2.17}$$

This is a second-order formula (page 73) because the variable symbol x represents a real number (an object of the universe) and f is a function symbol taking real numbers as arguments. Although this kind of reasoning is common to mathematics, it cannot be written as a first-order formula. This shows that there is a purpose to second-order logic. It is not a novelty.

■ **EXAMPLE 2.4.6**

To represent (2.16) as a first-order formula, let d be a unary function symbol representing the derivative and write

$$\forall x[d(x^2) = 2x].$$

Notice, however, that this formula does not convey the same meaning as (2.17).

Multiple Quantifiers

Let us take what we have learned concerning the universal and existential quantifiers and write paragraph proofs involving both. The first example is a simple one from algebra but will nicely illustrate our method.

■ **EXAMPLE 2.4.7**

Prove that for every real number x, there exists a real number y so that $x + y = 2$. This translates to

$$\forall x \exists y(x + y = 2)$$

with the universe equal to the real numbers. Remembering that a universal quantifier must apply to all objects of the universe and an existential quantifier means that we must find the desired object, we have the following:

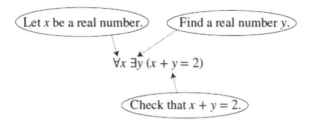

After taking an arbitrary x, our candidate will be $y = 2 - x$.

PROOF

Let x be a real number. Choose $y = 2 - x$ and calculate:

$$x + y = x + (2 - x) = 2. \blacksquare$$

Now let us switch the order of the quantifiers.

■ EXAMPLE 2.4.8

To see that there exists an integer x such that for all integers y, $x + y = y$, we use the following:

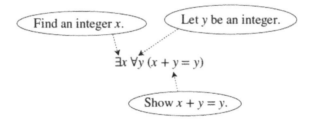

In the proof, the first goal is to identify a candidate. Then, we must show that it works with every real number.

PROOF

We claim that 0 is the sought after object. To see this, let y be an integer. Then $0 + y = y$. ■

The next example will involve two existential quantifiers. Therefore, we have to find two candidates.

■ EXAMPLE 2.4.9

We prove that there exist real numbers a and b so that for every real number x,

$$(a + 2b)x + (2a - b) = 2x - 6.$$

Translating we arrive at

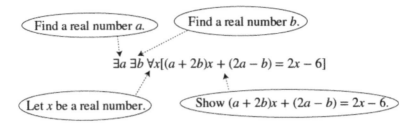

We have to choose two candidates, one for a and one for b, and then check by taking an arbitrary x.

PROOF

Solving the system,

$$a + 2b = 2,$$
$$2a - b = -6,$$

we choose $a = -2$ and $b = 2$. Let x be a real number. Then,

$$(a + 2b)x + (2a - b) = (-2 + 2 \cdot 2)x + (2 \cdot (-2) - 2) = 2x - 6. \blacksquare$$

Counterexamples

There are many times in mathematics when we must show that a proposition of the form $\forall x p(x)$ is false. This can be accomplished by proving $\exists x \neg p(x)$ is true, and this is done by showing that an object a exists in the universe such that $p(a)$ is false. This object is called a **counterexample** to $\forall x p(x)$.

■ **EXAMPLE 2.4.10**

Show false:

$$x + 2 = 7 \text{ for all real numbers } x.$$

To do this, we show that $\exists x(x + 2 \neq 7)$ is true by noting that the real number 0 satisfies $x + 2 \neq 7$. Hence, 0 is a counterexample to $\forall x(x + 2 = 7)$.

The idea of a counterexample can be generalized to cases with multiple universal quantifiers.

■ **EXAMPLE 2.4.11**

We know from algebra that *every quadratic polynomial with real coefficients has a real zero* is false. This can be symbolized as

$$\forall a \forall b \forall c \exists x(ax^2 + bx + c = 0),$$

where the universe is the collection of real numbers. The counterexample is found by demonstrating

$$\exists a \exists b \exists c \forall x(ax^2 + bx + c \neq 0).$$

Many polynomials could be chosen, but we select $x^2 + 1$. Its only zeros are i and $-i$. This means that $a = 1$, $b = 0$, and $c = 1$ is our counterexample.

Direct Proof

Direct Proof is the preferred method for proving implications. To use Direct Proof to write a paragraph proof identify the antecedent and consequent of the implication and then follow these steps:

- assume the antecedent,
- translate the antecedent,
- translate the consequent so that the goal of the proof is known,
- deduce the consequent.

Our first example will use Definitions 2.4.2 and 2.4.3. Notice the introductions in the proof.

■ **EXAMPLE 2.4.12**

We use Direct Proof to write a paragraph proof of the proposition

for all integers x, if 4 divides x, then x is even.

First, randomly choose an integer a and then identify the antecedent and consequent:

$$if \boxed{4 \ divides \ x} , \ then \boxed{x \ is \ even} .$$

Use these to identify the structure of the proof:

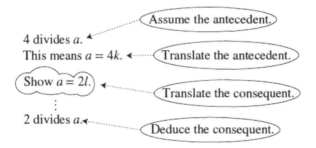

Now write the final version from this structure.

PROOF

Assume that 4 divides the integer a. This means $a = 4k$ for some integer k. We must show that $a = 2l$ for some integer l, but we know that $a = 4k = 2(2k)$. Hence, let $l = 2k$. ■

Sometimes it is difficult to prove a conditional directly. An alternative is to prove the contrapositive. This is sometimes easier or simply requires fewer lines. The next example shows this method in a paragraph proof.

■ EXAMPLE 2.4.13

Let us show that

> *for all integers n, if n^2 is odd, then n is odd.*

A direct proof of this is a problem. Instead, we prove its contrapositive,

> *if n is not odd, then n^2 is not odd.*

In other words, we prove that

> *if n is even, then n^2 is even.*

This will be done using Direct Proof:

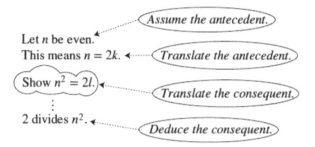

This leads to the final proof.

PROOF

Let n be an even integer. This means that $n = 2k$ for some integer k. To see that n^2 is even, calculate to find

$$n^2 = (2k)^2 = 4k^2 = 2(2k^2).\ ■$$

Notice that this proof is basically the same as the proof for *the square of every even integer is even* on page 99. This illustrates that there is a connection between universal proofs and direct proofs (Exercise 4).

Existence and Uniqueness

There will be times when we want to show that there exists exactly one object that satisfies a given predicate $p(x)$. In other words, there exists a **unique** object that satisfies $p(x)$. This is a two-step process.

- **Existence:** Show that there is at least one object that satisfies $p(x)$.

- **Uniqueness:** Show that there is at most one object that satisfies $p(x)$. This is usually done by assuming both a and b satisfy $p(x)$ and then proving $a = b$.

This means to prove that there *exists* a *unique* x such that $p(x)$, we prove

$$\exists x p(x) \wedge \forall x \forall y (p(x) \wedge p(y) \rightarrow x = y).$$

Use direct or indirect existential proof to demonstrate that an object exists. The next example illustrates proving uniqueness.

■ **EXAMPLE 2.4.14**

Let m and n be nonnegative integers with $m \neq 0$. To show that there is at most one pair of nonnegative integers r and q such that

$$r < m \text{ and } n = mq + r,$$

suppose in addition to r and q that there exists nonnegative integers r' and q' such that

$$r' < m \text{ and } n = mq' + r'.$$

Assume that $r' > r$. By Exercise 13, $q > q'$, so there exists $u, v > 0$ such that

$$r' = r + u \text{ and } q = q' + v.$$

Therefore,

$$m(q' + v) + r = mq' + r + u,$$
$$mq' + mv + r = mq' + r + u,$$
$$mv = u.$$

Since $v > 0$, there exists w such that $v = w + 1$. Hence,

$$mw + m = m(w + 1) = mv = u,$$

so $m \leq u$. However, since $r < m$ and $r' < m$, we have that $u < m$ (Exercise 13), which is impossible. Lastly, the assumption $r > r'$ leads to a similar contradiction. Therefore, $r = r'$, which implies that $q = q'$.

■ **EXAMPLE 2.4.15**

To prove $2x + 1 = 5$ has a unique real solution, we show

$$\exists x(2x + 1 = 5)$$

and

$$\forall x \forall y(2x + 1 = 5 \wedge 2y + 1 = 5 \rightarrow x = y).$$

• We know that $x = 2$ is a solution since $2(2) + 1 = 5$.

• Suppose that a and b are solutions. We know that both $2a + 1 = 5$ and $2b + 1 = 5$, so we calculate:

$$2a + 1 = 2b + 1,$$
$$2a = 2b,$$
$$a = b.$$

Indirect Proof

To use indirect proof, assume each premise and then assume the negation of the conclusion. Then proceed with the proof until a contradiction is reached. At this point, deduce the original conclusion.

■ EXAMPLE 2.4.16

Earlier we proved

for all integers n, if n^2 is odd, then n is odd

by showing its contrapositive. Here we nest an Indirect Proof within a Direct Proof:

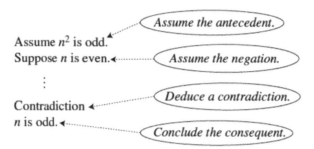

We use this structure to write the paragraph proof.

PROOF

Take an integer n and let n^2 be odd. In order to obtain a contradiction, assume that n is even. So, $n = 2k$ for some integer k. Substituting, we have

$$n^2 = (2k)^2 = 2(2k^2),$$

showing that n^2 is even. This is a contradiction. Therefore, n is an odd integer. ■

Indirect proof has been used to prove many famous mathematical results including the next example.

■ EXAMPLE 2.4.17

We show that $\sqrt{2}$ is an irrational number. Suppose instead that

$$\sqrt{2} = \frac{a}{b},$$

where a and b are integers, $b \neq 0$, and the fraction a/b has been reduced. Then,

$$2 = \frac{a^2}{b^2},$$

so $a^2 = 2b^2$. Therefore, $a = 2k$ for some integer k [Exercise 11(c)]. We conclude that $b^2 = 2k^2$, which implies that b is also even. However, this is impossible because the fraction was reduced.

There are times when it is difficult or impossible to find a candidate for an existential proof. When this happens, an **indirect existential proof** can sometimes help.

■ EXAMPLE 2.4.18

If n is an integer such that $n > 1$, then n is **prime** means that the only positive divisors of n are 1 and n; else n is **composite**. From the definition, if n is composite, there exist a and b such that $n = ab$ and $1 < a \leq b < n$. For example, 2, 11, and 97 are prime, and 4 and 87 are composite. Euclid proved that there are infinitely many prime numbers (*Elements* IX.20). Since it is impossible to find all of these numbers, we prove this theorem indirectly. Suppose that there are finitely many prime numbers and list them as

$$p_0, p_1, \ldots, p_{n-1}.$$

Consider

$$q = p_0 p_1 \cdots p_{n-1} + 1.$$

If q is prime, $q = p_i$ for some $i = 0, 1, \ldots, n - 1$, which would imply that p_i divides 1. Therefore, q is composite, but this is also impossible because q must have a prime factor, which again means that 1 would be divisible by a prime.

Biconditional Proof

The last three types of proofs that we will examine rely on direct proof. The first of these takes three forms, and they provide the usual method of proving biconditionals. The first form follows because $p \to q$ and $q \to p$ imply $(p \to q) \land (q \to p)$ by Conj, which implies $p \leftrightarrow q$ by Equiv.

■ INFERENCE RULE 2.4.19 [Biconditional Proof (BP)]

$$p \to q, q \to p \Rightarrow p \leftrightarrow q.$$

The rule states that a biconditional can be proved by showing both implications. Each is usually proved with direct proof. As is seen in the next example, the $p \to q$ subproof is introduced by ($\to$) and its converse with ($\leftarrow$). The conclusions of the two applications of direct proof are combined in line 7.

■ EXAMPLE 2.4.20

Prove: $p \to q \vdash p \land q \leftrightarrow p$

	1.	$p \to q$	Given
($\to$)	2.	$\rightharpoondown p \land q$	Assumption
	3.	p	2 Simp
($\leftarrow$)	4.	$\rightharpoondown p$	Assumption
	5.	q	1, 4 MP
	6.	$p \land q$	4, 5 Conj
	7.	$p \land q \leftrightarrow p$	2–6 BP

Sometimes the steps for one part are simply the steps for the other in reverse. When this happens, our work is cut in half, and we can use the **short rule of biconditional proof**. These proofs are simply a sequence of replacements with or without the reasons. This is a good method when only rules of replacement are used as in the next example. (Notice that there are no hypotheses to assume.)

■ **EXAMPLE 2.4.21**

Prove: $\vdash p \to q \leftrightarrow p \wedge \neg q \to \neg p$

$p \to q \Leftrightarrow \neg p \vee q$	Impl
$\Leftrightarrow \neg p \vee \neg p \vee q$	Idem
$\Leftrightarrow \neg p \vee (\neg p \vee q)$	Assoc
$\Leftrightarrow \neg p \vee q \vee \neg p$	Com
$\Leftrightarrow \neg p \vee \neg\neg q \vee \neg p$	DN
$\Leftrightarrow \neg(p \wedge \neg q) \vee \neg p$	DM
$\Leftrightarrow p \wedge \neg q \to \neg p$	Impl

■ **EXAMPLE 2.4.22**

Let us use biconditional proof to show that

for all integers n, n is even if and only if n^3 is even.

Since this is a biconditional, we must show both implications:

$$if \boxed{n \text{ is even}}, then \boxed{n^3 \text{ is even}},$$

and

$$if \boxed{n^3 \text{ is even}}, then \boxed{n \text{ is even}}.$$

To prove the second conditional, we prove its contrapositive. Therefore, using the pattern of the previous example, the structure is

$(\to)$	Assume n is even
	Then, $n = 2k$ for some integer k
	$\vdots$
	$n^3 = 2l$, with l being an integer
	n^3 is even
$(\leftarrow)$	Assume n is odd
	Then, $n = 2k + 1$ for some integer k
	$\vdots$
	$n^3 = 2l + 1$, with l being an integer
	n^3 is odd

PROOF

Let n be an integer.

- Assume n is even. Then, $n = 2k$ for some integer k. We show that n^3 is even. To do this, we calculate:

$$n^3 = (2k)^3 = 2(4k^3),$$

which means that n^3 is even.

- Now suppose that n is odd. This means that $n = 2k + 1$ for some integer k. To show that n^3 is odd, we again calculate:

$$n^3 = (2k+1)^3 = 8k^3 + 12k^2 + 6k + 1 = 2(4k^3 + 6k^2 + 3k) + 1.$$

Hence, n^3 is odd. ■

Notice that the words were chosen carefully to make the proof more readable. Furthermore, the example could have been written with the words *necessary* and *sufficient* introducing the two subproofs. The ($\rightarrow$) step could have been introduced with a phrase like

to show sufficiency,

and the ($\leftarrow$) could have opened with

as for necessity.

There will be times when we need to prove a sequence of biconditionals. The propositional forms $p_0, p_1, \ldots, p_{n-1}$ are **pairwise equivalent** if for all i, j,

$$p_i \leftrightarrow p_j.$$

In other words,

$$p_0 \leftrightarrow p_1, p_0 \leftrightarrow p_2, \ldots, p_1 \leftrightarrow p_2, p_1 \leftrightarrow p_3, \ldots, p_{n-2} \leftrightarrow p_{n-1}.$$

To prove all of these, we make use of the Hypothetical Syllogism. For example, if we know that $p_0 \rightarrow p_1$, $p_1 \rightarrow p_2$, and $p_2 \rightarrow p_0$, then

$$p_0 \rightarrow p_2 \text{ (because } p_0 \rightarrow p_1 \wedge p_1 \rightarrow p_2\text{)},$$
$$p_1 \rightarrow p_0 \text{ (because } p_1 \rightarrow p_2 \wedge p_2 \rightarrow p_0\text{)},$$
$$p_2 \rightarrow p_1 \text{ (because } p_2 \rightarrow p_0 \wedge p_0 \rightarrow p_1\text{)}.$$

The result is the equivalence rule.

■ **INFERENCE RULE 2.4.23 [Equivalence Rule]**

To prove that the propositional forms $p_0, p_1, \ldots, p_{n-1}$ are pairwise equivalent, prove:

$$p_0 \to p_1, p_1 \to p_2, \ldots, p_{n-2} \to p_{n-1}, p_{n-1} \to p_0.$$

In practice, the equivalence rule will typically be used to prove propositions that include the phrase

the following are equivalent.

■ **EXAMPLE 2.4.24**

Let

$$f(x) = a_n x^n + a_{n-1} x^{n-1} + \cdots + a_1 x + a_0$$

be a polynomial with real coefficients. That is, each a_i is a real number and n is a nonnegative integer. An integer r is a **zero** of $f(x)$ if

$$a_n r^n + a_{n-1} r^{n-1} + \cdots + a_1 r + a_0 = 0,$$

written as $f(r) = 0$, and a polynomial $g(x)$ is a **factor** of $f(x)$ if there is a polynomial $h(x)$ such that $f(x) = g(x)h(x)$. Whether $g(x)$ is a factor of $f(x)$ or not, there exist unique polynomials $q(x)$ and $r(x)$ such that

$$f(x) = g(x)q(x) + r(x) \tag{2.18}$$

and

$$\text{the degree of } r(x) \text{ is less than the degree of } g(x). \tag{2.19}$$

This result is called the **polynomial division algorithm,** The polynomial $q(x)$ is the **quotient** and $r(x)$ is the **remainder**. We prove that the following are equivalent:

- r is a zero of $f(x)$.

- r is a solution to $f(x) = 0$.

- $x - r$ is a factor of $f(x)$.

To do this, we prove three conditionals:

if $\boxed{r \text{ is a zero of } f(x)}$, then $\boxed{r \text{ is a solution to } f(x) = 0}$,

if $\boxed{r \text{ is a solution to } f(x) = 0}$, then $\boxed{x - r \text{ is a factor of } f(x)}$,

and

if $\boxed{x - r \text{ is a factor of } f(x)}$, then $\boxed{r \text{ is a zero of } f(x)}$.

We use direct proof on each.

PROOF

Let $a_n x^n + a_{n-1} x^{n-1} + \cdots + a_1 x + a_0$ be a polynomial and assume that the coefficients are real numbers. Denote the polynomial by $f(x)$.

- Let r be a zero of $f(x)$. By definition, this means $f(r) = 0$, so r is a solution to $f(x) = 0$.

- Suppose r is a solution to $f(x) = 0$. The polynomial division algorithm (2.18) gives polynomials $q(x)$ and $r(x)$ such that

$$f(x) = q(x)(x - r) + r(x)$$

  and the degree of $r(x)$ is less than 1 (2.19). Hence, $r(x)$ is a constant that we simply write as c. Now,

$$0 = f(r) = q(r)(r - r) + c = 0 + c = c.$$

  Therefore, $f(x) = q(x)(x - r)$, so $x - r$ is a factor of $f(x)$.

- Lastly, assume $x - r$ is a factor of $f(x)$. This means that there exists a polynomial $q(x)$ so that

$$f(x) = (x - r)q(x).$$

Thus,

$$f(r) = (r - r)q(r) = 0,$$

which means r is a zero of $f(x)$. ∎

Proof of Disjunctions

The second type of proof that relies on direct proof is the proof of a disjunction. To prove $p \lor q$, it is standard to assume $\neg p$ and show q. This means that we would be using direct proof to show $\neg p \to q$. This is what we want because

$$\neg p \to q \Leftrightarrow \neg\neg p \lor q \Leftrightarrow p \lor q.$$

The intuition behind the strategy goes like this. If we need to prove $p \lor q$ from some hypotheses, it is not reasonable to believe that we can simply prove p and then use Addition to conclude $p \lor q$. Indeed, if we could simply prove p, we would expect the conclusion to be stated as p and not $p \lor q$. Hence, we need to incorporate both disjuncts into the proof. We do this by assuming the negation of one of the disjuncts. If we can prove the other, the disjunction must be true.

■ **EXAMPLE 2.4.25**

To prove

for all integers a and b, if ab = 0, then a = 0 or b = 0,

we assume $ab = 0$ and show that $a \neq 0$ implies $b = 0$.

PROOF

Let a and b be integers. Let $ab = 0$ and suppose $a \neq 0$. Then, a^{-1} exists. Multiplying both sides of the equation by a^{-1} gives

$$a^{-1}ab = a^{-1} \cdot 0,$$

so $b = 0$. ■

Proof by Cases

The last type of proof that relies on direct proof is proof by cases. Suppose that we want to prove $p \to q$ and this is difficult for some reason. We notice, however, that p can be broken into cases. Namely, there exist $p_0, p_1, \ldots, p_{n-1}$ such that

$$p \Leftrightarrow p_0 \vee p_1 \vee \cdots \vee p_{n-1}.$$

If we can prove $p_i \to q$ for each i, we have proved $p \to q$. If $n = 2$, then $p \Leftrightarrow p_0 \vee p_1$, and the justification of this is as follows:

$$
\begin{aligned}
(p_0 \to q) \wedge (p_1 \to q) &\Leftrightarrow (\neg p_0 \vee q) \wedge (\neg p_1 \vee q) \\
&\Leftrightarrow (q \vee \neg p_0) \wedge (q \vee \neg p_1) \\
&\Leftrightarrow q \vee \neg p_0 \wedge \neg p_1 \\
&\Leftrightarrow \neg p_0 \wedge \neg p_1 \vee q \\
&\Leftrightarrow \neg(p_0 \vee p_1) \vee q \\
&\Leftrightarrow p_0 \vee p_1 \to q \\
&\Leftrightarrow p \to q.
\end{aligned}
$$

This generalizes to the next rule.

■ **INFERENCE RULE 2.4.26 [Proof by Cases (CP)]**

For every positive integer n, if $p \Leftrightarrow p_0 \vee p_1 \vee \cdots \vee p_{n-1}$, then

$$p_0 \to q, p_1 \to q, \ldots, p_{n-1} \to q \Rightarrow p \to q.$$

For example, since a is a real number if and only if $a > 0$, $a = 0$, or $a < 0$, if we needed to prove a proposition about an arbitrary real number, it would suffice to prove the result individually for $a > 0$, $a = 0$, and $a < 0$.

■ **EXAMPLE 2.4.27**

Our example of a proof by cases is a well-known one:

> *for all integers a and b, if a = ±b, then a divides b and b divides a.*

The antecedent means $a = b$ or $a = -b$, which are the two cases, so we have to show both

> *if a = b, then a divides b and b divides a*

and

> *if a = -b, then a divides b and b divides a.*

This leads to the following structure:

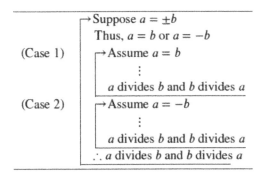

The final proof looks something like this.

PROOF

Let a and b be integers and suppose $a = \pm b$. To show a divides b and b divides a, we have two cases to prove.

- Assume $a = b$. Then, $a = b \cdot 1$ and $b = a \cdot 1$.
- Next assume $a = -b$. This means that $a = b \cdot (-1)$ and $b = a \cdot (-1)$.

In both the cases, we have proved that a divides b and b divides a. ■

Exercises

1. Let n be an integer. Demonstrate that each of the following are divisible by 6.
   - (a) 18
   - (b) −24
   - (c) 0
   - (d) $6n + 12$
   - (e) $2^3 \cdot 3^4 \cdot 7^5$
   - (f) $(2n + 2)(3n + 6)$

2. Let a be a nonzero integer. Write paragraph proofs for each proposition.

(a) 1 divides a.

(b) a divides 0.

(c) a divides a.

3. Write a universal proof for each proposition.

(a) For all real numbers x, $(x + 2)^2 = x^2 + 4x + 4$.

(b) For all integers x, $x - 1$ divides $x^3 - 1$.

(c) The square of every even integer is even.

4. Show why a proof of $(\forall x)p(x)$ is an application of direct proof.

5. Write existential proofs for each proposition.

(a) There exists a real number x such that $x - \pi = 9$.

(b) There exists an integer x such that $x^2 + 2x - 3 = 0$.

(c) The square of some integer is odd.

6. Prove each of the following by writing a paragraph proof.

(a) For all real numbers x, y, and z, $x^2 z + 2xyz + y^2 z = z(x + y)^2$.

(b) There exist real numbers u and v such that $2u + 5v = -29$.

(c) For all real numbers x, there exists a real number y so that $x - y = 10$.

(d) There exists an integer x such that for all integers y, $yx = x$.

(e) For all real numbers a, b, and c, there exists a complex number x such that $ax^2 + bx + c = 0$.

(f) There exists an integer that divides every integer.

7. Provide counterexamples for each of the following false propositions.

(a) Every integer is a solution to $x + 1 = 0$.

(b) For every integer x, there exists an integer y such that $xy = 1$.

(c) The product of any two integers is even.

(d) For every integer n, if n is even, then n^2 is a multiple of eight.

8. Assuming that a, b, c, and d are integers with $a \neq 0$ and $c \neq 0$, give paragraph proofs using direct proof for the following divisibility results.

(a) If a divides b, then a divides bd.

(b) If a divides b and a divides d, then a^2 divides bd.

(c) If a divides b and c divides d, then ac divides bd.

(d) If a divides b and b divides c, then a divides c.

9. Write paragraph proofs using direct proof.

(a) The sum of two even integers is even.

(b) The sum of two odd integers is even.

(c) The sum of an even and an odd is odd.

(d) The product of two even integers is even.

(e) The product of two odd integers is odd.

(f) The product of an even and an odd is even.

10. Let a and b be integers. Write paragraph proofs.

(a) If a and b are even, then $a^4 + b^4 + 32$ is divisible by 8.

(b) If a and b are odd, then 4 divides $a^4 + b^4 + 6$.

11. Prove the following by using direct proof to prove the contrapositive.

(a) For every integer n, if n^4 is even, then n is even.

(b) For all integers n, if $n^3 + n^2$ is odd, then n is odd.

(c) For all integers a and b, if ab is even, then a is even or b is even.

12. Write paragraph proofs.

(a) The equation $x - 10 = 23$ has a unique solution.

(b) The equation $\sqrt{2x - 5} = 2$ has a unique solution.

(c) For every real number y, the equation $2x + 5y = 10$ has a unique solution.

(d) The equation $x^2 + 5x + 6 = 0$ has at most two integer solutions.

13. From the proof of Example 2.4.14, prove that $q > q'$ and $u < m$.

14. Prove the results of Exercise 9 indirectly.

15. Prove using the method of biconditional proof.

(a) $p \vee q \to \neg r, s \to r, \neg p \vee q \to s \vdash p \leftrightarrow \neg s$

(b) $p \vee (\neg q \vee p), q \vee (\neg p \vee q) \vdash p \leftrightarrow q$

(c) $(p \to q) \wedge (r \to s), (p \to \neg s) \wedge (r \to \neg q), p \vee r \vdash q \leftrightarrow \neg s$

(d) $p \wedge r \to \neg(s \vee t), \neg s \vee \neg t \to p \wedge r \vdash s \leftrightarrow t$

16. Prove using the short rule of biconditional proof.

(a) $p \to q \wedge r \Leftrightarrow (p \to q) \wedge (p \to r)$

(b) $p \to q \vee r \Leftrightarrow p \wedge \neg q \to r$

(c) $p \vee q \to r \Leftrightarrow (p \to r) \wedge (q \to r)$

(d) $p \wedge q \to r \Leftrightarrow (p \to r) \vee (q \to r)$

17. Let a, b, c, and d be integers. Write paragraph proofs using biconditional proof.

(a) a is even if and only if a^2 is even.

(b) a is odd if and only if $a + 1$ is even.

(c) a is even if and only if $a + 2$ is even.

(d) $a^3 + a^2 + a$ is even if and only if a is even.

(e) If $c \neq 0$, then a divides b if and only if ac divides bc.

18. Suppose that a and b are integers. Prove that the following are equivalent.

- a divides b.
- a divides $-b$.
- $-a$ divides b.
- $-a$ divides $-b$.

19. Let a be an integer. Prove that the following are equivalent.

- a is divisible by 3.
- $3a$ is divisible by 9.
- $a + 3$ is divisible by 3.

20. Prove by using direct proof but do not use the contrapositive: for all integers a and b, if ab is even, then a is even or b is even.

21. Prove using proof by cases (Inference Rule 2.4.26).
    (a) For all integers a, if $a = 0$ or $b = 0$, then $ab = 0$.
    (b) The square of every odd integer is of the form $8k + 1$ for some integer k. (Hint: Square $2l + 1$. Then consider two cases: l is even and l is odd.)
    (c) $a^2 + a + 1$ is odd for every integer a.
    (d) The fourth power of every odd integer is of the form $16k + 1$ for some integer k.
    (e) Every nonhorizontal line intersects the x-axis.

22. Let a be an integer. Prove by cases.
    (a) 2 divides $a(a + 1)$
    (b) 3 divides $a(a + 1)(a + 2)$

23. For any real number c, the **absolute value** of c is defined as

$$|c| = \begin{cases} c & \text{if } c \geq 0, \\ -c & \text{if } c < 0. \end{cases}$$

Let a be a positive real number. Prove the following about absolute value for every real number x.
    (a) $|-x| = |x|$
    (b) $|x^2| = |x|^2$
    (c) $x \leq |x|$
    (d) $|xy| = |x||y|$
    (e) $|x| < a$ if and only if $-a < x < a$.
    (f) $|x| > a$ if and only if $x > a$ or $x < -a$.

24. Take a and b to be nonzero integers and prove the given propositions.
    (a) a divides 1 if and only if $a = \pm 1$.
    (b) If $a = \pm b$, then $|a| = |b|$.

CHAPTER 3

SET THEORY

3.1 SETS AND ELEMENTS

The development of logic that resulted in the work of Chapters 1 and 2 went through many stages and benefited from the work of various mathematicians and logicians through the centuries. Although modern logic can trace its roots to Descartes with his *mathesis universalis* and Gottfried Leibniz's *De Arte Combinatoria* (1666), the beginnings of modern symbolic logic is generally attributed to Augustus De Morgan [*Formal Logic* (1847)], George Boole [*Mathematical Analysis of Logic* (1847) and *An Investigation of the Laws of Thought* (1847)], and Frege [*Begriffsschrift* (1879), *Die Grundlagen der Arithmetik* (1884), and *Grundgesetze der Arithmetik* (1893)]. However, when it comes to set theory, it was Georg Cantor who, with his first paper, "Ueber eine Eigenschaft des Inbegriffs aller reellen algebraischen Zahlen" (1874), and over a decade of research, is the founder of the subject. For the next four chapters, Cantor's set theory will be our focus.

A **set** is a collection of objects known as **elements**. An element can be almost anything, such as numbers, functions, or lines. A set is a single object that can contain many elements. Think of it as a box with things inside. The box is the set, and the things are the elements. We use uppercase letters to label sets, and elements will usu-

A First Course in Mathematical Logic and Set Theory, First Edition. Michael L. O'Leary.

ally be represented by lowercase letters. The symbol $\in$ (fashioned after the Greek letter *epsilon*) is used to mean "element of," so if A is a set and a is an element of A, write $\in aA$ or, the more standard, $a \in A$. The notation $a, b \in A$ means $a \in A$ and $b \in A$. If c is not an element of A, write $c \notin A$. If A contains no elements, it is the **empty set**. It is represented by the symbol $\varnothing$. Think of the empty set as a box with no things inside.

Rosters

Since the elements are those that distinguish one set from another, one method that is used to write a set is to list its elements and surround them with braces. This is called the **roster method** of writing a set, and the list is known as a **roster**. The braces signify that a set has been defined. For example, the set of all integers between 1 and 10 inclusive is

$$\{1, 2, 3, 4, 5, 6, 7, 8, 9, 10\}.$$

Read this as "the set containing 1, 2, 3, 4, 5, 6, 7, 8, 9, and 10." The set of all integers between 1 and 10 exclusive is

$$\{2, 3, 4, 5, 6, 7, 8, 9\}.$$

If the roster is too long, use ellipses $(\ldots)$. When there is a pattern to the elements of the set, write down enough members so that the pattern is clear. Then use the ellipses to represent the continuing pattern. For example, the set of all integers inclusively between 1 and 1,000,000 can be written as

$$\{1, 2, 3, \ldots, 999,999, 1,000,000\}.$$

Follow this strategy to write infinite sets as rosters. For instance, the set of even integers can be written as

$$\{\ldots, -4, -2, 0, 2, 4, \ldots\}.$$

■ EXAMPLE 3.1.1

- As a roster, $\{\ \}$ denotes the empty set. *Warning:* Never write $\{\varnothing\}$ for the empty set. This set has one element in it.

- A set that contains exactly one element is called a **singleton**. Hence, the sets $\{1\}$, $\{f\}$, and $\{\varnothing\}$ are singletons written in roster form. Also, $1 \in \{1\}$, $f \in \{f\}$, and $\varnothing \in \{\varnothing\}$.

- The set of linear functions that intersect the origin with an integer slope can be written as:
$$\{\ldots, -2x, -x, 0, x, 2x, \ldots\}.$$
  (*Note:* Here 0 represents the function $f(x) = 0$.)

Let A and B be sets. These are **equal** if they contain exactly the same elements. The notation for this is $A = B$. What this means is if any element is in A, it is also in B, and conversely, if an element is in *B*, it is in *A*. To fully understand set equality, consider again the analogy between sets and boxes. Suppose that we have a box containing a carrot and a rabbit. We could describe it with the phrase *the box that contains the carrot and the rabbit*. Alternately, it could be referred to as *the box that contains the orange vegetable and the furry, cotton-tailed animal with long ears*. Although these are different descriptions, they do not refer to different boxes. Similarly, the set $\{1, 3\}$ and *the set containing the solutions of* $(x - 1)(x - 3) = 0$ are equal because they contain the same elements. Furthermore, the order in which the elements are listed does not matter. The box can just as easily be described as *the box with the rabbit and the carrot*. Likewise, $\{1, 3\} = \{3, 1\}$. Lastly, suppose that the box is described as containing *the carrot, the rabbit, and the carrot*, forgetting that the carrot had already been mentioned. This should not be confusing, for one understands that such mistakes are possible. It is similar with sets. A repeated element does not add to the set. Hence, $\{1, 3\} = \{1, 3, 1\}$.

Famous Sets

Although sets can contain many different types of elements, numbers are probably the most common for mathematics. For this reason particular important sets of numbers have been given their own symbols.

Symbol	Name
N	*The set of natural numbers*
Z	*The set of integers*
Q	*The set of rational numbers*
R	*The set of real numbers*
C	*The set of complex numbers*

As rosters,

$$N = \{0, 1, 2, \dots \}$$

and

$$Z = \{\dots, -2, -1, 0, 1, 2, \dots \}.$$

Notice that we define the set of natural numbers to include zero and do not make a distinction between counting numbers and whole numbers. Instead, write

$$Z^+ = \{1, 2, 3, \dots \}$$

and

$$Z^- = \{\dots, -3, -2, -1\}.$$

■ **EXAMPLE 3.1.2**

- $10 \in Z^+$, but $0 \notin Z^+$.

- $4 \in N$, but $-5 \notin N$.

- $-5 \in Z$, but $.65 \notin Z$.

- $.65 \in \mathbf{Q}$ and $1/2 \in \mathbf{Q}$, but $\pi \notin \mathbf{Q}$.

- $\pi \in \mathbf{R}$, but $3 - 2i \notin \mathbf{R}$.

- $3 - 2i \in \mathbf{C}$.

Of the sets mentioned above, the real numbers are probably the most familiar. It is the set of numbers most frequently used in calculus and is often represented by a number line. The line can be subdivided into **intervals**. Given two **endpoints**, an interval includes all real numbers between the endpoints and possibly the endpoints themselves. **Interval notation** is used to name these sets. A parenthesis next to an endpoint means that the endpoint is not included in the set, while a bracket means that the endpoint is included. If the endpoints are included, the interval is **closed**. If they are excluded, the interval is **open**. If one endpoint is included and the other is not, the interval is **half-open**. If the interval has only one endpoint, then the set is called a **ray** and is defined using the **infinity symbol** (∞), with or without the negative sign.

■ **DEFINITION 3.1.3**

Let $a, b \in \mathbf{R}$ such that $a < b$.

closed interval	$[a, b]$	*closed ray*	$[a, \infty)$
open interval	(a, b)	*closed ray*	$(-\infty, a]$
half-open interval	$[a, b)$	*open ray*	(a, ∞)
half-open interval	$(a, b]$	*open ray*	$(-\infty, a)$

■ **EXAMPLE 3.1.4**

We can describe $(4, 7)$ as

> *the interval of real numbers between 4 and 7 exclusive*

and $[4, 7]$ as

> *all real numbers between 4 and 7 inclusive.*

There is not a straightforward way to name the half-open intervals. For $(4, 7]$, we can try

> *the set of all real numbers x such that $4 < x \leq 7$*

or

> *the set of all real numbers greater than 4 and less than or equal to 7.*

The infinity symbol does not represent a real number, so a parenthesis must be used with it. Furthermore, the interval $(-\infty, \infty)$ can be used to denote $\mathbf{R}$.

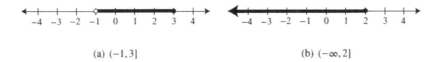

(a) $(-1, 3]$ (b) $(-\infty, 2]$

Figure 3.1 A half-open interval and a closed ray.

■ **EXAMPLE 3.1.5**

The interval $(-1, 3]$ contains all real numbers that are greater than -1 but less than or equal to 3 [Figure 3.1(a)]. A common mistake is to equate $(-1, 3]$ with $\{0, 1, 2, 3\}$. It is important to remember that $(-1, 3]$ includes all real numbers between -1 and 3. Hence, this set is infinite, as is $(-\infty, 2)$. It contains all real numbers less than 2 [Figure 3.1(b)].

■ **EXAMPLE 3.1.6**

Let $p(x) := x + 2 = 7$. Since $p(5)$ and there is no other real number a such that $p(a)$, there exists a unique $x \in \mathbf{R}$ such that $p(x)$. However, if a is an element of $\mathbf{Z}^-$ or $(-\infty, 5)$, then $\neg p(a)$.

■ **EXAMPLE 3.1.7**

If $q(x) := x \geq 10$, then $q(x)$ for all $x \in [20, 100]$, there exists $x \in \mathbf{Q}$ such that $\neg q(x)$, and there is no element a of $\{1, 2, 3\}$ such that $q(a)$.

Abstraction

When trying to write sets as rosters, we quickly discover issues with the technique. Since the rational numbers are defined using integers, we suspect that $\mathbf{Q}$ can be written as a roster, but when we try to begin a list, such as

$$1, \frac{1}{2}, \frac{1}{3}, \dots, 2, \frac{2}{3}, \dots,$$

we realize that there are complications with the pattern and are not quite sure that the we will exhaust them all. When considering $\mathbf{R}$, we know immediately that a roster is out of the question. We conclude that we need another method.

Fix a first-order alphabet with theory symbols S. Let A be a set and S-formula $p(x)$ have the property that for every a,

$$a \in A \Leftrightarrow p(a).$$

Notice that $p(x)$ completely describes the members of A. Namely, whenever we write $a \in A$, we can also write $p(a)$, and, conversely, whenever we write $p(a)$, we can also write $a \in A$. For example, let E be the set of even integers. As a roster,

$$E = \{\dots, -4, -2, 0, 2, 4, \dots\}.$$

Let $p(x)$ denote the formula

$$\exists n(n \in \mathbf{Z} \wedge x = 2n). \tag{3.1}$$

The even integers are exactly those numbers x such that $p(x)$. In particular, we have $p(2)$ and $p(-4)$ but not $p(5)$. Therefore, 2 and -4 are elements of E, but 5 is not.

■ **DEFINITION 3.1.8**

Let A be a first-order alphabet with theory symbols S. Let $p(x)$ be an S-formula and A be a set. Write $A = \{x : p(x)\}$ to mean

$$a \in A \Leftrightarrow p(a).$$

Using $\{x : p(x)\}$ to identify A is called the method of **abstraction**.

Using Definition 3.1.8 and (3.1), write E using abstraction as

$$E = \{x : \exists n(n \in \mathbf{Z} \wedge x = 2n)\}$$

or

$$E = \{x : p(x)\}.$$

Read this as "the set of x such that $p(x)$." Because $x = 2n$, it is customary to remove x from the definition of sets like E and write

$$E = \{2n : \in n\mathbf{Z}\},$$

or

$$E = \{2n : n \in \mathbf{Z}\}.$$

Read this as "the set of all $2n$ such that n is an integer." This simplified notation is still considered abstraction. Its form can be summarized as

$$\{\text{elements} : \text{condition}\}.$$

That is, what the elements look like come before the colon, and the condition that must be satisfied to be an element of the set comes after the colon.

■ **EXAMPLE 3.1.9**

Given the quadratic equation $x^2 - x - 2 = 0$, we know that its solutions are -1 and 2. Thus, its solution set is $A = \{-1, 2\}$. Using the method of abstraction, this can be written as

$$A = \{x : --\cdot xxx2 = 0 \wedge \in x\mathbf{R}\}.$$

However, as we have seen, it is customary to write the formula so that it is easier to read, so

$$A : x^2 - x - 2 = 0 \wedge x \in \mathbf{R}\},$$

or, using a common notation,

$$A\{x \in \mathbf{R} : x^2 - x - 2 = 0\}.$$

Therefore, given an arbitrary polynomial $f(x)$, its solution set over the real numbers is

$$\{x \in \mathbf{R} : f(x) = 0\}.$$

■ **EXAMPLE 3.1.10**

Since $x \notin \varnothing$ is always true, to write the empty set using abstraction, we use a formula like $x \neq x$ or a contradiction like $P \wedge \neg P$, where P is a propositional form. Then,

$$\varnothing = \{x \in \mathbf{R} : x \neq x\} = \{x : P \wedge \neg P\}.$$

■ **EXAMPLE 3.1.11**

Using the natural numbers as the starting point, $\mathbf{Z}$ and $\mathbf{Q}$ can be defined using the abstraction method by writing

$$\mathbf{Z} = \{n : n \in \mathbf{N} \vee -n \in \mathbf{N}\}$$

and

$$\mathbf{Q} = \left\{ \frac{a}{b} : a, b \in \mathbf{Z} \wedge b \neq 0 \right\}.$$

Notice the redundancy in the definition of $\mathbf{Q}$. The fraction $1/2$ is named multiple times like $2/4$ or $9/18$, but remember that this does not mean that the numbers appear infinitely many times in the set. They appear only once.

■ **EXAMPLE 3.1.12**

The open intervals can be defined using abstraction.

$$(a, b) = \{x \in \mathbf{R} : a < x < b\},$$
$$(a, \infty) = \{x \in \mathbf{R} : a < x\},$$
$$(-\infty, b) = \{x \in \mathbf{R} : x < a\}.$$

See Exercise 9 for the closed and half-open intervals. Also, as with $\mathbf{Z}^+$ and $\mathbf{Z}^-$, the superscript $+$ or $-$ is always used to denote the positive or negative numbers, respectively, of a set. For example,

$$\mathbf{R}^+ = \{x \in \mathbf{R} : x > 0\}$$

and

$$\mathbf{R}^- = \{x \in \mathbf{R} : x < 0\}.$$

■ **EXAMPLE 3.1.13**

Each of the following are written using the abstraction method and, where appropriate, as a roster.

- The set of all rational numbers with denominator equal to 3 in roster form is
$$\left\{ \ldots, -\frac{3}{3}, -\frac{2}{3}, -\frac{1}{3}, \frac{0}{3}, \frac{1}{3}, \frac{2}{3}, \frac{3}{3}, \ldots \right\}.$$
Using abstraction, it is
$$\left\{ \frac{n}{3} : n \in \mathbf{Z} \right\}.$$

- The set of all linear polynomials with integer coefficients and leading coefficient equal to 1 is
$$\{ \ldots, x - 2, x - 1, x, x + 1, x + 2, \ldots \}.$$
Using abstraction it is
$$\{ x + n : n \in \mathbf{Z} \}.$$

- The set of all polynomials of degree at most 5 can be written as
$$\{ a_5 x^5 + \cdots + a_1 x + a_0 : a_i \in \mathbf{R} \land i = 0, 1, \ldots, 5 \}.$$

Exercises

1. Determine whether the given propositions are true or false.
    (a) $0 \in \mathbf{N}$
    (b) $1/2 \in \mathbf{Z}$
    (c) $-4 \in \mathbf{Q}$
    (d) $4 + \pi \in \mathbf{R}$
    (e) $4.34534 \in \mathbf{C}$
    (f) $\{1, 2\} = \{2, 1\}$
    (g) $\{1, 2\} = \{1, 2, 1\}$
    (h) $[1, 2] = \{1, 2\}$
    (i) $(1, 3) = \{2\}$
    (j) $-1 \in (-\infty, -1)$
    (k) $-1 \in [-1, \infty)$
    (l) $\varnothing \in (-2, 2)$
    (m) $\varnothing \in \varnothing$
    (n) $0 \in \varnothing$

2. Write the given sets as rosters.
    (a) The set of all integers between 1 and 5 inclusive
    (b) The set of all odd integers
    (c) The set of all nonnegative integers

(d) The set of integers in the interval $(-3, 7]$

(e) The set of rational numbers in the interval $(0, 1)$ that can be represented with exactly two decimal places

3. If possible, find an element a in the given set such that $a + 3.14 = 0$.

    (a) **Z**

    (b) **R**

    (c) $\mathbf{R}^+$

    (d) $\mathbf{R}^-$

    (e) **Q**

    (f) **C**

    (g) $(0, 6)$

    (h) $(-\infty, -1)$

4. Determine whether the following are true or false.

    (a) $1 \in \{x : p(x)\}$ when $\neg p(1)$

    (b) $7 \in \{x \in \mathbf{R} : x^2 - 5x - 14 = 0\}$

    (c) $7x^2 - 0.5x \in \{a_2 x^2 + a_1 x + a_0 : a_i \in \mathbf{Q}\}$

    (d) $xy \in \{2k : k \in \mathbf{Z}\}$ if x is even and y is odd.

    (e) $\cos \theta \in \{a \cos \theta + b \sin \theta : a, b \in \mathbf{R}\}$

    (f) $\{1, 3\} = \{x : (x - 1)(x - 3) = 0\}$

    (g) $\{1, 3\} = \{x : (x - 1)(x - 3)^2 = 0\}$

    (h) $\left\{ \begin{bmatrix} a & 0 \\ 0 & 0 \end{bmatrix} : a \in \mathbf{R} \right\} = \left\{ \begin{bmatrix} x & 0 \\ 0 & y \end{bmatrix} : x \in \mathbf{R} \text{ and } y = 0 \right\}$

5. Write the following given sets in roster form.

    (a) $\{-3n : n \in \mathbf{Z}\}$

    (b) $\{0 \cdot n : n \in \mathbf{R}\}$

    (c) $\{n \cos x : n \in \mathbf{Z}\}$

    (d) $\{ax^2 + ax + a : a \in \mathbf{N}\}$

    (e) $\left\{ \begin{bmatrix} n & 0 \\ 0 & 0 \end{bmatrix} : n \in \mathbf{Z} \right\}$

6. Use a formula to uniquely describe the elements in the following sets. For example, $x \in \mathbf{N}$ if and only if $x \in \mathbf{Z} \wedge x \geq 0$.

    (a) $(0, 1)$

    (b) $(-3, 3]$

    (c) $[0, \infty)$

    (d) $\mathbf{Z}^+$

    (e) $\{\ldots, -2, -1, 0, 1, 2, \ldots\}$

    (f) $\{2a, 4a, 6a, \ldots\}$

7. Write each set using the method of abstraction.

    (a) All odd integers

    (b) All positive rational numbers

    (c) All integer multiples of 7

    (d) All integers that have a remainder of 1 when divided by 3

(e) All ordered pairs of real numbers in which the x-coordinate is positive and the y-coordinate is negative

(f) All complex numbers whose real part is 0

(g) All closed intervals that contain π

(h) All open intervals that do not contain a rational number

(i) All closed rays that contain no numbers in $(-\infty, 3]$

(j) All 2×2 matrices with real entries that have a diagonal sum of 0

(k) All polynomials of degree at most 3 with real coefficients

8. Write the given sets of real numbers using interval notation.

(a) The set of real numbers greater than 4

(b) The set of real numbers between -6 and -5 inclusive

(c) The set of real numbers x so that $x < 5$

(d) The set of real numbers x such that $10 < x \le 14$

9. Let $a, b \in \mathbf{R}$ with $a < b$. Write the given intervals using the abstraction method.

(a) $[a, b]$

(b) $[a, \infty)$

(c) $(-\infty, b]$

(d) $(a, b]$

3.2 SET OPERATIONS

We now use connectives to define the **set operations**. These allow us to build new sets from given ones.

Union and Intersection

The first set operation is defined using the $\vee$ connective.

■ **DEFINITION 3.2.1**

The **union** of A and B is

$$A \cup B = \{x : x \in A \vee x \in B\}.$$

The union of sets can be viewed as the combination of all elements from the sets. On the other hand, the next set operation is defined with $\wedge$ and can be considered as the overlap between the given sets.

■ **DEFINITION 3.2.2**

The **intersection** of A and B is

$$A \cap B = \{x : x \in A \wedge x \in B\}.$$

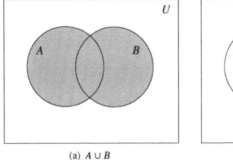 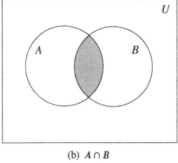

(a) $A \cup B$ (b) $A \cap B$

Figure 3.2   Venn diagrams for union and intersection.

For example, if $A = \{1, 2, 3, 4\}$ and $B = \{3, 4, 5, 6\}$, then

$$A \cup B = \{1, 2, 3, 4, 5, 6\}$$

and

$$A \cap B = \{3, 4\}.$$

The operations of union and intersection can be illustrated with pictures called **Venn diagrams**, named after the logician John Venn who used a variation of these drawings in his text on symbolic logic (Venn 1894). First, assume that all elements are members of a fixed universe U (page 97). In set theory, the universe is considered to be the set of all possible elements in a given situation. Use circles to represent sets and a rectangle to represent U. The space inside these shapes represent where elements might exist. Shading is used to represent where elements might exist after applying some set operations. The Venn diagram for union is in Figure 3.2(a) and the one for intersection is in Figure 3.2(b). If sets have no elements in their intersection, we can use the next definition to name them.

■ **DEFINITION 3.2.3**

The sets A and B are **disjoint** or **mutually exclusive** when $A \cap B = \varnothing$.

The sets $\{1, 2, 3\}$ and $\{6, 7\}$ are disjoint. A Venn diagram for two disjoint sets is given in Figure 3.3.

Set Difference

The next two set operations take all of the elements of one set that are not in another. They are defined using the *not* connective.

■ **DEFINITION 3.2.4**

The **set difference** of B from A is

$$A \setminus B = \{x : x \in A \land x \notin B\}.$$

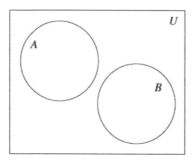

Figure 3.3   A Venn diagram for disjoint sets.

The **complement** of A is defined as

$$\overline{A} = U \setminus A = \{x : x \in U \wedge x \notin A\}.$$

Read $A \setminus B$ as "A minus B" or "A without B." See Figure 3.4(a) for the Venn diagram of the set difference of sets and Figure 3.4(b) for the complement of a set.

■ **EXAMPLE 3.2.5**

The following equalities use set difference.

- $\mathbf{N} = \mathbf{Z} \setminus \mathbf{Z}^-$

- The set of **irrational numbers** is $\mathbf{R} \setminus \mathbf{Q}$.

- $\mathbf{R} = \mathbf{C} \setminus \{a + bi : a, b \in \mathbf{R} \wedge b \neq 0\}$

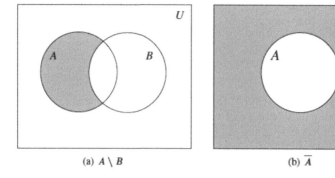

(a) $A \setminus B$ (b) $\overline{A}$

Figure 3.4   Venn diagrams for set difference and complement.

■ **EXAMPLE 3.2.6**

Let $U = \{1, 2, \ldots, 10\}$. Use a Venn diagram to find the results of the set operations on $A = \{1, 2, 3, 4, 5\}$ and $B = \{3, 4, 5, 6, 7, 8\}$. Each element will be represented as a point and labeled with a number.

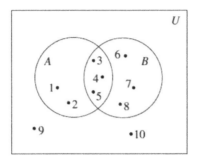

- $A \cup B = \{1, 2, 3, 4, 5, 6, 7, 8\}$

- $A \cap B = \{3, 4, 5\}$

- $A \setminus B = \{1, 2\}$

- $\overline{A} = \{6, 7, 8, 9, 10\}$

■ **EXAMPLE 3.2.7**

Let $C = (-4, 2)$, $D = [-1, 3]$, and $U = \mathbf{R}$. Use the diagram to perform the set operations.

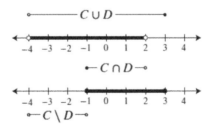

- $C \cup D = (-4, 3]$

- $C \cap D = [-1, 2)$

- $C \setminus D = (-4, -1)$

- $\overline{C} = (-\infty, -4] \cup [2, \infty)$

Cartesian Products

The last set operation is not related to the logic connectives as the others, but it is nonetheless very important to mathematics. Let A and B be sets. Given elements $a \in A$ and $b \in B$, we call (a, b) an **ordered pair**. In this context, a and b are called **coordinates**. It is similar to the set $\{a, b\}$ except that the order matters. The definition is due to Kazimierz Kuratowski (1921).

■ **DEFINITION 3.2.8**

If $a \in A$ and $b \in B$,
$$(a, b) = \{\{a\}, \{a, b\}\}.$$

Notice that $(a, b) = (a', b')$ means that
$$\{\{a\}, \{a, b\}\} = \{\{a'\}, \{a', b'\}\},$$
which implies that $a = a'$ and $b = b'$. Therefore,
$$(a, b) = (a', b') \text{ if and only if } a = a' \text{ and } b = b'.$$

The set of all ordered pairs with the first coordinate from A and the second from B is named after René Descartes.

■ **DEFINITION 3.2.9**

The **Cartesian product** of A and B is
$$A \times B = \{(a, b) : a \in A \wedge b \in B\}.$$

The product $\mathbf{R}^2 = \mathbf{R} \times \mathbf{R}$ is the set of ordered pairs of the **Cartesian plane**.

■ **EXAMPLE 3.2.10**

- Since $(1, 2) \neq (2, 1)$,
$$\{1, 2\} \times \{0, 1, 2\} = \{(1, 0), (1, 1), (1, 2), (2, 0), (2, 1), (2, 2)\}.$$

Even though we have a set definition for an ordered pair, we can still visually represent this set on a grid as in Figure 3.5(a).

- If $A = \{1, 2, 7\}$ and $B = \{\varnothing, \{1, 5\}\}$,
$$A \times B = \{(1, \varnothing), (1, \{1, 5\}), (2, \varnothing), (2, \{1, 5\}), (7, \varnothing), (7, \{1, 5\})\}.$$

See Figure 3.5(b).

- For any set A, $\varnothing \times A = A \times \varnothing = \varnothing$ because $\neg \exists x (x \in \varnothing \wedge y \in A)$.

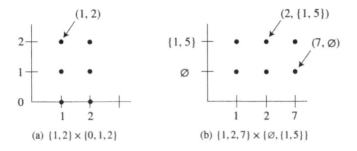

Figure 3.5 Two Cartesian products.

We generalize Definition 3.2.8 by defining

$$(a, b, c) = \{\{a\}, \{a, b\}, \{a, b, c\}\},$$

$$(a, b, c, d) = \{\{a\}, \{a, b\}, \{a, b, c\}, \{a, b, c, d\}\},$$

and for $n \in \mathbf{N}$,

$$(a_0, a_1, \dots, a_{n-1}) = \{\{a_0\}, \{a_0, a_1\}, \dots, \{a_0, a_1, \dots, a_{n-1}\}\},$$

which is called an **ordered n-tuple**. Then,

$$A \times B \times C = \{(a, b, c) : a \in A \wedge b \in B \wedge c \in C\},$$

$$A \times B \times C \times D = \{(a, b, c, d) : a \in A \wedge b \in B \wedge c \in C \wedge d \in D\},$$

and

$$A^n = \{(a_0, a_1, \dots, a_{n-1}) : a_i \in A \wedge i = 0, 1, \dots, n - 1\}.$$

Specifically, $\mathbf{R}^3 = \mathbf{R} \times \mathbf{R} \times \mathbf{R}$, and

$$\mathbf{R}^n = \underbrace{\mathbf{R} \times \mathbf{R} \times \cdots \times \mathbf{R}}_{n \text{ times}},$$

which is known as **Cartesian n-space**.

■ **EXAMPLE 3.2.11**

Let $A = \{1\}$, $B = \{2\}$, $C = \{3\}$, and $D = \{4\}$. Then,

$$A \times B \times C \times D = \{(1, 2, 3, 4)\}$$

is a singleton containing an ordered 4-tuple. Also,

$$(A \times B) \cup (C \times D) = \{(1, 2), (3, 4)\},$$

but

$$A \times (B \cup C) \times D = \{(1, 2, 4), (1, 3, 4)\}.$$

Order of Operations

As with the logical connectives, we need an order to make sense of expressions that involve many operations. To do this, we note the association between the set operations and certain logical connectives.

$$\left.\begin{array}{c} \overline{A} \\ A \setminus B \end{array}\right\} \quad \neg$$

$$\begin{array}{cc} A \cap B & \wedge \\ A \cup B & \vee \end{array}$$

From this we derive the order for the set operations.

■ **DEFINITION 3.2.12 [Order of Operations]**

To find a set determined by set operations, read from left to right and use the following precedence.

- sets within parentheses (innermost first),

- complements,

- set differences,

- intersections,

- unions.

■ **EXAMPLE 3.2.13**

If the universe is taken to be $\{1, 2, 3, 4, 5\}$, then

$$\{5\} \cup \overline{\{1, 2\}} \cap \{2, 3\} = \{5\} \cup \{3, 4, 5\} \cap \{2, 3\} = \{5\} \cup \{3\} = \{3, 5\}$$

This set can be written using parentheses as

$$\{5\} \cup (\overline{\{1, 2\}} \cap \{2, 3\}).$$

However,

$$(\{5\} \cup \overline{\{1, 2\}}) \cap \{2, 3\} = (\{5\} \cup \{3, 4, 5\}) \cap \{2, 3\} = \{3, 4, 5\} \cap \{2, 3\} = \{3\},$$

showing that

$$\{5\} \cup \overline{\{1, 2\}} \cap \{2, 3\} \neq (\{5\} \cup \overline{\{1, 2\}}) \cap \{2, 3\}.$$

■ **EXAMPLE 3.2.14**

Define $A = \{1\}$, $B = \{2\}$, $C = \{3\}$, and $D = \{4\}$. Then, we have

$$A \cup B \cup C \cup D = \{1, 2, 3, 4\}.$$

Written with parentheses and brackets,

$$A \cup B \cup C \cup D = ([(A \cup B) \cup C] \cup D).$$

With the given assignments, $A \cap B \cap C \cap D$ is empty.

Exercises

1. Each of the given propositions are false. Replace the underlined word with another word to make the proposition true.

   (a) Intersection is defined using a <u>disjunction</u>.

   (b) <u>Set</u> diagrams are used to illustrate set operations.

   (c) $\mathbf{R} \setminus (\mathbf{R} \setminus \mathbf{Q})$ is the set of <u>irrational</u> numbers.

   (d) Set difference has a higher order of precedence than <u>complements</u>.

   (e) The complement of A is equal to $\underline{\mathbf{R}}$ set minus A.

   (f) The intersection of two intervals is <u>always</u> an interval.

   (g) The union of two intervals is <u>never</u> an interval.

   (h) $A \times B$ is equal to $\varnothing$ if B does not contain <u>ordered pairs</u>.

2. Let $A = \{0, 2, 4, 6\}$, $B = \{3, 4, 5, 6\}$, $C = \{0, 1, 2\}$, and $U = \{0, 1, \ldots, 9, 10\}$. Write the given sets in roster notation.

   (a) $A \cup B$

   (b) $A \cap B$

   (c) $A \setminus B$

   (d) $B \setminus A$

   (e) $\overline{A}$

   (f) $A \times B$

   (g) $A \cup B \cap C$

   (h) $A \cap B \cup C$

   (i) $\overline{A} \cap B$

   (j) $A \cup \overline{A} \cap B$

   (k) $A \setminus B \setminus C$

   (l) $A \cup B \setminus A \cap C$

3. Write each of the given sets using interval notation.

   (a) $[2, 3] \cap (5/2, 7]$

   (b) $(-\infty, 4) \cup (-6, \infty)$

   (c) $(-2, 4) \cap [-6, \infty)$

   (d) $\varnothing \cup (4, 12]$

4. Identify each of the the given sets.
    (a) $[6, 17] \cap [17, 32)$
    (b) $[6, 17) \cap [17, 32)$
    (c) $[6, 17] \cup (17, 32)$
    (d) $[6, 17) \cup (17, 32)$

5. Draw Venn diagrams.
    (a) $A \cap \overline{B}$
    (b) $\overline{A \cup B}$
    (c) $A \cap \overline{B} \cup \overline{C}$
    (d) $(A \cup B) \setminus \overline{C}$
    (e) $A \cap C \cap \overline{B}$
    (f) $\overline{A \setminus B} \cap C$

6. Match each Venn diagram to as many sets as possible.

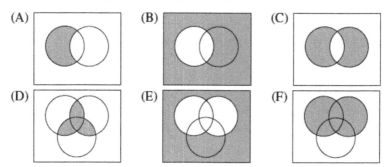

(A)   (B)   (C)

(D)   (E)   (F)

    (a) $A \cup B$
    (b) $A \setminus B$
    (c) $A \cap \overline{B}$
    (d) $(A \cup B) \setminus (A \cap B)$
    (e) $\overline{A \cap B \cup A \cap C \cup B \cap C}$
    (f) $\overline{A \cap B \cup A \setminus B}$
    (g) $(A \cup B) \cap (\overline{A} \cup \overline{B})$
    (h) $[(A \cup B) \cap \overline{C}] \cup [(A \cup B) \cap C]$
    (i) $[A \cap \overline{B} \cap C] \cup [\overline{A} \cap \overline{B} \cap \overline{C}]$
    (j) $\overline{A \setminus (B \cup C) \cap B \setminus (A \cup C) \cap C \setminus (A \cup B)}$

7. The ordered pair $(1, 2)$ is paired with the ordered pair $(2, 1)$ using a set operation. Write each resulting set as a roster.
    (a) $(1, 2) \cup (2, 1)$
    (b) $(1, 2) \cap (2, 1)$
    (c) $(1, 2) \setminus (2, 1)$

8. Let $p(x)$ be a formula. Prove the following.
    (a) $\forall x[x \in A \cup B \to p(x)] \Rightarrow \forall x[x \in A \cap B \to p(x)]$

(b) $\forall x[x \in A \cup B \rightarrow p(x)] \Leftrightarrow \forall x[x \in A \rightarrow p(x)] \wedge \forall x[x \in B \rightarrow p(x)]$

(c) $\exists x[x \in A \cap B \wedge p(x)] \Rightarrow \exists x[x \in A \wedge p(x)] \wedge \exists x[x \in B \wedge p(x)]$

(d) $\exists x[x \in A \cup B \wedge p(x)] \Leftrightarrow \exists x[x \in A \wedge p(x)] \vee \exists x[x \in B \wedge p(x)]$

9. Find a formula $p(x)$ and sets A and B to show that the following are false.

(a) $\forall x[x \in A \rightarrow p(x)] \vee \forall x[x \in B \rightarrow p(x)] \Rightarrow \forall x[x \in A \cup B \rightarrow p(x)]$

(b) $\exists x[x \in A \wedge p(x)] \wedge \exists x[x \in B \wedge p(x)] \Rightarrow \exists x[x \in A \cap B \wedge p(x)]$

3.3 SETS WITHIN SETS

An important relation between any two sets is when one is contained within another.

Subsets

Let A and B be sets. A is a **subset** of B exactly when every element of A is also an element of B, in symbols $A \subseteq B$. This is represented in a Venn diagram by the circle for A being within the circle for B [Figure 3.6(a)].

■ DEFINITION 3.3.1

For all sets A and B,

$$A \subseteq B \Leftrightarrow \forall x(x \in A \rightarrow x \in B).$$

If A is not a subset of B, write $A \nsubseteq B$. This is represented in a Venn diagram by A overlapping B with a point in A but not within B [Figure 3.6(b)]. Logically, this means

$$A \nsubseteq B \Leftrightarrow \neg\forall x(x \in A \rightarrow x \in B) \Leftrightarrow \exists x(x \in A \wedge x \notin B).$$

Thus, to show $A \nsubseteq B$ find an element in A that is not in B. For example, if we let $A = \{1, 2, 3\}$ and $B = \{1, 2, 5\}$, then $A \nsubseteq B$ because $3 \in A$ but $3 \notin B$.

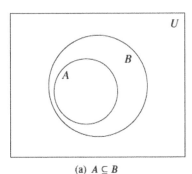

(a) $A \subseteq B$

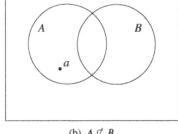

(b) $A \nsubseteq B$

Figure 3.6 Venn diagrams for a subset and a nonsubset.

■ EXAMPLE 3.3.2

The proposition *for all sets A and B, $A \cup B \subseteq A$* is false. To see this, we must prove that there exists A and B such that $A \cup B \nsubseteq A$. We take $A = \{1\}$ and $B = \{2\}$ as our candidates. Since $A \cup B = \{1, 2\}$ and $2 \notin A$, we have $A \cup B \nsubseteq A$.

Every set is the **improper subset** of itself. The notation $A \subset B$ means $A \subseteq B$ but $A \neq B$. In this case, A is a **proper subset** of B.

■ EXAMPLE 3.3.3

- $\{1, 2, 3\} \subset \{1, 2, 3, 4, 5\}$ and $\{1, 2, 3\} \subseteq \{1, 2, 3\}$

- $\mathbf{N} \subset \mathbf{Z} \subset \mathbf{Q} \subset \mathbf{R} \subset \mathbf{C}$

- $(4, 5) \subset (4, 5] \subset [4, 5]$

- $\{1, 2, 3\}$ is not a subset of $\{1, 2, 4\}$.

Our study of subsets begins with a fundamental theorem. It states that the empty set is a subset of every set.

■ THEOREM 3.3.4

$\varnothing \subseteq A$.

PROOF
Let A be a set. Since $x \notin \varnothing$, we have

$$\forall x(x \notin \varnothing \lor x \in A) \Leftrightarrow \forall x(x \in \varnothing \rightarrow x \in A) \Leftrightarrow \varnothing \subseteq A. ■$$

Proving that one set is a subset of another always means proving an implication, so Direct Proof is the primary tool in these proofs. That is, usually to prove that a set A is a subset of a set B, take an element $x \in A$ and show $x \in B$.

■ EXAMPLE 3.3.5

- Let $x \in \mathbf{Z} \setminus \mathbf{N}$. Then, $x \in \mathbf{Z}$ but $x \notin \mathbf{N}$, so $x \in \mathbf{Z}$ by Simp. Thus, $\mathbf{Z} \setminus \mathbf{N} \subseteq \mathbf{Z}$.

- Let
$$A = \{x : \exists n \in \mathbf{Z}\,[(x - 2n)(x - 2n - 2) = 0]\}$$
and
$$B = \{2n : n \in \mathbf{Z}\}.$$
Take $x \in A$. This means that
$$\exists n \in \mathbf{Z}\,[(x - 2n)(x - 2n - 2) = 0].$$
In other words,
$$(x - 2n)(x - 2n - 2) = 0$$

for some $n \in \mathbf{Z}$. Hence,

$$x - 2n = 0 \text{ or } x - 2n - 2 = 0.$$

We have two cases to check. If $x - 2n = 0$, then $x = 2n$. This means $x \in B$. If $x - 2n - 2 = 0$, then $x = 2n + 2 = 2(n + 1)$, which also means $x \in B$ since $n + 1$ is an integer. Hence, $A \subseteq B$.

- Fix $a, b \in \mathbf{Z}$. Let $x = na$, some $n \in \mathbf{Z}$. This means $x = na + 0b$, which implies that $\{na : n \in \mathbf{Z}\} \subseteq \{na + mb : n, m \in \mathbf{Z}\}$.

This next result is based only on the definitions and Inference Rules 1.2.10. As with Section 3.2, we see the close ties between set theory and logic.

■ **THEOREM 3.3.6**

- $A \subseteq A$.

- If $A \subseteq B$ and $B \subseteq C$, then $A \subseteq C$.

- If $A \subseteq B$ and $x \in A$, then $x \in B$.

- If $A \subseteq B$ and $x \notin B$, then $x \notin A$.

- Let $A \subseteq B$ and $C \subseteq D$. If $x \in A$ or $x \in C$, then $x \in B$ or $x \in D$.

- Let $A \subseteq B$ and $C \subseteq D$. If $x \notin B$ or $x \notin D$, then $x \notin A$ or $x \notin C$.

PROOF

Assume $A \subseteq B$ and $B \subseteq C$. By definition, $x \in A$ implies $x \in B$ and $x \in B$ implies $x \in C$. Therefore, by HS, if $x \in A$, then $x \in C$. In other words, $A \subseteq C$. The remaining parts are left to Exercise 4. ■

Note that it is not the case that for all sets A and B, $A \subseteq B$ implies that $B \subseteq A$. To prove this, choose $A = \varnothing$ and $B = \{0\}$. Then, $A \subseteq B$ yet $B \nsubseteq A$ because $0 \in B$ and $0 \notin \varnothing$.

Equality

For two sets to be equal, they must contain exactly the same elements (page 118 of Section 3.1). This can be stated more precisely using the idea of a subset.

■ **DEFINITION 3.3.7**

$A = B$ means $A \subseteq B$ and $B \subseteq A$.

To prove that two sets are equal, we show both inclusions. Let us do this to prove $\overline{A \cup B} = \overline{A} \cap \overline{B}$. This is one of **De Morgan's laws**. To prove it, we demonstrate both

$$\overline{A \cup B} \subseteq \overline{A} \cap \overline{B}$$

and

$$\overline{A \cap B} \subseteq \overline{A} \cup \overline{B}.$$

This amounts to proving a biconditional, which means we will use the rule of biconditional proof. Look at the first direction:

$x \in \overline{A \cup B}$	Given
$\neg(x \in A \cup B)$	Definition of complement
$\neg(x \in A \vee x \in B)$	Definition of union
$x \notin A \wedge x \notin B$	De Morgan's law
$x \in \overline{A} \wedge x \in \overline{B}$	Definition of complement
$x \in \overline{A} \cap \overline{B}$	Definition of intersection

Now read backward through those steps. Each follows logically when read in this order because only definitions and replacement rules were used. This means that the steps are reversible. Hence, we have a series of biconditionals, and we can use the short rule of biconditional proof (page 108):

$$x \in \overline{A \cup B} \Leftrightarrow \neg(x \in A \cup B)$$
$$\Leftrightarrow \neg(x \in A \vee x \in B)$$
$$\Leftrightarrow x \notin A \wedge x \notin B$$
$$\Leftrightarrow x \in \overline{A} \wedge x \in \overline{B}$$
$$\Leftrightarrow x \in \overline{A} \cap \overline{B}.$$

Hence, $\overline{A \cup B} = \overline{A} \cap \overline{B}$.

We must be careful when writing these types of proofs since it is easy to confuse the notation.

- The correct translation for $x \in A \cap B$ is $x \in A \wedge x \in B$. Common mistakes for this translation include using formulas with set operations:

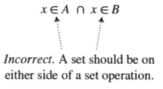

Incorrect. A set should be on either side of a set operation.

...and sets with connectives:

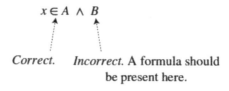

Correct.   *Incorrect.* A formula should be present here.

Remember that connectives connect formulas and set operations connect sets.

- Negations also pose problems. If a complement is used, first translate using

$$x \in \overline{A} \Leftrightarrow x \in U \wedge x \notin A \Leftrightarrow x \notin A$$

and then proceed with the proof. Similarly, the formula $x \notin A \cup B$ can be written as neither $x \notin A \cup x \notin B$ nor $x \notin A \vee x \notin B$. Instead, use DeM:

$$x \notin A \cup B \Leftrightarrow \neg(x \in A \cup B)$$
$$\Leftrightarrow \neg(x \in A \vee x \in B)$$
$$\Leftrightarrow x \notin A \wedge x \notin B$$
$$\Leftrightarrow x \in \overline{A} \cap \overline{B}.$$

We can now prove many basic properties about set operations. Notice how the following are closely related to their corresponding replacement rules (1.3.9).

■ **THEOREM 3.3.8**

- **Associative Laws**
  $A \cap B \cap C = A \cap (B \cap C)$
  $A \cup B \cup C = A \cup (B \cup C)$

- **Commutative Laws**
  $A \cap B = B \cap A$
  $A \cup B = B \cup A$

- **De Morgan's Laws**
  $\overline{A \cup B} = \overline{A} \cap \overline{B}$
  $\overline{A \cap B} = \overline{A} \cup \overline{B}$

- **Distributive Laws**
  $A \cap (B \cup C) = A \cap B \cup A \cap C$
  $A \cup B \cap C = (A \cup B) \cap (A \cup C)$

- **Idempotent Laws**
  $A \cap A = A$
  $A \cup A = A.$

■ **EXAMPLE 3.3.9**

We use the short rule of biconditional proof (page 108) to prove the equality $A \cap B \cap C = A \cap (B \cap C)$.

$$x \in A \cap B \cap C \Leftrightarrow x \in A \cap B \wedge x \in C$$
$$\Leftrightarrow x \in A \wedge x \in B \wedge x \in C$$
$$\Leftrightarrow x \in A \wedge (x \in B \wedge x \in C)$$
$$\Leftrightarrow x \in A \wedge (x \in B \cap C)$$
$$\Leftrightarrow x \in A \cap (B \cap C).$$

Another way to prove it is to use a chain of equal signs.

$$A \cap B \cap C = \{x : x \in A \cap B \wedge x \in C\}$$
$$= \{x : x \in A \wedge x \in B \wedge x \in C\}$$
$$= \{x : x \in A \wedge (x \in B \wedge x \in C)\}$$
$$= \{x : x \in A \wedge x \in B \cap C\}$$
$$= A \cap (B \cap C).$$

We have to be careful when proving equality. If two sets are equal, there are always proofs for both inclusions. However, the steps needed for the one implication might not simply be the steps for the converse in reverse. The next example illustrates this. It is always true that $A \cap B \subseteq A$. However, the premise is needed to show the other inclusion.

■ **EXAMPLE 3.3.10**

Let $A \subseteq B$. Prove $A \cap B = A$.

- Let $x \in A \cap B$. This means that $x \in A$ and $x \in B$. Then, $x \in A$ (Simp).

- Assume that x is an element of A. Since $A \subseteq B$, x is also an element of B. Hence, $x \in A \cap B$.

A more involved example of this uses the concept of divisibility. Let $a, b \in \mathbf{Z}$, not both equal to zero. A **common divisor** of a and b is c when $c \mid a$ and $c \mid b$. For example, 4 is a common divisor of 48 and 36, but it is not the largest.

■ **DEFINITION 3.3.11**

Let $a, b \in \mathbf{Z}$ with a and b not both zero. The integer g is the **greatest common divisor** of a and b if g is a common divisor of a and b and $e \le g$ for every common divisor $e \in \mathbf{Z}$. In this case, write $g = \gcd(a, b)$.

For example,

$$12 = \gcd(48, 36)$$

and

$$7 = \gcd(0, 7).$$

Notice that it is important that at least one of the integers is not zero. The $\gcd(0,0)$ is undefined since all a such that $a \ne 0$ divide 0. Further notice that the greatest common divisor is positive.

■ **EXAMPLE 3.3.12**

Let $a, b \in \mathbf{Z}$. Prove for all $n \in \mathbf{Z}$,

$$\gcd(a + nb, b) = \gcd(a, b).$$

If both pairs of numbers have the same common divisors, their greatest common divisors must be equal. So, define

$$S = \{k \in \mathbf{Z} : k \mid a + nb \wedge k \mid b\}$$

and

$$T = \{k \in \mathbf{Z} : k \mid a \wedge k \mid b\}.$$

To show that the greatest common divisors are equal, prove $S = T$.

- Let $d \in S$. Then $d \mid a + nb$ and $d \mid b$. This means $a + nb = dl$ and $b = dk$ for some $l, k \in \mathbf{Z}$. We are left to show $d \mid a$. By substitution, $a + ndk = dl$. Hence, $d \in T$ because

$$a = dl - ndk = d(l - nk).$$

- Now take $d \in T$. This means $d \mid a$ and $d \mid b$. Thus, there exists $l, k \in \mathbf{Z}$ such that $a = dl$ and $b = dk$. Then,

$$a + nb = dl + ndk = d(l + nk).$$

Therefore, $d \mid a + nb$ and $d \in S$.

As with subsets, let us now prove some results concerning the empty set and the universe. We use two strategies.

- Let A be a set. We know that

$$A = \varnothing \text{ if and only if } \forall x (x \notin A).$$

Therefore, to prove that A is empty, take an arbitrary a and show $a \notin A$. This can sometimes be done directly, but more often an indirect proof is better. That is, assume $a \in A$ and derive a contradiction. Since the contradiction arose simply by assuming $a \in A$, this formula must be the problem. Hence, A can have no elements.

- Let U be a universe. To prove $A = U$, we must show that $A \subseteq U$ and $U \subseteq A$. The first subset relation is always true. To prove the second, take an arbitrary element and show that it belongs to A. This works because U contains all possible elements.

■ **EXAMPLE 3.3.13**

- Suppose $x \in A \cap \varnothing$. Then $x \in \varnothing$, which is impossible. Therefore, $A \cap \varnothing = \varnothing$.

- Certainly, $A \subseteq A \cup \varnothing$, so to show the opposite inclusion take $x \in A \cup \varnothing$. Since $x \notin \varnothing$, it must be the case that $x \in A$. Thus, $A \cup \varnothing \subseteq A$, and we have that $A \cup \varnothing = A$.

- From Exercise 5(b), we know $A \cap U \subseteq A$, so let $x \in A$. This means that x must also belong to the universe. Hence, $x \in A \cap U$, so $A \cap U = A$.

- Certainly, $A \cup U \subseteq U$. Moreover, by Exercise 5(c), we have the other inclusion. Thus, $A \cup U = U$.

■ **EXAMPLE 3.3.14**

To prove that a set A is not equal to the empty set, show $\neg\forall x(x \notin A)$, but this is equivalent to $\exists x(x \in A)$. For instance, let

$$A = \{x \in \mathbf{R} : x^2 + 6x + 5 = 0\}.$$

We know that A is nonempty since $-1 \in A$.

■ **EXAMPLE 3.3.15**

Let

$$A = \{(a, b) \in \mathbf{R} \times \mathbf{R} : a + b = 0\}$$

and

$$B = \{(0, b) : b \in \mathbf{R}\}.$$

These sets are not disjoint since $(0, 0)$ is an element of both A and B. However, $A \neq B$ because $(1, -1) \in A$ but $(1, -1) \notin B$.

Let us combine the two strategies to show a relationship between $\varnothing$ and U.

■ **THEOREM 3.3.16**

For all sets A and B and universe U, the following are equivalent.

- $A \subseteq B$
- $\overline{A} \cup B = U$
- $A \cap \overline{B} = \varnothing.$

PROOF

- Assume $A \subseteq B$. Suppose $x \notin \overline{A}$. Then, $x \in A$, which implies that $x \in B$. Hence, for every element x, we have that $x \in \overline{A}$ or $x \in B$, and we conclude that $\overline{A} \cup B = U$.

- Suppose $\overline{A} \cup B = U$. In order to obtain a contradiction, take $x \in A \cap \overline{B}$. Then, $x \in \overline{B}$. Since $x \in A$, the supposition also gives $x \in B$, a contradiction. Therefore, $A \cap \overline{B} = \varnothing$.

- Let $A \cap \overline{B} = \varnothing$. Assume $x \in A$. By hypothesis, x cannot be a member of $\overline{B}$, otherwise the intersection would be nonempty. Hence, $x \in B$. ■

The following theorem is a generalization of the corresponding result concerning subsets. The proof could have been written using the short rule of biconditional proof or by appealing to Lemma 3.3.6 (Exercise 21).

■ **THEOREM 3.3.17**

If $A = B$ and $B = C$, then $A = C$.

PROOF

Assume $A = B$ and $B = C$. This means $A \subseteq B$, $B \subseteq A$, $B \subseteq C$, and $C \subseteq B$. We must show that $A = C$.

- Let $x \in A$. By hypothesis, x is then an element of B, which implies that $x \in C$.

- Let $x \in C$. Then, $x \in B$, from which $x \in A$ follows. ∎

The last result of the section involves the Cartesian product. The first part is illustrated in Figure 3.7. The sets B and C are illustrated along the vertical axis and A is illustrated along the horizontal axis. The Cartesian products are represented as boxes. The other parts of the theorem can be similarly visualized.

■ **THEOREM 3.3.18**

- $A \times (B \cap C) = (A \times B) \cap (A \times C)$.

- $A \times (B \cup C) = (A \times B) \cup (A \times C)$.

- $A \times (B \setminus C) = (A \times B) \setminus (A \times C)$.

- $(A \times B) \cap (C \times D) = (A \cap C) \times (B \cap D)$.

PROOF

We prove the first equation. The last three are left to Exercise 19. Take three sets A, B, and C. Then,

$$(a, b) \in A \times (B \cap C) \Leftrightarrow a \in A \wedge b \in B \cap C$$
$$\Leftrightarrow a \in A \wedge b \in B \wedge a \in A \wedge b \in C$$
$$\Leftrightarrow (a, b) \in A \times B \wedge (a, b) \in A \times C$$
$$\Leftrightarrow (a, b) \in (A \times B) \cap (A \times C). ∎$$

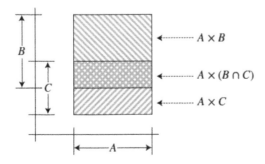

Figure 3.7 $A \times (B \cap C) = (A \times B) \cap (A \times C)$.

Inspired by the last result, we might try proving that $(A \times B) \cup (C \times D)$ is always equal to $(A \cup C) \times (B \cup D)$, but no such proof exists. To show this, take $A = B = \{1\}$ and $C = D = \{2\}$. Then

$$(A \cup C) \times (B \cup D) = \{1, 2\} \times \{1, 2\} = \{(1, 1), (1, 2), (2, 1), (2, 2)\}.$$

but

$$(A \times B) \cup (C \times D) = \{(1, 1)\} \cup \{(2, 2)\} = \{(1, 1), (2, 2)\},$$

Hence, $(A \cup C) \times (B \cup D) \not\subseteq (A \times B) \cup (C \times D)$. Notice, however, that the opposite inclusion is always true (Exercise 3.3.8).

Exercises

1. Answer true or false.
    (a) $\emptyset \in \emptyset$
    (b) $\emptyset \subseteq \{1\}$
    (c) $1 \in \mathbf{Z}$
    (d) $1 \subseteq \mathbf{Z}$
    (e) $1 \in \emptyset$
    (f) $\{1\} \subseteq \emptyset$
    (g) $0 \in \emptyset$
    (h) $\{1\} \in \mathbf{Z}$
    (i) $\emptyset \subseteq \overline{\emptyset}$

2. Answer true or false. For each false proposition, find one element that is in the first set but is not in the second.
    (a) $\mathbf{Z}^+ \subseteq \mathbf{C}$
    (b) $\mathbf{Q}^+ \subseteq \mathbf{Z}^+$
    (c) $\mathbf{Q} \setminus \mathbf{R} \subseteq \mathbf{Z}$
    (d) $\mathbf{R} \setminus \mathbf{Q} \subseteq \mathbf{Z}$
    (e) $\mathbf{Z} \cap (-1, 1) \subseteq \mathbf{Q}$
    (f) $(0, 1) \subseteq \mathbf{Q}^+$
    (g) $(0, 1) \subseteq \{0, 1, 2\}$
    (h) $(0, 1) \subseteq (0, 1]$

3. Prove.
    (a) $\{x \in \mathbf{R} : x^2 - 3x + 2 = 0\} \subseteq \mathbf{N}$
    (b) $(0, 1) \subseteq [0, 1]$
    (c) $[0, 1] \not\subseteq (0, 1)$
    (d) $\mathbf{Z} \times \mathbf{Z} \not\subseteq \mathbf{Z} \times \mathbf{N}$
    (e) $(0, 1) \cap \mathbf{Q} \not\subseteq [0, 1] \cap \mathbf{Z}$
    (f) $\mathbf{R} \subseteq \mathbf{C}$
    (g) $\{bi : b \in \mathbf{R}\} \subseteq \mathbf{C}$
    (h) $\mathbf{C} \not\subseteq \mathbf{R}$

4. Prove the remaining parts of Theorem 3.3.6.

5. Prove.
   (a) $A \subseteq U$
   (b) $A \cap B \subseteq A$
   (c) $A \subseteq A \cup B$
   (d) $A \setminus B \subseteq A$
   (e) If $A \subseteq B$, then $A \cup C \subseteq B \cup C$.
   (f) If $A \subseteq B$, then $A \cap C \subseteq B \cap C$.
   (g) If $A \subseteq B$, then $C \setminus B \subseteq C \setminus A$.
   (h) If $A \neq \varnothing$, then $A \not\subseteq \overline{A}$.
   (i) If $A \subseteq \overline{B}$, then $B \subseteq \overline{A}$.
   (j) If $A \subseteq B$, then $\{1\} \times A \subseteq \{1\} \times B$.
   (k) If $A \subseteq C$ and $B \subseteq D$, then $A \times B \subseteq C \times D$.
   (l) If $A \subseteq C$ and $B \subseteq D$, then $C \times D \subseteq A \times B$.

6. Prove that $A \subseteq B$ if and only if $\overline{B} \subseteq \overline{A}$.

7. Show that the given proposition is false:

   $$\text{for all sets } A \text{ and } B, \text{ if } A \cap B \neq \varnothing, \text{ then } A \not\subseteq A \cap B.$$

8. Prove: $(A \times B) \cup (C \times D) \subseteq (A \cup C) \times (B \cup D)$.

9. Take $a, b, c \in \mathbf{N}$. Let $A = \{n \in \mathbf{N} : n \mid a\}$ and $C = \{n \in \mathbf{N} : n \mid c\}$. Suppose $a \mid b$ and $b \mid c$. Prove $A \subseteq C$.

10. Prove.
    (a) $B \subseteq A$ and $C \subseteq A$ if and only if $B \cup C \subseteq A$.
    (b) $A \subseteq B$ and $A \subseteq C$ if and only if $A \subseteq B \cap C$.

11. Prove the unproven parts of Theorem 3.3.8.

12. Prove each equality.
    (a) $\overline{\varnothing} = U$
    (b) $\overline{U} = \varnothing$
    (c) $A \cap \overline{A} = \varnothing$
    (d) $A \cup \overline{A} = U$
    (e) $\overline{\overline{A}} = A$
    (f) $A \setminus A = \varnothing$
    (g) $A \setminus \varnothing = A$
    (h) $\overline{A} \cap B = B \setminus A$
    (i) $A \setminus B = A \cap \overline{B}$
    (j) $\overline{A \cup B} \cap B = \varnothing$
    (k) $A \cap B \setminus A = \varnothing$

13. Sketch a Venn diagram for each problem and then write a proof.

(a) $A = A \cap B \cup A \cap \overline{B}$

(b) $A \cup B = A \cup \overline{A} \cap B$

(c) $A \setminus (B \setminus C) = A \cap (\overline{B} \cup C)$

(d) $A \setminus (A \cap B) = A \setminus B$

(e) $A \cap B \cup A \cap \overline{B} \cup \overline{A} \cap B = A \cup B$

(f) $(A \cup B) \setminus C = A \setminus C \cup B \setminus C$

(g) $A \setminus (B \setminus C) = A \setminus B \cup A \cap C$

(h) $A \setminus B \setminus C = A \setminus (B \cup C)$

14. Prove.

(a) If $A \subseteq B$, then $A \setminus B = \varnothing$.

(b) If $A \subseteq \varnothing$, then $A = \varnothing$.

(c) Let U be a universe. If $U \subseteq A$, then $A = U$.

(d) If $A \subseteq B$, then $B \setminus (B \setminus A) = A$.

(e) $A \times B = \varnothing$ if and only if $A = \varnothing$ or

15. Let $a, c, m \in \mathbf{Z}$ and define $A = \{a + mk : k \in \mathbf{Z}\}$ and $B = \{a + m(c + k) : k \in \mathbf{Z}\}$. Show $A = B$.

16. Prove $A = B$, where

$$A = \left\{ \begin{bmatrix} a & 0 \\ 0 & b \end{bmatrix} : a + b = 0 \wedge a, b \in \mathbf{R} \right\}$$

and

$$B = \left\{ \begin{bmatrix} a & b \\ c & d \end{bmatrix} : a = -d \wedge b^2 + c^2 = 0 \wedge a, b, c, d \in \mathbf{R} \right\}.$$

17. Prove.

(a) $\mathbf{Q} \neq \mathbf{Z}$

(b) $\mathbf{C} \neq \mathbf{R}$

(c) $\{0\} \times \mathbf{Z} \neq \mathbf{Z}$

(d) $\mathbf{R} \times \mathbf{Z} \neq \mathbf{Z} \times \mathbf{R}$

(e) If $A = \{ax^3 + b : a, b \in \mathbf{R}\}$ and $B = \{x^3 + b : b \in \mathbf{R}\}$, then $A \neq B$.

(f) If $A = \{ax^3 + b : a, b \in \mathbf{Z}\}$ and $B = \{ax^3 + b : a, b \in \mathbf{C}\}$, then $A \neq B$.

(g) $A \neq B$, where

$$A = \left\{ \begin{bmatrix} a & 0 \\ 0 & b \end{bmatrix} : a, b \in \mathbf{R} \right\}$$

and

$$B = \left\{ \begin{bmatrix} a & b \\ c & d \end{bmatrix} : a, b, c, d \in \mathbf{R} \right\}.$$

18. Find A and B to illustrate the given inequalities.

(a) $A \setminus B \neq B \setminus A$

(b) $(A \times B) \times C \neq A \times (B \times C)$

(c) $A \times B \neq B \times A$

19. For the remaining parts of Theorem 3.3.18, draw diagrams as in Figure 3.7 and prove the results.

20. Is it possible for $A = \overline{A}$? Explain.

21. Prove Theorem 3.3.17 by first using the short rule of biconditional proof and then by directly appealing to Theorem 3.3.6.

22. Prove that the following are equivalent.
   - $A \subseteq B$
   - $A \cup B = B$
   - $A \setminus B = \varnothing$
   - $A \cap B = A$

23. Prove that the following are equivalent.
   - $A \cap B = \varnothing$
   - $A \setminus \overline{B} = \varnothing$
   - $A \subseteq \overline{B}$

24. Find an example of sets A, B, and C such that $A \cap B = A \cap C$ but $B \neq C$.

25. Does $A \cup B = A \cup C$ imply $B = C$ for all sets A and B? Explain.

26. Prove.
   (a) If $A \cup B \subseteq A \cap B$, then $A = B$.
   (b) If $A \cap B = A \cap C$ and $A \cup B = A \cup C$, then $B = C$.

27. Prove that there is a cancellation law with the Cartesian product. Namely, if $A \neq \varnothing$ and $A \times B = A \times C$, then $B = C$.

28. When does $A \times B = C \times D$ imply that $A = C$ and $B = D$?

29. Prove.
   (a) If $A \cup B \neq \varnothing$, then $A \neq \varnothing$ or $B \neq \varnothing$.
   (b) If $A \cap B \neq \varnothing$, then $A \neq \varnothing$ and $B \neq \varnothing$.

30. Find the greatest common divisors of each pair.
   (a) 12 and 18
   (b) 3 and 9
   (c) 14 and 0
   (d) 7 and 15

31. Let a be a positive integer. Find the following and prove the result.
   (a) $\gcd(a, a + 1)$
   (b) $\gcd(a, 2a)$
   (c) $\gcd(a, a^2)$
   (d) $\gcd(a, 0)$

3.4 FAMILIES OF SETS

The elements of a set can be sets themselves. We call such a collection a **family of sets** and often use capital script letters to name them. For example, let

$$\mathscr{E} = \{\{1,2,3\}, \{2,3,4\}, \{3,4,5\}\}. \tag{3.2}$$

The set $\mathscr{E}$ has three elements: $\{1,2,3\}$, $\{2,3,4\}$, and $\{3,4,5\}$.

■ EXAMPLE 3.4.1

- $\{1,2,3\} \in \{\{1,2,3\}, \{1,4,9\}\}$

- $1 \notin \{\{1,2,3\}, \{1,4,9\}\}$

- $\{1,2,3\} \nsubseteq \{\{1,2,3\}, \{1,4,9\}\}$

- $\{\{1,2,3\}\} \subseteq \{\{1,2,3\}, \{1,4,9\}\}$.

■ EXAMPLE 3.4.2

- $\varnothing \subseteq \{\varnothing, \{\varnothing\}\}$ by Theorem 3.3.4.

- $\{\varnothing\} \subseteq \{\varnothing, \{\varnothing\}\}$ because $\varnothing \in \{\varnothing, \{\varnothing\}\}$.

- $\{\{\varnothing\}\} \subseteq \{\varnothing, \{\varnothing\}\}$ because $\{\varnothing\} \in \{\varnothing, \{\varnothing\}\}$.

Families of sets can have infinitely many elements. For example, let

$$\mathscr{F} = \{[n, n+1] : n \in \mathbf{Z}\}. \tag{3.3}$$

In roster notation,

$$\mathscr{F} = \{\ldots, [-2, -1], [-1, 0], [0, 1], [1, 2], \ldots\}.$$

Notice that in this case, abstraction is more convenient. For each integer n, the closed interval $[n, n+1]$ is in $\mathscr{F}$. The set $\mathbf{Z}$ plays the role of an **index set**, a set whose only purpose is to enumerate the elements of the family. Each element of an index set is called an **index**. If we let $I = \mathbf{Z}$ and $A_i = [i, i+1]$, the family can be written as

$$\mathscr{F} = \{A_i : i \in I\}.$$

To write the family $\mathscr{E}$ (3.2) using an index set, let $I = \{0, 1, 2\}$ and define

$$A_0 = \{1, 2, 3\},$$
$$A_1 = \{2, 3, 4\},$$
$$A_2 = \{3, 4, 5\}.$$

Then, the family illustrated in Figure 3.8 is

$$\mathscr{E} = \{A_i : i \in I\} = \{A_1, A_2, A_3\}.$$

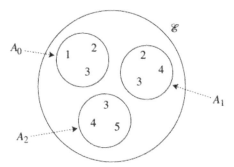

Figure 3.8   The family of sets $\mathscr{E} = \{A_i : i \in I\}$ with $I = \{1,2,3\}$.

There is no reason why I must be $\{0,1,2\}$. Any three-element set will do. The order in which the sets are defined is also irrelevant. For instance, we could have defined $I = \{w, \pi, 99\}$ and

$$A_w = \{3,4,5\},$$
$$A_\pi = \{2,3,4\},$$
$$A_{99} = \{1,2,3\}.$$

The goal is to have each set in the family referenced or **indexed** by at least one element of the set. We will still have $\mathscr{E} = \{A_i : i \in I\}$ with a similar diagram (Figure 3.9).

■ **EXAMPLE 3.4.3**

Write $\mathscr{F}$ (3.3) using **N** as the index set instead of **Z**. Use the even natural numbers to index the intervals with a nonnegative integer left-hand endpoint. The odd natural numbers will index the intervals with a negative integer left-hand endpoint. To do this, define the sets B_i as follows:

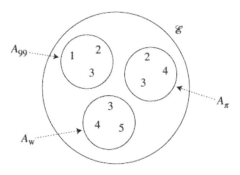

Figure 3.9   The family of sets $\mathscr{E} = \{A_i : i \in I\}$ with $I = \{w, \pi, 99\}$.

$$\vdots$$
$$B_5 = [-3, -2]$$
$$B_3 = [-2, -1]$$
$$B_1 = [-1, 0]$$
$$B_0 = [0, 1]$$
$$B_2 = [1, 2]$$
$$B_4 = [2, 3]$$
$$\vdots$$

Use $2n + 1$ to represent the odd natural numbers and $2n$ to represent the even natural numbers ($n \in \mathbf{N}$). Then,

$$\vdots$$
$$B_{2(2)+1} = [-2 - 1, -2]$$
$$B_{2(1)+1} = [-1 - 1, -1]$$
$$B_{2(0)+1} = [-0 - 1, -0]$$

and

$$B_{2(0)} = [0, 0 + 1]$$
$$B_{2(1)} = [1, 1 + 1]$$
$$B_{2(2)} = [2, 2 + 1]$$
$$\vdots$$

Therefore, define for all natural numbers n,

$$B_{2n+1} = [-n - 1, -n]$$

and

$$B_{2n} = [n, n + 1].$$

We have indexed the elements of $\mathscr{F}$ as

$$B_0 = [0, 1]$$
$$B_1 = [-1, 0]$$
$$B_2 = [1, 2]$$
$$B_3 = [-2, -1]$$
$$B_4 = [2, 3]$$
$$B_5 = [-3, -2]$$
$$\vdots$$

So under this definition, $\mathscr{F} = \{ B_i : i \in \mathbf{N} \}$.

Power Set

There is a natural way to form a family of sets. Take a set A. The collection of all subsets of A is called the **power set** of A. It is represented by $P(A)$.

■ DEFINITION 3.4.4

For any set A,
$$P(A) = \{B : B \subseteq A\}.$$

Notice that $\varnothing \in P(A)$ by Theorem 3.3.4.

■ EXAMPLE 3.4.5

- $P(\{1, 2, 3\}) = \{\varnothing, \{1\}, \{2\}, \{3\}, \{1, 2\}, \{1, 3\}, \{2, 3\}, \{1, 2, 3\}\}$
- $P(\{\varnothing, \{\varnothing\}\}) = \{\varnothing, \{\varnothing\}, \{\{\varnothing\}\}, \{\varnothing, \{\varnothing\}\}\}$
- $P(\mathbb{N}) = \{\varnothing, \{0\}, \{1\}, \ldots, \{0, 1\}, \{0, 2\}, \ldots\}$.

Consider $A = \{2, 6\}$ and $B = \{2, 6, 10\}$, so $A \subseteq B$. Examining the power sets of each, we find that
$$P(A) = \{\varnothing, \{2\}, \{6\}, \{2, 6\}\}$$
and
$$P(B) = \{\varnothing, \{2\}, \{6\}, \{10\}, \{2, 6\}, \{2, 10\}, \{6, 10\}, \{2, 6, 10\}\}.$$
Hence, $P(A) \subseteq P(B)$. This result is generalized in the next lemma.

■ LEMMA 3.4.6

$A \subseteq B$ if and only if $P(A) \subseteq P(B)$.

PROOF

- Let $A \subseteq B$. Assume $X \in P(A)$. Then, $X \subseteq A$, which gives $X \subseteq B$ by Theorem 3.3.6. Hence, $X \in P(B)$.

- Assume $P(A) \subseteq P(B)$. Let $x \in A$. In other words, $\{x\} \subseteq A$, but this means that $\{x\} \in P(A)$. Hence, $\{x\} \in P(B)$ by hypothesis, so $x \in B$. ■

The definition of set equality and Lemma 3.4.6 are used to prove the next theorem. Its proof is left to Exercise 9.

■ THEOREM 3.4.7

$A = B$ if and only if $P(A) = P(B)$.

Union and Intersection

We now generalize the set operations. Define the **union** of a family of sets to be the set of all elements that belong to some member of the family. This union is denoted by the same notation as in Definition 3.2.1 and can be defined using the abstraction method.

■ DEFINITION 3.4.8

Let $\mathscr{F}$ be a family of sets. Define

$$\bigcup \mathscr{F} = \{x : \exists A(A \in \mathscr{F} \wedge x \in A)\}.$$

If the family is indexed so that $\mathscr{F} = \{A_i : i \in I\}$, define

$$\bigcup_{i \in I} A_i = \{x : \exists i(i \in I \wedge x \in A_i)\}.$$

Observe that $\bigcup \mathscr{F} = \bigcup_{i \in I} A_i$ [Exercise 16(a)].

We generalize the notion of intersection similarly. The **intersection** of a family of sets is the set of all elements that belong to each member of the family.

■ DEFINITION 3.4.9

Let $\mathscr{F}$ be a family of sets.

$$\bigcap \mathscr{F} = \{x : \forall A(A \in \mathscr{F} \to x \in A)\}.$$

If the family is indexed so that $\mathscr{F} = \{A_i : i \in I\}$, define

$$\bigcap_{i \in I} A_i = \{x : \forall i(i \in I \to x \in A_i)\}.$$

Observe that $\bigcap \mathscr{F} = \bigcap_{i \in I} A_i$ [Exercise 16(b)].

Furthermore, notice that both definitions are indeed generalizations of the operations of Section 3.2 because as noted in Exercise 17,

$$\bigcup \{A, B\} = A \cup B,$$

and

$$\bigcap \{A, B\} = A \cap B.$$

■ EXAMPLE 3.4.10

Define $\mathscr{E} = \{[n, n+1] : n \in \mathbf{Z}\}$.

- When all of these intervals are combined, the result is the real line. This means that
$$\bigcup \mathscr{E} = \mathbf{R}.$$

- There is not one element that is common to all of the intervals. Hence,
$$\bigcap \mathscr{E} = \varnothing.$$

The next example illustrates how to write the union or intersection of a family of sets as a roster.

■ **EXAMPLE 3.4.11**

Let $\mathcal{F} = \{\{1, 2, 3\}, \{2, 3, 4\}, \{3, 4, 5\}\}$.

- Since 1 is in the first set of $\mathcal{F}$, $1 \in \bigcup \mathcal{F}$. The others can be explained similarly, so

$$\bigcup \mathcal{F} = \{1, 2, 3, 4, 5\}.$$

Notice that mechanically this amounts to removing the braces around the sets of the family and setting the union to the resulting set:

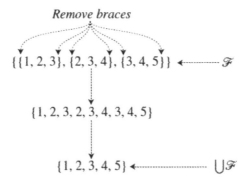

- The generalized intersection is simply the overlap of all of the sets. Hence, 3 is the only element of $\bigcap \mathcal{F}$. That is,

$$\bigcap \mathcal{F} = \{3\},$$

and this is illustrated by the following diagram:

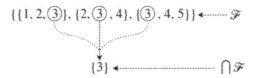

■ **EXAMPLE 3.4.12**

Since a family of sets can be empty, we must be able to take the union and intersection of the empty set. To prove that $\bigcup \emptyset = \emptyset$, take $x \in \bigcup \emptyset$. This means that there exists $A \in \emptyset$ such that $x \in A$, but this is impossible. We leave the fact that $\bigcap \emptyset$ is equal to the universe to Exercise 24.

The next theorem generalizes the Distributive Laws. Exercise 2.4.10 plays an important role in its proof. It allows us to move the quantifier.

■ **THEOREM 3.4.13**

Let A_i ($i \in I$) and B be sets.

- $B \cup \bigcap_{i \in I} A_i = \bigcap_{i \in I} (B \cup A_i)$

- $B \cap \bigcup_{i \in I} A_i = \bigcup_{i \in I} (B \cap A_i).$

PROOF

We leave the second part to Exercise 19. The first part is demonstrated by the following biconditional proof:

$$x \in B \cup \bigcap_{i \in I} A_i \Leftrightarrow x \in B \vee x \in \bigcap_{i \in I} A_i$$

$$\Leftrightarrow x \in B \vee \forall i (i \in I \rightarrow x \in A_i)$$
$$\Leftrightarrow x \in B \vee \forall i (i \notin I \vee x \in A_i)$$
$$\Leftrightarrow \forall i (x \in B \vee i \notin I \vee x \in A_i)$$
$$\Leftrightarrow \forall i (i \notin I \vee x \in B \vee x \in A_i)$$
$$\Leftrightarrow \forall i (i \in I \rightarrow x \in B \vee x \in A_i)$$
$$\Leftrightarrow \forall i (i \in I \rightarrow x \in B \cup A_i)$$
$$\Leftrightarrow x \in \bigcap_{i \in I} (B \cup A_i). \blacksquare$$

To understand the next result, consider the following example. Choose the universe to be $\{1, 2, 3, 4, 5\}$ and perform some set operations on the family

$$\{\{1, 2, 3\}, \{2, 3, 4\}, \{2, 3, 5\}\}.$$

First,

$$\overline{\bigcap \{\{1, 2, 3\}, \{2, 3, 4\}, \{2, 3, 5\}\}} = \overline{\{2, 3\}} = \{1, 4, 5\},$$

and second,

$$\bigcup \{\overline{\{1, 2, 3\}}, \overline{\{2, 3, 4\}}, \overline{\{2, 3, 5\}}\} = \bigcup \{\{4, 5\}, \{1, 5\}, \{1, 4\}\} = \{1, 4, 5\}.$$

This leads us to the next generalization of De Morgan's laws. Its proof is left to Exercise 23.

■ **THEOREM 3.4.14**

Let $\{A_i : i \in I\}$ be a family of sets.

- $\overline{\bigcap_{i \in I} A_i} = \bigcup_{i \in I} \overline{A_i}$

- $\overline{\bigcup_{i \in I} A_i} = \bigcap_{i \in I} \overline{A_i}.$

Disjoint and Pairwise Disjoint

What it means for two sets to be disjoint was defined in Section 3.2 (Definition 3.2.3). The next definition generalizes that notion to families of sets. Because a family can have more than two elements, it is appropriate to expand the concept of disjointness.

■ DEFINITION 3.4.15

Let $\mathcal{F}$ be a family of sets.

- $\mathcal{F}$ is **disjoint** when $\bigcap \mathcal{F} = \varnothing$.

- $\mathcal{F}$ is **pairwise disjoint** when for all $A, B \in \mathcal{F}$, if $A \neq B$, then $A \cap B = \varnothing$.

Observe that $\{\{1, 2\}, \{3, 4\}, \{5, 6\}\}$ is both disjoint and pairwise disjoint because its elements have no common members.

■ EXAMPLE 3.4.16

Let A be a set. We see that

$$\bigcup P(A) = \{x : \exists B[B \in P(A) \land x \in B]\} = A.$$

Although $P(A)$ is not a pairwise disjoint family of sets, it is a disjoint because $\varnothing \in P(A)$.

If the family is indexed, we can use another test to determine if it is pairwise disjoint. Let $\mathcal{F} = \{A_i : i \in I\}$ be a family of sets. If for all $i, j \in I$,

$$i \neq j \text{ implies } A_i \cap A_j = \varnothing, \tag{3.4}$$

then $\mathcal{F}$ is pairwise disjoint. To prove this, let $\mathcal{F}$ be a family that satisfies (3.4). Take $A_i, A_j \in \mathcal{F}$ for some $i, j \in I$ and assume $A_i \neq A_j$. Therefore, $i \neq j$, for otherwise they would be the equal. Hence, $A_i \cap A_j = \varnothing$ by Condition 3.4.

The next result illustrates the relationship between these two terms. One must be careful, though. The converse is false (Exercise 14).

■ THEOREM 3.4.17

Let $\mathcal{F}$ be a family of sets with at least two elements. If $\mathcal{F}$ is pairwise disjoint, then $\mathcal{F}$ is disjoint.

PROOF

Assume $\mathcal{F}$ is pairwise disjoint. Since $\mathcal{F}$ contains at least two sets, let $A, B \in \mathcal{F}$ such that $A \neq B$. Then, using Exercise 27, $\bigcap \mathcal{F} \subseteq A \cap B = \varnothing$. ■

Exercises

1. Let $I = \{1, 2, 3, 4, 5\}$, $A_1 = \{1, 2\}$, $A_2 = \{3, 4\}$, $A_3 = \{1, 4\}$, $A_4 = \{3, 4\}$, and $A_5 = \{1, 3\}$. Write the given families of sets as a rosters.

    (a) $\{A_i : i \in I\}$

    (b) $\{A_i : i \in \{2,4\}\}$

    (c) $\{A_i : i = 1\}$

    (d) $\{A_i : i = 1,2\}$

    (e) $\{A_i : i \in \varnothing\}$

    (f) $\{A_i : i \in A_5\}$

2. Answer true or false.

    (a) $1 \in \{\{1\}, \{2\}, \{1,2\}\}$

    (b) $\{1\} \in \{\{1\}, \{2\}, \{1,2\}\}$

    (c) $\{1\} \subseteq \{\{1\}, \{2\}, \{1,2\}\}$

    (d) $\{1,2\} \in \{\{\{1,2\}, \{3,4\}\}, \{1,2\}\}$

    (e) $\{1,2\} \subseteq \{\{\{1,2\}, \{3,4\}\}, \{1,2\}\}$

    (f) $\{3,4\} \in \{\{\{1,2\}, \{3,4\}\}, \{1,2\}\}$

    (g) $\{3,4\} \subseteq \{\{\{1,2\}, \{3,4\}\}, \{1,2\}\}$

    (h) $\varnothing \in \{\{\{1,2\}, \{3,4\}\}, \{1,2\}\}$

    (i) $\varnothing \subseteq \{\{\{1,2\}, \{3,4\}\}, \{1,2\}\}$

    (j) $\{\varnothing\} \in \{\varnothing, \{\varnothing, \{\varnothing\}\}\}$

    (k) $\{\varnothing\} \subseteq \{\varnothing, \{\varnothing, \{\varnothing\}\}\}$

    (l) $\varnothing \in \{\varnothing, \{\varnothing, \{\varnothing\}\}\}$

    (m) $\varnothing \subseteq \{\varnothing, \{\varnothing, \{\varnothing\}\}\}$

    (n) $\{\varnothing\} \subseteq \varnothing$

    (o) $\{\varnothing\} \subseteq \{\varnothing, \{\varnothing\}\}$

    (p) $\{\varnothing\} \subseteq \{\{\varnothing, \{\varnothing\}\}\}$

    (q) $\{\{\varnothing\}\} \subseteq \{\varnothing, \{\varnothing\}\}$

    (r) $\{1\} \in \mathbf{P}(\mathbf{Z})$

    (s) $\{1\} \subseteq \mathbf{P}(\mathbf{Z})$

    (t) $\varnothing \in \mathbf{P}(\varnothing)$

    (u) $\{\varnothing\} \in \mathbf{P}(\varnothing)$

    (v) $\{\varnothing\} \subseteq \mathbf{P}(\varnothing)$

3. Find sets such that $I \cap J = \varnothing$ but $\{A_i : i \in I\}$ and $\{A_j : j \in J\}$ are not disjoint.

4. Let $\{A_i : i \in K\}$ be a family of sets and let I and J be subsets of K. Define $\mathscr{E} = \{A_i : i \in I\}$ and $\mathscr{F} = \{A_j : j \in J\}$ and prove the following.

    (a) If $I \subseteq J$, then $\mathscr{E} \subseteq \mathscr{F}$.

    (b) $\mathscr{E} \cup \mathscr{F} = \{A_i : i \in I \cup J\}$

    (c) $\{A_i : i \in I \cap J\} \subseteq \mathscr{E} \cap \mathscr{F}$

5. Using the same notation as in the previous problem, find a family $\{A_i : i \in K\}$ and subsets I and J of K such that:

    (a) $\mathscr{E} \cap \mathscr{F} \not\subseteq \{A_i : i \in I \cap J\}$

    (b) $\{A_i : i \in I \setminus J\} \not\subseteq \mathscr{E} \setminus \mathscr{F}$

6. Show $\{A_i : i \in I\} = \varnothing$ if and only if $I = \varnothing$.

7. Let A be a finite set. How many elements are in $P(A)$ if A has n elements? Explain.

8. Find the given power sets.
    (a) $P(\{1,2\})$
    (b) $P(P(\{1,2\}))$
    (c) $P(\varnothing)$
    (d) $P(P(\varnothing))$
    (e) $P(\{\{\varnothing\}\})$
    (f) $P(\{\varnothing, \{\varnothing\}, \{\varnothing, \{\varnothing\}\}\})$

9. Prove Theorem 3.4.7.

10. For each of the given equalities, prove or show false. If one is false, prove any true inclusion.
    (a) $P(A \cup B) = P(A) \cup P(B)$
    (b) $P(A \cap B) = P(A) \cap P(B)$
    (c) $P(A \setminus B) = P(A) \setminus P(B)$
    (d) $P(A \times B) = P(A) \times P(B)$

11. Prove $P(A) \subseteq P(B)$ implies $A \subseteq B$ by using the fact that $A \in P(A)$.

12. Write the following sets in roster form.
    (a) $\bigcup\{\{1,2\}, \{1,2\}, \{1,3\}, \{1,4\}\}$
    (b) $\bigcap\{\{1,2\}, \{1,2\}, \{1,3\}, \{1,4\}\}$
    (c) $\bigcap P(\varnothing)$
    (d) $\bigcup P(\varnothing)$
    (e) $\bigcup\bigcup\{\{\{1\}\}, \{\{1,2\}\}, \{\{1,3\}\}, \{\{1,4\}\}\}$
    (f) $\bigcup\bigcap\{\{\{1\}\}, \{\{1,2\}\}, \{\{1,3\}\}, \{\{1,4\}\}\}$
    (g) $\bigcap\bigcup\{\{\{1\}\}, \{\{1,2\}\}, \{\{1,3\}\}, \{\{1,4\}\}\}$
    (h) $\bigcap\bigcap\{\{\{1\}\}, \{\{1,2\}\}, \{\{1,3\}\}, \{\{1,4\}\}\}$
    (i) $\bigcup\bigcup\varnothing$
    (j) $\bigcap\bigcup\varnothing$

13. Draw Venn diagrams for a disjoint family of sets that is not pairwise disjoint and for a pairwise disjoint family of sets.

14. Show by example that a disjoint family of sets might not be pairwise disjoint.

15. Given a family of sets $\{A_i : i \in I\}$, find a family $\mathscr{B} = \{B_i : i \in I\}$ such that $\{A_i \times B_i : i \in I\}$ is pairwise disjoint.

16. Let $\mathscr{F} = \{A_i : i \in I\}$ be a family of sets. Prove the given equations.
    (a) $\bigcup \mathscr{F} = \bigcup_{i \in I} A_i$
    (b) $\bigcap \mathscr{F} = \bigcap_{i \in I} A_i$

17. Prove for any sets A and B, $\bigcup\{A, B\} = A \cup B$ and $\bigcap\{A, B\} = A \cap B$.

18. Let $\mathscr{C} = \{(0, n) : n \in \mathbf{Z}^+\}$. Prove that $\bigcup \mathscr{C} = (0, \infty)$ and $\bigcap \mathscr{C} = (0, 1)$.

19. Prove the second part of Theorem 3.4.13.

20. Prove Theorem 3.4.17 indirectly.

21. Is Theorem 3.4.17 still true if the family of sets $\mathscr{F}$ has at most one element? Explain.

22. Let $\{A_i : i \in I\}$ be a family of sets and prove the following.
    - (a) If $B \subseteq A_i$ for some $i \in I$, then $B \subseteq \bigcup_{i \in I} A_i$.
    - (b) If $A_i \subseteq B$ for all $i \in I$, $\bigcap_{i \in I} A_i \subseteq B$.
    - (c) If $B \subseteq \bigcap_{i \in I} A_i$, then $B \subseteq A_i$ for all $i \in I$.

23. Prove Theorem 3.4.14.

24. Show $\bigcap \varnothing = U$ where U is a universe.

25. Let $\mathscr{F}$ be a family of sets such that $\varnothing \in \mathscr{F}$. Prove $\bigcap \mathscr{F} = \varnothing$.

26. Find families of sets $\mathscr{E}$ and $\mathscr{F}$ so that $\bigcup \mathscr{E} = \bigcup \mathscr{F}$ but $\mathscr{E} \neq \mathscr{F}$. Can this be repeated by replacing union with intersection?

27. Let $\mathscr{F}$ be a family of sets, and let $A \in \mathscr{F}$. Prove $\bigcap \mathscr{F} \subseteq A \subseteq \bigcup \mathscr{F}$.

28. Let $\mathscr{E}$ and $\mathscr{F}$ be families of sets. Show the following.
    - (a) $\bigcup \{\mathscr{F}\} = \mathscr{F}$
    - (b) $\bigcap \{\mathscr{F}\} = \mathscr{F}$
    - (c) If $\mathscr{E} \subseteq \mathscr{F}$, then $\bigcup \mathscr{E} \subseteq \bigcup \mathscr{F}$.
    - (d) If $\mathscr{E} \subseteq \mathscr{F}$, then $\bigcap \mathscr{F} \subseteq \bigcap \mathscr{E}$.
    - (e) $\bigcup (\mathscr{E} \cup \mathscr{F}) = \bigcup \mathscr{E} \cup \bigcup \mathscr{F}$
    - (f) $\bigcap (\mathscr{E} \cup \mathscr{F}) = \bigcap \mathscr{E} \cap \bigcap \mathscr{F}$

29. Find families of sets $\mathscr{E}$ and $\mathscr{F}$ that make the following false.
    - (a) $\bigcap (\mathscr{E} \cap \mathscr{F}) = \bigcap \mathscr{E} \cap \bigcap \mathscr{F}$
    - (b) $\bigcup (\mathscr{E} \cap \mathscr{F}) = \bigcup \mathscr{E} \cap \bigcup \mathscr{F}$

30. Let $\mathscr{F}$ be a family of sets. For each of the given equalities, prove true or show that it is false by finding a counter-example.
    - (a) $\bigcup P(\mathscr{F}) = \mathscr{F}$
    - (b) $\bigcap P(\mathscr{F}) = \mathscr{F}$
    - (c) $P(\bigcup \mathscr{F}) = \mathscr{F}$
    - (d) $P(\bigcap \mathscr{F}) = \mathscr{F}$

31. Prove these alternate forms of De Morgan's laws.
    - (a) $A \setminus \bigcup_{i \in I} B_i = \bigcap_{i \in I} A \setminus B_i$
    - (b) $A \setminus \bigcap_{i \in I} B_i = \bigcup_{i \in I} A \setminus B_i$

32. Assume that the universe U contains only sets such that for all $A \in U$, $A \subseteq U$ and $P(A) \in U$. Prove the following equalities.
    - (a) $\bigcup U = U$
    - (b) $\bigcap U = \varnothing$

33. Let $I_n = [n, n+1]$ and $J_n = [n, n+1]$, $n \in \mathbf{Z}$. Define

$$\mathcal{F} = \{I_n \times J_m : n, m \in \mathbf{Z}\}.$$

Show $\bigcup \mathcal{F} = \mathbf{R}^2$ and $\bigcap \mathcal{F} = \emptyset$. Is $\mathcal{F}$ pairwise disjoint?

CHAPTER 4

RELATIONS AND FUNCTIONS

4.1 RELATIONS

A relation is an association between objects. A book on a table is an example of the relation of one object being *on* another. It is especially common to speak of relations among people. For example, one person could be the niece of another. In mathematics, there are many relations such as equals and less-than that describe associations between numbers. To formalize this idea, we make the next definition.

■ **DEFINITION 4.1.1**

A set R is an (***n*-ary**) **relation** if there exist sets $A_0, A_1, \ldots, A_{n-1}$ such that

$$R \subseteq A_0 \times A_1 \times \cdots \times A_{n-1}.$$

In particular, R is a **unary** relation if $n = 1$ and a **binary** relation if $n = 2$. If $R \subseteq A \times A$ for some set A, then R is a **relation on** A and we write (A, R).

The relation *on* can be represented as a subset of the Cartesian product of the set of all books and the set of all tables. We could then write (*dictionary*, *desk*) to mean that the

A First Course in Mathematical Logic and Set Theory, First Edition. Michael L. O'Leary.
© 2016 John Wiley & Sons, Inc. Published 2016 by John Wiley & Sons, Inc.

dictionary is on the desk. Similarly, the set $\{(2,4),(7,3),(0,0)\}$ is a relation because it is a subset of $\mathbf{Z} \times \mathbf{Z}$. The ordered pair $(2,4)$ means that 2 is related to 4. Likewise, $\mathbf{R} \times \mathbf{Q}$ is a relation where every real number is related to every rational number, and according to Definition 4.1.1, the empty set is also a relation because $\varnothing = \varnothing \times \varnothing$.

■ **EXAMPLE 4.1.2**

For any set A, define
$$I_A = \{(a,a) : a \in A\}.$$
Call this set the **identity** on A. In particular, the identity on $\mathbf{R}$ is
$$I_\mathbf{R} = \{(x,x) : x \in \mathbf{R}\},$$
and the identity on $\mathbf{Z}$ is
$$I_\mathbf{Z} = \{(x,x) : x \in \mathbf{Z}\}.$$
Notice that $\varnothing$ is the identity on $\varnothing$.

■ **EXAMPLE 4.1.3**

The less-than relation on $\mathbf{Z}$ is defined as
$$L = \{(a,b) : a, b \in \mathbf{Z} \wedge a < b\}.$$
Another approach is to use membership in the set of positive integers as our condition. That is,
$$L = \{(a,b) : a, b \in \mathbf{Z} \wedge b - a \in \mathbf{Z}^+\}.$$
Hence, $(4,7) \in L$ because $7 - 4 \in \mathbf{Z}^+$. See Exercise 1 for another definition of L.

When a relation $R \subseteq A \times B$ is defined, all elements in A or B might not be used. For this reason, it is important to identify the sets that comprise all possible values for the two coordinates of the relation.

■ **DEFINITION 4.1.4**

Let $R \subseteq A \times B$. The **domain** of R is the set
$$\mathrm{dom}(R) = \{x \in A : \exists y(y \in B \wedge (x,y) \in R)\},$$
and the **range** of R is the set
$$\mathrm{ran}(R) = \{y \in B : \exists x(x \in A \wedge (x,y) \in R)\}.$$

■ **EXAMPLE 4.1.5**

If $R = \{(1,3),(2,4),(2,5)\}$, then $\mathrm{dom}(R) = \{1,2\}$ and $\mathrm{ran}(R) = \{3,4,5\}$. We represent this in Figure 4.1 where A and B are sets so that $\{1,2\} \subseteq A$ and $\{3,4,5\} \subseteq B$. The ordered pair $(1,3)$ is denoted by the arrow pointing from 1 to 3. Also, both $\mathrm{dom}(R)$ and $\mathrm{ran}(R)$ are shaded.

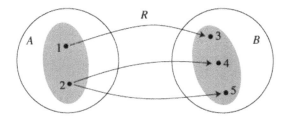

Figure 4.1 $R = \{(1,3),(2,4),(2,5)\}$.

■ **EXAMPLE 4.1.6**

Let $S = \{(x,y) : |x| = y \wedge x, y \in \mathbf{R}\}$. Notice that both $(2,2)$ and $(-2,2)$ are elements of S. Furthermore, $\text{dom}(S) = \mathbf{R}$ and $\text{ran}(S) = [0,\infty)$.

The domain and range of a relation can be the same set as in the next two examples.

■ **EXAMPLE 4.1.7**

If $R = \{(0,1),(0,2),(1,0),(2,0)\}$, then R is a relation on the set $\{0,1,2\}$ with $\text{dom}(R) = \text{ran}(R) = \{0,1,2\}$.

■ **EXAMPLE 4.1.8**

For the relation
$$S = \left\{(x,y) \in \mathbf{R}^2 : \sqrt{x^2 + y^2} \geq 1\right\},$$

both the domain and range equal $\mathbf{R}$. First, note that it is clear by the definition of S that both $\text{dom}(S)$ and $\text{ran}(S)$ are subsets of $\mathbf{R}$. For the other inclusion, take $x \in \mathbf{R}$. Since $(x,1) \in S$, $x \in \text{dom}(S)$, and since $(1,y) \in S$, $y \in \text{ran}(S)$.

Composition

Given the relations R and S, let us define a new relation. Suppose that $(a,b) \in S$ and $(b,c) \in R$. Therefore, a is related to c through b. The new relation will contain the ordered pair (a,c) to represent this relationship.

■ **DEFINITION 4.1.9**

Let $R \subseteq A \times B$ and $S \subseteq B \times C$. The **composition** of S and R is the subset of $A \times C$ defined as

$$S \circ R = \{(x,z) : \exists y(y \in B \wedge (x,y) \in R \wedge (y,z) \in S)\}.$$

As illustrated in Figure 4.2, the reason that $(a,c) \in R \circ S$ is because $(a,b) \in S$ and $(b,c) \in R$. That is, a is related to c via $R \circ S$ because a is related to b via S and then b

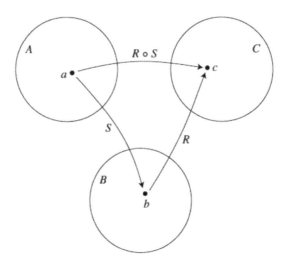

Figure 4.2 A composition of relations.

is related to c via R. The composition can be viewed as the direct path from a to c that does not require the intermediary b.

■ **EXAMPLE 4.1.10**

To clarify the definition, let

$$R = \{(2,4), (1,3), (2,5)\}$$

and

$$S = \{(0,1), (1,0), (0,2), (2,0)\}.$$

We have that $R \circ S = \{(0,3), (0,4), (0,5)\}$. Notice that $(0,3) \in R \circ S$ because $(0,1) \in S$ and $(1,3) \in R$. However, $S \circ R$ is empty since $\mathrm{ran}(R)$ and $\mathrm{dom}(S)$ are disjoint.

■ **EXAMPLE 4.1.11**

Define

$$R = \{(x, y) \in \mathbf{R}^2 : x^2 + y^2 = 1\}$$

and

$$S = \{(x, y) \in \mathbf{R}^2 : y = x + 1\}.$$

Notice that R is the unit circle and S is the line with slope of 1 and y-intercept of $(0, 1)$. Let us find $R \circ S$.

$$\begin{aligned}
R \circ S &= \{(x, z) \in \mathbf{R}^2 : \exists y (y \in \mathbf{R} \wedge (x, y) \in S \wedge (y, z) \in R)\} \\
&= \{(x, z) \in \mathbf{R}^2 : \exists y (y \in \mathbf{R} \wedge y = x + 1 \wedge y^2 + z^2 = 1)\} \\
&= \{(x, z) \in \mathbf{R}^2 : (x + 1)^2 + z^2 = 1\}.
\end{aligned}$$

Therefore, $R \circ S$ is the circle with center $(-1, 0)$ and radius 1.

Example 4.1.10 shows that it is possible that $S \circ R \neq R \circ S$. However, we can change the order of the composition.

■ THEOREM 4.1.12

If $R \subseteq A \times B$, $S \subseteq B \times C$, and $T \subseteq C \times D$, then $T \circ (S \circ R) = (T \circ S) \circ R$.

PROOF

Assume that $R \subseteq A \times B$, $S \subseteq B \times C$, and $T \subseteq C \times D$. Then,

$(a, d) \in T \circ (S \circ R)$

$\Leftrightarrow \exists c (c \in C \wedge (a, c) \in S \circ R \wedge (c, d) \in T)$

$\Leftrightarrow \exists c (c \in C \wedge \exists b (b \in B \wedge (a, b) \in R \wedge (b, c) \in S) \wedge (c, d) \in T)$

$\Leftrightarrow \exists c \exists b (c \in C \wedge b \in B \wedge (a, b) \in R \wedge (b, c) \in S \wedge (c, d) \in T)$

$\Leftrightarrow \exists b \exists c (b \in B \wedge (a, b) \in R \wedge c \in C \wedge (b, c) \in S \wedge (c, d) \in T)$

$\Leftrightarrow \exists b (b \in B \wedge (a, b) \in R \wedge \exists c [c \in C \wedge (b, c) \in S \wedge (c, d) \in T])$

$\Leftrightarrow \exists b (b \in B \wedge (a, b) \in R \wedge (b, d) \in T \circ S)$

$\Leftrightarrow (a, d) \in (T \circ S) \circ R.$ ■

Inverses

Let $R \subseteq A \times B$. We know that $R \circ I_A = R$ and $I_B \circ R = R$ [Exercise 10(a)]. If we want $I_A = I_B$, we need $A = B$ so that R is a relation on A. Then,

$$R \circ I_A = I_A \circ R = R.$$

For example, if we again define $R = \{(2, 4), (1, 3), (2, 5)\}$ and view it as a relation on $\mathbb{Z}$, then R composed on either side by $I_\mathbb{Z}$ yields R. To illustrate, consider the ordered pair $(1, 3)$. It is an element of $R \circ I_\mathbb{Z}$ because

$$(1, 1) \in I_\mathbb{Z} \wedge (1, 3) \in R,$$

and it is also an element of $I_\mathbb{Z} \circ R$ because

$$(1, 3) \in R \wedge (3, 3) \in I_\mathbb{Z}.$$

Notice that not every identity relation will have this property. Using the same definition of R as above,

$$R \circ I_{\{1,2\}} = R,$$

but

$$I_{\{1,2\}} \circ R = \varnothing.$$

Now let us change the problem. Given a relation R on A, can we find a relation S on A such that $R \circ S = S \circ R = I_A$? The next definition is used to try to answer this question.

■ **DEFINITION 4.1.13**

Let R be a binary relation. The **inverse** of R is the set

$$R^{-1} = \{(y, x) : (x, y) \in R\}.$$

For a relation S, we say that R and S are **inverse relations** if $R^{-1} = S$.

■ **EXAMPLE 4.1.14**

Let L be the less-than relation on $\mathbf{R}$ (Example 4.1.3). Then,

$$
\begin{aligned}
L^{-1} &= \{(y, x) \in \mathbf{R}^2 : (x, y) \in L\} \\
&= \{(y, x) \in \mathbf{R}^2 : x < y\} \\
&= \{(y, x) \in \mathbf{R}^2 : y > x\}.
\end{aligned}
$$

This shows that less-than and greater-than are inverse relations.

We now check whether $R \circ R^{-1} = R^{-1} \circ R = I_A$ for any relation R on A. Consider $R = \{(2, 1), (4, 3)\}$, which is a relation on $\{1, 2, 3, 4\}$. Then,

$$R^{-1} = \{(1, 2), (3, 4)\},$$

and we see that composing does not yield the identity on $\{1, 2, 3, 4\}$ because

$$R \circ R^{-1} = \{(1, 1), (3, 3)\} = I_{\{1,3\}}$$

and

$$R^{-1} \circ R = \{(2, 2), (4, 4)\} = I_{\{2,4\}}.$$

The situation is worse when we define $S = \{(2, 1), (2, 3), (4, 3)\}$. In this case, we have that

$$S^{-1} = \{(1, 2), (3, 2), (3, 4)\}$$

but

$$S \circ S^{-1} = \{(1, 1), (1, 3), (3, 1), (3, 3)\}$$

and

$$S^{-1} \circ S = \{(2, 2), (2, 4), (4, 4)\}.$$

Neither of these compositions leads to an identity, but at least we have that

$$\{(1, 1), (3, 3)\} \subseteq S \circ S^{-1}$$

and

$$\{(2, 2), (4, 4)\} \subseteq S^{-1} \circ S.$$

This can be generalized.

■ **THEOREM 4.1.15**

$I_{\mathrm{ran}(R)} \subseteq R \circ R^{-1}$ and $I_{\mathrm{dom}(R)} \subseteq R^{-1} \circ R$ for any binary relation R.

PROOF

The first inclusion is proved in Exercise 12. To see the second inclusion, let $x \in \text{dom}(R)$. By definition, there exists $y \in \text{ran}(R)$ so that $(x, y) \in R$. Hence, $(y, x) \in R^{-1}$, which implies that $(x, x) \in R^{-1} \circ R$. ■

Exercises

1. Let $L \subseteq \mathbf{Z} \times \mathbf{Z}$ be the less-than relation as defined in Example 4.1.3. Prove that

$$L = \{(a, b) \in \mathbf{Z} \times \mathbf{Z} : a - b \in \mathbf{Z}^-\}.$$

2. Find the domain and range of the given relations.
   - (a) $\{(0, 1), (2, 3), (4, 5), (6, 7)\}$
   - (b) $\{((a, b), 1), ((a, c), 2), ((a, d), 3)\}$
   - (c) $\mathbf{R} \times \mathbf{Z}$
   - (d) $\varnothing \times \varnothing$
   - (e) $\mathbf{Q} \times \varnothing$
   - (f) $\{(x, y) : x, y \in [0, 1] \wedge x < y\}$
   - (g) $\{(x, y) \in \mathbf{R}^2 : y = 3\}$
   - (h) $\{(x, y) \in \mathbf{R}^2 : y = |x|\}$
   - (i) $\{(x, y) \in \mathbf{R}^2 : x^2 + y^2 = 4\}$
   - (j) $\{(x, y) \in \mathbf{R}^2 : y \leq \sqrt{x} \wedge x \geq 0\}$
   - (k) $\{(f, g) : \exists a \in \mathbf{R}(f(x) = e^x \wedge g(x) = ax)\}$
   - (l) $\{((a, b), a + b) : a, b \in \mathbf{Z}\}$

3. Write $R \circ S$ as a roster.
   - (a) $R = \{(1, 0), (2, 3), (4, 6)\}$, $S = \{(1, 2), (2, 3), (3, 4)\}$
   - (b) $R = \{(1, 3), (2, 5), (3, 1)\}$, $S = \{(1, 3), (3, 1), (5, 2)\}$
   - (c) $R = \{(1, 2), (3, 4), (5, 6)\}$, $S = \{(1, 2), (3, 4), (5, 6)\}$
   - (d) $R = \{(1, 2), (3, 4), (5, 6)\}$, $S = \{(2, 1), (3, 5), (5, 7)\}$

4. Write $R \circ S$ using abstraction.
   - (a) $R = \{(x, y) \in \mathbf{R}^2 : x^2 + y^2 = 1\}$
     $S = \mathbf{R}^2$
   - (b) $R = \{(x, y) \in \mathbf{R}^2 : x^2 + y^2 = 1\}$
     $S = \mathbf{Z} \times \mathbf{Z}$
   - (c) $R = \{(x, y) \in \mathbf{R}^2 : x^2 + y^2 = 1\}$
     $S = \{(x, y) \in \mathbf{R}^2 : (x - 2)^2 + y^2 = 1\}$
   - (d) $R = \{(x, y) \in \mathbf{R}^2 : x^2 + y^2 = 1\}$
     $S = \{(x, y) \in \mathbf{R}^2 : y = 2x - 1\}$

5. Write the inverse of each relation. Use the abstraction method where appropriate.
   - (a) $\varnothing$
   - (b) $I_\mathbf{Z}$
   - (c) $\{(1, 0), (2, 3), (4, 6)\}$

(d) $\{(1,0),(1,1),(2,1)\}$

(e) $\mathbf{Z} \times \mathbf{R}$

(f) $\{(x, \sin x) : x \in \mathbf{R}\}$

(g) $\{(x, y) \in \mathbf{R}^2 : x + y = 1\}$

(h) $\{(x, y) \in \mathbf{R}^2 : x^2 + y^2 = 1\}$

6. Let $R \subseteq A \times B$ and $S \subseteq B \times C$. Show the following.

   (a) $R^{-1} \circ S^{-1} \subseteq C \times A$.

   (b) $\text{dom}(R) = \text{ran}(R^{-1})$.

   (c) $\text{ran}(R) = \text{dom}(R^{-1})$.

7. Prove that if R is a binary relation, $(R^{-1})^{-1} = R$.

8. Prove that $(S \circ R)^{-1} = R^{-1} \circ S^{-1}$ if $R \subseteq A \times B$ and $S \subseteq B \times C$.

9. Let $R, S \subseteq A \times B$. Prove the following.

   (a) If $R \subseteq S$, then $R^{-1} \subseteq S^{-1}$.

   (b) $(R \cup S)^{-1} = R^{-1} \cup S^{-1}$.

   (c) $(R \cap S)^{-1} = R^{-1} \cap S^{-1}$.

10. Let $R \subseteq A \times B$.

    (a) Prove $R \circ I_A = R$ and $I_B \circ R = R$.

    (b) Show that if there exists a set C such that A and B are subsets of C, then $R \circ I_C = I_C \circ R = R$.

11. Let $R \subseteq A \times B$ and $S \subseteq B \times C$. Show that $S \circ R = \varnothing$ if and only if $\text{dom}(S)$ and $\text{ran}(R)$ are disjoint.

12. For any relation R, prove $I_{\text{ran}(R)} \subseteq R \circ R^{-1}$.

13. Let $R \subseteq A \times B$. Prove.

    (a) $\bigcup_{b \in B} \{x \in A : (x, b) \in R\} = \text{dom}(R)$.

    (b) $\bigcup_{a \in A} \{y \in B : (a, y) \in R\} = \text{ran}(R)$.

4.2 EQUIVALENCE RELATIONS

In practice we usually do not write relations as sets of ordered pairs. We instead write propositions like $4 = 4$ or $3 < 9$. To copy this, we will introduce an alternate notation.

■ DEFINITION 4.2.1

Let R be a relation on A. For all $a, b \in A$,

$$a \, R \, b \text{ if and only if } (a, b) \in R,$$

and

$$a \, \cancel{R} \, b \text{ if and only if } (a, b) \notin R.$$

For example, the less-than relation L (Example 4.1.3) is usually denoted by $<$, and we write $2 < 3$ instead of $(2, 3) \in L$ or $(2, 3) \in <$.

■ **EXAMPLE 4.2.2**

Define the relation R on $\mathbf{Z}$ by

$$R = \{(a, b) \in \mathbf{Z} \times \mathbf{Z} : \exists c \in \mathbf{Z}(b = ac \wedge a \neq 0)\}.$$

Therefore, for all $a, b \in \mathbf{Z}$, $a R b$ if and only if a divides b. Therefore, $4 R 8$ but $8 \not R 4$.

Relations can have different properties depending on their definitions. Here are three important examples using relations on $A = \{1, 2, 3\}$.

- $\{(1, 1), (2, 2), (3, 3)\}$ has the property that every element of A is related to itself.

- $\{(1, 2), (2, 1), (2, 3), (3, 2)\}$ has the property that if a is related to b, then b is related to a.

- $\{(1, 2), (2, 3), (1, 3)\}$ has the property that if a is related to b and b is related to c, then a is related to c.

These examples lead to the following definitions.

■ **DEFINITION 4.2.3**

Let R be a relation on A.

- R is **reflexive** if $a R a$ for all $a \in A$.

- R is **symmetric** when for all $a, b \in A$, if $a R b$, then $b R a$.

- R is **transitive** means that for all $a, b, c \in A$, if $a R b$ and $b R c$, then $a R c$.

Notice that the relation in Example 4.2.2 is not reflexive because 0 does not divide 0 and is not symmetric because 4 divides 8 but 8 does not divide 4, but it is transitive because if a divides b and b divides c, then a divides c.

When a relation is reflexive, symmetric, and transitive, it behaves very much like an identity relation (Example 4.1.2). Such relations play an important role in mathematics, so we name them.

■ **DEFINITION 4.2.4**

A relation R on A is an **equivalence relation** if R is reflexive, symmetric, and transitive.

Observe that the relation in Example 4.2.2 is not an equivalence relation. However, any identity relation is an equivalence relation. We see this assumption at work in the next example.

■ **EXAMPLE 4.2.5**

Let R be a relation on $\mathbf{Z} \times (\mathbf{Z} \setminus \{0\})$ so that for all $a, c \in \mathbf{Z}$ and $b, d \in \mathbf{Z} \setminus \{0\}$,

$$(a, b) \, R \, (c, d) \text{ if and only if } ad = bc.$$

To see that this is an equivalence relation, let (a, b), (c, d), and (e, f) be elements of $\mathbf{Z} \times (\mathbf{Z} \setminus \{0\})$.

- $(a, b) \, R \, (a, b)$ since $ab = ab$.

- Assume $(a, b) \, R \, (c, d)$. Then, $ad = bc$. This implies that $cb = da$, so $(c, d) \, R \, (a, b)$.

- Let $(a, b) \, R \, (c, d)$ and $(c, d) \, R \, (e, f)$. This gives $ad = bc$ and $cf = de$. Therefore, $(a, b) \, R \, (e, f)$ because

$$af = \frac{bcf}{d} = be.$$

■ **EXAMPLE 4.2.6**

Take $m \in \mathbf{Z}^+$ and let a, b, and c be integers. Define a to be **congruent** to b **modulo** m and write

$$a \equiv b \pmod{m} \text{ if and only if } m \mid a - b.$$

That is,

$$a \equiv b \pmod{m} \text{ if and only if } a = b + mk \text{ for some } k \in \mathbf{Z}.$$

For example, we have that $7 \equiv 1 \pmod{3}$, $1 \equiv 13 \pmod{3}$, and $27 \equiv 0 \pmod{3}$, but $2 \not\equiv 9 \pmod{3}$ and $25 \not\equiv 0 \pmod{3}$. Congruence modulo m defines the relation

$$R_m = \{(a, b) : a \equiv b \pmod{m}\}.$$

Observe that

$$R_m = \{(a, b) : m \mid a - b\} = \{(a, b) : \exists k (k \in \mathbf{Z} \wedge a = b + mk)\}.$$

Prove that R_m is an equivalence relation.

- $(a, a) \in R_m$ because $a \equiv a \pmod{m}$.

- Assume $(a, b) \in R_m$. This implies that $m \mid a - b$. By Exercise 2.4.18, $m \mid b - a$. Hence, $(b, a) \in R_m$.

- Let $(a, b), (b, c) \in R_m$. Then $a = b + mk$ and $b = c + ml$ for some $k, l \in \mathbf{Z}$. Substitution yields

$$a = c + ml + mk = c + m(l + k).$$

Since the sum of two integers is an integer, $(a, c) \in R_m$.

Equivalence Classes

Let R be a relation on $\{1, 2, 3, 4\}$ such that

$$R = \{(1,2),(1,3),(2,4)\}. \tag{4.1}$$

Observe that 1 is related to 2 and 3, 2 is related to 4, and 3 and 4 are not related to any number. Combining the elements that are related to a particular element results in a set named by the next definition.

■ **DEFINITION 4.2.7**

Let R be a relation on A with $a \in A$. The **class** of a with respect to R is the set

$$[a]_R = \{x \in A : a\,R\,x\}.$$

If R is an equivalence relation, $[a]_R$ is called an **equivalence class**. We often denote $[a]_R$ by $[a]$ if the relation is clear from context.

Using R as defined in (4.1),

$$[1]_R = \{2, 3\}, [2]_R = \{4\}, \text{ and } [3]_R = [4]_R = \varnothing.$$

If R had been an equivalence relation on a set A, then $[a]_R$ would be nonempty for all $a \in A$ because a would be an element of $[a]_R$ (Exercise 17).

■ **EXAMPLE 4.2.8**

Let R be the equivalence relation from Example 4.2.5. We prove that

$$[(1,3)] = \{(n, 3n) : n \in \mathbf{Z} \setminus \{0\}\}.$$

To see this, take $(a, b) \in [(1, 3)]$. This means that $(1, 3)\,R\,(a, b)$, so $b = 3a$ and $a \neq 0$ because $b \neq 0$. Hence, $(a, b) = (a, 3a)$. Conversely, let $n \neq 0$. Then, $(1, 3)\,R\,(n, 3n)$ because $1 \cdot 3n = 3 \cdot n$. Thus, $(n, 3n) \in [(1, 3)]$.

■ **EXAMPLE 4.2.9**

Using the notation of Example 4.2.6, let R_m be the relation defined by congruence modulo m. For all $n \in \mathbf{Z}$, define

$$[n]_m = R_m(n).$$

Therefore, when $m = 5$, the equivalence classes are:

$$[0]_5 = \{\ldots, -10, -5, 0, 5, 10, \ldots\},$$
$$[1]_5 = \{\ldots, -9, -4, 1, 6, 11, \ldots\},$$
$$[2]_5 = \{\ldots, -8, -3, 2, 7, 12, \ldots\},$$
$$[3]_5 = \{\ldots, -7, -2, 3, 8, 13, \ldots\},$$
$$[4]_5 = \{\ldots, -6, -1, 4, 9, 14, \ldots\}.$$

In addition,

$$[a]_5 = [b]_5 \text{ if and only if } a \equiv b \pmod 5.$$

The collection of all equivalence classes of a relation is a set named by the next definition.

■ DEFINITION 4.2.10

Let R be an equivalence relation on A. The **quotient set** of A **modulo** R is

$$A/R = \{[a]_R : a \in A\}.$$

Observe by Exercise 3 that it is always the case that

$$A = \bigcup A/R. \tag{4.2}$$

■ EXAMPLE 4.2.11

Let $m \in \mathbf{Z}^+$. The quotient set $\mathbf{Z}/R_m$ is denoted by $\mathbf{Z}_m$. That is,

$$\mathbf{Z}_m = \{[0]_m, [1]_m, \dots, [m-1]_m\}.$$

■ EXAMPLE 4.2.12

Define the relation R on $\mathbf{R}^2$ by

$$(a, b) \, R \, (c, d) \text{ if and only if } b - a = d - c.$$

R is an equivalence relation by Exercise 2. We note that for any $(a, b) \in \mathbf{R}^2$,

$$\begin{aligned}
[(a, b)] &= \{(x, y) : (x, y) \, R \, (a, b)\} \\
&= \{(x, y) : y - x = b - a\} \\
&= \{(x, y) : y = x + (b - a)\}.
\end{aligned}$$

Therefore, the equivalence class of (a, b) is the line with a slope of 1 and a y-intercept equal to $(0, b - a)$. The equivalence classes of $(0, 1.5)$ and $(0, -1)$ are illustrated in the graph in Figure 4.3. The quotient set $\mathbf{R}^2/R$ is the collection of all such lines. Notice that

$$\mathbf{R}^2 = \bigcup \mathbf{R}^2/R.$$

Partitions

In Example 4.2.12, we saw that $\mathbf{R}^2$ is the union of all the lines with slope equal to 1, and since the lines are parallel, they form a pairwise disjoint set. These properties can be observed in the other equivalence relations that we have seen. Each set is equal to the union of the equivalence classes, and the quotient set is pairwise disjoint. Generalizing these two properties leads to the next definition.

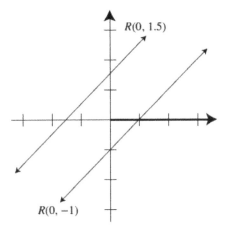

Figure 4.3   Two equivalence classes in $\mathbf{R}^2$ when $(a, b) R (c, d)$ if and only if $b - a = d - c$.

■ **DEFINITION 4.2.13**

Let A be a nonempty set. The family $\mathscr{P}$ is a **partition** of A if and only if

- $\mathscr{P} \subseteq \mathbf{P}(A)$,

- $\bigcup \mathscr{P} = A$,

- $\mathscr{P}$ is pairwise disjoint.

To illustrate the definition, let $A = \{1, 2, 3, 4, 5, 6, 7\}$ and define the elements of the partition to be $A_0 = \{1, 2, 5\}$, $A_1 = \{3\}$, and $A_2 = \{4, 6, 7\}$. The family

$$\mathscr{P} = \{A_1, A_2, A_3\}$$

is a subset of $\mathbf{P}(A)$, $A = A_0 \cup A_1 \cup A_2$, and $\mathscr{P}$ is pairwise disjoint. Therefore, $\mathscr{P}$ is a partition of A. This is illustrated in Figure 4.4.

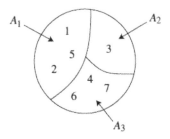

Figure 4.4   A partition of the set $A = \{1, 2, 3, 4, 5, 6, 7\}$.

■ **EXAMPLE 4.2.14**

For each real number $r \geq 0$, define C_r to be the circle with radius r centered at the origin. Namely,

$$C_r = \left\{ (x, y) : \sqrt{x^2 + y^2} = r \right\}.$$

Let $\mathscr{C} = \{C_r : r \in [0, \infty)\}$. We claim that $\mathscr{C}$ is a partition of $\mathbf{R}^2$.

- $\mathscr{C} \subseteq P(\mathbf{R}^2)$ because $C_r \subseteq \mathbf{R}^2$ for all $r \geq 0$.

- To prove that $\mathbf{R}^2 = \bigcup_{r \in [0,\infty)} C_r$, it suffices to show that $\mathbf{R}^2 \subseteq \bigcup_{r \in [0,\infty)} C_r$, but this follows because if $(a, b) \in \mathbf{R}^2$, then

$$(a, b) \in C_{\sqrt{a^2+b^2}}.$$

- To see that $\mathscr{C}$ is pairwise disjoint, let $r, s \geq 0$ and assume that (a, b) is an element of $C_r \cap C_s$. Then,

$$r = \sqrt{a^2 + b^2} = s,$$

which implies that $C_r = C_s$.

The set $\mathbf{Z}_5$ is a family of subsets of $\mathbf{Z}$, has the property that $\bigcup \mathbf{Z}_5 = \mathbf{Z}$, and is pairwise disjoint. Hence, $\mathbf{Z}_5$ is a partition for $\mathbf{Z}$. We generalize this result to the next theorem. It uses an arbitrary equivalence relation on a given set to define a partition for that set. In this case, we say that the equivalence relation **induces** the partition.

■ **THEOREM 4.2.15**

If R is an equivalence relation on A, then A/R is a partition of A.

PROOF

Take a set A with an equivalence relation R.

- Since an equivalence class is a subset of A, we have $A/R \subseteq P(A)$.

- $\bigcup A/R = A$ is (4.2).

- Let $[a], [b] \in A/R$ and assume that there exists $y \in [a] \cap [b]$. In other words, $a\, R\, y$ and $b\, R\, y$. Now take $x \in [a]$. This means that $a\, R\, x$. Since $x\, R\, a$ and $y\, R\, b$ by symmetry, we have that $x\, R\, y$, and then $x\, R\, b$ by transitivity. Thus, $x \in [b]$, which shows $[a] \subseteq [b]$. Similarly, $[b] \subseteq [a]$, so $[a] = [b]$. ■

The collection of equivalence relations forms a partition of a set. Conversely, if we have a partition of a set, the partition gives rise to an equivalence relation on the set. To see this, take any set A and a partition $\mathscr{P}$ of A. For all $a, b \in A$, define

$a\, R\, b$ if and only if there exists $C \in \mathscr{P}$ such that $a, b \in C$.

To show that R is an equivalence relation, take a, b, and c in A.

- Since $a \in A$ and $A = \bigcup \mathcal{P}$, there exists $C \in \mathcal{P}$ such that $a \in C$. Therefore, $a \, R \, a$.

- Assume $a \, R \, b$. This means that $a, b \in C$ for some $C \in \mathcal{P}$. This, of course, is the same as $b, a \in C$. Hence, $b \, R \, a$.

- Suppose $a \, R \, b$ and $b \, R \, c$. Then, there are sets C and D in $\mathcal{P}$ so that $a, b \in C$ and $b, c \in D$. This means that $C \cap D \neq \varnothing$. Since $\mathcal{P}$ is pairwise disjoint, $C = D$. So, a and c are elements of C, and we have $a \, R \, c$.

This equivalence relation is said to be **induced** from the partition.

■ **EXAMPLE 4.2.16**

The sets

$$\{\dots, -10, -5, 0, 5, 10, \dots \},$$
$$\{\dots, -9, -4, 1, 6, 11, \dots \},$$
$$\{\dots, -8, -3, 2, 7, 12, \dots \},$$
$$\{\dots, -7, -2, 3, 8, 13, \dots \},$$
$$\{\dots, -6, -1, 4, 9, 14, \dots \},$$

form a collection that is a partition of $\mathbf{Z}$. The equivalence relation that is induced from this partition is congruence modulo 5 (Example 4.2.9).

Exercises

1. For all $a, b \in \mathbf{R} \setminus \{0\}$, let $a \, R \, b$ if and only if $ab > 0$.
    (a) Show that R is an equivalence relation on $\mathbf{R} \setminus \{0\}$.
    (b) Find $[1]$ and $[-3]$.

2. Define the relation S on $\mathbf{R}^2$ by $(a, b) \, S \, (c, d)$ if and only if $b - a = d - c$. Prove that S is an equivalence relation.

3. Let S be an equivalence relation on A. Prove that $A = \bigcup_{a \in A} [a]$.

4. Prove that if C is an equivalence class for some equivalence relation R and $a \in C$, then $C = [a]$.

5. For all $a, b \in \mathbf{Z}$, let $a \, R \, b$ if and only if $|a| = |b|$.
    (a) Prove R is an equivalence relation on $\mathbf{Z}$.
    (b) Sketch the partition of $\mathbf{Z}$ induced by this equivalence relation.

6. For all $(a, b), (c, d) \in \mathbf{Z} \times \mathbf{Z}$, define $(a, b) \, S \, (c, d)$ if and only if $ab = cd$.
    (a) Show that S is an equivalence relation on $\mathbf{Z} \times \mathbf{Z}$.
    (b) What is the equivalence class of $(1, 2)$?
    (c) Sketch the partition of $\mathbf{Z} \times \mathbf{Z}$ induced by this equivalence relation.

7. Let A be a set and $a \in A$. Show that the given relations are not equivalence relations on $\mathbf{P}(A)$.

(a) For all $C, D \subseteq A$, define $C \, R \, D$ if and only if $C \cap D \neq \varnothing$.

(b) For all $C, D \subseteq A$, define $C \, S \, D$ if and only if $a \in C \cap D$.

8. Let $z, z' \in \mathbf{C}$ and write $z = a + bi$ and $z' = a' + b'i$. Define $z = z'$ to mean that $a = a'$ and $b = b'$. Prove that R is an equivalence relation.

9. Find.

(a) $[3]_5$

(b) $[12]_6$

(c) $[2]_5 \cup [27]_5$

(d) $[4]_7 \cap [5]_7$

10. Let r be the remainder obtained when n is divided by m. Prove $[n]_m = [r]_m$.

11. Let $c, m \in \mathbf{Z}$ and suppose that $\gcd(c, m) = 1$. Prove.

(a) There exists b such that $bc \equiv 1 \pmod{m}$. (Notice that b is that multiplicative inverse of c modulo m.)

(b) If $ca \equiv cb \pmod{m}$, then $a \equiv b \pmod{m}$.

(c) Prove that the previous implication is false if $\gcd(c, m) \neq 1$.

12. Prove that $\{(n, n + 1] : n \in \mathbf{Z}\}$ is a partition of $\mathbf{R}$.

13. Prove that the following are partitions of $\mathbf{R}^2$.

(a) $\mathcal{P} = \{\{(a, b)\} : a, b \in \mathbf{R}\}$.

(b) $\mathcal{V} = \{\{(r, y) : y \in \mathbf{R}\} : r \in \mathbf{R}\}$.

(c) $\mathcal{H} = \{\mathbf{R} \times (n, n + 1] : n \in \mathbf{Z}\}$.

14. Is $\{[n, n + 1] \times (n, n + 1) : n \in \mathbf{Z}\}$ a partition of $\mathbf{R}^2$? Explain.

15. Let R be a relation on A and show the following:

(a) R is reflexive if and only if R^{-1} is reflexive.

(b) R is symmetric if and only if $R = R^{-1}$.

(c) R is symmetric if and only if $(A \times A) \setminus R$ is symmetric.

16. Let R and S be equivalence relations on A. Prove or show false.

(a) $R \cup S$ is an equivalence relation on A.

(b) $R \cap S$ is an equivalence relation on A.

17. Prove for all relations R on A.

(a) R is reflexive if and only if $\forall a (a \in [a])$.

(b) R is symmetric if and only if $\forall a \forall b (a \in [b] \leftrightarrow b \in [a])$.

(c) R is transitive if and only if $\forall a \forall b \forall c ([b \in [a] \wedge c \in [b]] \rightarrow c \in [a])$.

(d) R is an equivalence relation if and only if $\forall a \forall b ((a, b) \in R \leftrightarrow [a] = [b])$.

18. Let R be a relation on A with the property that if $a \, R \, b$ and $b \, R \, c$, then $c \, R \, a$. Prove that if R is also reflexive, R is an equivalence relation.

19. Define the relation R on $\mathbf{C}$ by $a + bi \, R \, c + di$ if and only if

$$\sqrt{a^2 + b^2} = \sqrt{c^2 + d^2}.$$

(a) Prove R is an equivalence relation on $\mathbf{C}$.

(b) Graph $[1 + i]$ in the complex plane.

(c) Describe the partition that R induces on $\mathbf{C}$.

20. Let R and S be relations on A. The **symmetric closure** of R is S if $R \subseteq S$ and for all symmetric relations T on A such that $R \subseteq T$, then $S \subseteq T$. Prove the following.

(a) $R \cup R^{-1}$ is the symmetric closure of R.

(b) A symmetric closure is unique.

4.3   PARTIAL ORDERS

While equivalence relations resemble equality, there are other common relations in mathematics that we can model. To study some of their attributes, we expand Definition 4.2.3 with three more properties.

■ **DEFINITION 4.3.1**

Let R be a relation on A.

- R is **irreflexive** if $a \not\!\!R\, a$ for all $a \in A$.

- R is **asymmetric** when for all $a, b \in A$, if $a \, R \, b$, then $b \not\!\!R\, a$.

- R is **antisymmetric** means that for all $a, b \in A$, if $a \, R \, b$ and $b \, R \, a$, then $a = b$.

Notice that a relation on a nonempty set cannot be both reflexive and irreflexive. However, many relations have neither property. For example, consider the relation $R = \{(1, 1)\}$ on $\{1, 2\}$. Since $(1, 1) \in R$, the relation R is not irreflexive, and R is not reflexive because $(2, 2) \notin R$. Likewise, a relation on a nonempty set cannot be both symmetric and asymmetric.

■ **EXAMPLE 4.3.2**

The less-than relation on $\mathbf{Z}$ is irreflexive and asymmetric. It is also antisymmetric. To see this, let $a, b \in \mathbf{Z}$. Since $a < b$ and $b < a$ is false, the implication

$$\text{if } a < b \text{ and } b < a, \text{ then } a = b$$

is true. The $\leq$ relation is also antisymmetric. However, $\leq$ is neither irreflexive nor asymmetric since $3 \leq 3$.

■ **EXAMPLE 4.3.3**

Let $R = \{(1, 2)\}$ and $S = \{(1, 2), (2, 1)\}$. Both are relations on $\{1, 2\}$. The first relation is asymmetric since $1 \, R \, 2$ but $2 \not\!\!R\, 1$. It is also antisymmetric, but S is not antisymmetric because $2 \, S \, 1$ and $1 \, S \, 2$. Both the relations are irreflexive.

■ **EXAMPLE 4.3.4**

Let R be a relation on a set A. We prove that

$$R \text{ is antisymmetric if and only if } R \cap R^{-1} \subseteq I_A.$$

- Assume that R is antisymmetric and take $(a, b) \in R \cap R^{-1}$. This means that $(a, b) \in R$ and $(a, b) \in R^{-1}$. Therefore, $(b, a) \in R$, and since R is antisymmetric, $a = b$.

- Now suppose $R \cap R^{-1} \subseteq I_A$. Let $(a, b), (b, a) \in R$. We conclude that $(a, b) \in R^{-1}$, which implies that $(a, b) \in R \cap R^{-1}$. Then, $(a, b) \in I_A$. Hence, $a = b$.

As an equivalence relation is a generalization of an identity relation, the following relation is a generalization of $\leq$ on **N**. For this reason, instead of naming the relation R, it is denoted by the symbol $\preccurlyeq$.

■ **DEFINITION 4.3.5**

If a relation $\preccurlyeq$ on a set A is reflexive, antisymmetric, and transitive, $\preccurlyeq$ is a **partial order** on A and the ordered pair $(A, \preccurlyeq)$ is called a **partially ordered set** (or simply a **poset**). Furthermore, for all $a, b \in A$, the notation $a \prec b$ means $a \preccurlyeq b$ but $a \neq b$.

For example, $\leq$ and $=$ are partial orders on **R**, but $<$ is not a partial order on **R** because the relation $<$ is not reflexive. Although $=$ is a partial order on any set, in general an equivalence relation is not a partial order (Example 4.2.6).

■ **EXAMPLE 4.3.6**

Divisibility (Definition 2.4.2) is a partial order on $\mathbf{Z}^+$. To prove this, let a, b, and c be positive integers.

- $a \mid a$ since $a = a \cdot 1$ and $a \neq 0$.

- Suppose that $a \mid b$ and $b \mid a$. This means that $b = ak$ and $a = bl$, for some $k, l \in \mathbf{Z}^+$. Hence, $b = blk$, so $lk = 1$. Since k and l are positive integers, $l = k = 1$. That is, $a = b$.

- Assume that we have $a \mid b$ and $b \mid c$. This means that $b = al$ and $c = bk$ for some $l, k \in \mathbf{Z}^+$. By substitution, $c = (al)k = a(lk)$. Hence, $a \mid c$.

■ **EXAMPLE 4.3.7**

Let $\mathbb{A}$ be a collection of symbols and let $\mathbb{A}^*$ denote the set of all strings over $\mathbb{A}$ (as on page 5). Use the symbol $\square$ to denote the empty string, the string of length zero. As with the empty set, the empty string is always an element of $\mathbb{A}^*$. For example, if $\mathbb{A} = \{a, b, c\}$, then abc, $aaabbb$, c, and $\square$ are elements of $\mathbb{A}^*$. Now, take $\sigma, \tau \in \mathbb{A}^*$. The **concatenation** of σ and τ is denoted by $\sigma^\frown \tau$ and is the

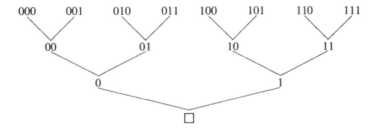

Figure 4.5 A partial order defined on $\{0, 1\}^*$.

string consisting of the elements of σ followed by those of τ. For example, if $\sigma = 011$ and $\tau = 1010$, then $\sigma^\frown \tau = 0111010$. Finally, for all $\sigma, \tau \in \mathbb{A}^*$, define

$$\sigma \preccurlyeq \tau \text{ if and only if there exists } v \in \mathbb{A}^* \text{ such that } \tau = \sigma^\frown v.$$

It can be shown that $\preccurlyeq$ is a partial order on $\mathbb{A}^*$ (Exercise 8) with the structure seen in Figure 4.5 for $\mathbb{A} = \{0, 1\}$.

The partial order $\leq$ on $\mathbf{R}$ has the property that for all $a, b \in \mathbf{R}$, either $a < b$, $b < a$, or $a = b$. As we see in Figure 4.5, this is not the case for every partially ordered set. We do, however, have the following slightly weaker property.

■ **THEOREM 4.3.8 [Weak-Trichotomy Law]**

If $\preccurlyeq$ is a partial order on A, for all $a, b \in A$, at most one of the following are true: $a < b$, $b < a$, or $a = b$.

PROOF

Let $a, b \in A$. We have three cases to consider.

- Suppose $a < b$. This means that $a \preccurlyeq b$ and $a \neq b$. If in addition $b < a$, by transitivity $a < a$, which is a contradiction.

- That $b < a$ precludes both $a < b$ and $a = b$ is proved like the first case.

- If $a = b$, then by definition of $<$ it is impossible for $a < b$ or $b < a$ to be true. ■

Technically, subset is not a relation, but in a natural way, it can be considered as one. Let $\mathscr{F}$ be a family of sets. Define

$$S = \{(A, B) : A \subseteq B \wedge A, B \in \mathscr{F}\}.$$

Associate $\subseteq$ with the relation S.

■ **EXAMPLE 4.3.9**

Let A be a nonempty set. We show that $(\mathbf{P}(A), \subseteq)$ is a partially ordered set. Let B, C, and D be subsets of A.

- Since $B \subseteq B$, the relation $\subseteq$ is reflexive.

- Since $B \subseteq C$ and $C \subseteq B$ implies that $B = C$ (Definition 3.3.7), $\subseteq$ is antisymmetric.

- Since $B \subseteq C$ and $C \subseteq D$ implies $B \subseteq D$ (Theorem 3.3.6), we see that $\subseteq$ is transitive.

This example is in line with what we know about subsets. For instance, if $A \subset B$, then we conclude that $B \not\subset A$ and $A \neq B$, which is what we expect from the weak-trichotomy law (4.3.8).

Bounds

Let $\preccurlyeq$ be a partial order on A with elements m and m' such that $a \preccurlyeq m$ and $a \preccurlyeq m'$ for all $a \in A$. In particular, this implies that $m \preccurlyeq m'$ and $m' \preccurlyeq m$, so since $\preccurlyeq$ is antisymmetric, we conclude that $m = m'$. Similarly, if $m \preccurlyeq a$ and $m' \preccurlyeq a$ for all $a \in A$, then $m = m'$. This argument justifies the use of the word *the* in the next definition.

■ **DEFINITION 4.3.10**

Let $(A, \preccurlyeq)$ be a poset and $m \in A$.

- m is the **least element** of A (with respect to $\preccurlyeq$) if $m \preccurlyeq a$ for all $a \in A$.

- m is the **greatest element** of A (with respect to $\preccurlyeq$) if $a \preccurlyeq m$ for all $a \in A$.

There is no guarantee that a partially ordered set will have a least or a greatest element. In Example 4.3.9, the greatest element of $\mathbf{P}(A)$ with respect to $\subseteq$ is A and the least element is $\varnothing$. However, in Example 4.3.7 (Figure 4.5), the least element of $\mathbb{A}^*$ is $\square$, but there is no greatest element.

■ **EXAMPLE 4.3.11**

If A is a finite set of real numbers, A has a least and a greatest element with respect to $\leq$. Under the partial order $\leq$, the set $\mathbf{Z}^+$ also has a least element but no greatest element, and both $\{5n : n \in \mathbf{Z}^-\}$ and $\mathbf{Z}^-$ have greatest elements but no least elements. To show that $B = \{5n : n \in \mathbf{Z}^+\}$ has a least element with respect to $\leq$, use the fact that the least element of $\mathbf{Z}^+$ is 1. Therefore, the least element of B is 5 because $5 \in B$ and $5(1) \leq 5n$ for all $n \in \mathbf{Z}^+$.

Some sets will not have a least or greatest element with a given partial order but there will still be elements that are considered greater or lesser than every element of the set.

■ **DEFINITION 4.3.12**

Let $\preccurlyeq$ be a partial order on A and $B \subseteq A$.

- $u \in A$ is an **upper bound** of B if $b \preccurlyeq u$ for all $b \in B$. The element u is the **least upper bound** of B if it is an upper bound and for all upper bounds u' of $B, u \preccurlyeq u'$.

- $l \in A$ is a **lower bound** of B if $l \preccurlyeq b$ for all $b \in B$. The element l is the **greatest lower bound** of B if it is a lower bound and for all lower bounds l' of B, $l' \preccurlyeq l$.

In the definition we can write the word *the* because the order is antisymmetric.

■ **EXAMPLE 4.3.13**

The interval $(3, 5)$ is a subset of $\mathbf{R}$. Under the partial order $\leq$, both 5 and 10 are upper bounds of this interval, while 5 is a least upper bound. Also, 3 and $-\pi$ are lower bounds, but 3 is the greatest lower bound.

■ **EXAMPLE 4.3.14**

Assume that $\mathscr{C} \subseteq \mathbf{P}(\mathbf{Z})$. Since the elements of $\mathscr{C}$ are subsets of $\mathbf{Z}$, we conclude that $\bigcup \mathscr{C} \in \mathbf{P}(\mathbf{Z})$ (Definition 3.4.8). If $A \in \mathscr{C}$, then $A \subseteq \bigcup \mathscr{C}$. Therefore, $\bigcup \mathscr{C}$ is an upper bound of $\mathscr{C}$ with respect to $\subseteq$. To see that it is the least upper bound, let U be any upper bound of $\mathscr{C}$. Take $x \in \bigcup \mathscr{C}$. This means that there exists $D \in \mathscr{C}$ such that $x \in D$. Since U is an upper bound of $\mathscr{C}$, $D \subseteq U$. Hence, $x \in U$, and we conclude that $\bigcup \mathscr{C} \subseteq U$.

Comparable and Compatible Elements

In Figure 4.5, we see that $01 \preccurlyeq 010$ and $01 \preccurlyeq 011$. However, $010 \npreccurlyeq 011$ and $011 \npreccurlyeq 010$. This means that in the poset of Example 4.3.7, there are pairs of elements that are related to each other and there are other pairs that are not.

■ **DEFINITION 4.3.15**

Let $(A, \preccurlyeq)$ be a poset and $a, b \in A$. If $a \preccurlyeq b$ or $b \preccurlyeq a$, then a and b are **comparable** with respect to $\preccurlyeq$. Elements of A are **incomparable** if they are not comparable.

Continuing our review of the partially ordered set of Example 4.3.7, we note that the element $\square$ has the property that no element is less than it, but as seen in Figure 4.5, for every element of $\mathbb{A}^*$, there exists an element of $\mathbb{A}^*$ that is greater. However, that same relation defined on

$$A = \{\sigma \in \mathbb{A}^* : \sigma \text{ has at most 3 characters}\} \tag{4.3}$$

has the property that $000, 001, 010, 011, 100, 101, 110, 111$ have no elements greater than them. This leads to the next definition.

■ **DEFINITION 4.3.16**

Let $(A, \preccurlyeq)$ be a poset and $m \in A$.

- m is a **minimal element** of A (with respect to $\preccurlyeq$) if $a \npreccurlyeq m$ for all $a \in A$.
- m is a **maximal element** of A (with respect to $\preccurlyeq$) if $m \npreccurlyeq a$ for all $a \in A$.

Therefore, the empty string is a minimal element of A (4.3), and 000, 001, 010, 011, 100, 101, 110, 111 are maximal. Notice that every least element is minimal and every greatest element is maximal.

Although not every pair of elements is comparable in the partially ordered set of Example 4.3.7, there are infinite sequences of comparable elements, such as

$$\Box \preccurlyeq 0 \preccurlyeq 01 \preccurlyeq 001 \preccurlyeq 0001 \preccurlyeq \cdots .$$

■ **DEFINITION 4.3.17**

A subset C of the poset $(A, \preccurlyeq)$ is a **chain** with respect to $\preccurlyeq$ if a is comparable to b for all $a, b \in C$.

When $\mathbf{Z}$ is partially ordered by $\leq$, the sets $\{0, 1, 2, 3, \dots\}$, $\{\dots, -3, -2, -1, 0\}$, and $\{\dots, -2, 0, 2, 4, \dots\}$ are chains. In $\mathbf{P(Z)}$, both

$$\{\{1, 2\}, \{1, 2, 3, 4\}, \{1, 2, 3, 4, 5, 6\}\}$$

and

$$\{\varnothing, \{0\}, \{0, 1\}, \{0, 1, 2\}, \dots\}$$

are chains with respect to $\subseteq$.

■ **EXAMPLE 4.3.18**

To see that $\{A_k : k \in \mathbf{Z}^+\}$ where $A_k = \{x \in \mathbf{Z} : (x-1)(x-2)\cdots(x-k) = 0\}$ is a chain with respect to $\subseteq$, take $m, n \in \mathbf{Z}^+$. By definition,

$$A_m = \{1, 2, \dots, m\}$$

and

$$A_n = \{1, 2, \dots, n\}.$$

If $m \leq n$, then $A_m \subseteq A_n$, otherwise $A_n \subseteq A_m$.

■ **EXAMPLE 4.3.19**

Let C_0 and C_1 be chains of A with respect to $\preccurlyeq$. Take $a, b \in C_0 \cap C_1$. Then, $a, b \in C_0$, so $a \preccurlyeq b$ or $b \preccurlyeq a$. Therefore, $C_0 \cap C_1$ is a chain. However, the union of two chains might not be a chain. For example, $\{\{1\}, \{1, 2\}\}$ and $\{\{1\}, \{1, 3\}\}$ are chains in $(\mathbf{P(Z)}, \subseteq)$, but $\{\{1\}, \{1, 2\}, \{1, 3\}\}$ is not a chain because $\{1, 2\} \nsubseteq \{1, 3\}$ and $\{1, 3\} \nsubseteq \{1, 2\}$.

The sets $\mathbf{Z}$, $\mathbf{Q}$, and $\mathbf{R}$ are chains of $\mathbf{R}$ with respect to $\leq$. In fact, any subset of $\mathbf{R}$ is a chain of $\mathbf{R}$ because subsets of chains are chains. This motivates the next definition.

■ **DEFINITION 4.3.20**

The poset $(A, \preccurlyeq)$ is a **linearly ordered set** and $\preccurlyeq$ is a **linear order** if A is a chain with respect to $\preccurlyeq$.

Since every subset A of $\mathbf{R}$ is a chain with respect to $\leq$, the relation $\leq$ is a linear order on A. Furthermore, since every pair of elements in a linear order are comparable, Theorem 4.3.8 can be strengthened.

■ **THEOREM 4.3.21 [Trichotomy Law]**

If $\preccurlyeq$ is a linear order on A, for all $a, b \in A$, exactly one of the following are true: $a \prec b$, $b \prec a$, or $a = b$.

Although it is not the case that every pair of elements of the poset defined in Example 4.3.7 is comparable, it is the case that for any given pair of elements, there exists another element that is related to the given elements. For example, for the pair 100 and 110, $1 \preccurlyeq 100$ and $1 \preccurlyeq 110$. Also, for the pair 101 and 10, $10 \preccurlyeq 101$ and $10 \preccurlyeq 10$.

■ **DEFINITION 4.3.22**

Let $(A, \preccurlyeq)$ be a poset. The elements $a, b \in A$ are **compatible** if there exists $c \in A$ such that $c \preccurlyeq a$ and $c \preccurlyeq b$. If a and b are not compatible, they are **incompatible** and we write $a \perp b$.

Observe that if a and b are comparable, they are also compatible. On the other hand, it takes some work to define a relation in which every pair of elements is incompatible.

■ **DEFINITION 4.3.23**

A subset D of a poset $(A, \preccurlyeq)$ is an **antichain** with respect to $\preccurlyeq$ when for all $a, b \in D$, if $a \neq b$, then $a \perp b$.

The sets

$$\{\{n\} : n \in \mathbf{Z}\}$$

and

$$\{\{1\}, \{2, 3\}, \{4, 5, 6\}, \{7, 8, 9, 10\}, \dots \}$$

are antichains of $\mathbf{P}(\mathbf{Z})$ with respect to $\subseteq$.

Well-Ordered Sets

The notion of a linear order incorporates many of the properties of $\leq$ on $\mathbf{N}$ since $\leq$ is reflexive, antisymmetric, and transitive, and $\mathbf{N}$ is a chain with respect to $\leq$. However, there is one important property of $(\mathbf{N}, \leq)$ that is not included among those of a linear order. Because of the nature of the natural numbers, every subset of $\mathbf{N}$ that is nonempty has a least element. For example, 5 is the least element of $\{5, 7, 32, 99\}$ and 2 is the least element of $\{2, 4, 6, 8, \dots \}$. We want to be able to identify those partial orders that also have this property.

■ **DEFINITION 4.3.24**

The linearly ordered set $(A, \preccurlyeq)$ is a **well-ordered set** and $\preccurlyeq$ is a **well-order** if every nonempty subset of A has a least element with respect to $\preccurlyeq$.

According to Definition 4.3.24, $(N, \leq)$ is a well-ordered set, but this fact about the natural numbers cannot be proved without making an assumption. Therefore, so that we have at least one well-ordered set with which to work, we assume the following.

■ AXIOM 4.3.25

$(N, \leq)$ is a well-ordered set.

Coupling Axiom 4.3.25 with the next theorem will yield infinitely many well-ordered sets.

■ THEOREM 4.3.26

If $(A, \preccurlyeq)$ is well-ordered and B is a nonempty subset of A, then $(B, \preccurlyeq)$ is well-ordered.

PROOF

Let $B \subseteq A$ and $B \neq \varnothing$. To prove that B is well-ordered, let $C \subseteq B$ and $C \neq \varnothing$. Then, $C \subseteq A$. Since $\preccurlyeq$ well-orders A, we know that C has a least element with respect to $\preccurlyeq$. ■

Because $Z \cap [5, \infty) \subseteq Z^+ \subseteq N$, by Axiom 4.3.25 and Theorem 4.3.26, both $(Z^+, \leq)$ and $(Z \cap [5, \infty), \leq)$ are well-ordered sets.

■ EXAMPLE 4.3.27

Let $A = \{n\pi : n \in N\}$. To prove that A is well-ordered by $\leq$, let $B \subseteq A$ such that $B \neq \varnothing$. This means that there exists a nonempty subset I of N such that $B = \{n\pi : n \in I\}$. Since N is well-ordered, I has least element m. We claim that $m\pi$ is the least element of B. To see this, take $b \in B$. Then, $b = i\pi$ for some $i \in I$. Since m is the least element of I, $m \leq i$. Therefore, $m\pi \leq i\pi = b$.

To prove that a set A is not well-ordered by $\preccurlyeq$, we must find a nonempty subset B of A that does not have a least element. This means that for every $b \in B$, there exists $c \in B$ such that $c \prec b$. That is, there are elements $b_n \in B$ ($n \in N$) such that

$$\cdots \prec b_2 \prec b_1 \prec b_0.$$

This informs the next definition.

■ DEFINITION 4.3.28

Let $\preccurlyeq$ be a partial order on A and $B = \{a_i : i \in N\}$ be a subset of A.

- B is **increasing** means $i < j$ implies $a_i \prec a_j$ for all $i, j \in N$.

- B is **decreasing** means $i < j$ implies $a_j \prec a_i$ for all $i, j \in N$.

If a set is well-ordered, it has a least element, but the converse is not true. To see this, consider $A = \{0, 1/2, 1/3, 1/4, \dots\}$. It has a least element, namely, 0, but A also contains the decreasing set

$$\left\{ \frac{1}{2}, \frac{1}{3}, \frac{1}{4}, \frac{1}{5}, \cdots \right\}.$$ (4.4)

Therefore, A has a subset without a least element, so A is not well-ordered. We summarize this observation with the following theorem, and leave its proof to Exercise 23.

■ **THEOREM 4.3.29**

$(A, \preccurlyeq)$ is not a well-ordered set if and only if $(A, \preccurlyeq)$ does not have a decreasing subset.

Theorem 4.3.29 implies that any finite linear order is well-ordered.

■ **EXAMPLE 4.3.30**

The decreasing sequence (4.4) with Theorem 4.3.29 shows that the sets $(0, 1)$, $[0, 1]$, $\mathbf{Q}$, and $\{1/n : n \in \mathbf{Z}^+\}$ are not well-ordered by $\leq$.

We close this section by proving two important results from number theory. Their proofs use Axiom 4.3.25. The strategy is to define a nonempty subset of a well-ordered set. Its least element r will be a number that we want. This least element also needs to have a particular property, say $p(r)$. To show that it has the property, assume $\neg p(r)$ and use this to find another element of the set that is less than r. This contradicts the minimality of r allowing us to conclude $p(r)$.

■ **THEOREM 4.3.31 [Division Algorithm]**

If $m, n \in \mathbf{N}$ with $m \neq 0$, there exist unique $q, r \in \mathbf{N}$ such that $r < m$ and $n = mq + r$.

PROOF
Uniqueness is proved in Example 2.4.14. To prove existence, take $m, n \in \mathbf{N}$ and define

$$S = \{k \in \mathbf{N} : \exists l(l \in \mathbf{N} \wedge n = ml + k)\}.$$

Notice that $n \in S$, so $S \neq \varnothing$. Therefore, S has a least element by Axiom 4.3.25. Call it r and write $n = mq + r$ for some natural number q. Assume $r \geq m$, which implies that $r - m \geq 0$. Also,

$$n = m(q - 1) + r - m$$

because

$$r - m = n - mq - m = n - m(q + 1),$$

so $r - m \in S$. Since $r > r - m$ because m is positive, r cannot be the minimum of S, a contradiction. ■

The value q of the division algorithm (Theorem 4.3.31) is called the **quotient** and r is the **remainder**. For example, if we divide 5 into 17, the division algorithm returns a quotient of 3 and a remainder of 2, so we can write that $17 = 5(3) + 2$. Notice that $2 < 3$.

Call $n \in \mathbf{Z}$ a **linear combination** of the integers a and b if $n = ua + vb$ for some $u, v \in \mathbf{Z}$. Since $37 = 5(2) + 3(9)$, we see that 37 is a linear combination of 2 and 9. Furthermore, if $d \mid a$ and $d \mid b$, then $d \mid ua + vb$. To see this, write $a = dl$ and $b = dk$ for some $l, k \in \mathbf{Z}$. Then,

$$ua + vb = udl + vdk = d(ul + vk),$$

and this means $d \mid ua + vb$.

■ THEOREM 4.3.32

Let $a, b \in \mathbf{Z}$ with not both equal to 0. If $c = \gcd(a, b)$, there exists $m, n \in \mathbf{Z}$ such that $c = ma + nb$.

PROOF
Define
$$T = \{z \in \mathbf{Z}^+ : \exists x \exists y (x, y \in \mathbf{Z} \land z = xa + yb)\}.$$

Notice that T is not empty because $a^2 + b^2 \in T$. By Axiom 4.3.25, T has a least element d, so write $d = ma + nb$ for some $m, n \in \mathbf{Z}$.

- Since $d > 0$, the division algorithm (4.3.31) yields $a = dq + r$ for some natural numbers q and r with $r < d$. Then,

$$r = a - dq = a - (ma - nb)q = (1 - mq)a + (nq)b.$$

If $r > 0$, then $r \in T$, which is impossible because d is the least element of T. Therefore, $r = 0$ and $d \mid a$. Similarly, $d \mid b$.

- To show that d is the greatest of the common divisors, suppose $s \mid a$ and $s \mid b$ with $s \in \mathbf{Z}^+$. By definition, $a = sk$ and $b = sl$ for some $k, l \in \mathbf{Z}$. Hence,

$$d = m(sk) + n(sl) = s(mk - nl).$$

Thus, $s \leq d$ because s is nonzero and s divides d (Exercise 25). ■

Exercises

1. Is $(\varnothing, \subseteq)$ a partial order, linear order, or well order? Explain.

2. For each relation on $\{1, 2\}$, determine if it is reflexive, irreflexive, symmetric, asymmetric, antisymmetric, or transitive.
    (a) $\{(1, 2)\}$
    (b) $\{(1, 2), (2, 1)\}$
    (c) $\{(1, 1), (1, 2), (2, 1)\}$
    (d) $\{(1, 1), (1, 2), (2, 2)\}$
    (e) $\{(1, 1), (1, 2), (2, 1), (2, 2)\}$
    (f) $\varnothing$

3. Give an example of a relation that is neither symmetric nor asymmetric.

4. Let R be a relation on A. Prove that R is reflexive if and only if $(A \times A) \setminus R$ is irreflexive.

5. Show that a relation R on A is asymmetric if and only if $R \cap R^{-1} = \varnothing$.

6. Let $(A, \preccurlyeq)$ and $(B, \leq)$ be posets. Define $\sim$ on $A \times B$ by

$$(a, b) \sim (a', b') \text{ if and only if } a \preccurlyeq a' \text{ and } b \leq b'.$$

Show that $\sim$ is a partial order on $A \times B$.

7. For any alphabet $\mathbb{A}$, prove the following.
   (a) For all $\sigma, \tau, v \in \mathbb{A}^*$, $\sigma^\frown(\tau^\frown v) = (\sigma^\frown \tau)^\frown v$.
   (b) There exists $\sigma, \tau \in \mathbb{A}^*$ such that $\sigma^\frown \tau \neq \tau^\frown \sigma$.
   (c) $\mathbb{A}^*$ has an identity with respect to $\frown$, but for all $\sigma \in \mathbb{A}^*$, there is no inverse for σ if $\sigma \neq \square$.

8. Prove that $\mathbb{A}^*$ from Example 4.3.7 is partially ordered by $\preccurlyeq$.

9. Prove that $(\mathbf{P}(A), \subseteq)$ is not a linear order if A has at least three elements.

10. Show that $(\mathbb{A}^*, \preccurlyeq)$ is not a linear order if $\mathbb{A}$ has at least two elements.

11. Prove that the following families of sets are chains with respect to $\subseteq$.
    (a) $\{[0, n] : n \in \mathbf{Z}^+\}$
    (b) $\{(2^n)\mathbf{Z} : n \in \mathbf{N}\}$ where $(2^n)\mathbf{Z} = \{2^n \cdot k : k \in \mathbf{Z}\}$
    (c) $\{B_n : n \in \mathbf{N}\}$ where $B_n = \bigcup\{A_i : i \in \mathbf{N} \wedge i \leq n\}$ and A_i is a set for all $i \in \mathbf{N}$

12. Can a chain be disjoint or pairwise disjoint? Explain.

13. Suppose that $\{A_i : i \in \mathbf{N}\}$ is a chain of sets such that for all $i \leq j$, $A_i \subseteq A_j$. Prove for all $k \in \mathbf{N}$.
    (a) $\bigcup\{A_i : i \in \mathbf{N} \wedge i \leq k\} = A_k$
    (b) $\bigcap\{A_i : i \in \mathbf{N} \wedge i \leq k\} = A_0$

14. Let $\{A_n : n \in \mathbf{N}\}$ be a family of sets. For every $m \in \mathbf{N}$, define $B_m = \bigcup_{i=0}^{m} A_i$. Show that $\{B_m : m \in \mathbf{N}\}$ is a chain.

15. Let $\mathscr{C}_i$ be a chain of the poset $(A, \preccurlyeq)$ for all $i \in I$. Prove that $\bigcap_{i \in I} \mathscr{C}_i$ is a chain. Is $\bigcup_{i \in I} \mathscr{C}_i$ necessarily a chain? Explain.

16. Let $(A, \preccurlyeq)$ and $(B, \leq)$ be linear orders. Define $\sim$ on $A \times B$ by

$$(a, b) \sim (a', b')$$

if and only if

$$a \prec a' \text{ or } a = a' \text{ and } b \leq b'.$$

This relation is called a **lexicographical order** since it copies the order of a dictionary.
    (a) Prove that $(A \times B, \sim)$ is a linear order.
    (b) Suppose that (a, b) is a maximal element of $A \times B$ with respect to $\sim$. Show that a is a maximal element of A with respect to $\preccurlyeq$.

17. Let A be a set. Prove that $B \subseteq P(A) \setminus \{\varnothing\}$ is an antichain with respect to $\subseteq$ if and only if B is pairwise disjoint.

18. Prove the following true or false.
    (a) Every well-ordered set contains a least element.
    (b) Every well-ordered set contains a greatest element.
    (c) Every subset of a well-ordered set contains a least element.
    (d) Every subset of a well-ordered set contains a greatest element.
    (e) Every well-ordered set has a decreasing subset.
    (f) Every well-ordered set has a increasing subset.

19. For each of the given sets, indicate whether or not it is well-ordered by $\leq$. If it is, prove it. If it is not, find a decreasing sequence of elements of the set.

    (a) $\{\sqrt{2}, 5, 6, 10.56, 17, -100\}$
    (b) $\{2n : n \in \mathbf{N}\}$
    (c) $\{\pi/n : n \in \mathbf{Z}^+\}$
    (d) $\{\pi/n : n \in \mathbf{Z}^-\}$
    (e) $\{-4, -3, -2, -1, \dots\}$
    (f) $\mathbf{Z} \cap (\pi, \infty)$
    (g) $\mathbf{Z} \cap (-7, \infty)$

20. Prove that $(\mathbf{Z} \cap (x, \infty), \leq)$ is a well-ordered set for all $x \in \mathbf{R}$.

21. Show that a well-ordered set has a unique least element.

22. Let $(A, \preccurlyeq)$ be a well-ordered set. If $B \subseteq A$ and there is an upper bound for B in A, then B has a greatest element.

23. Prove Theorem 4.3.29.

24. Where does the proof of Theorem 4.3.31 go wrong if $m = 0$?

25. Prove that if $a \mid b$, then $a \leq b$ for all $a, b \in \mathbf{Z}$.

26. Let $a, b, c \in \mathbf{Z}$. Show that if $\gcd(a, c) = \gcd(b, c) = 1$, then $\gcd(a, bc) = 1$.

27. Let $a, b \in \mathbf{Z}$ and assume that $\gcd(a, b) = 1$. Prove.
    (a) If $a \mid n$ and $b \mid n$, then $ab \mid n$.
    (b) $\gcd(a + b, b) = \gcd(a + b, a) = 1$.
    (c) $\gcd(a + b, a - b) = 1$ or $\gcd(a + b, a - b) = 2$.
    (d) If $c \mid a$, then $\gcd(b, c) = 1$.
    (e) If $c \mid a + b$, then $\gcd(a, c) = \gcd(b, c) = 1$.
    (f) If $d \mid ac$ and $d \mid bc$, then $d \mid c$.

28. Let $a, b \in \mathbf{Z}$, where at least one is nonzero. Prove that $S = T$ if

$$S = \{x : \exists l [l \in \mathbf{Z} \wedge x = l \gcd(a, b)]\}$$

and

$$T = \{x : x > 0 \wedge \exists u \exists v (u, v \in \mathbf{Z} \wedge x = ua + vb)\}.$$

29. Prove that if $d \mid a$ and $d \mid b$, then $d \mid \gcd(a, b)$ for all $a, b, d \in \mathbf{Z}$.

4.4 FUNCTIONS

From algebra and calculus, we know what a function is. It is a rule that assigns to each possible input value a unique output value. The common picture is that of a machine that when a certain button is pushed, the same result always happens. In basic algebra a relation can be graphed in the Cartesian plane. Such a relation will be a function if and only if every vertical line intersects its graphs at most once. This is known as the **vertical line test** (Figure 4.6). This criteria is generalized in the next definition.

■ **DEFINITION 4.4.1**

Let A and B be sets. A relation $f \subseteq A \times B$ is a **function** means that for all $(x, y), (x', y') \in A \times B$,

$$\text{if } x = x', \text{ then } y = y'.$$

The function f is an **n-ary** function if there exists sets $A_0, A_1, \ldots, A_{n-1}$ such that $A = A_0 \times A_1 \times \cdots \times A_{n-1}$. If $n = 1$, then f is a **unary** function, and if $n = 2$, then f is a **binary** function,

■ **EXAMPLE 4.4.2**

The set $\{(1, 2), (4, 5), (6, 5)\}$ is a function, but $\{(1, 2), (1, 5), (6, 5)\}$ is not since it contains $(1, 2)$ and $(1, 5)$. Also, $\varnothing$ is a function (Exercise 16).

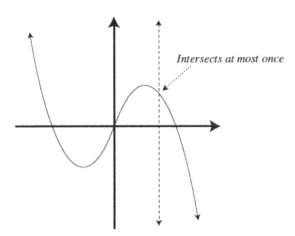

Intersects at most once

Figure 4.6 Passing the vertical line test.

■ **EXAMPLE 4.4.3**

Define $f = \{(u, 1 + 2\cos\pi u) : u \in \mathbf{R}\}$. Assume that both $(x, 1 + 2\cos\pi x)$ and $(x', 1 + 2\cos\pi x')$ are elements of f. Then, because cosine is a function,

$$
\begin{aligned}
x = x' &\Rightarrow \pi x = \pi x' \\
&\Rightarrow \cos\pi x = \cos\pi x' \\
&\Rightarrow 2\cos\pi x = 2\cos\pi x' \\
&\Rightarrow 1 + 2\cos\pi x = 1 + 2\cos\pi x'.
\end{aligned}
$$

Therefore, f is a function.

■ **EXAMPLE 4.4.4**

The standard arithmetic operations are functions. For example, taking a square root is a unary function, while addition, subtraction, multiplication, and division are binary functions. To illustrate this, addition on $\mathbf{Z}$ is the set

$$A = \{((a, b), a + b) : a, b \in \mathbf{Z}\}. \tag{4.5}$$

Let $f \subseteq A \times B$ be a function. This implies that for all $a \in A$, either $[a]_f = \varnothing$ or $[a]_f$ is a singleton (Definition 4.2.7). For example, if $f = \{(1, 2), (4, 5), (6, 5)\}$, then $[1]_f = \{2\}$ and $[3]_f = \varnothing$. Because $[a]_f$ can contain at most one element, we typically simplify the notation.

■ **DEFINITION 4.4.5**

Let f be a function. For all $a \in A$, define

$$f(a) = b \text{ if and only if } [a]_f = \{b\}$$

and write that $f(a)$ is **undefined** if $[a]_f = \varnothing$.

For example, using (4.5),

$$A((5, 7)) = 12,$$

but $A(5, \pi)$ is undefined. With ordered n-tuples, the outer parentheses are usually eliminated so that we write

$$A(5, 7) = 12.$$

Moreover, if $D \subseteq A$ is the domain of the function f, write the **function notation**

$$f : D \to B$$

and call B a **codomain** of f (Figure 4.7). Because functions are often represented by arrows that "send" one element to another, a function can be called a **map**. If $f(x) = y$, we can say that "f maps x to y." We can also say that y is the **image** of x under f and x is a **pre-image** of y. For example, if

$$f = \{(3, 1), (4, 2), (5, 2)\},$$

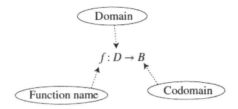

Figure 4.7   Function notation.

f maps 3 to 1, 2 is the image of 4, and 5 is a pre-image of 2 (Figure 4.8). If g is also a function with domain D and codomain B, we can use the abbreviation

$$f, g : D \rightarrow B$$

to represent both functions. An alternate choice of notation involves referring to the functions as $D \rightarrow B$.

■ **EXAMPLE 4.4.6**

If $f(x) = \cos x$, then f maps π to -1, 0 is the image of $\pi/2$, and $\pi/4$ is a pre-image of $\sqrt{2}/2$.

■ **EXAMPLE 4.4.7**

Let A be any set. The identity relation on A (Definition 4.1.2) is a function, so call I_A the **identity map** and write

$$I_A(x) = x.$$

Sometimes a function f is defined using a rule that pairs an element of the function's domain with an element of its codomain. When this is done successfully, f is said to be **well-defined**. Observe that proving that f is well-defined is the same as proving it to be a function.

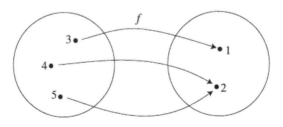

Figure 4.8   The map $f = \{(3, 1), (4, 2), (5, 2)\}$.

■ EXAMPLE 4.4.8

Let $f : \mathbf{R} \to \mathbf{R}$ be defined by $f(x) = 1 + 2 \cos \pi x$. This is the function

$$f = \{(x, 1 + 2 \cos \pi x) : x \in \mathbf{R}\}.$$

The work of Example 4.4.3 shows that f is well-defined.

Before we examine another example, let us set a convention on naming functions. It is partly for aesthetics, but it does help in organizing functions based on the type of elements in their domains and ranges.

- Use English letters (usually, f, g, and h) for naming functions that involve numbers. Typically, these will be lowercase, but there are occasions when we will choose them to be uppercase.

- Use Greek letters (often φ or ψ) for general functions or those with domains not consisting of numbers. They are also usually lowercase, but uppercase Greek letters like Φ and Ψ are sometimes appropriate. (See the appendix for the Greek alphabet.)

■ EXAMPLE 4.4.9

Let $n, m \in \mathbf{Z}^+$ such that $m \mid n$. Define $\varphi([a]_n) = [a]_m$ for all $a \in \mathbf{Z}$. This means that

$$\varphi = \{([a]_n, [a]_m) : a \in \mathbf{Z}\}.$$

It is not clear that φ is well-defined since an equivalence class can have many representatives, so assume that $[a]_n = [b]_n$ for $a, b \in \mathbf{Z}$. Therefore, $n \mid a - b$. Then, by hypothesis, $m \mid a - b$, and this yields

$$\varphi([a]_n) = [a]_m = [b]_m = \varphi([b]_n).$$

■ EXAMPLE 4.4.10

Let $x \in \mathbf{R}$. Define the **greatest integer function** as

$$[\![x]\!] = \text{the greatest integer} \le x.$$

For example, $[\![5]\!] = 5$, $[\![1.4]\!] = 1$, and $[\![-3.4]\!] = -4$. The greatest integer function is a function $\mathbf{R} \to \mathbf{Z}$. It is well-defined because the relation $\le$ well-orders $\mathbf{Z}$ (Exercise 4.3.22).

If a relation on $\mathbf{R}$ is not a function, it will fail the vertical line test (Figure 4.9). To generalize, let $\varphi \subseteq A \times B$. To show that φ is not a function, we must show that there exists $(x, y_1), (x, y_2) \in \varphi$ such that $y_1 \ne y_2$.

■ EXAMPLE 4.4.11

The relation $f = I_{\mathbf{R}} \cup \{(x, -x) : x \in \mathbf{R}\}$ is not a function since $(4, 4) \in f$ and $(4, -4) \in f$.

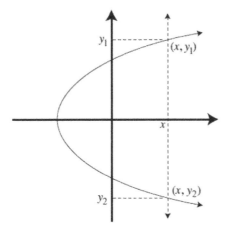

Figure 4.9   The relation is not a function.

■ **EXAMPLE 4.4.12**

Define $\varphi \subseteq \mathbf{Z}_2 \times \mathbf{Z}_3$ by $\varphi([a]_2) = [a]_3$ for all $a \in \mathbf{Z}$. Since 3 does not divide 2, φ is not well-defined. This is proved by noting that $[0]_2 = [2]_2$ but $[0]_3 \neq [2]_3$.

There will be times when we want to examine sets of functions. If each function is to have the same domain and codomain, we use the following notation.

■ **DEFINITION 4.4.13**

If A and B are sets,

$$^AB = \{\varphi : \varphi \text{ is a function } A \to B\}.$$

For instance, ^AB is a set of **real-valued** functions if $A, B \subseteq \mathbf{R}$.

■ **EXAMPLE 4.4.14**

If a_n is a sequence of real numbers with $n = 0, 1, 2, \ldots$, the sequence a_n is an element of $^N\mathbf{R}$. Illustrating this, the sequence

$$a_n = (-1/2)^n$$

is a function and can be graphed as in Figure 4.10.

■ **EXAMPLE 4.4.15**

Let $A \subseteq \mathbf{R}$ and fix $a \in A$. The **evaluation map**,

$$\epsilon_a : {}^AA \to A,$$

is defined as

$$\epsilon_a(f) = f(a).$$

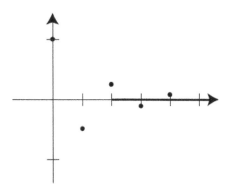

Figure 4.10 $a_n = (-1/2)^n$ is a function.

For example, if $g(x) = x^2$, then $\epsilon_3(g) = 9$. Observe that the evaluation map is an element of $^{(^AA)}A$.

Equality

Since functions are sets, we already know that two functions are equal when they contain the same ordered pairs. However, there is a common test to determine function equality other than a direct appeal to Definition 3.3.7.

■ THEOREM 4.4.16

Functions $\varphi, \psi : A \to B$ are equal if and only if $\varphi(x) = \psi(x)$ for all $x \in A$.

PROOF

Sufficiency is clear, so to prove necessity suppose that $\varphi(x) = \psi(x)$ for every $x \in A$. Take $(a, b) \in \varphi$. This means that $\varphi(a) = b$. By hypothesis, $\psi(a) = b$, which implies that $(a, b) \in \psi$. Hence, $\varphi \subseteq g$. The proof of $\psi \subseteq \varphi$ is similar, so $\varphi = g$. ■

We use Theorem 4.4.16 in the next examples.

■ EXAMPLE 4.4.17

Let f and g be functions $\mathbf{R} \to \mathbf{R}$ defined by

$$f(x) = (x - 3)^2 + 2$$

and

$$g(x) = x^2 - 6x + 11.$$

We show that $f = g$ by taking $x \in \mathbf{R}$ and calculating

$$f(x) = (x - 3)^2 + 2 = (x^2 - 6x + 9) + 2 = x^2 - 6x + 11 = g(x).$$

■ **EXAMPLE 4.4.18**

Let $\varphi, \psi : \mathbf{Z} \to \mathbf{Z}_6$ be functions such that

$$\varphi(n) = [n]_6$$

and

$$\psi(n) = [n + 12]_6.$$

Take $n \in \mathbf{Z}$. We show that $[n + 12]_6 = [n]_6$ by proceeding as follows:

$$\begin{aligned} x \in [n + 12]_6 &\Leftrightarrow \exists k \, [k \in \mathbf{Z} \wedge x = n + 12 + 6k] \\ &\Leftrightarrow \exists k \, [k \in \mathbf{Z} \wedge x = n + 6(2 + k)] \\ &\Leftrightarrow x \in [n]_6. \end{aligned}$$

Therefore, $\varphi = \psi$.

■ **EXAMPLE 4.4.19**

Define

$$\psi : \mathbf{R} \to {}^{(^{\mathbf{R}}\mathbf{R})}\mathbf{R}$$

by $\psi(x) = \epsilon_x$ for all $x \in \mathbf{R}$ (Example 4.4.15). We show that ψ is well-defined. Take $a, b \in \mathbf{R}$ and assume that $a = b$. To show that $\psi(a) = \psi(b)$, we prove $\epsilon_a = \epsilon_b$. Therefore, let $f \in {}^{\mathbf{R}}\mathbf{R}$. Since f is a function and $a, b \in \mathrm{dom}(f)$, we have that $f(a) = f(b)$. Thus,

$$\epsilon_a(f) = f(a) = f(b) = \epsilon_b(f).$$

By Theorem 4.4.16, we see that two functions f and g are not equal when either $\mathrm{dom}(f) \neq \mathrm{dom}(g)$ or $f(x) \neq g(x)$ for some x in their common domain. For example, $f(x) = x^2$ and $g(x) = 2x$ are not equal because $f(3) = 9$ and $g(3) = 6$. Although these two functions differ for every $x \neq 0$ and $x \neq 2$, it only takes one inequality to prove that the functions are not equal. For example, if we define

$$h(x) = \begin{cases} x^2 & \text{if } x \neq 0, \\ 7 & \text{if } x = 0, \end{cases}$$

then $f \neq h$ since $f(0) = 0$ and $h(0) = 7$.

Composition

We now consider the composition of relations when those relations are functions.

■ **THEOREM 4.4.20**

If $\varphi : A \to B$ and $\psi : C \to D$ are functions such that $\mathrm{ran}(\varphi) \subseteq C$, then $\psi \circ \varphi$ is a function $A \to D$ and $(\psi \circ \varphi)(x) = \psi(\varphi(x))$.

PROOF

Because the range of φ is a subset of C, we know that $\varphi \subseteq A \times C$ and $\psi \subseteq C \times D$. Let $(a, d_1), (a, d_2) \in \psi \circ \varphi$. This means by Definition 4.1.9 that there exists $c_1, c_2 \in C$ such that $(a, c_1), (a, c_2) \in \varphi$ and $(c_1, d_1), (c_2, d_2) \in \psi$. Since φ is a function, $c_1 = c_2$, and then since ψ is a function, $d_1 = d_2$. Therefore, $\psi \circ \varphi$ is a function, which is clearly $A \to D$. Furthermore,

$$
\begin{aligned}
\psi \circ \varphi &= \{(x, z) : \exists y [y \in C \wedge (x, y) \in \varphi \wedge (y, z) \in \psi]\} \\
&= \{(x, z) : \exists y [y \in C \wedge \varphi(x) = y \wedge \psi(y) = z]\} \\
&= \{(x, z) : x \in A \wedge \psi(\varphi(x)) = z\} \\
&= \{(x, \psi(\varphi(x))) : x \in A\}.
\end{aligned}
$$

Hence, $(\psi \circ \varphi)(x) = \psi(\varphi(x))$. ■

The $\mathrm{ran}(\varphi) \subseteq C$ condition is important to Theorem 4.4.20. For example, take the real-valued functions $f(x) = x$ and $g(x) = \sqrt{x}$. Since $f(-1) = -1$ but $g(-1) \notin \mathbf{R}$, we conclude that $(g \circ f)(-1)$ is undefined.

■ **EXAMPLE 4.4.21**

Define the two functions $f : \mathbf{R} \to \mathbf{Z}$ and $g : \mathbf{R} \setminus \{0\} \to \mathbf{R}$ by $f(x) = [\![x]\!]$ and $g(x) = 1/x$. Since $\mathrm{ran}(f) = \mathbf{Z} \nsubseteq \mathrm{dom}(g)$, there are elements of $\mathbf{R}$ for which $g \circ f$ is undefined. However,

$$
\mathrm{ran}(g) = \mathbf{R} \setminus \{0\} \subseteq \mathbf{R} = \mathrm{dom}(f),
$$

so $f \circ g$ is defined and for all $x \in \mathbf{R}$,

$$
(f \circ g)(x) = f(g(x)) = f(1/x) = [\![1/x]\!].
$$

■ **EXAMPLE 4.4.22**

Let $\psi : {}^{\mathbf{Z}}\mathbf{Z} \to \mathbf{Z}$ be defined by $\psi(f) = \epsilon_3(f)$ and also let $\varphi : \mathbf{Z} \to \mathbf{Z}_7$ be $\varphi(n) = [n]_7$. Since $\mathrm{ran}(\psi) \subseteq \mathrm{dom}(\varphi)$, $\varphi \circ \psi$ is defined. Thus, if $g : \mathbf{Z} \to \mathbf{Z}$ is defined as $g(n) = 3n$,

$$
(\varphi \circ \psi)(g) = \varphi(\psi(g)) = \varphi(g(3)) = \varphi(9) = [9]_7 = [2]_7.
$$

We should note that function composition is not a binary operation unless both functions are $A \to A$ for some set A. In this case, function composition is a binary operation on ${}^A A$.

Restrictions and Extensions

There are times when a subset of a given function is required. For example, consider

$$
f = \{(x, x^2) : x \in \mathbf{R}\}.
$$

If only positive values of x are required, we can define

$$g = \{(x, x^2) : x \in (0, \infty)\}$$

so that $g \subseteq f$. We have notation for this.

■ **DEFINITION 4.4.23**

Let $\varphi : A \to B$ be a function and $C \subseteq A$.

- The **restriction** of φ to C is the function $\varphi \restriction C : C \to B$ so that

$$(\varphi \restriction C)(x) = \varphi(x) \text{ for all } x \in C.$$

- The function $\psi : D \to E$ is an **extension** of φ if $A \subseteq D$, $B \subseteq E$, and $\psi \restriction A = \varphi$.

■ **EXAMPLE 4.4.24**

Let $f = \{(1, 2), (2, 3), (3, 4), (4, 1)\}$ and $g = \{(1, 2), (2, 3)\}$. We conclude that $g = f \restriction \{1, 2\}$, and f is an extension of g.

■ **EXAMPLE 4.4.25**

Let $\varphi : U \to V$ be a function and $A, B \subseteq U$. We conclude that

$$\varphi \restriction (A \cup B) = (\varphi \restriction A) \cup (\varphi \restriction B)$$

because

$$
\begin{aligned}
(x, y) \in \varphi \restriction (A \cup B) &\Leftrightarrow y = \varphi(x) \wedge x \in A \cup B \\
&\Leftrightarrow y = \varphi(x) \wedge (x \in A \vee x \in B) \\
&\Leftrightarrow y = \varphi(x) \wedge x \in A \vee y = \varphi(x) \wedge x \in B \\
&\Leftrightarrow (x, y) \in \varphi \restriction A \vee (x, y) \in \varphi \restriction B \\
&\Leftrightarrow (x, y) \in (\varphi \restriction A) \cup (\varphi \restriction B).
\end{aligned}
$$

Binary Operations

Standard addition and multiplication of real numbers are functions $\mathbf{R} \times \mathbf{R} \to \mathbf{R}$ (Example 4.4.4). This means two things. First, given any two real numbers, their sum or product will always be the same number. For instance, $3+5$ is 8 and never another number. Second, given any two real numbers, their sum or product is also a real number. Notice that subtraction also has these two properties when it is considered an operation involving real numbers, but when we restrict substraction to $\mathbf{Z}^+$, it no longer has the second property because the difference of two positive integers might not be a positive integer. That is, subtraction is not a function $\mathbf{Z}^+ \times \mathbf{Z}^+ \to \mathbf{Z}^+$.

■ **DEFINITION 4.4.26**

A **binary operation** $*$ on the nonempty set A is a function $A \times A \to A$.

The symbol that represents the addition function is $+$. It can be viewed as a function $\mathbf{R} \to \mathbf{R}$. Therefore, using function notation, $+(3, 5) = 8$. However, we usually write this as $3 + 5 = 8$. Similarly, since $*$ represents an operation like addition, instead of writing $*(a, b)$, we usually write $a * b$,

To prove that a relation $*$ is a binary operation on A, we must show that it satisfies Definition 4.4.26. To do this, take $a, a', b, b' \in A$, and prove:

- $a = a'$ and $b = b'$ implies $a * b = a' * b'$,

- A is **closed** under $*$, that is $a * b \in A$.

■ **EXAMPLE 4.4.27**

Define $x * y = 2x - y$ and take $a, a', b, b' \in \mathbf{Z}$.

- Assume $a = a'$ and $b = b'$. Then,

$$a * b = 2a - b = 2a' - b' = a' * b'.$$

  The second equality holds because multiplication and subtraction are binary operations on $\mathbf{Z}$.

- Because the product and difference of two integers is an integer, we have that $a * b \in \mathbf{Z}$, so $\mathbf{Z}$ is closed under $*$.

Thus, $*$ is a binary operation on $\mathbf{Z}$.

■ **EXAMPLE 4.4.28**

Let $S = \{e, a, b, c\}$ and define $*$ by the following table:

$*$	e	a	b	c
e	e	a	b	c
a	a	b	c	e
b	b	c	e	a
c	c	e	a	b

The table is read from left to right, so $b * c = a$. The table makes $*$ into a binary operation since every pair of elements of S is assigned a unique element of S.

■ **EXAMPLE 4.4.29**

Fix a set A. For any $X, Y \in P(A)$, define $X * Y = X \cup Y$.

- Let $X_1, X_2, Y_1, Y_2 \in P(A)$. If we assume that $X_1 = X_2$ and $Y_1 = Y_2$, we have $X_1 \cup Y_1 = X_2 \cup Y_2$. Hence, $*$ is well-defined.

- To show that $\mathbf{P}(A)$ is closed under $*$, let B and C be subsets of A. Then $B * C = B \cup C \in \mathbf{P}(A)$ because $B \cup C \subseteq A$ [Exercise 3.3.10(a)].

This shows that $*$ is a binary operation on $\mathbf{P}(A)$. Notice that $*$ is a subset of $[\mathbf{P}(A) \times \mathbf{P}(A)] \times \mathbf{P}(A)$.

■ **EXAMPLE 4.4.30**

Let $m \in \mathbf{Z}^+$ and define $[a]_m + [b]_m = [a + b]_m$. We show that this is a binary operation on $\mathbf{Z}_m$.

- Let $a_1, a_2, b_1, b_2 \in \mathbf{Z}$. Suppose $[a_1]_m = [a_2]_m$ and $[b_1]_m = [b_2]_m$. This means that $a_1 = a_2 + nk$ and $b_1 = b_2 + nl$ for some $k, l \in \mathbf{Z}$. Hence, $a_1 + b_1 = a_2 + b_2 + n(k + l)$, and we have $[a_1 + b_1]_m = [a_2 + b_2]_m$.

- For closure, let $[a]_m, [b]_m \in \mathbf{Z}_m$ where a and b are integers. Then, we have that $[a]_m + [b]_m = [a + b]_m \in \mathbf{Z}_m$ since $a + b$ is an integer.

Many binary operations share similar properties with the operations of $+$ and $\times$ on $\mathbf{R}$. The next definition gives four of these properties.

■ **DEFINITION 4.4.31**

Let $*$ be a binary operation on A.

- $*$ is **associative** means that $(a * b) * c = a * (b * c)$ for all $a, b, c \in A$.

- $*$ is **commutative** means that $a * b = b * a$ for all $a, b \in A$.

- The element e is an **identity** of A with respect to $*$ when $e \in A$ and $e * a = a * e = a$ for all $a \in A$.

- Suppose that A has an identity e with respect to $*$ and let $a \in A$. The element $a' \in A$ is an **inverse** of a with respect to $*$ if $a * a' = a' * a = e$.

Notice that the identity, if it exists, must be unique. To prove this, suppose that both e and e' are identities. These must be equal because $e = e * e' = e'$. So, if a set has an identity with respect to an operation, we can refer to it as *the* identity of the set. Similarly, we can write *the* inverse if it exists for associative binary operations (Exercise 20).

■ **EXAMPLE 4.4.32**

We assume that $+$ and $\times$ are both associative and commutative on $\mathbf{C}$ and all subsets of $\mathbf{C}$, that 0 is the identity with respect to $+$ (the **additive identity**) and 1 is the identity with respect to $\times$ (the **multiplicative identity**), and that every complex number has an inverse with respect to $+$ (an **additive inverse**) and every nonzero complex number has an inverse with respect to $\times$ (a **multiplicative inverse**).

■ **EXAMPLE 4.4.33**

The binary operation defined in Example 4.4.30 is both associative and commutative. To see that it is commutative, let $a, b \in \mathbf{Z}$. Then,

$$[a]_m + [b]_m = [a + b]_m = [b + a]_m = [b]_m + [a]_m, \tag{4.6}$$

where the second equality holds because $+$ is commutative on $\mathbf{Z}$. Its identity is $[0]_m$ [let $b = 0$ in (4.6)] and the additive inverse of $[a]_m$ is $[-a]_m$.

■ **EXAMPLE 4.4.34**

Since $A \cup B = B \cup A$ and $(A \cup B) \cup C = A \cup (B \cup C)$ for all sets A, B, and C, the binary operation in Example 4.4.29 is both associative and commutative. Its identity is $\varnothing$, and only $\varnothing$ has an inverse.

Exercises

1. Indicate whether each of the given relations are functions. If a relation is not a function, find an element of its domain that is paired with two elements of its range.
   (a) $\{(1, 2), (2, 3), (3, 4), (4, 5), (5, 1)\}$
   (b) $\{(1, 1), (1, 2), (1, 3), (1, 4), (1, 5)\}$
   (c) $\{(x, \sqrt{|x|}) : x \in \mathbf{R}\}$
   (d) $\{(x, \pm\sqrt{|x|}) : x \in \mathbf{R}\}$
   (e) $\{(x, x^2) : x \in \mathbf{R}\}$
   (f) $\{([a]_5, b) : \exists k \in \mathbf{Z}(a = b + 5k)\}$
   (g) $\psi : \mathbf{Z} \to \mathbf{Z}_5$ if $\psi(a) = [a]_5$

2. Prove that the given relations are functions.
   (a) $\{(x, 1/x) : x \in \mathbf{R} \setminus \{0\}\}$
   (b) $\{(x, x + 1) : x \in \mathbf{Z}\}$
   (c) $\{(x, |x|) : x \in \mathbf{R}\}$
   (d) $\{(x, \sqrt{x}) : x \in [0, \infty)\}$

3. Let $f = \{(x, y) \in \mathbf{R}^2 : 2x + y = 1\}$. Show that f is a function with domain and codomain equal to the set of real numbers.

4. Let $f, g : \mathbf{R} \to \mathbf{R}$ be functions. Prove that $\varphi(x, y) = (f(x), g(y))$ is a function with domain and codomain equal to $\mathbf{R} \times \mathbf{R}$.

5. Let A be a set and define $\psi(A) = \mathbf{P}(A)$. Show that ψ is a function.

6. Define

$$f(x) = \begin{cases} x^2 & \text{if } x \geq 0, \\ 5 & \text{if } x < 0. \end{cases}$$

Show that f is well-defined with domain equal to $\mathbf{R}$. What is ran(f)?

7. Let x be in the domain and y in the range of each relation. Explain why each of the given equations does not describe a function.
    (a) $y = 5 \pm x$
    (b) $x^2 + y^2 = 1$
    (c) $x = 4y^2 - 1$
    (d) $y^2 - x^2 = 9$

8. Let $\varphi : \mathbf{Z} \to \mathbf{Z}_7$ be defined by $\varphi(a) = [a]_7$. Write the given images as rosters.
    (a) $\varphi(0)$
    (b) $\varphi(7)$
    (c) $\varphi(3)$
    (d) $\varphi(-3)$

9. Define $\varphi([a]_n) = [a]_m$ for all $a \in \mathbf{Z}$. Is φ being function sufficient for $m \mid n$? Explain.

10. Give an example of a function that is an element of the given sets.
    (a) $^{\mathbf{R}}\mathbf{R}$
    (b) $^{\mathbf{R}}\mathbf{Z}$
    (c) $^{\mathbf{N}}\mathbf{R}$
    (d) $^{\mathbf{R}}[0, \infty)$
    (e) $^{\mathbf{Z}}(\mathbf{Z}_5)$

11. Evaluate the indicated expressions.
    (a) $\epsilon_4(f)$ if $f(x) = 9x + 2$
    (b) $\epsilon_\pi(g)$ if $g(\theta) = \sin \theta$

12. Since functions are sets, we can perform set operations on them. Let $f(x) = x^2$ and $g(x) = -x$. Find the following.
    (a) $f \cup g$
    (b) $f \cap g$
    (c) $f \setminus g$
    (d) $g \setminus f$

13. Let $\mathscr{C}$ be a chain of functions with respect to $\subseteq$. Prove that $\bigcup \mathscr{C}$ is a function.

14. Let f and g be functions. Prove the following.
    (a) If f and g are functions, $f \cap g$ is a function.
    (b) $f \cup g$ is a function if and only if $f(x) = g(x)$ for all $x \in \mathrm{dom}(f) \cap \mathrm{dom}(g)$.

15. Let $f : A \to B$ be a function. Define a relation S on A by $a\, S\, b$ if and only if $f(a) = f(b)$.
    (a) Show S is an equivalence relation.
    (b) Find $[3]_S$ if $f : \mathbf{Z} \to \mathbf{Z}$ is defined by $f(n) = 2n$.
    (c) Find $[2]_S$ if $f : \mathbf{Z} \to \mathbf{Z}_5$ is given by $f(n) = [n]_5$.

16. Show that $\varnothing$ is a function and find its domain and range.

17. Define $*$ by $x * y = x + y + 2$ for all $x, y \in \mathbf{Z}$.
    (a) Show that $*$ is a binary operation on $\mathbf{Z}$.
    (b) Prove that -2 is the identity of $\mathbf{Z}$ with respect to $*$.
    (c) For every $n \in \mathbf{Z}$, show that $-n - 4$ is the inverse of n with respect to $*$.

18. Define the binary operation $*$ by $x * y = 2x - y$ for all $x, y \in \mathbf{Z}$.
    (a) Is there an integer that serves as an identity with respect to $*$?
    (b) Does every integer have an inverse with respect to $*$?

19. Let $f, g : \mathbf{R} \to \mathbf{R}$ be functions. Prove that $f \circ g$ is well-defined.

20. For an associative binary operation, prove that the inverse of an element is unique if it exists. Show that this might not be the case if the binary operation is not associative.

21. Prove that the given pairs of functions are equal.
    (a) $f(x) = (x - 1)(x - 2)(x + 3)$ and $g(x) = x^3 - 7x + 6$
        where $f, g : \mathbf{R} \to \mathbf{R}$
    (b) $\varphi(a, b) = a + b$ and $\psi(a, b) = b + a$
        where $\varphi, \psi : \mathbf{Z} \times \mathbf{Z} \to \mathbf{Z}$
    (c) $\varphi(a, b) = ([a]_5, [b + 7]_5)$ and $\psi(a, b) = ([a + 5]_5, [b - 3]_5)$
        where $\varphi, \psi : \mathbf{Z} \times \mathbf{Z} \to \mathbf{Z}_5 \times \mathbf{Z}_5$
    (d) $\varphi(f) = f \restriction \mathbf{Z}$ and $\psi(f) = \{(n, f(n)) : n \in \mathbf{Z}\}$
        where $\varphi, \psi : {}^{\mathbf{R}}\mathbf{R} \to {}^{\mathbf{Z}}\mathbf{R}$

22. Show that the given pairs of functions are not equal.
    (a) $f(x) = x$ and $g(x) = 2x$ where $f, g : \mathbf{R} \to \mathbf{R}$
    (b) $f(x) = x - 3$ and $g(x) = x + 3$ where $f, g : \mathbf{R} \to \mathbf{R}$
    (c) $\varphi(a) = [a]_5$ and $\psi(a) = [a]_4$ where $\varphi, \psi : \mathbf{Z} \to \mathbf{Z}_4 \cup \mathbf{Z}_5$
    (d) $\varphi(A) = A \setminus \{0\}$ and $\psi(A) = A \cap \{1, 2, 3\}$ where $\varphi, \psi : P(\mathbf{Z}) \to P(\mathbf{Z})$

23. Let $\psi : \mathbf{R} \to {}^{\mathbf{R}}\mathbf{R}$ be defined by $\psi(a) = f_a$ where f_a is the function $f_a : \mathbf{R} \to \mathbf{R}$ with $f_a(x) = ax$. Prove that ψ is well-defined.

24. For each pair of functions, find the indicated values when possible.
    (a) $f : \mathbf{R} \to \mathbf{R}$ and $f(x) = 2x^3$
        $g : \mathbf{R} \to \mathbf{R}$ and $g(x) = x + 1$
        $(f \circ g)(2)$
        $(g \circ f)(0)$
    (b) $f : [0, \infty) \to \mathbf{R}$ and $f(x) = \sqrt{x}$
        $g : \mathbf{R} \to \mathbf{R}$ and $g(x) = |x| - 1$
        $(f \circ g)(0)$
        $(g \circ f)(4)$
    (c) $\varphi : \mathbf{Z} \to \mathbf{Z}_5$ and $\varphi(a) = [a]_5$
        $\psi : {}^{\mathbf{R}}\mathbf{R} \to \mathbf{R}$ and $\psi(f) = f(0)$
        $(\varphi \circ \psi)(.5x + 1)$
        $(\psi \circ \varphi)(2)$

25. For each of the given functions, find the composition of the function with itself. For example, find $f \circ f$ for part (a).

    (a) $f : \mathbf{R} \to \mathbf{R}$ with $f(x) = x^2$
    (b) $g : \mathbf{R} \to \mathbf{R}$ with $g(x) = 3x + 1$
    (c) $\varphi : \mathbf{Z} \times \mathbf{Z} \to \mathbf{Z} \times \mathbf{Z}$ with $\varphi(x, y) = (2y, 5x - y)$
    (d) $\psi : \mathbf{Z}_m \to \mathbf{Z}_m$ with $\psi([n]_m) = [n + 2]_m$

26. Let $\varphi : A \to B$ be a function and $\psi = \varphi \restriction C$ where $C \subseteq A$. Prove that if $\iota : C \to A$ is defined by $\iota(c) = c$ (known as the **inclusion map**), then $\psi = \varphi \circ \iota$.

27. Write the given restrictions as rosters.

    (a) $\{(1, 2), (2, 2), (3, 4), (4, 7)\} \restriction \{1, 3\}$
    (b) $f \restriction \{0, 1, 2, 3\}$ where $f(x) = 7x - 1$ and $\mathrm{dom}(f) = \mathbf{R}$
    (c) $(g + h) \restriction \{-3.3, 1.2, 7\}$ where $g(x) = [\![x]\!]$, $h(x) = x + 1$, and both $\mathrm{dom}(g)$ and $\mathrm{dom}(h)$ equal $\mathbf{R}$

28. For functions f and g such that $A, B \subseteq \mathrm{dom}(f)$, prove the following.

    (a) $f \restriction A = f \cap [A \times \mathrm{ran}(f)]$
    (b) $f \restriction (A \cap B) = (f \restriction A) \cap (f \restriction B)$
    (c) $f \restriction (A \setminus B) = (f \restriction A) \setminus (f \restriction B)$
    (d) $(g \circ f) \restriction A = g \circ (f \restriction A)$

29. Let $f : U \to V$ be a function. Prove that if $A \subseteq U$, then $f \restriction A = f \circ I_A$.

30. Let $\varphi : {}^A C \to {}^B C$ be defined by $\varphi(f) = f \restriction B$. Prove that φ is well-defined.

31. A real-valued function f is **periodic** if there exists $k > 0$ so that $f(x) = f(x + k)$ for all $x \in \mathrm{dom}(f)$. Let $g, h : \mathbf{R} \to \mathbf{R}$ be functions with period k. Prove that $g \circ h$ is periodic with period k.

32. Let $(A, \preccurlyeq)$ be a poset. A function $\varphi : A \to A$ is **increasing** means for all $x, y \in A$, if $x \prec y$, then $\varphi(x) \prec \varphi(y)$. A **decreasing** function is defined similarly. Suppose that σ and τ are increasing. Prove that $\sigma \circ \tau$ is increasing.

4.5 INJECTIONS AND SURJECTIONS

When looking at relations, we studied the concept of an inverse relation. Given R, obtain R^{-1} by exchanging the x- and y-coordinates. The same can be done with functions, but the inverse might not be a function. For example, given

$$f = \{(1, 2), (2, 3), (3, 2)\},$$

its inverse is

$$f^{-1} = \{(2, 1), (3, 2), (2, 3)\}.$$

However, if the original relation is a function, we often want the inverse also to be a function. This leads to the next definition.

■ **DEFINITION 4.5.1**

$\varphi : A \to B$ is **invertible** means that φ^{-1} is a function $B \to A$.

An immediate consequence of the definition is the next result.

■ **LEMMA 4.5.2**

Let φ be invertible. Then, $\varphi(x) = y$ if and only if $\varphi^{-1}(y) = x$ for all $x \in \text{dom}(\varphi)$.

PROOF
Suppose that $\varphi(x) = y$. This means that $(x, y) \in \varphi$, so $(y, x) \in \varphi^{-1}$. Since φ is invertible, φ^{-1} is a function, so write $\varphi^{-1}(y) = x$. The converse is proved similarly. ■

We use Lemma 4.5.2 in the proof of the next theorem, which gives conditions for when a function is invertible.

■ **THEOREM 4.5.3**

$\varphi : A \to B$ is invertible if and only if $\varphi^{-1} \circ \varphi = I_A$ and $\varphi \circ \varphi^{-1} = I_B$.

PROOF
Take a function $\varphi : A \to B$. Then, $\varphi^{-1} \subseteq B \times A$.

- Assume that φ^{-1} is a function $B \to A$. Let $x \in A$ and $y \in B$. By assumption, we have $x_0 \in A$ and $y_0 \in B$ such that $\varphi(x) = y_0$ and $\varphi^{-1}(y) = x_0$. This implies that $\varphi^{-1}(y_0) = x$ and $\varphi(x_0) = y$ by Lemma 4.5.2. Therefore,

$$(\varphi^{-1} \circ \varphi)(x) = \varphi^{-1}(\varphi(x)) = \varphi^{-1}(y_0) = x,$$

and

$$(\varphi \circ \varphi^{-1})(y) = \varphi(\varphi^{-1}(y)) = \varphi(x_0) = y.$$

- Now assume $\varphi^{-1} \circ \varphi = I_A$ and $\varphi \circ \varphi^{-1} = I_B$. To show that φ^{-1} is a function, take $(y, x), (y, x') \in \varphi^{-1}$. From this, we know that $(x, y) \in \varphi$. Therefore,

$$(x, x') \in \varphi^{-1} \circ \varphi = I_A,$$

so $x = x'$. In addition, we know that $\text{dom}(\varphi^{-1}) \subseteq B$, so to prove equality, let $y \in B$. Then,

$$(y, y) \in I_B = \varphi \circ \varphi^{-1}.$$

Thus, there exists $x \in A$ such that $(y, x) \in \varphi^{-1}$, so $y \in \text{dom}(\varphi^{-1})$. ■

■ **EXAMPLE 4.5.4**

- Let $f : \mathbf{R} \to \mathbf{R}$ be the function given by $f(x) = x + 2$. Its inverse is $g(x) = x - 2$ by Theorem 4.5.3. This is because

$$(g \circ f)(x) = g(x + 2) = (x + 2) - 2 = x$$

and

$$(f \circ g)(x) = f(x - 2) = (x - 2) + 2 = x.$$

- If the function $g : [0, \infty) \rightarrow [0, \infty)$ is defined by $g(x) = x^2$, then g^{-1} is a function $[0, \infty) \rightarrow [0, \infty)$ and is defined by $g^{-1}(x) = \sqrt{x}$.

- Let $h : \mathbf{R} \rightarrow (0, \infty)$ be defined as $h(x) = e^x$. By Theorem 4.5.3, we know that $h^{-1}(x) = \ln x$ because $e^{\ln x} = x$ for all $x \in [0, \infty)$ and $\ln e^x = x$ for all $x \in \mathbf{R}$.

Injections

Theorem 4.5.3 can be improved by finding a condition for the invertibility of a function based only on the given function. Consider the following. In order for a relation to be a function, it cannot look like Figure 4.9. Since the inverse exchanges the roles of the two coordinates, in order for an inverse to be a function, the original function cannot look like the graph in Figure 4.11. In other words, if f^{-1} is to be a function, there cannot exist x_1 and x_2 so that $x_1 \neq x_2$ and $f(x_1) = f(x_2)$. But then,

$$\neg \exists x_1 \exists x_2 [x_1 \neq x_2 \wedge f(x_1) = f(x_2)]$$

is equivalent to

$$\forall x_1 \forall x_2 [x_1 = x_2 \vee f(x_1) \neq f(x_2)],$$

which in turn is equivalent to

$$\forall x_1 \forall x_2 [f(x_1) = f(x_2) \rightarrow x_1 = x_2].$$

Hence, f being an invertible function implies that for every $x_1, x_2 \in \text{dom}(f)$,

$$x_1 = x_2 \text{ if and only if } f(x_1) = f(x_2). \tag{4.7}$$

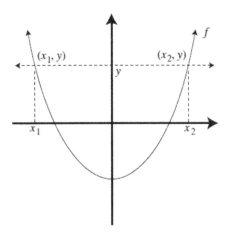

Figure 4.11   The inverse of a function might not be a function.

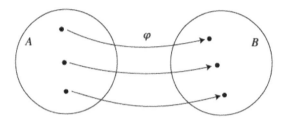

Figure 4.12 φ is a one-to-one function.

This means that the elements of the domain of f and the elements of the range of f form pairs of elements as illustrated in Figure 4.12. The sufficiency of (4.7) is the definition of a function, while necessity is the next definition.

■ **DEFINITION 4.5.5**

The function $\varphi : A \to B$ is **one-to-one** if and only if for all $x_1, x_2 \in A$,

$$\text{if } \varphi(x_1) = \varphi(x_2), \text{ then } x_1 = x_2.$$

A one-to-one function is sometimes called an **injection**.

■ **EXAMPLE 4.5.6**

Define $f : \mathbf{R} \to \mathbf{R}$ by $f(x) = 5x+1$. To show that f is one-to-one, let $x_1, x_2 \in \mathbf{R}$ and assume $f(x_1) = f(x_2)$. Then,

$$5x_1 + 1 = 5x_2 + 1,$$
$$5x_1 = 5x_2,$$
$$x_1 = x_2.$$

■ **EXAMPLE 4.5.7**

Let $\varphi : \mathbf{Z} \times \mathbf{Z} \to \mathbf{Z} \times \mathbf{Z} \times \mathbf{Z}$ be the function

$$\varphi(a, b) = (a, b, 0).$$

For any $(a_1, b_1), (a_2, b_2) \in \mathbf{Z} \times \mathbf{Z}$, assume

$$\varphi(a_1, b_1) = \varphi(a_2, b_2).$$

This means that

$$(a_1, b_1, 0) = (a_2, b_2, 0).$$

Hence, $a_1 = a_2$ and $b_1 = b_2$, and this yields $(a_1, b_1) = (a_2, b_2)$.

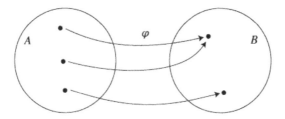

Figure 4.13 φ is not a one-to-one function.

If a function is not one-to-one, there must be an element of the range that has at least two pre-images (Figure 4.13). An example of a function that is not one-to-one is $f(x) = x^2$ where both the domain and codomain of f are **R**. This is because $f(2) = 4$ and $f(-2) = 4$. Another example is $g : \mathbf{R} \to \mathbf{R}$ defined by $g(\theta) = \cos\theta$. It is not an injection because $g(0) = g(2\pi) = 1$.

Although the original function might not be one-to-one, we can always restrict the function to a subset of its domain so that the resulting function is one-to-one. This is illustrated in the next two examples.

■ **EXAMPLE 4.5.8**

Let f be the function $\{(1,5),(2,8),(3,8),(4,6)\}$. We observe that f is not one-to-one, but both

$$f \upharpoonright \{1,2\} = \{(1,5),(2,8)\}$$

and

$$f \upharpoonright \{3,4\} = \{(3,8),(4,6)\}$$

are one-to-one as in Figure 4.14.

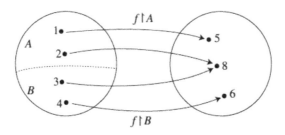

Figure 4.14   Restrictions of f to A and B are one-to-one.

■ **EXAMPLE 4.5.9**

Let $g : \mathbf{R} \to \mathbf{R}$ be the function $g(x) = x^2$. This function is not one-to-one, but $g \upharpoonright [0, \infty)$ and $g \upharpoonright (-10, -5)$ are one-to-one.

Let $f(x) = 3x + 6$ and $g(x) = \sqrt{5x - 8}$. Both are injections. Notice that

$$(f \circ g)(x) = 3\sqrt{5x - 8} + 6$$

and

$$(g \circ f)(x) = \sqrt{15x + 22}$$

are also injections. We can generalize this result to the next theorem.

■ **THEOREM 4.5.10**

If $\varphi : A \to B$ and $\psi : B \to C$ are injections, $\psi \circ \varphi$ is an injection.

PROOF

Assume that $\varphi : A \to B$ and $\psi : B \to C$ are one-to-one. Let $a_1, a_2 \in A$ and assume $(\psi \circ \varphi)(a_1) = (\psi \circ \varphi)(a_2)$. Then,

$$\psi(\varphi(a_1)) = \psi(\varphi(a_2)).$$

Since ψ is one-to-one,

$$\varphi(a_1) = \varphi(a_2),$$

and since φ is one-to-one, $a_1 = a_2$. ■

Surjections

The function being an injection is not sufficient for it to be invertible since it is possible that not every element of the codomain will have a pre-image. In this case, the codomain cannot be the domain of the inverse. To prevent this situation, we will need the function to satisfy the next definition.

■ **DEFINITION 4.5.11**

A function $\varphi : A \to B$ is **onto** if and only if for every $y \in B$, there exists $x \in A$ such that $\varphi(x) = y$. An onto function is also called a **surjection**.

This definition is related to the range (or **image**) of the function. The range of the function $\varphi : A \to B$ is

$$\text{ran}(\varphi) = \{y : \exists x(x \in A \land (x, y) \in \varphi)\} = \{\varphi(x) : x \in \text{dom}(\varphi)\}$$

as illustrated in Figure 4.15. Thus, φ is onto if and only if $\text{ran}(\varphi) = B$.

■ **EXAMPLE 4.5.12**

Define $f : [0, \infty) \to [0, \infty)$ by $f(x) = \sqrt{x}$. Its range is also $[0, \infty)$, so it is onto.

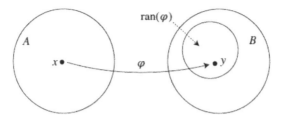

Figure 4.15   The range of $\varphi : A \to B$ with $y = \varphi(x)$.

■ **EXAMPLE 4.5.13**

The ranges of the following functions are different from their codomains, so they are not onto.

- Let $g : \mathbf{R} \to \mathbf{R}$ be defined by $g(x) = |x|$. Then, $\mathrm{ran}(g) = [0, \infty)$.
- Define $h : \mathbf{Z} \to \mathbf{Z}$ by $h(n) = 2n$. Here, $\mathrm{ran}(h) = \{2n : n \in \mathbf{Z}\}$.

The functions illustrated in Figures 4.12, 4.13, and 4.14 are onto functions as are those in the next examples.

■ **EXAMPLE 4.5.14**

Any linear function $f : \mathbf{R} \to \mathbf{R}$ that is not a horizontal line is a surjection. To see this, let $f(x) = ax + b$ for some $a \neq 0$. Take $y \in \mathbf{R}$. We need to find $x \in \mathbf{R}$ so that $ax + b = y$. Choose

$$x = \frac{y - b}{a}.$$

Then,

$$f(x) = a\left(\frac{y - b}{a}\right) + b = y.$$

The approach in the example is typical. To show that a function is onto, take an arbitrary element of the codomain and search for a candidate to serve as its pre-image. When found, check it.

■ **EXAMPLE 4.5.15**

Take a positive integer m and let $\varphi : \mathbf{Z} \to \mathbf{Z}_m$ be defined as $\varphi(k) = [k]_m$. To see that φ is onto, take $[l]_m \in \mathbf{Z}_m$ for some $l \in \mathbf{Z}$. We then find that $\varphi(l) = [l]_m$.

■ **EXAMPLE 4.5.16**

Let $m, n \in \mathbf{N}$ with $m > n$. A function $\pi : \mathbf{R}^m \to \mathbf{R}^n$ defined by

$$\pi(x_0, x_1, \dots, x_{n-1}, x_n, \dots, x_{m-1}) = (x_0, x_1, \dots, x_{n-1})$$

is called a **projection.**. Such functions are not one-to-one, but they are onto. For instance, define $\pi : \mathbf{R} \times \mathbf{R} \times \mathbf{R} \to \mathbf{R} \times \mathbf{R}$ by

$$\pi(x, y, z) = (x, y).$$

It is not one-to-one because $\pi(1, 2, 3) = \pi(1, 2, 4) = (1, 2)$. However, if we take $(a, b) \in \mathbf{R} \times \mathbf{R}$, then $\pi(a, b, 0) = (a, b)$, so π is onto.

If a function is not onto, it has a diagram like that of Figure 4.16. Therefore, to show that a function is not a surjection, we must find an element of the codomain that does not have a pre-image.

■ **EXAMPLE 4.5.17**

Define $f : \mathbf{Z} \to \mathbf{Z}$ by $f(n) = 3n$. This function is not onto because 5 does not have a pre-image in $\mathbf{Z}$.

■ **EXAMPLE 4.5.18**

The function

$$\varphi : \mathbf{Z} \times \mathbf{Z} \to \mathbf{Z} \times \mathbf{Z} \times \mathbf{Z}$$

defined by $\varphi(a, b) = (a, b, 0)$ is not onto because $(1, 1, 1)$ does not have a pre-image in $\mathbf{Z} \times \mathbf{Z}$.

We have the following analog of Theorem 4.5.10 for surjections.

■ **THEOREM 4.5.19**

If $\varphi : A \to B$ and $\psi : B \to C$ are surjections, $\psi \circ \varphi$ is a surjection.

PROOF

Assume that $\varphi : A \to B$ and $\psi : B \to C$ are surjections. Take $c \in C$. Then, there exists $b \in B$ so that $\psi(b) = c$ and $a \in A$ such that $\varphi(a) = b$. Therefore,

$$(\psi \circ \varphi)(a) = \psi(\varphi(a)) = \psi(b) = c. ∎$$

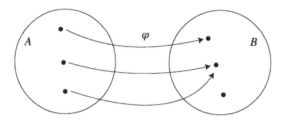

Figure 4.16 φ is not a onto function.

Bijections

If we have a function that is both one-to-one and onto, then it is called a **bijection** or a **one-to-one correspondence**. Observe that $\varnothing$ is a bijection.

■ EXAMPLE 4.5.20

As illustrated in Examples 4.5.6 and 4.5.14, every linear $f : \mathbf{R} \to \mathbf{R}$ with nonzero slope is a bijection.

■ EXAMPLE 4.5.21

Both $g : (-\pi/2, \pi/2) \to \mathbf{R}$ such that $g(\theta) = \tan\theta$ and $h : \mathbf{R} \to (0, \infty)$ where $h(x) = e^x$ are bijections.

We are now ready to give the standard test for invertibility. Its proof requires both Lemma 4.5.2 and Theorem 4.5.3. The benefit of this theorem is that it provides a test for invertibility in which the given function is examined instead of its inverse.

■ THEOREM 4.5.22

A function is invertible if and only if it is a bijection.

PROOF

Let $\varphi : A \to B$ be a function.

- Suppose that φ is invertible. To show that φ is one-to-one, let $x_1, x_2 \in A$ and assume that $\varphi(x_1) = \varphi(x_2)$. Then, by Theorem 4.5.3,

$$x_1 = \varphi^{-1}(\varphi(x_1)) = \varphi^{-1}(\varphi(x_2)) = x_2.$$

  To see that φ is onto, take $y \in B$. Then, there exists $x \in A$ such that $\varphi^{-1}(y) = x$. Hence, $\varphi(x) = y$ by Lemma 4.5.2.

- Assume that φ is both one-to-one and onto. To show that φ^{-1} is a function, let $(y, x), (y, x') \in \varphi^{-1}$. This implies that $\varphi(x) = y = \varphi(x')$. Since φ is one-to-one, $x = x'$. To prove that the domain of φ^{-1} is B, take $y \in B$. Since φ is onto, there exists $x \in A$ such that $\varphi(x) = y$. By Lemma 4.5.2, we have that $\varphi^{-1}(y) = x$, so $y \in \text{dom}(\varphi^{-1})$. ■

By Theorem 4.5.22 the functions of Examples 4.5.20 and 4.5.21 are invertible.

■ THEOREM 4.5.23

If $\varphi : A \to B$ and $\psi : B \to C$ are bijections, then φ^{-1} and $\psi \circ \varphi$ are bijections.

PROOF

Suppose φ is a bijection. By Theorem 4.5.22, it is invertible, so φ^{-1} is a function that has φ as its inverse. Therefore, φ^{-1} is a bijection by Theorem 4.5.22. Combining the proofs of Theorems 4.5.10 and 4.5.19 show that $\psi \circ \varphi$ is a bijection when ψ is a bijection. ■

Using the functions g and h from Example 4.5.21, we conclude from Theorem 4.5.23 that $h \circ g$ is a bijection with domain $(-\pi/2, \pi/2)$ and range $(0, \infty)$.

Order Isomorphims

Consider the function

$$\varphi : \mathbf{Z} \times \{0\} \to \{0\} \times \mathbf{Z} \tag{4.8}$$

defined by $\varphi(m, 0) = (0, m)$. It can be shown that φ is a bijection (compare Exercise 12). Define the linear orders $\preccurlyeq$ on $\mathbf{Z} \times \{0\}$ and $\preccurlyeq'$ on $\{0\} \times \mathbf{Z}$ by

$$(m, 0) \preccurlyeq (n, 0) \text{ if and only if } m \le n$$

and

$$(0, m) \preccurlyeq' (0, n) \text{ if and only if } m \le n.$$

Notice that $(3, 0) \preccurlyeq (5, 0)$ and $(0, 3) \preccurlyeq' (0, 5)$ because $3 \le 5$. We can generalize this to conclude that

$$(m, 0) \preccurlyeq (n, 0) \text{ if and only if } (0, m) \preccurlyeq' (0, n),$$

and this implies that

$$(m, 0) \preccurlyeq (n, 0) \text{ if and only if } \varphi(m, 0) \preccurlyeq' \varphi(n, 0).$$

This leads to the next definition.

■ DEFINITION 4.5.24

Let R be a relation on A and S be a relation on B.

- $\varphi : A \to B$ is an **order-preserving** function if for all $a_1, a_2 \in A$,

$$(a_1, a_2) \in R \text{ if and only if } (\varphi(a_1), \varphi(a_2)) \in S$$

  and we say that φ **preserves** R with S.

- An order-preserving bijection is an **order isomorphism**.

- (A, R) is **order isomorphic** to (B, S) and we write $(A, R) \cong (B, S)$ if there exists an order isomorphism $\varphi : A \to B$ preserving R with S. If $(A, R) \cong (B, S)$ and the relations R and S are clear from context, we can write $A \cong B$. Sometimes (A, R) and (B, S) are said to have the same **order type** when they are order isomorphic.

An isomorphism pairs elements from two sets in such a way that the orders on the two sets appear to be the same.

■ **EXAMPLE 4.5.25**

Define $f : \mathbf{R}^+ \to \mathbf{R}^-$ by $f(x) = -x$ (Example 3.1.12). Clearly, f is a bijection. Moreover, f preserves $\leq$ with $\geq$. To prove this, let $x_1, x_2 \in \mathbf{R}^+$. Then,

$$x_1 \leq x_1 \Leftrightarrow -x_1 \geq -x_2 \Leftrightarrow f(x_1) \geq f(x_2).$$

Therefore, $(\mathbf{R}^+, \leq) \cong (\mathbf{R}^-, \geq)$.

Observe that the inverse of (4.8) is the function

$$\varphi^{-1} : \{0\} \times \mathbf{Z} \to \mathbf{Z} \times \{0\}$$

such that $\varphi^{-1}(0, m) = (m, 0)$. This function preserves $\preccurlyeq'$ with $\preccurlyeq$. This result is generalized and proved in the next theorem.

■ **THEOREM 4.5.26**

The inverse of an order isomorphism preserving R with S is an order isomorphism preserving S with R.

PROOF

Let $\varphi : A \to B$ be an order isomorphism preserving R with S. By Theorem 4.5.23, φ^{-1} is a bijection. Suppose that $(b_1, b_2) \in S$. Since φ is onto, there exists $a_1, a_2 \in A$ such that $\varphi(a_1) = b_1$ and $\varphi(a_2) = b_2$. This implies that $(\varphi(a_1), \varphi(a_2)) \in S$. Since φ is an isomorphism preserving R with S, we have that $(a_1, a_2) \in R$. However, $a_1 = \varphi^{-1}(b_1)$ and $a_2 = \varphi^{-1}(b_2)$, so $(\varphi^{-1}(b_1), \varphi^{-1}(b_2)) \in R$. Therefore, $\varphi^{-1} : B \to A$ is an isomorphism preserving S with R. ■

Also, observe that $g : \mathbf{R} \to (0, \infty)$ defined by $g(x) = e^x$ is an order isomorphism preserving $\leq$ with $\leq$. Using f from Example 4.5.25, the composition $f \circ g$ is an order isomorphism $\mathbf{R} \to (-\infty, 0)$ such that $(f \circ g)(x) = -e^x$. That this happens in general is the next theorem. Its proof is left to Exercise 25.

■ **THEOREM 4.5.27**

If $\varphi : A \to B$ is an isomorphism preserving R with S and $\psi : B \to C$ is an isomorphism preserving S with T, then $\psi \circ \varphi : A \to C$ is an isomorphism preserving R with T.

■ **EXAMPLE 4.5.28**

Let R be a relation on A, S be a relation on B, and T be a relation on C.

- Since the identity map is an order isomorphism, $(A, R) \cong (A, R)$.

- Suppose $(A, R) \cong (B, S)$. This means that there exists an order isomorphism $\varphi : A \to B$ preserving R with S. By Theorem 4.5.26, φ^{-1} is an order isomorphism preserving S with R. Therefore, $(B, S) \cong (A, R)$.

- Let $(A, R) \cong (B, S)$ and $(B, S) \cong (C, T)$. By Theorem 4.5.27, we conclude that $(A, R) \cong (C, T)$.

If an order-preserving function is one-to-one, even it is not a surjection, the function still provides an order isomorphism between its domain and range. This concept is named by the next definition.

■ **DEFINITION 4.5.29**

$\varphi : A \to B$ is an **embedding** if φ is an order isomorphism $A \to \mathrm{ran}(\varphi)$.

For example, $f : \mathbf{Z} \to \mathbf{Q}$ such that $f(n) = n$ is an embedding preserving $\leq$ and $\pi : \mathbf{R}^2 \to \mathbf{R}^3$ such that $\psi(x, y) = (x, y, 0)$ is an embedding preserving the lexicographical order (Exercise 4.3.16). Although $\mathbf{R}^2$ is not a subset of $\mathbf{R}^3$, we view the image of ψ as a **copy** of $\mathbf{R}^2$ in $\mathbf{R}^3$ that preserves the orders.

Exercises

1. Show that the given pairs of functions are inverses.
    (a) $f(x) = 3x + 2$ and $g(x) = \frac{1}{3}x - \frac{2}{3}$
    (b) $\varphi(a, b) = (2a, b + 2)$ and $\psi(a, b) = (\frac{1}{2}a, b - 2)$
    (c) $f(x) = a^x$ and $g(x) = \log_a x$, where $a > 0$

2. For each function, graph the indicated restriction.
    (a) $f \upharpoonright (0, \infty)$, $f(x) = x^2$
    (b) $g \upharpoonright [-5, -2]$, $g(x) = |x|$
    (c) $h \upharpoonright [0, \pi/2]$, $h(x) = \cos x$

3. Prove that the given functions are one-to-one.
    (a) $f : \mathbf{R} \to \mathbf{R}$, $f(x) = 2x + 1$
    (b) $g : \mathbf{R}^2 \to \mathbf{R}^2$, $g(x, y) = (3y, 2x)$
    (c) $h : \mathbf{R} \setminus \{9\} \to \mathbf{R} \setminus \{0\}$, $h(x) = 1/(x - 9)$
    (d) $\varphi : \mathbf{Z} \times \mathbf{R} \to \mathbf{Z} \times (0, \infty)$, $\varphi(n, x) = (3n, e^x)$
    (e) $\psi : P(A) \to P(B)$, $\psi(C) = C \cup \{b\}$ where $A \subset B$ and $b \in B \setminus A$

4. Let $f : (a, b) \to (c, d)$ be defined by

$$f(x) = \frac{d - c}{b - a}(x - a) + c.$$

Graph f and show that it is a bijection.

5. Let f and g be functions such that $\mathrm{ran}(g) \subseteq \mathrm{dom}(f)$.
    (a) Prove that if $f \circ g$ is one-to-one, g is one-to-one.
    (b) Give an example of functions f and g such that $f \circ g$ is one-to-one, but f is not one-to-one.

6. Define $\varphi : \mathbf{Z} \to \mathbf{Z}_m$ by $\varphi(k) = [k]_m$. Show that φ is not one-to-one.

7. Show that the given functions are not one-to-one.
    (a) $f : \mathbf{R} \to \mathbf{R}$, $f(x) = x^4 + 3$
    (b) $g : \mathbf{R} \to \mathbf{R}$, $g(x) = |x - 2| + 4$
    (c) $\varphi : P(A) \to \{\{a\}, \varnothing\}$, $\varphi(B) = B \cap \{a\}$, where $a \in A$ and A has at least two elements
    (d) $\epsilon_5 : {}^{\mathbf{R}}\mathbf{R} \to \mathbf{R}$, $\epsilon_5(f) = f(5)$

8. Let $f : \mathbf{R} \to \mathbf{R}$ be periodic (Exercise 4.4.31). Prove that f is not one-to-one.

9. Show that the given functions are onto.
    (a) $f : \mathbf{R} \to \mathbf{R}$, $f(x) = 2x + 1$
    (b) $g : \mathbf{R} \to (0, \infty)$, $g(x) = e^x$
    (c) $h : \mathbf{R} \setminus \{0\} \to \mathbf{R} \setminus \{0\}$, $h(x) = 1/x$
    (d) $\varphi : \mathbf{Z} \times \mathbf{Z} \to \mathbf{Z}$, $\varphi(a, b) = a + b$
    (e) $\epsilon_5 : {}^{\mathbf{R}}\mathbf{R} \to \mathbf{R}$, $\epsilon_5(f) = f(5)$

10. Show that the given functions are not onto.
    (a) $f : \mathbf{R} \to \mathbf{R}$, $f(x) = e^x$
    (b) $g : \mathbf{R} \to \mathbf{R}$, $g(x) = |x|$
    (c) $\varphi : \mathbf{Z} \times \mathbf{Z} \to \mathbf{Z} \times \mathbf{Z}$, $\varphi(a, b) = (3a, b^2)$
    (d) $\psi : \mathbf{R} \to {}^{\mathbf{R}}\mathbf{R}$, $\psi(a) = f$, where $f(x) = a$ for all $x \in \mathbf{R}$

11. Let f and g be functions such that $\mathrm{ran}(g) \subseteq \mathrm{dom}(f)$.
    (a) Prove that if $f \circ g$ is onto, then f is onto.
    (b) Give an example of functions f and g such that $f \circ g$ is onto, but g is not onto.

12. Define $\varphi : \mathbf{Q} \times \mathbf{Z} \to \mathbf{Z} \times \mathbf{Q}$ by $\varphi(x, y) = (y, x)$. Show that φ is a bijection.

13. Show that the function $\gamma : A \times B \to C \times D$ defined by

$$\gamma(a, b) = (\varphi(a), \psi(b))$$

is a bijection if both $\varphi : A \to C$ and $\psi : B \to D$ are bijections.

14. Define $\gamma : A \times (B \times C) \to (A \times B) \times C$ by

$$\gamma(a, (b, c)) = ((a, b), c).$$

Prove γ is a bijection.

15. Demonstrate that the inverse of a bijection is a bijection.

16. Prove that the empty set is a bijection with domain and range equal to $\varnothing$.

17. Let $A \subseteq \mathbf{R}$ and define $\varphi : {}^{\mathbf{R}}\mathbf{R} \to {}^{A}\mathbf{R}$ by $\varphi(f) = f \restriction A$. Is φ always one-to-one? Is it always onto? Explain.

18. A function $f : A \to B$ has a **left inverse** if there exists a function $g : B \to A$ such that $g \circ f = I_A$. Prove that a function is one-to-one if and only if it has a left inverse.

19. A function $f : A \to B$ has a **right inverse** if there exists a function $g : B \to A$ so that $f \circ g = I_B$. Prove a function is onto if and only if it has a right inverse.

20. Let $(A, \preccurlyeq)$ be a poset. For every bijection $\varphi : A \to A$, φ is increasing (Exercise 4.4.32) if and only if φ^{-1} is increasing.

21. Show that if a real-valued function is increasing, it is one-to-one.

22. Prove or show false this modification of Theorem 4.5.23: If $\varphi : A \to B$ and $\psi : C \to D$ are bijections with ran$(\varphi) \subseteq C$, then $\psi \circ \varphi$ is a bijection.

23. Let $A, B \subseteq \mathbf{R}$ be two sets ordered by $\leq$. Let A be well-ordered by $\leq$. Prove that if $f : A \to B$ is an order-preserving surjection, B is well-ordered by $\leq$.

24. Let $\varphi : A \to B$ be an isomorphism preserving R with R'. Let $C \subseteq A$ and $D \subseteq B$. Prove the following.
    (a) $(C, R \cap [C \times C]) \cong (\varphi [C], R' \cap [\varphi [C] \times \varphi [C]])$
    (b) $(D, R' \cap [D \times D]) \cong (\varphi^{-1} [D], R \cap [\varphi^{-1} [D] \times \varphi^{-1} [D]])$

25. Prove Theorem 4.5.27.

26. Define $f : \mathbf{R} \to \mathbf{R}$ by $f(x) = 2x + 1$. Prove that f is an order isomorphism preserving $<$ with $<$.

27. Suppose that (A, R) and (B, S) are posets. Let $\varphi : A \to B$ be an order isomorphism preserving R with S and $C \subseteq A$. Prove that if m is the least element of C with respect to R, then $\varphi(m)$ is the least element of $\varphi [C]$ with respect to S.

28. Find linear orders $(A, \preccurlyeq)$ and $(B, \preccurlyeq')$ such that each is isomorphic to a subset of the other but $(A, \preccurlyeq)$ is not isomorphic to $(B, \preccurlyeq')$.

29. Let $(A, \preccurlyeq)$ be a poset. Prove that there exists $B \subseteq P(A)$ such that $(A, \preccurlyeq) \cong (B, \subseteq)$.

4.6   IMAGES AND INVERSE IMAGES

So far we have focused on the image of single element in the domain of a function. Sometimes we will need to examine a set of images.

■ **DEFINITION 4.6.1**

Let $\varphi : A \to B$ be a function and $C \subseteq A$. The **image** of C (under φ) is

$$\varphi [C] = \{\varphi(x) : x \in C\}.$$

Notice that $\varphi [C] \subseteq B$ (Figure 4.17) and $\varphi [A] = $ ran(φ).
    A similar definition can be made with subsets of the codomain.

■ **DEFINITION 4.6.2**

Let $\varphi : A \to B$ be a function and $D \subseteq B$. The **inverse image** of D (under φ) is

$$\varphi^{-1} [D] = \{x \in A : \varphi(x) \in D\}.$$

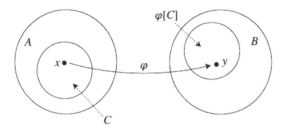

Figure 4.17   The image of C under φ.

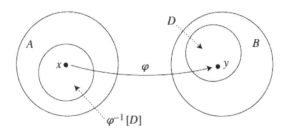

Figure 4.18   The inverse image of D under φ.

Observe that $\varphi^{-1}[D] \subseteq A$ (Figure 4.18) and $\varphi^{-1}[\mathrm{ran}(\varphi)] = A$.

■ **EXAMPLE 4.6.3**

Let $f = \{(1,2),(2,4),(3,5),(4,5)\}$. This set is a function. Its domain is $\{1,2,3,4\}$, and its range is $\{2,4,5\}$. Then,

$$f[\{1,3\}] = \{2,5\}$$

and

$$f^{-1}[\{5\}] = \{3,4\}.$$

■ **EXAMPLE 4.6.4**

Define $f : \mathbf{R} \to \mathbf{R}$ by $f(x) = x^2 + 1$.

- To prove $f[(1,2)] = (2,5)$, we show both inclusions. Let $y \in f[(1,2)]$. Then, $y = x^2 + 1$ for some $x \in (1,2)$. By a little algebra, we see that $2 < x^2 + 1 < 5$. Hence, $y \in (2,5)$. Conversely, let $y \in (2,5)$. Because

$$2 < y < 5 \Leftrightarrow 1 < y - 1 < 4 \Leftrightarrow 1 < \sqrt{y-1} < 2,$$

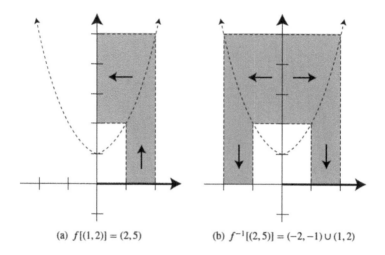

(a) $f[(1,2)] = (2,5)$ (b) $f^{-1}[(2,5)] = (-2,-1) \cup (1,2)$

Figure 4.19 The image and inverse image of a set under f.

$\sqrt{y-1} \in (1,2)$. Furthermore,

$$f(\sqrt{y-1}) = f(\sqrt{y-1}) = (\sqrt{y-1})^2 + 1 = y.$$

Therefore, $y \in f[(1,2)]$. This is illustrated in Figure 4.19(a).

- Simply because $f[(1,2)] = (2,5)$, we cannot conclude $f^{-1}[(2,5)]$ equals $(1,2)$. Instead, $f^{-1}[(2,5)] = (-2,-1) \cup (1,2)$, as seen in Figure 4.19(b), because

$$x \in (-2,-1) \cup (1,2) \Leftrightarrow -2 < x < -1 \text{ or } 1 < x < 2$$
$$\Leftrightarrow 1 < x^2 < 4$$
$$\Leftrightarrow 2 < x^2 + 1 < 5$$
$$\Leftrightarrow f(x) \in (2,5)$$
$$\Leftrightarrow x \in f^{-1}[(2,5)].$$

- To show that $f^{-1}[(-2,-1)]$ is empty, take $x \in f^{-1}[(-2,-1)]$. This means that $-2 < f(x) < -1$, but this is impossible because $f(x)$ is positive.

Let $f = \{(1,3),(2,3),(3,4),(4,5)\}$. This is a function $\{1,2,3,4\} \to \{3,4,5\}$. Notice that

$$f[\{1\} \cup \{2,3\}] = \{3,4\} = f[\{1\}] \cup f[\{2,3\}],$$
$$f[\{1,2\} \cap \{3\}] = f[\varnothing] = \varnothing \subseteq \{3\} = f[\{1\}] \cap f[\{2,3\}],$$
$$f^{-1}[\{3,4\} \cup \{5\}] = \{1,2,3,4\} = f^{-1}[\{3,4\}] \cup f^{-1}[\{5\}],$$

and

$$f^{-1}[\{3,4\} \cap \{5\}] = \varnothing = f^{-1}[\{3,4\}] \cap f^{-1}[\{5\}].$$

This result concerning the interaction of images and inverse images with union and intersection is generalized in the next theorem.

■ **THEOREM 4.6.5**

Let $\varphi : A \to B$ be a function with $C, D \subseteq A$ and $E, F \subseteq B$.

- $\varphi[C \cup D] = \varphi[C] \cup \varphi[D]$.

- $\varphi[C \cap D] \subseteq \varphi[C] \cap \varphi[D]$.

- $\varphi^{-1}[E \cup F] = \varphi^{-1}[E] \cup \varphi^{-1}[F]$.

- $\varphi^{-1}[E \cap F] = \varphi^{-1}[E] \cap \varphi^{-1}[F]$.

PROOF
We prove the first and third parts, leaving the others for Exercise 7. By Exercise 3.2.8(d),

$$
\begin{aligned}
y \in \varphi[C \cup D] &\Leftrightarrow \exists x (x \in C \cup D \wedge \varphi(x) = y) \\
&\Leftrightarrow \exists x (x \in C \wedge \varphi(x) = y) \vee \exists x (x \in D \wedge \varphi(x) = y) \\
&\Leftrightarrow y \in \varphi[C] \vee y \in \varphi[D] \\
&\Leftrightarrow y \in \varphi[C] \cup \varphi[D].
\end{aligned}
$$

In addition,

$$
\begin{aligned}
x \in \varphi^{-1}[E \cup F] &\Leftrightarrow \varphi(x) \in E \cup F \\
&\Leftrightarrow \varphi(x) \in E \vee \varphi(x) \in F \\
&\Leftrightarrow x \in \varphi^{-1}[E] \vee x \in \varphi^{-1}[F] \\
&\Leftrightarrow x \in \varphi^{-1}[E] \cup \varphi^{-1}[F]. \blacksquare
\end{aligned}
$$

It might seem surprising that we only have an inclusion in the second part of Theorem 4.6.5. To see that the other inclusion is false, let $f = \{(1,3),(2,3)\}$. Then,

$$f[\{1\} \cap \{2\}] = f[\varnothing] = \varnothing,$$

but

$$f[\{1\}] \cap f[\{2\}] = \{3\} \cap \{3\} = \{3\}.$$

Hence, $f[\{1\}] \cap f[\{2\}] \not\subseteq f[\{1\} \cap \{2\}]$. However, if f had been a bijection, the inclusion would hold (Exercise 13).

■ **EXAMPLE 4.6.6**

Let $f(x) = x^2 + 1$. We check the union results of Theorem 4.6.5.

- We have already seen in Example 4.6.4 that $f[(1, 2)] = (2, 5)$. Since $(1, 2) = (1, 1.5] \cup [1.5, 2)$, apply f to both of these intervals and find that

$$f[(1, 1.5]] = (2, 3.25],$$
$$f[[1.5, 2)] = [3.25, 5).$$

Therefore, $f[(1, 2)] = f[(1, 1.5]] \cup f[[1.5, 2)]$.

- Also, from Example 4.6.4, $f^{-1}[(2, 5)] = (-2, -1) \cup (1, 2)$. We can write $(2, 5)$ as the union of $(2, 4)$ and $(3, 5)$. Since

$$f^{-1}[(2, 4)] = \left(-\sqrt{3}, -1\right) \cup \left(1, \sqrt{3}\right),$$
$$f^{-1}[(3, 5)] = \left(-2, -\sqrt{2}\right) \cup \left(\sqrt{2}, 2\right),$$

we have that

$$f^{-1}[(2, 5)] = f^{-1}[(2, 4)] \cup f^{-1}[(3, 5)].$$

Theorem 4.6.5 can be modified to arbitrary unions and intersections. The proof is left to Excercise 17.

■ **THEOREM 4.6.7**

Let $\{A_i : i \in I\}$ be a family of sets.

- $\varphi\left[\bigcup_{i \in I} A_i\right] = \bigcup_{i \in I} \varphi[A_i]$

- $\varphi\left[\bigcap_{i \in I} A_i\right] \subseteq \bigcap_{i \in I} \varphi[A_i]$

- $\varphi^{-1}\left[\bigcup_{i \in I} A_i\right] = \bigcup_{i \in I} \varphi^{-1}[A_i]$

- $\varphi^{-1}\left[\bigcap_{i \in I} A_i\right] = \bigcap_{i \in I} \varphi^{-1}[A_i].$

We know that the composition of a function and its inverse equals the identity map (Theorem 4.5.3). The last two results of the section show a similar result involving the image and inverse image using functions that are either one-to-one or onto.

■ **THEOREM 4.6.8**

Let $\varphi : A \to B$ be a function. Suppose $C \subseteq A$ and $D \subseteq B$.

- If φ is one-to-one, then $\varphi^{-1}[\varphi[C]] = C$.
- If φ is onto, then $\varphi[\varphi^{-1}[D]] = D$.

PROOF

We prove the first part and leave the second to Exercise 8. Suppose that φ is an injection. We show that $\varphi^{-1}[\varphi[C]] = C$.

- Let $x \in \varphi^{-1}[\varphi[C]]$. This means that there exists $y \in \varphi[C]$ such that $\varphi(x) = y$. Furthermore, there is a $z \in C$ so that $\varphi(z) = y$. Therefore, since φ is one-to-one, $x = z$, which means that $x \in C$.

- This step will work for any function. Take $x \in C$, so $\varphi(x) \in \varphi[C]$. By definition,
$$\varphi^{-1}[\varphi[C]] = \{z \in A : \varphi(z) \in \varphi[C]\}.$$
Since x is also an element of A, we conclude that $x \in \varphi^{-1}[\varphi[C]]$. ■

Because we used the one-to-one condition only to prove $\varphi^{-1}[\varphi[C]] \subseteq C$, we suspect that this inclusion is false if the function is not one-to-one. To see this, let $f : \mathbf{R} \to \mathbf{R}$ be defined as $f(x) = x^2$. We know that f is not one-to-one. Choose $C = \{1\}$. Then,

$$f^{-1}[f[C]] = f^{-1}[\{1\}] = \{-1, 1\}.$$

Therefore, $f^{-1}[f[C]] \not\subseteq C$.

When examining the example, we might conjecture that the function being one-to-one is necessary for equality. That this is the case is the following theorem.

■ **THEOREM 4.6.9**

Let $\varphi : A \to B$ be a function.

- If $\varphi^{-1}[\varphi[C]] = C$ for all $C \subseteq A$, then φ is one-to-one.
- If $\varphi[\varphi^{-1}[D]] = D$ for all $D \subseteq B$, then φ is onto.

PROOF

As with the previous theorem, we prove only the first part. The second part is Exercise 8. Assume

$$\varphi^{-1}[\varphi[C]] = C \text{ for all } C \subseteq A.$$

Take $x_1, x_2 \in A$ and let $\varphi(x_1) = \varphi(x_2)$. Now,

$$\{x_1\} = \varphi^{-1}[\varphi[\{x_1\}]] = \{z \in A : \varphi(z) = \varphi(x_1)\}.$$

Because $\varphi(x_1) = \varphi(x_2)$, we have that $\{x_1, x_2\} \subseteq \{z \in A : \varphi(z) = \varphi(x_1)\}$, so we must have $x_1 = x_2$. ■

Exercises

1. Let $f : \mathbf{R} \to \mathbf{R}$ be defined by $f(x) = 2x + 1$. Find the images and inverse images.
   - (a) $f[(1, 3]]$
   - (b) $f[(-\infty, 0)]$
   - (c) $f^{-1}[(-1, 1)]$
   - (d) $f^{-1}[(0, 2) \cup (5, 8)]$

2. Let $g : \mathbf{R} \to \mathbf{R}$ be the function $g(x) = x^4 - 1$. Find the images and inverse images.
   - (a) $g[\{0\}]$
   - (b) $g[\mathbf{Z}]$
   - (c) $g^{-1}[\{0, 15\}]$
   - (d) $g^{-1}[[-9, -5] \cup [0, 5]]$

3. Define $\varphi : \mathbf{R} \times \mathbf{R} \to \mathbf{Z}$ by $\varphi(a, b) = \llbracket a \rrbracket + \llbracket b \rrbracket$. Find the images and inverse images.
   - (a) $\varphi[\{0\} \times \mathbf{R}]$
   - (b) $\varphi[(0, 1) \times (0, 1)]$
   - (c) $\varphi^{-1}[\{2, 4\}]$
   - (d) $\varphi^{-1}[\mathbf{N}]$

4. Let $\psi : \mathbf{Z} \to \mathbf{Z}$ be a function and define $\gamma : P(\mathbf{Z}) \to P(\mathbf{Z})$ by $\gamma(C) = \psi[C]$.
   - (a) Prove that γ is well-defined.
   - (b) Let $\psi(n) = 2n$. Find $\gamma[\{1, 2, 3\}]$, $\gamma[\mathbf{Z}]$, $\gamma^{-1}[\{1, 2, 3\}]$, and $\gamma^{-1}[\mathbf{Z}]$.
   - (c) Under what conditions is γ one-to-one?
   - (d) Under what conditions is γ onto?

5. For any function ψ, show that $\psi[\varnothing] = \varnothing$ and $\psi^{-1}[\varnothing] = \varnothing$.

6. Prove for every $B \subseteq A$, $I_A[B] = B$ and $(I_A)^{-1}[B] = B$.

7. Prove the remaining parts of Theorem 4.6.5.

8. Prove the unproven parts of Theorems 4.6.8 and 4.6.9.

9. Let $\varphi : A \to B$ be an injection and $C \subseteq A$.
   - (a) Prove $\varphi(x) \in \varphi[C]$ if and only if $x \in C$.
   - (b) Show that Exercise 9(a) is false if the function is not one-to-one.

10. If ψ is a function, $A \subseteq B \subseteq \text{dom}(\psi)$, and $C \subseteq D \subseteq \text{ran}(\psi)$, show that both $\psi[A] \subseteq \psi[B]$ and $\psi^{-1}[C] \subseteq \psi^{-1}[D]$.

11. Let $\varphi : A \to B$ be a function and take disjoint sets U and V.
    - (a) Prove false: If $U, V \subseteq A$, then $\varphi[U] \cap \varphi[V] = \varnothing$.
    - (b) Prove false: If $U, V \subseteq B$, then $\varphi^{-1}[U] \cap \varphi^{-1}[V] = \varnothing$.
    - (c) What additional assumption is needed to prove both of the implications?

12. Assume that φ and ψ are functions such that $\text{ran}(\psi) \subseteq \text{dom}(\varphi)$. Let A be a subset of $\text{dom}(\psi)$. Prove or show false with a counterexample: $(\varphi \circ \psi)[A] = \varphi[\psi[A]]$.

13. Let $\varphi : A \to B$ be one-to-one. Prove the following.

(a) $\varphi[A] \cap \varphi[B] \subseteq \varphi[A \cap B]$.

(b) If $C \subseteq A$ and $D \subseteq B$, then $\varphi[C] = D$ if and only if $\varphi^{-1}[D] = C$.

14. Prove that if $\varphi : A \to B$ is a bijection and $C \subseteq A$, then $\varphi[A \setminus C] = B \setminus \varphi[C]$.

15. Find a function $\varphi : A \to B$ and a set $D \subseteq B$ such that $D \not\subseteq \varphi[\varphi^{-1}[D]]$.

16. Let ψ be an injection with $A \subseteq \mathrm{dom}(\psi)$ and $B \subseteq \mathrm{ran}(\psi)$. Prove that

$$B \subseteq \psi[A] \text{ if and only if } \psi^{-1}[B] \subseteq A.$$

17. Prove Theorem 4.6.7.

CHAPTER 5

AXIOMATIC SET THEORY

5.1 AXIOMS

When we began studying set theory in Chapter 3, we made several assumptions regarding which things are sets. For example, we assumed that collections of numbers, like $\mathbf{N}$, $\mathbf{R}$, or $(0, \infty)$ are sets. We supposed that operating with given sets to form new collections, as with union or intersection, resulted in sets. We also assumed that formulas could be used to describe certain sets. All of this seemed perfectly reasonable, but since all of these assumptions were made without a carefully thought-out system, we would be wise to pause and investigate if we have introduced any problems.

Consider the following question. Given a formula $p(x)$, is there a set of the form $\{x : p(x)\}$? Consistent with the attitude of our previous work, we might quickly answer in the affirmative. Mathematicians, including Cantor, also initially thought that this was the case. However, it was shown independently by Bertrand Russell and Ernst Zermelo that not every formula can be used to define a set. For example, let $p(x) := x \notin x$ and $A = \{x : p(x)\}$ and consider whether A is an element of itself. If $A \in A$, then due to the definition of $p(x)$, $A \notin A$, and if $A \notin A$, then $A \in A$. Because of this built-in contradiction, it is impossible for A to be a set. This is known as **Russell's paradox**, and it was a serious challenge to set theory. One solution would have been to dismiss

A First Course in Mathematical Logic and Set Theory, First Edition. Michael L. O'Leary.
© 2016 John Wiley & Sons, Inc. Published 2016 by John Wiley & Sons, Inc.

set theory altogether. The problem was that this new subject combined with advances in logic appeared to promise a framework in which to study foundational questions of mathematics. David Hilbert famously supported set theory by remarking that "no one will drive us out of this paradise that Cantor has created for us," so dismissal was not an option. In order to prevent contradictions such as Russell's paradox from appearing, mathematicians settled on the method of Euclid as the solution, but instead of assuming geometric postulates, over a period of time, certain set-theoretic axioms were chosen. Their purpose was to define a system by which one could determine whether a given collection should be considered a set in such a manner that prevented any contradictions from arising. In this chapter we identify the axioms and then redefine **N**, **R**, and other collections so that we are confident in our previous assumptions regarding them being sets.

Equality Axioms

We begin with the basics. Although not officially among the set axioms, $=$ is always assumed to satisfy the following rules. They are defined to replicate the standard reasoning of equality that we have been using in previous chapters.

■ **AXIOMS 5.1.1 [Equality]**

Let x, y, and z be variable symbols from theory symbols S.

- **[E1]** $x = x$.

- **[E2]** $x = y \Leftrightarrow y = x$.

- **[E3]** $x = y, y = z \Rightarrow x = z$.

Let $x_0, x_1, \ldots, x_{n-1}$ and $y_0, y_1, \ldots, y_{n-1}$ be variable symbols from S.

- **[E4]** For any n-ary function symbol f of S,

$$x_0 = y_0 \wedge x_1 = y_1 \wedge \cdots \wedge x_{n-1} = y_{n-1}$$
$$\Rightarrow f(x_0, x_1, \ldots, x_{n-1}) = f(y_0, y_1, \ldots, y_{n-1}).$$

- **[E5]** For any S-formula $p(u_0, u_1, \ldots, u_{n-1})$,

$$x_0 = y_0 \wedge x_1 = y_1 \wedge \cdots \wedge x_{n-1} = y_{n-1}$$
$$\Rightarrow p(x_0, x_1, \ldots, x_{n-1}) \leftrightarrow p(y_0, y_1, \ldots, y_{n-1}).$$

Axioms **E1**, **E2**, and **E3** give $=$ the behavior of an equivalence relation (Definition 4.2.4). For example, we can use **E2** to prove that for any constant symbols c_0 and c_1,

$$\vdash c_0 = c_1 \leftrightarrow c_1 = c_0.$$

The proof goes as follows:

$$c_0 = c_1 \Leftrightarrow \left[(x_0 = x_1)\frac{c_0}{x_0}\right]\frac{c_1}{x_1}$$

$$\Leftrightarrow \left[(x_1 = x_0)\frac{c_0}{x_0}\right]\frac{c_1}{x_1}$$

$$\Leftrightarrow c_1 = c_0.$$

Axiom **E4** allows a function symbol to be used in a proof as a function, and **E5** allows equal terms to be substituted into formulas with the result being equivalent formulas. For example, given the NT-term $u + v$, by **E4**,

$$x_0 = y_0 \wedge x_1 = y_1 \Rightarrow x_0 + x_1 = y_0 + y_1,$$

and given the NT-formula $u + v = v + u$, by **E5**,

$$x_0 = y_0 \wedge x_1 = y_1 \Rightarrow x_0 + x_1 = x_1 + x_0 \leftrightarrow y_0 + y_1 = y_1 + y_0.$$

Formal proofs that require a deduction on an equality need to reference one of the equality axioms from Axioms 5.1.1.

Existence and Uniqueness Axioms

The axioms will be ST-formulas (Example 2.1.3), where all terms represent sets. We begin by assuming the existence of two sets.

■ **AXIOM 5.1.2 [Empty Set]**

$$\exists x \forall y (\neg\, y \in x).$$

In ST-formulas, a witness for the empty set axiom is denoted by $\{\ \}$, although it is usually written as $\varnothing$.

■ **AXIOM 5.1.3 [Infinity]**

$$\exists x(\{\ \} \in x \wedge \forall u[u \in x \rightarrow \exists y(y \in x \wedge u \in y \wedge \forall v[v \in u \rightarrow v \in y])]).$$

The standard interpretation of the infinity axiom is that there exists a set that contains $\varnothing$ and $a \cup \{a\}$ is an element of the set if a is an element of the set.

In Chapter 3, we noted that the elements of a set determine the set. For example, $\{1,2\} = \{1,2,2\}$. This principle is codified by the next axiom.

■ **AXIOM 5.1.4 [Extensionality]**

$$\forall x \forall y (\forall u[u \in x \leftrightarrow u \in y] \rightarrow x = y).$$

Suppose A_1 and A_2 are witnesses to the empty set axiom (5.1.2). Since $x \notin A_1$ and $x \notin A_2$ for every x, we conclude that

$$x \in A_1 \leftrightarrow x \in A_2.$$

Therefore, $A_1 = A_2$ by extensionality (Axiom 5.1.4), which means that $\varnothing$ is *the* witness to the empty set axiom. This uniqueness result does not appear to extend to the infinity axiom because both

$$\{\varnothing, \{\varnothing\}, \{\varnothing, \{\varnothing\}\}, \{\varnothing, \{\varnothing\}, \{\varnothing, \{\varnothing\}\}\}, \dots \}$$

and

$$\{\varnothing, \{\varnothing\}, \{\varnothing, \{\varnothing\}\}, \{\varnothing, \{\varnothing\}, \{\varnothing, \{\varnothing\}\}\},$$
$$\dots, \{\varnothing, \{\varnothing\}, \{\varnothing, \{\varnothing\}\}, \{\varnothing, \{\varnothing\}, \{\varnothing, \{\varnothing\}\}\}\}, \dots \}$$

are witnesses, provided that they are sets.

Construction Axioms

Now to build some sets. The next four axioms allow us to do this.

■ AXIOM 5.1.5 [Pairing]

$\forall u \, \forall v \, \exists x \, \forall w \, (w \in x \leftrightarrow w = u \lor w = v).$

Suppose that M and N are sets. Since $\{M, N\}$ is a witness of

$$\exists x \, \forall w \, (w \in x \leftrightarrow w = M \lor w = N),$$

$\{M, N\}$ is a set by the pairing axiom, and from this, we conclude that $\{M, \{M, N\}\}$, $\{N, \{M, N\}\}$, and $\{\{M, N\}\} = \{\{M, N\}, \{M, N\}\}$ are sets. Because $\{M, M\}$ equals $\{M\}$, pairing along with extensionality prove the existence of singletons. For example, if W is a witness to the Infinity Axiom, $\{\varnothing, W\}$, $\{\varnothing\}$, and $\{\varnothing, \{\varnothing, W\}\}$ are sets.

■ AXIOM 5.1.6 [Union]

$\forall x \exists y \forall u (u \in y \leftrightarrow \exists v [v \in x \land u \in v]).$

By the union axiom, $\bigcup M$ is a set, and since $M \cup N = \bigcup \{M, N\}$, we conclude that $M \cup N$ is a set. Furthermore, the empty set, union, and pairing axioms can be used to prove that for any $n \in \mathbf{N}$, there exists a set of the form $\{a_0, a_1, \dots, a_n\}$ (Exercise 3).

■ AXIOM 5.1.7 [Power Set]

$\forall x \exists y \forall u (u \in y \leftrightarrow \forall v [v \in u \rightarrow v \in x]).$

Because of Definition 3.3.1, the power set axiom can be written as

$$\forall x \exists y \forall u (u \in y \leftrightarrow u \subseteq x).$$

We conclude that for every set M, $\mathbf{P}(M)$ is a set by the power set axiom, and by extensionality, $\mathbf{P}(M)$ is the unique set of subsets of M.

The next axiom is actually what is called an **axiom scheme**, infinitely many axioms, one for every formula. They are sometimes called the **separation axioms**.

■ **AXIOMS 5.1.8 [Subset]**

For every ST-formula $p(u)$ not containing the symbol y, the following is an axiom:

$$\forall x \exists y \forall u \, [u \in y \leftrightarrow u \in x \wedge p(u)] .$$

The formula $p(u)$ in the subset axioms cannot contain the symbol y because the axioms yield the existence of this set. If y was among its symbols, the existence of y would depend on y.

The subset axioms yield many familiar sets.

- Let $\mathscr{F}$ be a set. By a subset axiom, there exists a set C such that

$$x \in C \leftrightarrow x \in \bigcup \mathscr{F} \wedge \forall c (c \in \mathscr{F} \rightarrow x \in c).$$

Observe that the symbol C does not appear in the formula

$$\forall c (c \in A \rightarrow x \in c).$$

Also, observe that the set C is the intersection of $\mathscr{F}$. Hence, $\bigcap \mathscr{F}$ is a set, and since $M \cap N = \bigcap \{M, N\}$, we conclude that $M \cap N$ is a set.

- By a subset axiom, there exists a set D such that

$$x \in D \leftrightarrow x \in M \wedge x \notin N,$$

so $M \setminus N$ is a set.

Replacement Axioms

Given sets A and B, the function $\varphi : A \rightarrow B$ is a set because $\varphi \subseteq A \times B$ and $A \times B$ is a set. Suppose, instead, that the function is defined using a formula $p(x, y)$ and that its domain is given by a set A. It cannot be concluded from Axioms 5.1.4–5.1.10 that the range $\{y : x \in A \wedge p(x, y)\}$ is a set. However, it appears reasonable that it is. For example, define

$$p(x, y) := y \in \mathbf{Z} \wedge y \leq x \wedge \forall z (z \in \mathbf{Z} \wedge y \leq z \rightarrow x < z).$$

An examination of the formula shows $p(3.4, 4)$ and $p(-7.1, -8)$. Since

$$p(x, y_1) \wedge p(x, y_2) \rightarrow y_1 = y_2,$$

the formula $p(x, y)$ defines a function (Definition 4.4.10). If the domain is given to be $[0, \infty)$, its range is the set $[0, \infty) \cap \mathbf{Z}$. Generalizing to an arbitrary $p(x, y)$, it is expected that the range would be a set if the domain is a set. The next axiom scheme guarantees this. It was first found in correspondence between Cantor and Richard Dedekind (Cantor 1932) and Dmitry Mirimanoff (1917) with formal versions by Abraham Fraenkel (1922) and Thoralf Skolem (1922).

■ **AXIOMS 5.1.9 [Replacement]**

For every ST-formula $p(t, w)$ not containing the symbol y, the following is an axiom:

$$\forall x[\forall u \forall v_1 \forall v_2(u \in x \wedge p(u, v_1) \wedge p(u, v_2) \to v_1 = v_2)$$
$$\to \exists y \forall w(w \in y \leftrightarrow \exists t[t \in x \wedge p(t, w)])].$$

As an example, every indexed family of sets is a set. To prove this, let I be a set and A_i be a set for all $i \in I$. Define

$$p(i, y) := y = A_i.$$

Observe that by **E2** and **E3** (Axioms 5.1.1),

$$y_1 = A_i \wedge y_2 = A_i \Rightarrow y_1 = y_2.$$

Therefore, by a replacement axiom (5.1.9), there exists a set $\mathscr{F}$ such that

$$w \in \mathscr{F} \leftrightarrow \exists i (i \in I \wedge w = A_i),$$

so $\mathscr{F} = \{A_i : i \in I\}$ is a set, which implies that the union and intersection of families of sets are sets.

Suppose $a \in M$ and $b \in N$. By the subset and pairing axioms, we conclude that $\{a, b\}$, $(a, b) = \{\{a\}, \{a, b\}\}$ (Definition 3.2.8), and $\{(a, b)\}$ are sets. Therefore, fixing b,

$$\{\{(a, b)\} : a \in M\}$$

is a family of sets, which implies that it is a set. Likewise,

$$\{\{\{(a, b)\} : a \in M\} : b \in N\}$$

is a set. Hence, by the union axiom (5.1.6),

$$M \times N = \bigcup \bigcup \{\{\{(a, b)\} : a \in M\} : b \in N\}$$

is a set. This implies, using a subset axiom (5.1.8), that any binary relation R such that $\text{dom}(R) \subseteq M$ and $\text{ran}(R) \subseteq N$ is a set.

Axiom of Choice

Suppose that we are given the pairwise disjoint family of sets

$$\mathscr{F} = \{\{1, 3, 5\}, \{2, 9, 11\}, \{7, 8, 13\}\}.$$

It is easy to find a set S such that

$$S \cap A \text{ is a singleton for every } A \in \mathscr{F}. \tag{5.1}$$

Simply run through the elements of $\mathscr{F}$ and choose an element from each set and put it in S. Since $\mathscr{F}$ is pairwise disjoint, each choice will differ from the others. For example, it might be that

$$S = \{1, 9, 13\}.$$

However, what happens if $\mathscr{F}$ is an infinite set? If there was not a systematic way where elements could be chosen from the sets of $\mathscr{F}$, we would be left with making infinitely many choices, which is something that we cannot do. Nonetheless, it appears reasonable that there is a set S that intersects each member of $\mathscr{F}$ exactly once. Such a set cannot be proved to exist from Axioms 5.1.2 to 5.1.9, so we need another axiom. It is called the **axiom of choice**. We will have to use it every time that an infinite number of arbitrary choices need to be made.

■ AXIOM 5.1.10 [Choice]

If $\mathscr{F}$ is a family of pairwise disjoint, nonempty sets, there exists $S \subseteq \bigcup \mathscr{F}$ such that $S \cap A$ is a singleton for all $A \in \mathscr{F}$.

The statement of the axiom of choice can be written as an ST-formula (Exercise 12). Also, notice that S in Axiom 5.1.10 is a function (Exercise 13). It is called a **selector**.

The next follows quickly from the axiom of choice. In fact, the proposition is equivalent to the axiom (Exercise 15), so this corollary is often used as a replacement for it.

■ COROLLARY 5.1.11

For every binary relation R, there exists a function φ such that $\varphi \subseteq R$ and $\text{dom}(\varphi) = \text{dom}(R)$.

PROOF

Let $R \subseteq A \times B$. Define $\mathscr{F} = \{\{a\} \times [a]_R : a \in A\}$, which is a set by a replacement axiom (Exercise 14). Since $\mathscr{F}$ is pairwise disjoint and the set $\{a\} \times [a]_R \neq \varnothing$ for all $a \in \text{dom}(R)$, the axiom of choice (5.1.10) implies that there exists a selector S such that $S \subseteq \bigcup \mathscr{F}$ and $S \cap (\{a\} \times [a]_R)$ is a singleton for all $a \in \text{dom}(R)$. Thus, $S \subseteq R$, and for all $a \in \text{dom}(R)$, there exists a unique $b \in B$ such that $a R b$. This implies that S is the desired function. ■

Given a family of sets $\mathscr{F}$, define a relation $R \subseteq \mathscr{F} \times \bigcup \mathscr{F}$ by

$$R = \{(A, a) : A \in \mathscr{F} \wedge a \in A\}.$$

By Corollary 5.1.11, there exists a function $\xi : \mathscr{F} \to \bigcup \mathscr{F}$ such that $\xi(A) \in A$ for all $A \in \mathscr{F}$. The function ξ is called a **choice function**.

■ EXAMPLE 5.1.12

Let $\mathscr{A} = \{A_i : i \in I\}$ be a family of nonempty sets. We want to define a family of singletons $\mathscr{B}$ such that for all $i \in I$,

$$\text{if } \{a_i\} \in \mathscr{B}, \text{ then } a_i \in A_i.$$

By Corollary 5.1.11, there exists a choice function $\xi : \mathcal{A} \to \bigcup \mathcal{A}$. The family $\mathcal{B} = \{\{\xi(A_i)\} : i \in I\}$ is the desired set because $\xi(A_i) \in A_i$ for all $i \in I$.

There are many theorems equivalent to the axiom of choice (5.1.10). One such result involves families of sets. Take $n \in \mathbf{N}$ and define

$$\mathcal{A}_n = \{\{0\}, \{0, 1\}, \dots, \{0, 1, \dots, n\}\}.$$

Observe that $\mathcal{A}_n$ is a chain with respect to $\subseteq$ and contains a maximal element (Definition 4.3.16), which is $\{0, 1, \dots, n\}$. However, the chain

$$\mathcal{A} = \{\{0\}, \{0, 1\}, \dots, \{0, 1, \dots, n\}, \dots\}$$

has no maximal element. There are many sets that can be added to $\mathcal{A}$ to give it a maximal element, but the natural choice is to add the union of $\mathcal{A}$ to the family giving

$$\mathcal{A}' = \{\{0\}, \{0, 1\}, \dots, \{0, 1, \dots, n\}, \dots, \mathbf{N}\}.$$

$\mathcal{A}'$ has a maximal element, namely, $\mathbf{N}$. The generalization of this result to any family of sets was first proved by Kuratowski (1922) and then independently by Zorn (1935) for whom the theorem is named despite Kuratowski's priority. The proof given is essentially due to Zermelo (Halmos 1960).

■ **THEOREM 5.1.13 [Zorn's Lemma]**

Let $\mathcal{A}$ be a family of sets. If $\bigcup \mathscr{C} \in \mathcal{A}$ for every chain $\mathscr{C}$ of $\mathcal{A}$ with respect to $\subseteq$, there exists $M \in \mathcal{A}$ such that $M \not\subset A$ for all $A \in \mathcal{A}$.

PROOF

Let $\xi : \mathbf{P}(\mathcal{A}) \setminus \{\varnothing\} \to \mathcal{A}$ be a choice function (Corollary 5.1.11). For every chain $\mathscr{C}$ of $\mathcal{A}$, define

$$\bar{\mathscr{C}} = \{A \in \mathcal{A} : \mathscr{C} \cup \{A\} \text{ is a chain}\}.$$

Notice that $\bar{\mathscr{C}}$ is the set of elements of $\mathcal{A}$ that when added to $\mathscr{C}$ yields a chain. Let $\mathrm{Ch}(\mathcal{A})$ be defined as the set of all chains of $\mathcal{A}$. That both $\bar{\mathscr{C}}$ and $\mathrm{Ch}(\mathcal{A})$ are sets is left to Exercise 17(a). Define

$$X : \mathrm{Ch}(\mathcal{A}) \to \mathrm{Ch}(\mathcal{A})$$

by

$$X(\mathscr{C}) = \begin{cases} \mathscr{C} \cup \{\xi(\bar{\mathscr{C}} \setminus \mathscr{C})\} & \text{if } \bar{\mathscr{C}} \setminus \mathscr{C} \neq \varnothing, \\ \mathscr{C} & \text{if } \bar{\mathscr{C}} \setminus \mathscr{C} = \varnothing. \end{cases}$$

Next, suppose that $\mathscr{C}_0$ is a chain of $\mathcal{A}$ such that $X(\mathscr{C}_0) = \mathscr{C}_0$. By the assumption on $\mathcal{A}$, we have that $\bigcup \mathscr{C}_0 \in \mathcal{A}$. To prove that $\bigcup \mathscr{C}_0$ is a maximal element of $\mathcal{A}$, take $A \in \mathcal{A}$ such that $\bigcup \mathscr{C}_0 \subset A$. Since A has an element that is not in any of the elements of $\mathscr{C}_0$, the union $\mathscr{C}_0 \cup \{A\}$ is a chain properly containing $\mathscr{C}_0$, which is impossible. We conclude that the theorem is proved if $\mathscr{C}_0$ is shown to exist.

To accomplish this, we begin with a definition. A subset $\mathscr{T}$ of $\mathrm{Ch}(\mathscr{A})$ is called a **tower** when

- $\varnothing \in \mathscr{T}$,

- $X(\mathscr{C}) \in \mathscr{T}$ for all $\mathscr{C} \in \mathscr{T}$,

- $\bigcup \mathscr{D} \in \mathscr{T}$ for every chain $\mathscr{D} \subseteq \mathscr{T}$.

Define $\mathrm{To}(\mathscr{A})$ to be the set of all towers of $\mathscr{A}$. Observe that $\mathrm{Ch}(\mathscr{C})$ and $\bigcap \mathrm{To}(\mathscr{A})$ are both towers [Exercise 17(b)].

Take $C \in \bigcap \mathrm{To}(\mathscr{A})$ such that C is comparable with respect to inclusion to every element of $\bigcap \mathrm{To}(\mathscr{A})$. Such a set C exists since $\varnothing$ is an element of $\mathrm{To}(\mathscr{A})$ and comparable to every element of $\bigcap \mathrm{To}(\mathscr{A})$. Suppose $A \in \bigcap \mathrm{To}(\mathscr{A})$ such that $A \subset C$. Because $\bigcap \mathrm{To}(\mathscr{A})$ is a tower, $X(A) \in \bigcap \mathrm{To}(\mathscr{A})$. If $C \subset X(A)$, then $X(A) \setminus A$ has at least two elements, which is impossible. Therefore,

$$\text{if } A \in \bigcap \mathrm{To}(\mathscr{A}) \text{ and } A \subset C, \text{ then } X(A) \subseteq C. \tag{5.2}$$

Define

$$T = \{ A \in \bigcap \mathrm{To}(\mathscr{A}) : A \subseteq C \vee X(C) \subseteq A \}.$$

If $A \subseteq C$, then $A \subseteq X(C)$ since $C \subseteq X(C)$. Hence, every element of T is comparable to $X(C)$. Also, T is a tower because of the following:

- $\varnothing \in T$ because $\varnothing \in \bigcap \mathrm{To}(\mathscr{A})$ and $\varnothing \subseteq C$.

- Let $B \in T$. Since $\bigcap \mathrm{To}(\mathscr{A})$ is a tower, $X(B) \in \bigcap \mathrm{To}(\mathscr{A})$. If $B \subset C$, then $X(B) \subseteq C$ by (5.2). If $C \subseteq B$, then $X(C) \subseteq X(B)$. In both cases, $X(B) \in T$.

- Let $\mathscr{C} \subseteq T$ be a chain. Then, $\bigcup \mathscr{C} \in \bigcap \mathrm{To}(\mathscr{A})$. Suppose there exists $C_0 \in \mathscr{C}$ such that C_0 is not a subset of C. This implies that $X(C) \subseteq C_0$. Thus, $X(C) \subseteq \bigcup \mathscr{C}$, so $\bigcup \mathscr{C} \in T$.

Hence, $T = \bigcap \mathrm{To}(\mathscr{A})$ because $T \subseteq \bigcap \mathrm{To}(\mathscr{A})$. Therefore, $\bigcap \mathrm{To}(\mathscr{A})$ is a chain because C is arbitrary [Exercise 17(c)]. Since $\bigcap \mathrm{To}(\mathscr{A})$ is a tower,

$$\bigcup \bigcap \mathrm{To}(\mathscr{A}) \in \bigcap \mathrm{To}(\mathscr{A})$$

and then

$$X \left(\bigcup \bigcap \mathrm{To}(\mathscr{A}) \right) \in \bigcap \mathrm{To}(\mathscr{A}).$$

Hence, since $\bigcup \bigcap \mathrm{To}(\mathscr{A})$ is an upper bound of $\bigcap \mathrm{To}(\mathscr{A})$,

$$X \left(\bigcup \bigcap \mathrm{To}(\mathscr{A}) \right) \subseteq \bigcup \bigcap \mathrm{To}(\mathscr{A}),$$

which yields

$$X \left(\bigcup \bigcap \mathrm{To}(\mathscr{A}) \right) = \bigcup \bigcap \mathrm{To}(\mathscr{A}). \ \blacksquare$$

There are many equivalents to the axiom of choice (Rubin and Rubin 1985, Jech 1973). One of them is Zorn's lemma.

■ **THEOREM 5.1.14**

The axiom of choice is equivalent to Zorn's lemma.

PROOF
Since we have already proved that the axiom of choice (5.1.10) implies Zorn's lemma (Theorem 5.1.13), we only need to prove the converse. We must be careful to only use Axioms 5.1.2–5.1.9 and not Axiom 5.1.10.

Assume Zorn's lemma and let R be a relation. Define

$$\mathscr{A} = \{\varphi : \varphi \subseteq R \wedge \varphi \text{ is a function}\}.$$

Let $\mathscr{C}$ be a chain of elements of $\mathscr{A}$.

- Take $\psi \in \bigcup \mathscr{C}$, so $\psi \in C$ for some $C \in \mathscr{C}$. This implies that C is a subset of R. Therefore, $\psi \in R$, proving that $\bigcup \mathscr{C} \subseteq R$.

- $\bigcup \mathscr{C}$ is a function by Exercise 4.4.13.

We conclude that $\bigcup \mathscr{C} \in \mathscr{A}$. Thus, by Zorn's lemma, there exists a maximal element $\Phi \in \mathscr{A}$. This means that $\Phi \subseteq R$ and $\text{dom}(\Phi) \subseteq \text{dom}(R)$. To prove that $\text{dom}(R) \subseteq \text{dom}(\Phi)$, let $(x, y) \in R$ and suppose that $x \notin \text{dom}(\Phi)$. This implies that $\Phi \cup \{(x, y)\} \subseteq R$ is a function. Hence, $\Phi \cup \{(x, y)\} \in \mathscr{A}$, contradicting the maximality of Φ. We conclude that Φ is the desired function, and the axiom of choice follows (Exercise 15). ■

There was a controversy regarding the axiom of choice when Zermelo first proposed it. Despite mathematicians having previously used it implicitly, some objected to its nonconstructive nature. The other axioms yield distinct results, but the axiom of choice results in a set with elements that are not clearly identified. Over time, however, most objections have faded. This is because the majority of mathematicians regard it as reasonable and generally those who question the axiom of choice realize that eliminating it would lead to serious problems because many proofs in various fields of mathematics rely on the axiom.

Axiom of Regularity

The ideas that led to the next axiom (also known as the **axiom of foundation**) can be found in Mirimanoff (1917), while the statement of the axiom is credited to Skolem (1922) and John von Neumann (1923).

■ **AXIOM 5.1.15 [Regularity]**

$\forall x(x \neq \{\} \rightarrow \exists y[y \in x \wedge \neg \exists u(u \in y \wedge u \in x)]).$

The main result of the regularity axiom is that it prevents sets from being elements of themselves. Suppose there exists a set A such that $A \in A$. Then, $A \cap \{A\} \neq \varnothing$, but the regularity axiom implies that $A \cap \{A\}$ should be empty.

■ THEOREM 5.1.16

No set is an element of itself.

If $V = \{x : x \text{ is a set}\}$ is a set, $V \in V$, contradicting Theorem 5.1.16. Thus, the theorem's corollary quickly follows.

■ COROLLARY 5.1.17

There is no set of all sets.

Because of the regularity axiom (5.1.15), $A \notin A$ for all sets A. Therefore, $\{x : x \notin x\}$ is not a set by, which prevents Russell's paradox from being deduced from the axioms, provided that they are consistent (Theorem 1.5.2).

The axiom of regularity is the final axiom of our chosen collection of axioms. It is believed that they do not prove any contradictions, which implies that the axioms prevent the construction of $\{x : x \notin x\}$ as a set. Therefore, we write the following definition. The collections of axioms are named after the mathematician who was primarily responsible for their selection (Zermelo 1908).

■ DEFINITION 5.1.18

- Axioms 5.1.2–5.1.8 are the **Zermelo axioms**. This collection of sentences is denoted by **Z**.

- Axioms 5.1.2–5.1.8 combined with replacement and regularity (Axioms 5.1.9 and 5.1.15) are the **Zermelo–Fraenkel axioms**, denoted by **ZF**.

- The Zermelo–Fraenkel axioms with the axiom of choice is denoted by **ZFC**.

The nonempty sets that follow from **ZFC** have the property that all of their elements are sets. Sets with this property are called **hereditary** or **pure**. Assuming **ZFC** does not prevent us from working with different types of sets, such as sets of symbols or formulas, but we must remember that such nonhereditary sets are not products of **ZFC**, so they must be handled with care because we do not want to fall into a paradox.

Exercises

1. Let S be a set of theory symbols. Let $c_1, c_2, c_3, c_4 \in S$ be constant symbols and $f \in S$ be a binary function symbol. Suppose that $p(x, y)$ is an S-formula. Use the Equality Axioms (5.1.1) to prove the following.

  (a) $\vdash c_1 = c_1$
  (b) $\vdash c_1 = c_2 \wedge c_2 = c_3 \rightarrow c_1 = c_3$
  (c) $\vdash c_1 = c_2 \wedge c_3 = c_4 \rightarrow f(c_1, c_3) = f(c_2, c_4)$
  (d) $\vdash c_1 = c_2 \wedge c_3 = c_4 \rightarrow [p(f(c_1), c_3) \leftrightarrow p(f(c_2), c_4)]$

2. Prove $\forall x \forall y (x = y \rightarrow \forall u [u \in x \leftrightarrow u \in y])$.

3. For any $n \in \mathbb{N}$ with $n > 0$, prove that there exists a set of the form $\{a_0, a_1, \ldots, a_{n-1}\}$.

4. Let $\mathscr{F}$ be a family of sets. Prove that $\mathbf{P}(\bigcup \bigcap \mathscr{F})$ is a set.

5. Use a subset axiom (5.1.8) to prove that there is no set of all sets. This proves that Russell's paradox does not follow from the axiom of Z.

6. Prove that there is no set that has every singleton as an element.

7. Let A and B be sets. Define the **symmetric difference** of A and B by

$$A \bigtriangleup B = A \setminus B \cup B \setminus A.$$

Prove that $A \bigtriangleup B$ is a set.

8. Prove the given equations for all sets A, B, and C.
   (a) $A \bigtriangleup \varnothing = A$
   (b) $A \bigtriangleup U = \overline{A}$
   (c) $A \bigtriangleup B = B \bigtriangleup A$
   (d) $A \bigtriangleup (B \bigtriangleup C) = (A \bigtriangleup B) \bigtriangleup C$
   (e) $A \bigtriangleup B = (A \cup B) \setminus (B \cup A)$
   (f) $(A \bigtriangleup B) \cap C = (A \cap C) \bigtriangleup (B \cap C)$

9. Given sets I and A_i for all $i \in I$, show that $\bigcup_{i \in I} A_i$ and $\bigcap_{i \in I} A_i$ are sets.

10. Prove that $A_0 \times A_1 \times \cdots \times A_{n-1}$ is a set if $A_0, A_1, \ldots, A_{n-1}$ are sets.

11. Show that the Cartesian product of a nonempty family of nonempty sets is not empty.

12. Write the Axiom of Choice (5.1.10) as an ST-formula.

13. Demonstrate that a selector is a function.

14. In the proof of Corollary 5.1.11, prove that $\{\{a\} \times [a]_R : a \in A\}$ is a set.

15. Prove that Corollary 5.1.11 implies Axiom 5.1.10.

16. Let $R = \mathbf{R} \times \{0, 1\}$. Find a function $F : \mathbf{R} \to \{0, 1\}$ such that $F(a) = 0$ for all $a \in \mathbf{R}$.

17. Prove the following parts of the proof of Zorn's lemma (5.1.13).
   (a) $\overline{\mathscr{C}}$ and $\mathrm{Ch}(\mathscr{A})$ are sets.
   (b) $\mathrm{Ch}(\mathscr{C})$ and $\cap \mathrm{To}(\mathscr{A})$ are towers.
   (c) $\bigcap \mathrm{To}(\mathscr{A})$ is a chain.

18. Prove that ZF and Zorn's lemma imply the axiom of choice.

19. Use Zorn's lemma to prove that for every function $\varphi : A \to B$, there exists a maximal $C \subseteq A$ such that $\varphi \upharpoonright C$ is one-to-one.

20. Prove that if $\mathscr{A}$ is a collection of sets such that $\bigcup \mathscr{C} \in \mathscr{A}$ for every chain $\mathscr{C} \subseteq \mathscr{A}$, then $\mathscr{A}$ contains a maximal element.

21. Let (A, R) be a partial order. Prove that there exists an order S on A such that $R \subseteq S$ and S is linear.

22. Use the regularity axiom (5.1.15) to prove that if $\{a, \{a, b\}\} = \{c, \{c, d\}\}$, then $a = c$ and $b = d$. This gives an alternative to Kuratowski's definition of an ordered pair (Definition 3.2.8).

23. Prove that there does not exist an infinite sequence of sets $A_0, A_1, A_2, \ldots$ such that $A_{i+1} \in A_i$ for all $i = 0, 1, 2, \ldots$.

24. Prove that the empty set axiom (5.1.2) can be proved from the other axioms of **ZFC**.

25. Use Axiom 5.1.9 to prove that the subset axioms (5.1.8) can be proved from the other axioms of **ZFC**.

26. Let (A, R) be a poset. Prove that every chain of A is contained in a maximal chain with respect to $\subseteq$. This is called the **Hausdorff maximal princple**.

27. Prove that the Hausdorff maximal principle implies Zorn's lemma, so is equivalent to the axiom of choice.

5.2 NATURAL NUMBERS

In order to study mathematics itself, as opposed to studying the contents of mathematics as when we study calculus or Euclidean geometry, we need to first develop a system in which all of mathematics can be interpreted. In such a system, we should be able to precisely define mathematical concepts, like functions and relations; construct examples of them; and write statements about them using a very precise language. **ZFC** with first-order logic seems like a natural choice for such an endeavor. However, in order to be a success, this system must have the ability to represent the most basic objects of mathematical study. Namely, it must be able to model numbers. Therefore, within **ZFC** families of sets that copy the properties of $\mathbf{N}, \mathbf{Z}, \mathbf{Q}, \mathbf{R}$, and $\mathbf{C}$ will be constructed. Since we can do this, we conclude that $\mathbf{N}, \mathbf{Z}, \mathbf{Q}, \mathbf{R}$, and $\mathbf{C}$ are themselves sets, and we also conclude that what we discover about their analogs are true about them. We begin with the natural numbers.

■ **DEFINITION 5.2.1**

For every set a, the **successor** of a is the set $a^+ = a \cup \{a\}$. If a is the successor of b, then b is the **predecessor** of a and we write $b = a^-$.

For example, we have that $\{3, 5, 7\}^+ = \{3, 5, 7, \{3, 5, 7\}\}$ and $\varnothing^+ = \{\varnothing\}$. For convenience, write a^{++} for $(a^+)^+$, so $\varnothing^{++} = \{\varnothing, \{\varnothing\}\}$. Also, the predecessor of the set $\{\varnothing, \{\varnothing\}, \{\varnothing, \{\varnothing\}\}\}$ is $\{\varnothing, \{\varnothing\}\}$, so write $\{\varnothing, \{\varnothing\}, \{\varnothing, \{\varnothing\}\}\}^- = \{\varnothing, \{\varnothing\}\}$. Furthermore, since $\varnothing$ contains no elements, we have the following.

■ **THEOREM 5.2.2**

$\varnothing$ does not have a predecessor.

Although every set has a successor, we are primarily concerned with certain sets that have a particular property.

■ **DEFINITION 5.2.3**

The set A is **inductive** if $\varnothing \in A$ and $a^+ \in A$ for all $a \in A$.

Definition 5.2.3 implies that if A is inductive, A contains the sets

$$\varnothing, \{\varnothing\}, \{\varnothing, \{\varnothing\}\}, \{\varnothing, \{\varnothing\}, \{\varnothing, \{\varnothing\}\}\}, \ldots$$

because

$$\varnothing^+ = \{\varnothing\},$$
$$\{\varnothing\}^+ = \{\varnothing, \{\varnothing\}\},$$
$$\{\varnothing, \{\varnothing\}\}^+ = \{\varnothing, \{\varnothing\}, \{\varnothing, \{\varnothing\}\}\},$$
$$\{\varnothing, \{\varnothing\}, \{\varnothing, \{\varnothing\}\}\}^+ = \{\varnothing, \{\varnothing\}, \{\varnothing, \{\varnothing\}\}, \{\varnothing, \{\varnothing\}, \{\varnothing, \{\varnothing\}\}\}\},$$
$$\vdots$$

Of course, Definition 5.2.3 does not guarantee the existence of an inductive set. The infinity axiom (5.1.3) does that. Then, by a subset axiom (5.1.8),

$$\forall x \exists y \forall u (u \in y \leftrightarrow u \in x \wedge x \text{ is inductive} \wedge \forall w[w \text{ is inductive} \rightarrow u \in w]),$$

where w *is inductive* can be written as

$$\varnothing \in w \wedge \forall a(a \in w \rightarrow a^+ \in w).$$

Therefore, the collection that contains the elements that are common to all inductive sets is a set, and we can make the next definition (von Neumann 1923).

■ **DEFINITION 5.2.4**

An element that is a member of every inductive set is called a **natural number**. Let ω denote the set of natural numbers. That is,

$$\omega = \{\varnothing, \{\varnothing\}, \{\varnothing, \{\varnothing\}\}, \{\varnothing, \{\varnothing\}, \{\varnothing, \{\varnothing\}\}\}, \ldots \}.$$

Definition 5.2.4 suggests that the elements of ω will be interpreted to represent the elements of **N**, so represent each natural number with the appropriate element of **N**:

$$0 = \varnothing,$$
$$1 = \{\varnothing\},$$
$$2 = \{\varnothing, \{\varnothing\}\},$$
$$3 = \{\varnothing, \{\varnothing\}, \{\varnothing, \{\varnothing\}\}\},$$
$$4 = \{\varnothing, \{\varnothing\}, \{\varnothing, \{\varnothing\}\}, \{\varnothing, \{\varnothing\}, \{\varnothing, \{\varnothing\}\}\}\},$$
$$\vdots$$

Although the choice of which sets should represent the numbers of $\mathbf{N}$ is arbitrary, our choice does have some fortunate properties. For example, the number of elements in each natural number and the number of $\mathbf{N}$ that it represents are the same. Moreover, notice that each natural number can be written as

$$0 = \{\ \},$$
$$1 = \{0\},$$
$$2 = \{0, 1\},$$
$$3 = \{0, 1, 2\},$$
$$4 = \{0, 1, 2, 3\},$$
$$\vdots$$

and also note that

$$\bigcup 1 = 0,$$
$$\bigcup 2 = 1,$$
$$\bigcup 3 = 2,$$
$$\bigcup 4 = 3,$$
$$\vdots$$

The empty set is an element of ω by definition, so suppose $n \in \omega$ and take A to be an inductive set. Since n is a natural number, $n \in A$, and since A is inductive, $n^+ \in A$. Because A is arbitrary, we conclude that n^+ is an element of every inductive set, so $n^+ \in \omega$ (Definition 5.2.4). This proves the next theorem.

■ **THEOREM 5.2.5**

ω is inductive.

The proof of the next theorem is left to Exercise 5.

■ **THEOREM 5.2.6**

If A is an inductive set and $A \subseteq \omega$, then $A = \omega$.

Order

Because we want to interpret ω so that it represents $\mathbf{N}$, we should be able to order ω as $\mathbf{N}$ is ordered by $\leq$. That is, we need to find a partial order on ω that is also a well-order. To do this, we start with a definition.

■ **DEFINITION 5.2.7**

A set A is **transitive** means for all a and b, if $b \in a$ and $a \in A$, then $b \in A$.

Observe that $\varnothing$ is transitive because $a \in \varnothing$ is always false. More transitive sets are found in the next theorem.

■ THEOREM 5.2.8

- Every natural number is transitive.

- ω is transitive.

PROOF

The proof of the second part is left to Exercise 4. We prove the first by defining

$$A = \{n \in \omega : n \text{ is transitive}\}.$$

We have already noted that $0 \in A$, so assume that $n \in A$ and let $b \in a$ and $a \in n \cup \{n\}$. If $a \in \{n\}$, then $b \in n$. If $a \in n$, then by hypothesis, $b \in n$. In either case, $b \in n^+$ since $n \subseteq n^+$, so $n^+ \in A$. Hence, A is inductive, so $A = \omega$ by Theorem 5.2.6. ■

In addition to being transitive, each element of ω has another useful property that provides another reason why their choice was a wise one.

■ LEMMA 5.2.9

If $m, n \in \omega$, then $m \subset n$ if and only if $m \in n$.

PROOF

Take $m, n \in \omega$. If $m \in n$, then $m \subset n$ since n is transitive (Exercise 3). To prove the converse, define

$$A = \{k \in \omega : \forall l(l \in \omega \wedge [l \subset k \rightarrow l \in k])\}.$$

We show that A is inductive. Trivially, $0 \in A$. Now take $n \in A$ and let $m \in \omega$ such that $m \subset n^+$. We have two cases to consider.

- Suppose $n \notin m$. Then, $m \subseteq n$. If $m \subset n$, then $m \in n$ by hypothesis, which implies that $m \in n^+$. If $m = n$, then $m \in n^+$.

- Next assume that $n \in m$. Take $x \in n$. Since m is transitive by Theorem 5.2.8, we have that $x \in m$. This implies that $n \cup \{n\} \subseteq m$, but this is impossible since $m \subset n^+$, so $n \notin m$. ■

For example, we see that $2 \in 3$ and $2 \subset 3$ because

$$\{\varnothing, \{\varnothing\}\} \in \{\varnothing, \{\varnothing\}, \{\varnothing, \{\varnothing\}\}\}$$

and

$$\{\varnothing, \{\varnothing\}\} \subset \{\varnothing, \{\varnothing\}, \{\varnothing, \{\varnothing\}\}\}.$$

Now use Lemma 5.2.9 to define an order on ω. Instead of using $\preccurlyeq$, we use $\leq$ to copy the order on **N**.

■ DEFINITION 5.2.10

For all $m, n \in \omega$, let

$$m \leq n \text{ if and only if } m \subseteq n.$$

Define $m < n$ to mean $m \leq n$ but $m \neq n$.

Lemma 5.2.9 in conjunction with Definition 5.2.10 implies that for all $m, n \in \omega$, we have that

$$m < n \text{ if and only if } m \subset n \text{ if and only if } m \in n.$$

The order of Definition 5.2.10 makes ω a chain with $\varnothing$ as its least element.

■ THEOREM 5.2.11

$(\omega, \leq)$ is a linear order.

PROOF

That $(\omega, \leq)$ is a poset follows as in Example 4.3.9. To show that ω is a chain under $\leq$, define

$$A = \{k \in \omega : \forall l (l \in \omega \wedge [k \leq l \vee l \leq k])\}.$$

We prove that A is inductive.

- Since $\varnothing$ is a subset of every set, $0 \in A$.

- Suppose that $n \in A$ and let $m \in \omega$. We have two cases to check. First, assume that $n \leq m$. If $n < m$, then $n^+ \leq m$, while if $n = m$, then $m < n^+$. Now suppose that $m \leq n$, but this implies that $m < n^+$. In either case, $n^+ \in A$. ■

We can now quickly prove the following.

■ COROLLARY 5.2.12

For all $m, n \in \omega$, if $m^+ = n^+$, then $m = n$.

PROOF

Let m and n be natural numbers and assume that

$$m \cup \{m\} = n \cup \{n\}. \tag{5.3}$$

Take $x \in m$. Then, $x \in n$ or $x = n$. If $x \in n$, then $m \subseteq n$, so suppose that $x = n$. This implies that $n \in m$, so $m \neq n$ by Theorem 5.1.16. However, we then have $m \in n$ by (5.3), which contradicts the trichotomy law (Theorem 4.3.21). Similarly, we can prove that $n \subseteq m$. ■

Since the successor defines a function $S : \omega \to \omega$ where $S(n) = n^+$, Corollary 5.2.12 implies that S is one-to-one.

Thereom 5.2.11 shows that $(\omega, \leq)$ is a linear order as $(\mathbf{N}, \leq)$ is a linear order. The next theorem shows that the similarity goes further.

■ **THEOREM 5.2.13**

$(\omega, \leq)$ is a well-ordered set.

PROOF

By Theorem 5.2.11, $(\omega, \leq)$ is a linear order. To prove that it is well-ordered, suppose that $A \subseteq \omega$ such that A does not have a least element. Define

$$B = \{k \in \omega : \{0, 1, \ldots, k\} \cap A = \varnothing\}.$$

We prove that B is inductive.

- If $0 \in A$, then A has a least element because 0 is the least element of ω. Hence, $\{0\} \cap A = \varnothing$, so $0 \in B$.

- Let $n \in B$. This implies that $\{0, 1, \ldots, n\} \cap A$ is empty. Thus, n^+ cannot be an element of A for then it would be the least element of A. Hence, $\{0, 1, \ldots, n, n^+\} \cap A = \varnothing$ proving that $n^+ \in B$.

Therefore, $B = \omega$, which implies that A is empty. ■

Recursion

The familiar **factorial** is defined recursively as

$$0! = 1, \tag{5.4}$$

$$(n + 1)! = (n + 1)n! \quad (n \in \mathbf{N}). \tag{5.5}$$

A **recursive definition** is one that is given in terms of itself. This is illustrated in (5.5) where the factorial is defined using the factorial. It appears that the factorial is a function $\mathbf{N} \to \mathbf{N}$, but a function is a set, so why do (5.4) and (5.5) define a set? Such a definition is not found among the axioms of Section 5.1 or the methods of Section 3.1. That they do define a function requires an important theorem.

■ **THEOREM 5.2.14 [Recursion]**

Let A be a set and $a \in A$. If g is a function $A \to A$, there exists a unique function $f : \omega \to A$ such that

- $f(0) = a$,

- $f(n^+) = g(f(n))$ for all $n \in \omega$.

PROOF

Let $g : A \to A$ be a function. Define

$$\mathscr{F} = \{h : h \subseteq \omega \times A \wedge (0, a) \in h \\ \wedge \forall n \forall y [(n, y) \in h \to (n^+, g(y)) \in h])\}. \tag{5.6}$$

Note that $\mathscr{F}$ is a set by a subset axiom (5.1.8) and $\mathscr{F}$ is nonempty because $\omega \times A \in \mathscr{F}$. Let

$$f = \bigcap \mathscr{F}.$$

Observe that $f \in \mathscr{F}$ (Exercise 8). Define

$$D = \{n \in \omega : \exists z[(n, z) \in f] \land \forall y \forall y'[(n, y) \in f \land (n, y') \in f \to y = y']\}.$$

We prove that D is inductive.

- Since $\mathscr{F} \neq \varnothing$, we know that $(0, a) \in f$ by (5.6), so let $(0, b)$ also be an element of f. If we assume that $a \neq b$, then $f \setminus \{(0, b)\} \in \mathscr{F}$, which is impossible because it implies that $(0, b) \notin \bigcap \mathscr{F}$. Hence, $0 \in D$.

- Suppose that $n \in D$. This means that $(n, z) \in f$ for some $z \in A$. Thus, $(n^+, g(z)) \in f$ by (5.6). Assume that $(n^+, y) \in f$. If $y \neq g(z)$, then $f \setminus \{(n^+, y)\} \in \mathscr{F}$, which again leads to the contradictory $(n^+, y) \notin \bigcap \mathscr{F}$.

Hence, f is a function $\omega \to A$ (Theorem 5.2.6). We confirm that f has the desired properties.

- $f(0) = a$ because $(0, a) \in \bigcap \mathscr{F}$.

- Take $n \in \omega$ and write $y = f(n)$. This implies that $(n, y) \in f$, so we have that $(n^+, g(y)) \in f$. Therefore, $f(n^+) = g(y) = g(f(n))$.

To prove that f is unique, let $f' : \omega \to A$ be a function such that $f'(0) = a$ and $f'(n^+) = g(f'(n))$ for all $n \in \omega$. Let

$$E = \{n \in \omega : f(n) = f'(n)\}.$$

The set E is inductive because $f(0) = a = f'(0)$ and assuming $n \in \omega$ we have that $f(n^+) = g(f(n)) = g(f'(n)) = f'(n^+)$. ∎

The factorial function has domain $\mathbf{N}$ but the recursion theorem (5.2.14) gives a function with domain ω and uses the successor of Definition 5.2.1. We need a connection between $\mathbf{N}$ and ω. We do this by defining two operations on ω that we designate by $+$ and $\cdot$ and then showing that the basic properties of ω under these two operations are the same as the basic properties of $\mathbf{N}$ under standard addition and multiplication.

Arithmetic

We begin with addition. Let $g : \omega \to \omega$ be defined by $g(n) = n^+$. For every $m \in \omega$, by Theorem 5.2.14, there exists a unique function $f_m : \omega \to \omega$ such that

$$f_m(0) = m$$

and for all $n \in \omega$,

$$f_m(n^+) = g(f_m(n)) = [f_m(n)]^+.$$

Define

$$a = \{((m, n), f_m(n)) : m, n \in \omega\}.$$

Since f_m is a function for every $n \in \omega$, the set a is a binary operation (Definition 4.4.26). Observe that for all $m, n \in \omega$,

- $a(m, 0) = m$,

- $a(m, n^+) = a(m, n)^+$.

We know that for every $k, l \in \mathbf{N}$, we have that $k + 0 = k$ and to add $k + l$, one simply adds 1 a total of l times to k. This is essentially what a does to the natural numbers. For example, for $1, 3, 4 \in \mathbf{N}$,

$$1 + 3 = ([(1 + 1) + 1] + 1) = 4$$

and for $1, 2, 3, 4 \in \omega$ (page 238),

$$
\begin{aligned}
a(1, 3) &= a(1, 2^+) \\
&= a(1, 2)^+ \\
&= a(1, 1^+)^+ \\
&= a(1, 1)^{++} \\
&= a(1, 0^+)^{++} \\
&= a(1, 0)^{+++} \\
&= 1^{+++} \\
&= 2^{++} \\
&= 3^+ \\
&= 4.
\end{aligned}
$$

Therefore, we choose a to be addition on ω. To define the addition, we only need to cite the two properties given in Theorem 5.2.14. Since each f_m is unique, there is no other function to which the definition could be referring. Therefore, we can define addition recursively.

■ DEFINITION 5.2.15

For all $m, n \in \omega$,

- $m + 0 = m$,

- $m + n^+ = (m + n)^+$.

Notice that $m^+ = (m + 0)^+ = m + 0^+ = m + 1$. Furthermore, using the notation from page 238, we conclude that $1 + 3 = 4$ because $a(1, 3) = 4$.

The following lemma shows that the addition given in Definition 5.2.15 has a property similar to that of commutativity.

■ LEMMA 5.2.16

For all $m, n \in \omega$,

- $0 + n = n$.

- $m^+ + n = (m + n)^+$.

PROOF

Let $m, n \in \omega$. To prove that $0 + n = n$, we show that

$$A = \{k \in \omega : 0 + k = k\}$$

is inductive.

- $0 \in A$ because $0 + 0 = 0$.

- Let $n \in A$. Then, $0 + n^+ = (0 + n)^+ = n^+$, so $n^+ \in A$.

To prove that $m^+ + n = (m + n)^+$, we show that

$$B = \{k \in \omega : m^+ + k = (m + k)^+\}$$

is inductive.

- Again, $0 \in B$ because $0 + 0 = 0$.

- Suppose that $n \in B$. We have that

$$m^+ + n^+ = (m^+ + n)^+ = (m + n)^{++} = (m + n^+)^+,$$

   where the first and third equality follow by Definition 5.2.15 and the second follows because $n \in B$. Thus, $n^+ \in B$. ■

Now to see that $+$ behaves on ω as $+$ behaves on **N**, we use Definition 4.4.31.

■ THEOREM 5.2.17

- The binary operation $+$ on ω is associative and commutative.

- 0 is the additive identity for ω.

PROOF

0 is the additive identity by Definition 5.2.15 and Lemma 5.2.16, and that $+$ is associative on ω is Exercise 14. To show that $+$ is commutative, let $m \in \omega$ and define

$$A = \{k \in \omega : m + k = k + m\}.$$

As has been our strategy, we show that A is inductive.

- $m + 0 = m$ by Definition 5.2.15, and $0 + m = m$ by Lemma 5.2.16, so $0 \in A$.

- Let $n \in A$. Therefore, $m + n = n + m$, which implies that

$$m + n^+ = (m + n)^+ = (n + m)^+ = n^+ + m.$$

   Hence, $n^+ \in A$. ■

Multiplication on **N** can be viewed as iterated addition. For example,

$$3 \cdot 4 = 3 + 3 + 3 + 3,$$

so we define multiplication recursively along these lines. As with addition, the result is a binary operation by Theorem 5.2.14.

■ **DEFINITION 5.2.18**

For all $m, n \in \omega$,

- $m \cdot 0 = 0$,
- $m \cdot n^+ = m \cdot n + m$.

For example, $3 \cdot 4 = 12$ because

$$
\begin{aligned}
3 \cdot 4 &= 3 \cdot 3 + 3 \\
&= (3 \cdot 2 + 3) + 3 \\
&= ([3 \cdot 1 + 3] + 3) + 3 \\
&= ([(3 \cdot 0 + 3) + 3] + 3) + 3 \\
&= ([3 + 3] + 3) + 3 \\
&= 12,
\end{aligned}
$$

where $([3 + 3] + 3) + 3 = 12$ is left to Exercise 13.

The next result is analogous to Lemma 5.2.16. Its proof is left to Exercise 9.

■ **LEMMA 5.2.19**

For all $m, n \in \omega$,

- $0 \cdot m = 0$,
- $n^+ \cdot m = n \cdot m + m$.

We now prove that $\cdot$ on ω behaves as $\cdot$ on $\mathbf{N}$ and that $+$ and $\cdot$ on ω interact with each other via the distributive law as the two operations on $\mathbf{N}$. For the proof, we introduce two common conventions for these two operations.

- So that we lessen the use of parentheses, define $\cdot$ to have precedence over $+$, and read from left to right. That is,

$$
m \cdot n + o = (m \cdot n) + o \text{ and } m + n \cdot o = m + (n \cdot o),
$$

and

$$
m + n + o = (m + n) + o \text{ and } m \cdot n \cdot o = (m \cdot n) \cdot o.
$$

- Define $mn = m \cdot n$.

■ **THEOREM 5.2.20**

- The binary operation $\cdot$ on ω is associative and commutative.
- 1 is the multiplicative identity for ω.
- The **distributive law** holds for ω. This means that for all $m, n, o \in \omega$,

$$
m(n + o) = mn + mo.
$$

PROOF

That the associative and commutative properties hold is Exercise 12. To prove the other parts of the theorem, let $m, n, o \in \omega$. Since $0^+ = 1$, by Definition 5.2.18 and Lemma 5.2.16,

$$m1 = m0^+ = m0 + m = 0 + m = m,$$

and by Lemma 5.2.19 and Definition 5.2.15, we have that $1m = m$. To show that the distributive law holds, define

$$A = \{k \in \omega : k(n + o) = kn + ko\}.$$

Since $0(n + o) = 0$ and $0n + 0o = 0 + 0 = 0$, we have that $0 \in A$. Now suppose that $m \in A$. That is,

$$m(n + o) = mn + mo.$$

Therefore, $m^+ \in A$ because

$$
\begin{aligned}
m^+(n + o) &= m(n + o) + (n + o) \\
&= (mn + mo) + (n + o) \\
&= (mn + n) + (mo + o) \\
&= m^+n + m^+o. \quad \blacksquare
\end{aligned}
$$

■ **EXAMPLE 5.2.21**

We revisit the factorial function. Since addition and multiplication are now defined on ω, let $g : \omega \times \omega \to \omega \times \omega$ be the function

$$g(k, l) = (k + 1, (k + 1)l).$$

Theorem 5.2.14 implies that there exists a unique function $f : \omega \to \omega \times \omega$ such that

$$f(0) = (0, 1)$$

and

$$f(n + 1) = g(f(n)).$$

Let π be the projection map $\pi(k, l) = l$. By Theorem 5.2.6 and Exercise 10, we conclude that for all $n \in \omega$,

$$f(n + 1) = (n + 1, (n + 1)(\pi \circ f)(n)). \tag{5.7}$$

Therefore, the factorial function $n!$, which is defined recursively by

$$0! = 1,$$
$$(n + 1)! = (n + 1)n! \quad (n \in \omega),$$

is $\pi \circ f$ by the uniqueness of f.

To solve an equation like $6 + 2x = 14$ where the coefficients are from **N**, we can use the cancellation law and write

$$6 + 2x = 14,$$
$$6 + 2x = 6 + 8,$$
$$2x = 8,$$
$$2x = 2 \cdot 4,$$
$$x = 4.$$

For equations with coefficients in ω, we need a similar law.

■ THEOREM 5.2.22 [Cancellation]

Let $a, b, c \in \omega$.

- If $a + b = a + c$, then $b = c$.

- If $ab = ac$ and $a \neq 0$, then $b = c$.

PROOF

The proof for multiplication is Exercise 15. For addition, define

$$A = \{k \in \omega : \forall l \forall m (l \in \omega \wedge m \in \omega \wedge [k + l = k + m \to l = m])\}.$$

Clearly, $0 \in A$, so assume $n \in A$. To prove that $n^+ \in A$, let $b, c \in \omega$ and suppose that $n^+ + b = n^+ + c$. Then, $(n+b)^+ = (n+c)^+$ by Lemma 5.2.16, so $n+b = n+c$ by Corollary 5.2.12. Hence, $b = c$ because $n \in A$. ■

Exercises

1. For every set A, show that A^+ is a set.

2. For every $n \in \omega$, show that $n^{+++++} = n + 5$.

3. Show that the following are equivalent.
   - A is transitive.
   - If $a \in A$, then $a \subset A$.
   - $A \subseteq P(A)$.
   - $\bigcup A \subseteq A$.
   - $\bigcup(A^+) = A$.

4. Prove that ω is transitive.

5. Prove Theorem 5.2.6.

6. Let $A \subseteq \omega$. Show that if $\bigcup A = A$, then $A = \omega$.

7. Take $u, v, x, y \in \omega$ and assume that $u + x = v + y$. Prove that $u \in v$ if and only if $y \in x$.

8. Prove that $f \in \mathscr{F}$ in the proof of Theorem 5.2.14.

9. Prove Lemma 5.2.19.

10. Prove (5.7) from Example 5.2.21.

11. Let A be a set and $\varphi : A \to A$ be a one-to-one function. Take $a \in A \setminus \text{ran}(\varphi)$. Recursively define $f : \omega \to A$ such that

$$f(0) = a,$$
$$f(n^+) = \varphi(f(n)).$$

Prove that f is one-to-one.

12. Let $m, n, o \in \omega$. Prove the given equations.
   (a) $mn = nm$.
   (b) $m(n + o) = mn + mo$.

13. Show that $([3 + 3] + 3) + 3 = 12$.

14. Prove that addition on ω is associative.

15. For all $a, b, c \in \omega$ with $a \neq 0$, prove that if $ab = ac$, then $b = c$.

16. Show that for all $m, n \in \omega$, if $mn = 0$, then $m = 0$ or $n = 0$.

17. We define exponentiation on ω. For all $n \in \omega$,

$$n^0 = 1,$$
$$n^{k^+} = n^k \cdot n.$$

   (a) Use the recursion (Theorem 5.2.14) to prove that this defines a function $\omega \times \omega \to \omega$.
   (b) Show that exponentiation on ω is one-to-one.

18. Let $x, y, x \in \omega$. Use the definition of Exercise 17 to prove the given equations.
   (a) $x^{y+z} = x^y x^z$.
   (b) $(xy)^z = x^z y^z$.
   (c) $(x^y)^z = x^{yz}$.

19. Let $m, n, k \in \omega$. Assume that $m \leq n$ and $0 \leq k$. Demonstrate the following.
   (a) $m + k \leq n + k$.
   (b) $mk \leq nk$.
   (c) $m^k \leq n^k$.

5.3 INTEGERS AND RATIONAL NUMBERS

Now that we have defined within set theory the set of natural numbers and confirmed its basic properties, we wish to continue this with other sets of numbers. We begin with the integers.

Integers

We build the integers using the natural numbers. The problem is how to define the negative integers. We need to decide how to represent the adjoining of the negative sign to a natural number. One option that might work is to use ordered pairs. These are always good options when extra information needs to be included with each element of a set. For example, $(4, 0)$ could represent 4 because $4 - 0 = 4$, and $(0, 4)$ could represent -4 because $0 - 4 = -4$. However, this is a problem because we did not define subtraction on ω, so we need another solution. Our decision is to generalize this idea of subtracting coordinates to the set $\omega \times \omega$, but use addition to do it. Since there are infinitely many pairs (m, n) such that $m - n = 4$, we equate them by using the equivalence relation of Exercise 4.2.2.

■ **DEFINITION 5.3.1**

Let R be the equivalence relation on $\omega \times \omega$ defined by

$$(m, n) \; R \; (m', n') \text{ if and only if } m + n' = m' + n.$$

Define $\mathbb{Z} = (\omega \times \omega)/R$ to be the set of **integers**.

Using Definition 5.3.1 we associate elements of **Z** with elements of $\mathbb{Z}$:

Z	$\mathbb{Z}$
$\vdots$	$\vdots$
-2	$[(0, 2)]$
-1	$[(0, 1)]$
0	$[(0, 0)]$
1	$[(1, 0)]$
2	$[(2, 0)]$
$\vdots$	$\vdots$

Notice that because 0 and 2 are elements of ω, the equivalence class $[(0, 2)]$ is the name for $[(\varnothing, \{\varnothing, \{\varnothing\}\})]$. Also,

$$[(0, 2)] = [(1, 3)] = [(2, 4)] = \cdots .$$

The ordering of $\mathbb{Z}$ is defined in terms of the ordering on ω. Be careful to note that the symbol $\leq$ will represent two different orders, one on $\mathbb{Z}$ and one on ω (Definition 5.2.10). This overuse of the symbol will not lead to confusion because the order will be clear from context. For example, in the next definition, the first $\leq$ is the order on $\mathbb{Z}$, and the second $\leq$ is the order on ω.

■ **DEFINITION 5.3.2**

For all $[(m, n)]$, $[(m', n')] \in \mathbb{Z}$, define

$$[(m, n)] \leq [(m', n')] \text{ if and only if } m + n' \leq m' + n$$

and

$$[(m, n)] < [(m', n')] \text{ if and only if } m + n' < m' + n.$$

Since $-4 = [(0, 4)]$ and $3 = [(5, 2)]$, we have that $-4 < 3$ because $0 + 2 < 5 + 4$.

Since $(\mathbb{Z}, \leq)$ has no least element (Exercise 3), it is not well-ordered. However, it is a chain. Note that in the proof the symbol $\leq$ is again overused.

■ THEOREM 5.3.3

$(\mathbb{Z}, \leq)$ is a linear order.

PROOF

We use the fact that the order on ω is a linear order (Theorem 5.2.11). Let $a, b \in \mathbb{Z}$, so $a = (m, n)$ and $b = (m', n')$ for some $m, n, m', n' \in \omega$.

- Because $m + n \leq m + n$, we have $a \leq a$.

- Suppose that $a \leq b$ and $b \leq a$. This implies that $m + n' \leq m' + n$ and $m' + n \leq m + n'$. Since $\leq$ is antisymmetric on ω, $m + n' = m' + n$, which implies that $a = b$, so $\leq$ is antisymmetric on $\mathbb{Z}$.

- Using a similar strategy, it can be shown that $\leq$ is transitive on $\mathbb{Z}$ (Exercise 4). Thus, $(\mathbb{Z}, \leq)$ is a poset.

- Since $(\omega, \leq)$ is a linear order, $m + n' \leq m' + n$ or $m' + n \leq m + n'$. This implies that $a \leq b$ or $b \leq a$, so $\mathbb{Z}$ is a chain under $\leq$. ■

Although ω is not a subset of $\mathbb{Z}$ like $\mathbf{N}$ is a subset of $\mathbf{Z}$, the set ω can be embedded in $\mathbb{Z}$ (Definition 4.5.29). This is shown by the function $\varphi : \omega \to \mathbb{Z}$ defined by

$$\varphi(n) = [(n, 0)]. \tag{5.8}$$

We see that φ is one-to-one because $[(m, 0)] = [(n, 0)]$ implies that $m = n$. The function φ is then an order isomorphism using the relation from Definition 5.3.2 (Exercise 1).

As with ω, we define what it means to add and multiply two integers.

■ DEFINITION 5.3.4

Let $[(m, n)], [(u, v)] \in \mathbb{Z}$.

- $[(m, n)] + [(u, v)] = [(m + u, n + v)]$.

- $[(m, n)] \cdot [(u, v)] = [(mu + nv, mv + un)]$.

For example, the equation in $\mathbb{Z}$,

$$[(5, 2)] + [(7, 1)] = [(12, 3)],$$

corresponds to the equation in $\mathbf{Z}$,

$$3 + 6 = 9,$$

and the equation

$$[(5, 2)] \cdot [(7, 1)] = [(35 + 2, 5 + 14)] = [(37, 19)]$$

corresponds to the equation

$$3 \cdot 6 = 18.$$

Before we check the properties of $+$ and $\cdot$ on $\mathbb{Z}$, we should confirm that these are well-defined. Exercise 5 is for addition. To prove that multiplication is well-defined, let $[(m, n)] = [(m', n')]$ and $[(u, v)] = [(u', v')]$. Then, $m + n' = m' + n$ and $u + v' = u' + v$. Hence, by Theorem 5.2.20 in ω we have that

$$mu + n'u = m'u + nu,$$
$$mv + n'v = m'v + nv,$$
$$m'u + m'v' = m'u' + m'v,$$
$$n'u + n'v' = n'u' + n'v.$$

Therefore,

$$mu + n'u + m'v + nv + m'u + m'v' + n'u' + n'v$$

equals

$$m'u + nu + mv + n'v + m'u' + m'v + n'u + n'v',$$

so by Cancellation (Theorem 5.2.22),

$$mu + nv + m'v' + n'u' = mv + nu + m'u' + n'v'.$$

This implies by Definition 5.3.1 that

$$[(mu + nv, mv + nu)] = [(m'u' + n'v', m'v' + n'u')],$$

and this by Definition 5.3.4 implies that

$$[(m, n)] \cdot [(u, v)] = [(m', n')] \cdot [(u', v')].$$

We follow the same order of operations with $+$ and $\cdot$ on $\mathbb{Z}$ as on ω and also write mn for $m \cdot n$.

■ THEOREM 5.3.5

- The binary operations $+$ and $\cdot$ on $\mathbb{Z}$ are associative and commutative.

- $[(0, 0)]$ is the additive identity for $\mathbb{Z}$, and $[(1, 0)]$ is its multiplicative identity.

- For every $a \in \mathbb{Z}$, there exists an additive inverse of a.

- For all $a, b, c \in \mathbb{Z}$, the distributive law holds.

PROOF

We will prove parts of the first and third properties and leave the remaining properties to Exercise 6. Let $a, b, c \in \mathbb{Z}$. This means that there exist natural numbers $m, n, r, s, u,$ and v such that $a = [(m, n)]$, $b = [(r, s)]$, and $c = [(u, v)]$. Then,

$$
\begin{aligned}
a + b + c &= [(m, n)] + [(r, s)] + [(u, v)] \\
&= [(m + r, n + s)] + [(u, v)] \\
&= [(m + r + u, n + s + v)] \\
&= [(m + [r + u], n + [s + v])] \\
&= [(m, n)] + [(r + u, s + v)] \\
&= [(m, n)] + ([(r, s)] + [(u, v)]) \\
&= a + (b + c)
\end{aligned}
$$

and

$$
\begin{aligned}
abc &= [(m, n)] \cdot [(r, s)] \cdot [(u, v)] \\
&= [(mr + ns, ms + nr)] \cdot [(u, v)] \\
&= [(u[mr + ns] + v[ms + nr], v[mr + ns] + u[ms + nr])] \\
&= [(umr + uns + vms + vnr, vmr + vns + ums + unr)] \\
&= [(m[ru + sv] + n[su + rv], m[su + rv] + n[ru + sv])] \\
&= [(m, n)] \cdot [(ru + sv, su + rv)] \\
&= [(m, n)] \cdot ([(r, s)] \cdot [(u, v)]) \\
&= a(bc).
\end{aligned}
$$

To prove that every element of $\mathbb{Z}$ has an additive inverse, notice that

$$
[(m, n)] + [(n, m)] = [(m + n, n + m)] = [(0, 0)].
$$

Therefore, if $a = [(m, n)]$, the additive inverse of a is $[(n, m)]$. ∎

For all $n \in \mathbb{Z}$, denote the additive inverse of n by $-n$, and for all $m, n, r, s \in \omega$, define

$$
[(m, n)] - [(r, s)] = [(m, n)] + [(s, r)].
$$

Rational Numbers

As the integers were built using ω, so the set of rational numbers will be built using $\mathbb{Z}$. Its definition is motivated by the behavior of fractions in $\mathbf{Q}$. For instance,

$$
\frac{2}{3} = \frac{8}{12}
$$

because $2 \cdot 12 = 3 \cdot 8$. Imagining that the ordered pair (m, n) represents the fraction m/n, we define an equivalence relation. Notice that this is essentially the relation from Example 4.2.5. Notice that $\mathbb{Z}$ is defined using addition (Definition 5.3.1) while the rational numbers are defined using multiplication.

■ **DEFINITION 5.3.6**

Let S be the equivalence relation on $\mathbb{Z} \times (\mathbb{Z} \setminus \{0\})$ defined by

$$(m, n) \, S \, (m', n') \text{ if and only if } mn' = m'n.$$

Define $\mathbb{Q} = [\mathbb{Z} \times (\mathbb{Z} \setminus \{0\})]/S$ to be the set of **rational numbers**.

Using Definition 5.3.6, we associate elements of $\mathbb{Q}$ with elements of Q:

$\mathbb{Q}$	Q		$\mathbb{Q}$	Q
$\vdots$	$\vdots$		$\vdots$	$\vdots$
-2	$[(-2, 1)]$		$1/2$	$[(1, 2)]$
-1	$[(-1, 1)]$		$2/3$	$[(2, 3)]$
0	$[(0, 1)]$		$3/4$	$[(3, 4)]$
1	$[(1, 1)]$		$4/5$	$[(4, 5)]$
2	$[(2, 1)]$		$5/6$	$[(5, 6)]$
$\vdots$	$\vdots$		$\vdots$	$\vdots$

Notice that since $1, 2 \in \mathbb{Z}$, the equivalence class $[(1, 2)]$ is the name for

$$[[(1, 0)]_R, [(2, 0)]_R]_S = [[(\{\varnothing\}, \varnothing)]_R, [(\{\varnothing, \{\varnothing\}\}, \varnothing)]_R]_S,$$

where R is the relation of Definition 5.3.1. Also,

$$[(1, 2)] = [(2, 4)] = [(3, 6)] = \cdots.$$

We next define a partial order on $\mathbb{Q}$.

■ **DEFINITION 5.3.7**

For all $[(m, n)], [(m', n')] \in \mathbb{Q}$, define

$$[(m, n)] \leq [(m', n')] \text{ if and only if } mn' \leq nm'$$

and

$$[(m, n)] < [(m', n')] \text{ if and only if } mn' < nm'.$$

When working with $\mathbb{Q}$, we often denote $[(m, n)]$ by m/n or $\frac{m}{n}$ and $[(m, 1)]$ by m. Thus, since $2/3 = [(2, 3)]$ and $7/8 = [(7, 8)]$, we conclude that $2/3 < 7/8$ because $2 \cdot 8 < 3 \cdot 7$.

As ω can be embedded in $\mathbb{Z}$, so $\mathbb{Z}$ can be embedded in $\mathbb{Q}$. The function that can be used is

$$\psi : \mathbb{Z} \to [\mathbb{Z} \times (\mathbb{Z} \setminus \{0\})]/S$$

defined by

$$\psi(n) = [(n, 1)]. \tag{5.9}$$

(See Exercise 2) Using the function φ (5.8), we see that ω can be embedded in $\mathbb{Q}$ via the order isomorphism $\psi \circ \varphi$.

Since $(\mathbb{Q}, \leq)$ has no least element (Exercise 3), it is not well-ordered. However, it is a chain (Exercise 11).

■ **THEOREM 5.3.8**

$(\mathbb{Q}, \leq)$ is a linear order.

Lastly, we define two operations on $\mathbb{Q}$ that represent the standard operations of $+$ and $\cdot$ on **Q**.

■ **DEFINITION 5.3.9**

Let $[(m, n)], [(m', n')] \in \mathbb{Q}$.

- $[(m, n)] + [(m', n')] = [(mn' + m'n, nn')]$.

- $[(m, n)] \cdot [(m', n')] = [(mm', nn')]$.

That $+$ and $\cdot$ as given in Definition 5.3.9 are binary operations is left to Exercise 13.
   As examples of these two operations, the equation in $\mathbb{Q}$

$$[(1, 3)] + [(3, 4)] = [(1 \cdot 4 + 3 \cdot 3, 3 \cdot 4)] = [(13, 12)]$$

corresponds to the equation in **Q**

$$\frac{1}{3} + \frac{3}{4} = \frac{4}{12} + \frac{9}{12} = \frac{13}{12}$$

and the equation in $\mathbb{Q}$

$$[(1, 3)] \cdot [(3, 4)] = [(1 \cdot 3, 3 \cdot 4)] = [(3, 12)] = [(1, 4)]$$

corresponds to the equation in **Q**

$$\frac{1}{3} \cdot \frac{3}{4} = \frac{3}{12} = \frac{1}{4}.$$

We follow the same order of operations with $+$ and $\cdot$ on $\mathbb{Q}$ as on $\mathbb{Z}$ and also write mn for $m \cdot n$.

■ **THEOREM 5.3.10**

- The binary operations $+$ and $\cdot$ on $\mathbb{Q}$ are associative and commutative.

- $[(0, 1)]$ is the additive identity, and $[(1, 1)]$ is the multiplicative identity.

- Every rational number has an additive inverse, and every element of $\mathbb{Q} \setminus \{[(0, 1)]\}$ has a multiplicative inverse.

- The distributive law holds.

PROOF

Since $[(m, n)] \cdot [(1, 1)] = [(m \cdot 1, n \cdot 1)] = [(m, n)]$ for all $[(m, n)] \in \mathbb{Q}$ and multiplication is commutative, the multiplicative identity of $\mathbb{Q}$ is $[(1, 1)]$. Also, because $m \neq 0$ and $n \neq 0$ implies that

$$[(m, n)] \cdot [(n, m)] = [(mn, nm)] = [(1, 1)],$$

every element of $\mathbb{Q} \setminus [(0, 1)]$ has a multiplicative inverse. The other properties are left to Exercise 14. ∎

For all $a, b \in \mathbb{Q}$ with $b \neq 0$, denote the additive inverse of a by $-a$ and the multiplicative inverse of b by b^{-1}.

Actual Numbers

We now use the elements of ω, $\mathbb{Z}$, and $\mathbb{Q}$ as if they were the actual elements of **N**, **Z**, and **Q**. For example, we understand the formula $n \in \mathbb{Z}$ to mean that n is an integer of Definition 5.3.1 with all of the properties of $n \in \mathbf{Z}$. Also, when we write that the formula $p(n)$ is satisfied by some rational number a, we interpret this to mean that $p(a)$ with $a \in \mathbb{Q}$ where a has all of the properties of $a \in \mathbf{Q}$. This allows us to use the set properties of a natural number, integer, or rational number when needed yet also use the results we know concerning the actual numbers. That the partial orders and binary operations defined on ω, $\mathbb{Z}$, and $\mathbb{Q}$ have essentially the same properties as those on **N**, **Z**, and **Q** allows for this association of properties to be legitimate.

Exercises

1. Prove that the function φ (5.8) is an order isomorphism using the order of Definition 5.3.2.

2. Prove that the function ψ (5.9) is an order isomorphism using the order of Definition 5.3.7.

3. Show that $(\mathbb{Z}, \leq)$ and $(\mathbb{Q}, \leq)$ do not have least elements.

4. Prove that $\leq$ is transitive on $\mathbb{Z}$.

5. Prove that $+$ is well-defined on $\mathbb{Z}$.

6. Complete the remaining proofs of the properties of Theorem 5.3.5.

7. Show that every nonempty subset of $\mathbb{Z}^- = \{n : n \in \mathbb{Z} \wedge n < 0\}$ has a greatest element.

8. Let $m, n \in \mathbb{Z}$ and $k \in \mathbb{Z}^-$ (Exercise 7) Prove that if $m \leq n$, then $nk \leq mk$.

9. Prove that for every $m, n \in \mathbb{Z}$, if $mn = 0$, then $m = 0$ or $n = 0$. Show that this same result also holds for $\mathbb{Q}$.

10. The **cancellation law** for the integers states that for all $m, n, k \in \mathbb{Z}$,
   - if $m + k = n + k$, then $m = n$,
   - if $mk = nk$ and $k \neq 0$, then $m = n$.

Prove that the cancellation law holds for $\mathbb{Z}$. Prove that a similar law holds for $\mathbb{Q}$.

11. Prove that $(\mathbb{Q}, \leq)$ is a chain.

12. Demonstrate that every nonempty set of integers with a lower bound with respect to $\leq$ has a least element.

13. Show that $+$ and $\cdot$ as given in Definition 5.3.9 are binary operations on $\mathbb{Q}$.

14. Prove the remaining parts of Theorem 5.3.10.

15. Let $a, b, c, d \in \mathbb{Q}$. Prove.
  (a) If $0 \leq a \leq c$ and $0 \leq b \leq d$, then $ab \leq cd$.
  (b) If $0 \leq a < c$ and $0 \leq b < d$, then $ab < cd$.

16. Let $a, b \in \mathbb{Q}$. Prove that if $b < 0$, then $a + b < a$.

17. Let $a, b \in \mathbb{Q}$. Prove that if $a \geq 0$ and $b < 1$, then $ab \leq a$.

18. Prove that between any two rational numbers is a rational number. This means that $\mathbb{Q}$ is **dense**. Show that this is not the case for $\mathbb{Z}$.

19. Generalize the definition of exponentiation given in Exercise 5.2.17 to $\mathbb{Z}$. Prove that this defines a one-to-one function $\mathbb{Z} \times (\mathbb{Z}^+ \cup \{0\}) \to \mathbb{Z}$. Does the proof require recursion (Theorem 5.2.14)?

20. Let $m, n, k \in \mathbb{Z}$. Assume that $m \leq n$. Demonstrate the following.
  (a) $m + k \leq n + k$.
  (b) $mk \leq nk$ if $0 \leq k$.
  (c) $nk \leq mk$ if $k < 0$.
  (d) $m^k \leq n^k$ if $k \geq 0$.

21. Let $\varphi : \omega \to \mathbb{Z}$ be the embedding defined by (5.8). Define $f : \omega \times \omega \to \omega$ by $f(m, n) = m^n$ and $g : \mathbb{Z} \times (\mathbb{Z}^+ \cup \{0\}) \to \mathbb{Z}$ by $g(u, v) = u^v$. Prove that for all $m, n \in \omega$, we have that $g(\varphi(m), \varphi(n)) = \varphi(f(m, n))$. Explain the significance of this result.

5.4   MATHEMATICAL INDUCTION

Suppose that we want to prove $p(n)$ for all integers n greater than or equal to some n_0. Our previous method (Section 2.4) is to take an arbitrary $n \geq n_0$ and try to prove $p(n)$. If this is not possible, we might be tempted to try proving $p(n)$ for each n individually. This is impossible because it would take infinitely many steps. Instead, we combine the results of Sections 5.2 and 5.3 and use the next theorem.

■ **THEOREM 5.4.1 [Mathematical Induction 1]**

Let $p(k)$ be a formula. For any $n_0 \in \mathbb{Z}$, if

$$p(n_0) \wedge \forall n \in \mathbb{Z}[n \geq n_0 \wedge p(n) \to p(n + 1)],$$

then

$$\forall n \in \mathbb{Z}[n \geq n_0 \to p(n)].$$

PROOF

Assume $p(n_0)$ and that $p(k)$ implies $p(k+1)$ for all integers $k \geq n_0$. Define

$$A = \{k \in \omega : p(n_0 + k)\}.$$

Notice that $0 \in A \Leftrightarrow p(n_0)$, $1 \in A \Leftrightarrow p(n_0 + 1)$, and so on. To prove $p(k)$ for all integers $k \geq n_0$, we show that A is inductive.

- By hypothesis, $0 \in A$ because $p(n_0)$.

- Assume $n \in A$. This implies that $p(n_0 + n)$. Therefore, $p(n_0 + n + 1)$, so $n + 1 \in A$. ∎

Theorem 5.4.1 gives rise to a standard proof technique known as **mathematical induction**. First, prove $p(n_0)$. Then, show that $p(n)$ implies $p(n + 1)$ for every integer $n \geq n_0$. Often an analogy of dominoes is used to explain this. Proving $p(n_0)$ is like tipping over the first domino, and then proving the implication shows that the dominoes have been set properly. This means that by *modus ponens*, if $p(n_0 + 1)$ is true, then $p(n_0 + 2)$ is true, and so forth, each falling like dominoes:

$$p(n_0)$$
$$p(n_0) \to p(n_0 + 1) \quad \therefore p(n_0 + 1)$$
$$p(n_0 + 1) \to p(n_0 + 2) \quad \therefore p(n_0 + 2) \cdots$$

This two-step process is characteristic of proofs by mathematical induction. So much so, the two stages have their own terminology.

- Proving $p(n_0)$ is called the **basis case**. It is typically the easiest part of the proof but should, nonetheless, be explicitly shown.

- Proving that $p(n)$ implies $p(n + 1)$ is the **induction step**. For this, we typically use direct proof, assuming $p(n)$ to show $p(n + 1)$. The assumption is called the **induction hypothesis**.

Often induction is performed to prove a formula for all positive integers, so to represent this set, define

$$\mathbb{Z}^+ = \{1, 2, 3, \dots\}.$$

■ **EXAMPLE 5.4.2**

Prove $p(k)$ for all $k \in \mathbb{Z}^+$, where

$$p(k) := 1^2 + 2^2 + \cdots + k^2 = \frac{k(k+1)(2k+1)}{6}.$$

Proceed by mathematical induction.

- The basis case $p(1)$ holds because

$$\frac{1(1+1)(2 \cdot 1 + 1)}{6} = \frac{1(2)(3)}{6} = 1^2.$$

- Now for the induction step, assume

$$1^2 + 2^2 + \cdots + n^2 = \frac{n(n+1)(2n+1)}{6}. \tag{5.10}$$

This is the induction hypothesis. We must show that the equation holds for $n + 1$. Adding $(n + 1)^2$ to both sides of (5.10) gives

$$1^2 + 2^2 + \cdots + n^2 + (n+1)^2 = \frac{n(n+1)(2n+1)}{6} + (n+1)^2$$

$$= \frac{(n+1)[n(2n+1) + 6(n+1)]}{6}$$

$$= \frac{(n+1)(2n^2 + 7n + 6)}{6}$$

$$= \frac{(n+1)(n+2)(2n+3)}{6}$$

$$= \frac{(n+1)([n+1] + 1)(2[n+1] + 1)}{6}.$$

■ EXAMPLE 5.4.3

Let x be a positive rational number. To prove that for any $k \in \mathbb{Z}^+$,

$$(x + 1)^k \geq x^k + 1,$$

we proceed by mathematical induction.

- $(x + 1)^1 = x + 1 = x^1 + 1$.

- Let $n \in \mathbb{Z}^+$ and assume that $(x + 1)^n \geq x^n + 1$. By multiplying both sides of the given inequality by $x + 1$, we have

$$(x + 1)^n(x + 1) \geq (x^n + 1)(x + 1)$$

$$= x^{n+1} + x^n + x + 1$$

$$\geq x^{n+1} + 1.$$

The first inequality is true by induction (that is, by appealing to the induction hypothesis) and since $x + 1$ is positive, and the last one holds because $x \geq 0$.

■ EXAMPLE 5.4.4

Recursively define a sequence of numbers,

$$a_1 = 3,$$
$$a_n = 2a_{n-1} \text{ for all } n \in \mathbb{Z} \text{ such that } n > 1.$$

The sequence is

$$a_1 = 3, a_2 = 6, a_3 = 12, a_4 = 24, a_5 = 48, \ldots,$$

and we conjecture that $a_k = 3 \cdot 2^{k-1}$ for all positive integers k. We prove this by mathematical induction.

- For the basis case, $a_1 = 3 \cdot 2^0 = 3$.

- Let $n > 1$ and assume $a_n = 3 \cdot 2^{n-1}$. Then,

$$a_{n+1} = 2a_n = 2 \cdot 3 \cdot 2^{n-1} = 3 \cdot 2^n.$$

■ EXAMPLE 5.4.5

Use mathematical induction to prove that $k^3 < k!$ for all integers, $k \geq 6$.

- First, show that the inequality holds for $n = 6$:

$$6^3 = 216 < 720 = 6!.$$

- Assume $n^3 < n!$ with $n \geq 6$. The induction hypothesis yields three inequalities. Namely,

$$3n^2 < n \cdot n^2 \leq n!,$$

$$3n < n \cdot n < n^3 \leq n!,$$

and

$$1 < n!.$$

Therefore,

$$(n+1)^3 = n^3 + 3n^2 + 3n + 1$$
$$< n! + n! + n! + n!$$
$$= 4n!$$
$$< (n+1)n!$$
$$= (n+1)!.$$

Combinatorics

We now use mathematical induction to prove some basic results from two areas of mathematics. The first is **combinatorics**, the study of the properties that sets have based purely on their size.

A **permutation** of a given set is an arrangement of the elements of the set. For example, the number of permutations of $\{a, b, c, d, e, f\}$ is 720. If we were to write all of the permutations in a list, it would look like the following:

$$
\begin{array}{cccccc}
a & b & c & d & e & f \\
a & b & c & d & f & e \\
a & b & c & e & d & f \\
& & \vdots & & & \\
f & e & d & c & a & b \\
f & e & d & c & b & a \\
\end{array}
$$

We observe that $6! = 720$ and hypothesize that the number of permutations of a set with k elements is $k!$. To prove it, we use mathematical induction.

- There is only one way to write the elements of a singleton. Since $1! = 1$, we have proved the basis case.

- Assume that the number of permutations of a set with $n \geq 1$ elements is $n!$. Let $A = \{a_1, a_2, \ldots, a_{n+1}\}$ be a set with $n + 1$ elements. By induction, there are $n!$ permutations of the set $\{a_1, a_2, \ldots, a_n\}$. After writing the permutations in a list, notice that there are $n + 1$ columns before, between, and after each element of the permutations:

$$
\left| \quad \left| \begin{array}{c} a_1 \\ a_1 \\ \vdots \\ a_n \end{array} \right| \quad \left| \begin{array}{c} a_2 \\ a_2 \\ \vdots \\ a_{n-1} \end{array} \right| \quad \left| \begin{array}{c} \cdots \\ \cdots \\ \end{array} \right| \quad \left| \begin{array}{c} a_n \\ a_{n-1} \\ \vdots \\ a_1 \end{array} \right| \quad \right|
$$

To form the permutations of A, place a_{n+1} into the positions of each empty column. For example, if a_{n+1} is put into the first column, the following permutations are obtained:

$$
\begin{array}{ccccc}
a_{n+1} & a_1 & a_2 & \cdots & a_n \\
a_{n+1} & a_1 & a_2 & \cdots & a_{n-1} \\
\vdots & \vdots & \vdots & & \vdots \\
a_{n+1} & a_n & a_{n-1} & \cdots & a_1 \\
\end{array}
$$

Since there are $n!$ rows with $n + 1$ ways to add a_{n+1} to each row, we conclude that there are $(n + 1)n! = (n + 1)!$ permutations of A.

This argument proves the first theorem.

■ **THEOREM 5.4.6**

Let $n \in \mathbb{Z}^+$. The number of permutations of a set with n elements is $n!$.

Suppose that we do not want to rearrange the entire set but only subsets of it. For example, let $A = \{a, b, c, d, e\}$. To see all three-element permutations of A, look at the following list:

$$
\begin{array}{cccccc}
abc & acb & bac & bca & cab & cba \\
abd & adb & bad & bda & dab & dba \\
abe & aeb & bae & bea & eab & eba \\
acd & adc & cad & cda & dac & dca \\
ace & aec & cae & cea & eac & eca \\
ade & aed & dae & dea & ead & eda \\
bcd & bdc & cbd & cdb & dbc & dcb \\
bce & bec & ceb & cbe & ebc & ecb \\
bde & bed & dbe & deb & ebd & edb \\
cde & ced & dce & dec & ecd & edc
\end{array}
$$

There are 60 arrangements because there are 5 choices for the first entry. Once that is chosen, there are only 4 left for the second, and then 3 for the last. We calculate that as

$$
60 = 5 \cdot 4 \cdot 3 = \frac{5 \cdot 4 \cdot 3 \cdot 2 \cdot 1}{2 \cdot 1} = \frac{5!}{(5-3)!}.
$$

Generalizing, we define for all $n, r \in \omega$,

$$
{}_nP_r = \frac{n!}{(n-r)!},
$$

and we conclude the following theorem.

■ **THEOREM 5.4.7**

Let $r, n \in \mathbb{Z}^+$. The number of permutations of r elements from a set with n elements is ${}_nP_r$.

Now suppose that we only want to count subsets. For example, $A = \{a, b, c, d, e\}$ has 10 subsets of three elements. They are the following:

$$
\begin{array}{ccccc}
\{a,b,c\} & \{a,b,d\} & \{a,b,e\} & \{a,c,d\} & \{a,c,e\} \\
\{a,d,e\} & \{b,c,d\} & \{b,c,e\} & \{b,d,e\} & \{c,d,e\}
\end{array}
$$

The number of subsets can be calculated by considering the next grid.

$$
\left.
\begin{array}{cccccc}
abc & acb & bac & bca & cab & cba \\
abd & adb & bad & bda & dab & dba \\
abe & aeb & bae & bea & eab & eba \\
acd & adc & cad & cda & dac & dca \\
ace & aec & cae & cea & eac & eca \\
ade & aed & dae & dea & ead & eda \\
bcd & bdc & cbd & cdb & dbc & dcb \\
bce & bec & ceb & cbe & ebc & ecb \\
bde & bed & dbe & deb & ebd & edb \\
cde & ced & dce & dec & ecd & edc
\end{array}
\right\} \text{10 rows}
$$

$$
\underbrace{\qquad\qquad\qquad\qquad\qquad\qquad}_{\text{3! columns}}
$$

There are $_5P_3$ permutations with three elements from A. They are found as the entries in the grid. However, since we are looking at subsets, we do not want to count abc as different from acb because $\{a, b, c\} = \{a, c, b\}$. For this reason, all elements in any given row of the grid are considered as one subset. Each row has $6 = 3!$ entries because that is the number of permutations of a set with three elements. Hence, multiplying the number of rows by the number of columns gives

$$_5P_3 = 10(3!).$$

Therefore,

$$\frac{_5P_3}{3!} = \frac{5!}{3!(5-3)!} = \frac{5!}{3!2!} = 10.$$

A generalization of this calculation leads to the formula for the arbitrary **binomial coefficient**,

$$\binom{n}{r} = \frac{n!}{r!(n-r)!},$$

where $n, r \in \omega$. Read $\binom{n}{r}$ as "n choose r." A generalization of the argument leads to the next theorem.

■ **THEOREM 5.4.8**

Let $n, r \in \mathbb{Z}^+$. The number of subsets of r elements from a set with n elements is $\binom{n}{r}$.

When we expand $(x + 1)^3$, we find that

$$(x + 1)^3 = x^3 + 3x^2 + 3x^2 + 1 = \sum_{r=0}^{3} \binom{3}{r} x^{3-r} 1^r.$$

To prove this for any binomial $(x + y)^n$, we need the following equation. It was proved by Blaise Pascal (1653). The proof is Exercise 7.

■ **LEMMA 5.4.9 [Pascal's Identity]**

If $n, r \in \omega$ so that $n \geq r$,

$$\binom{n}{r} + \binom{n}{r-1} = \binom{n+1}{r}.$$

■ **THEOREM 5.4.10 [Binomial Theorem]**

Let $n \in \mathbb{Z}^+$. Then,

$$(x + y)^n = \sum_{r=0}^{n} \binom{n}{r} x^{n-r} y^r.$$

PROOF

- Since $\binom{1}{0} = \binom{1}{1} = 1$,

$$(x + y)^1 = \binom{1}{0}x + \binom{1}{1}y = \sum_{r=0}^{1}\binom{1}{r}x^{1-r}y^r.$$

- Assume for $k \in \mathbb{Z}^+$,

$$(x + y)^k = \sum_{r=0}^{k}\binom{k}{r}x^{k-r}y^r.$$

Then,

$$(x + y)^{k+1} = (x + y)(x + y)^k = (x + y)\sum_{r=0}^{k}\binom{k}{r}x^{k-r}y^r.$$

Multiplying the $(x + y)$ term through the summation yields

$$\sum_{r=0}^{k}\binom{k}{r}x^{k-r+1}y^r + \sum_{r=0}^{k}\binom{k}{r}x^{k-r}y^{r+1}.$$

Taking out the $(k + 1)$-degree terms and shifting the index on the second summation gives

$$x^{k+1} + \sum_{r=1}^{k}\binom{k}{r}x^{k-r+1}y^r + \sum_{r=1}^{k}\binom{k}{r-1}x^{k-r+1}y^r + y^{n+1},$$

which using Pascal's identity (Lemma 5.4.9) equals

$$x^{k+1} + \sum_{r=1}^{k}\binom{k+1}{r}x^{k-r+1}y^r + y^{k+1},$$

and this is

$$\sum_{r=0}^{k+1}\binom{k+1}{r}x^{k+1-r}y^r. \blacksquare$$

Euclid's Lemma

Our second application of mathematical induction comes from number theory. It is the study of the greatest common divisor (Definition 3.3.11). We begin with a lemma.

■ **LEMMA 5.4.11**

Let $a, b, c \in \mathbb{Z}$ such that $a \neq 0$ or $b \neq 0$. If $a \mid bc$ and $\gcd(a, b) = 1$, then $a \mid c$.

PROOF

Assume $a \mid bc$ and $\gcd(a, b) = 1$. Then, $bc = ak$ for some $k \in \mathbb{Z}$. By Theorem 4.3.32, there exist $m, n \in \mathbb{Z}$ such that

$$1 = ma + nb.$$

Therefore, $a \mid c$ because

$$c = cma + cnb = cma + nak = a(cm + nk). \blacksquare$$

Suppose that $p \in \mathbb{Z}$ is a prime (Example 2.4.18) that does not divide a. We show that $\gcd(a, p) = 1$. Take $d > 0$ and assume $d \mid a$ and $d \mid p$. Since p is prime, $d = 1$ or $d = p$. Since $p \nmid a$, we conclude that d must equal 1, which means $\gcd(a, p) = 1$. Use this to prove the next result attributed to Euclid (*Elements* VII.30).

■ **THEOREM 5.4.12 [Euclid's Lemma]**

An integer $p > 1$ is prime if and only if $p \mid ab$ implies $p \mid a$ or $p \mid b$ for all $a, b \in \mathbb{Z}$.

PROOF

- Let p be prime. Suppose $p \mid ab$ but $p \nmid a$. Then, $\gcd(a, p) = 1$. Therefore, $p \mid b$ by Theorem 5.4.11.

- Let $p > 1$. Suppose p satisfies the condition,

$$\forall a \forall b (p \mid ab \to p \mid a \vee p \mid b).$$

Assume p is not prime. This means that there are integers c and d so that $p = cd$ with $1 < c \le d < p$. Hence, $p \mid cd$. By hypothesis, $p \mid c$ or $p \mid d$. However, since $c, d < p$, p can divide neither c nor d. This is a contradiction. Hence, p must be prime. ■

Since 6 divides $3 \cdot 4$ but $6 \nmid 3$ and $6 \nmid 4$, the lemma tells us that 6 is not prime. On the other hand, if p is a prime that divides 12, then p divides 4 or 3. This means that $p = 2$ or $p = 3$.

The next theorem is a generalization of Euclid's lemma. Its proof uses mathematical induction.

■ **THEOREM 5.4.13**

Let p be prime and $a_i \in \mathbb{Z}$ for $i = 0, 1, \ldots, n - 1$. If $p \mid a_0 a_1 \cdots a_{n-1}$, then $p \mid a_j$ for some $j = 0, 1, \ldots, n - 1$.

PROOF

- The case when $n = 1$ is trivial because p divides a_0 by definition of the product.

- Assume if $p \mid a_0 a_1 \cdots a_{n-1}$, then $p \mid a_j$ for some $j = 0, 1, \ldots, n - 1$. Suppose $p \mid a_0 a_1 \cdots a_n$. Then, by Lemma 5.4.12,

$$p \mid a_0 a_1 \cdots a_{n-1} \text{ or } p \mid a_n.$$

If $p \mid a_n$, we are done. Otherwise, p divides $a_0 a_1 \cdots a_{n-1}$. Hence, p divides one of the a_i by induction. ∎

Exercises

1. Let $n \in \mathbb{Z}^+$. Prove.

   (a) $1 + 2 + 3 + \cdots + n = \dfrac{n(n+1)}{2}$

   (b) $1 + 3 + 5 + \cdots + (2n-1) = n^2$

   (c) $1^2 + 3^2 + 5^2 + \cdots + (2n-1)^2 = \dfrac{n(2n-1)(2n+1)}{3}$

   (d) $1^3 + 2^3 + 3^3 + \cdots + n^3 = \left[\dfrac{n(n+1)}{2} \right]^2$

   (e) $1 + r + r^2 + \cdots + r^n = \dfrac{1 - r^{n+1}}{1 - r}$ $(r \neq 1)$

   (f) $1 \cdot 1! + 2 \cdot 2! + \cdots + n \cdot n! = (n+1)! - 1$

   (g) $\dfrac{1}{2!} + \dfrac{2}{3!} + \cdots + \dfrac{n}{(n+1)!} = 1 - \dfrac{1}{(n+1)!}$

   (h) $2 \cdot 6 \cdot 10 \cdot 14 \cdots \cdots (4n-2) = \dfrac{(2n)!}{n!}$

2. Prove for all positive integers n.

   (a) $\displaystyle\sum_{i=1}^{n} i(i+1) = \dfrac{n(n+1)(n+2)}{3}$

   (b) $\displaystyle\sum_{i=1}^{n} \dfrac{1}{(2i-1)(2i+1)} = \dfrac{n}{2n+1}$

3. Let $n \in \mathbb{Z}^+$. Prove.

   (a) $n < 2^n$

   (b) $n! \leq n^n$

   (c) $\displaystyle\sum_{i=1}^{n} \dfrac{1}{i^2} \leq 2 - \dfrac{1}{n}$

   (d) $\dfrac{1}{2} + \dfrac{2}{2^2} + \dfrac{3}{2^3} + \cdots + \dfrac{n}{2^n} \leq 2 - \dfrac{n}{2^n}$

4. Let $n \in \mathbb{Z}$. Prove.

   (a) $n^2 < 2^n$ for all $n \geq 5$.

   (b) $2^n < n!$ for all $n \geq 4$.

   (c) $n^2 < n!$ for all $n \geq 4$.

5. For $n \in \mathbb{Z}^+$, prove that if A has n elements, $P(A)$ has 2^n elements.

6. Prove that the number of lines in a truth table with n propositional variables is 2^n.

7. Demonstrate Pascal's identity (Lemma 5.4.9).

8. For all integers $n \geq r \geq 0$, prove the given equations.

(a) $\binom{n}{0} = \binom{n}{n} = 1$.

(b) $\binom{n}{r} = \binom{n}{n-r}$.

9. Let $n, r \in \mathbb{Z}^+$ with $n \geq r$. Prove.

(a) $\binom{r}{r} + \binom{r+1}{r} + \cdots + \binom{n}{r} = \binom{n+1}{r+1}$

(b) $1^2 + 3^2 + 5^2 + \cdots + (2n-1)^2 = \binom{2n+1}{3}$

10. Let $n \geq 2$ be an integer and prove the given equations.

(a) $\sum_{r=1}^{n} r \binom{n}{r} = n2^{n-1}$

(b) $\sum_{r=1}^{n} (-1)^{r-1} r \binom{n}{r} = 0$

11. Let n and r be positive integers and $n \geq r$. Use induction to show the given equations.

(a) $\binom{r}{r} + \binom{r+1}{r} + \cdots + \binom{n}{r} = \binom{n+1}{r+1}$

(b) $1^2 + 3^2 + 5^2 + \cdots + (2n-1)^2 = \binom{2n+1}{3}$

12. Prove the following for all $n \in \omega$.
   (a) $5 \mid n^5 - n$
   (b) $9 \mid n^3 + (n+1)^3 + (n+2)^3$
   (c) $8 \mid 5^{2n} + 7$
   (d) $5 \mid 3^{3n+1} + 2^{n+1}$

13. If p is prime and a and b are positive integers such that $a + b = p$, prove that $\gcd(a, b) = 1$.

14. Prove for all $n \in \mathbb{Z}^+$, there exist n consecutive composite integers (Example 2.4.18) by showing that $(n+1)! + 2, (n+1)! + 3, \ldots, (n+1)! + n + 1$ are composite.

15. Prove.
   (a) If $a \neq 0$, then $a \cdot \gcd(b, c) = \gcd(ab, ac)$.
   (b) Prove if $\gcd(a_i, b) = 1$ for $i = 1, \ldots, n$, then $\gcd(a_1 \cdot a_2 \cdots a_n, b) = 1$).

16. For $k \in \mathbb{Z}^+$, let $a_0, a_1, \ldots, a_{k-1} \in \omega$, not all equal to zero. Define

$$g = \gcd(a_0, a_1, \ldots, a_{k-1})$$

to mean that g is the greatest integer such that $g \mid a_i$ for all $i = 0, 1, \ldots, k-1$. Assuming $k \geq 3$, prove the given equations.

(a) $\gcd(a_0, a_1, \ldots, a_{k-1}) = \gcd(a_0, a_1, \ldots, a_{k-3}, \gcd(a_{k-2}, a_{k-1}))$

(b) $\gcd(ca_0, ca_1, \ldots, ca_{k-1}) = c \gcd(a_0, a_1, \ldots, a_{k-1})$ for all integers $c \neq 0$

5.5 STRONG INDUCTION

Suppose we want to find an equation for the terms of a sequence defined recursively in which each term is based on two or more previous terms. To prove that such an equation is correct, we modify mathematical induction. Remember the domino picture that we used to explain how mathematical induction works (page 258). The first domino is tipped causing the second to fall, which in turn causes the third to fall. By the time the sequence of falls reaches the $n + 1$ domino, n dominoes have fallen. This means that sentences $p(1)$ through $p(n)$ have been proved true. It is at this point that $p(n + 1)$ is proved. This is the intuition behind the next theorem. It is sometimes called **strong induction**.

■ **THEOREM 5.5.1 [Mathematical Induction 2]**

Let $p(k)$ be a formula. For any $n_0 \in \mathbb{Z}$, if

$$p(n_0) \wedge \forall k(k \in \omega \wedge [\forall l(l \in \omega \wedge [l \leq k \rightarrow p(n_0 + l)]) \rightarrow p(n_0 + k + 1)]),$$

then

$$\forall k[k \in \mathbb{Z} \wedge k \geq n_0 \rightarrow p(k)].$$

PROOF

Assume $p(n_0)$ and

$$p(n_0) \wedge p(n_0 + 1) \wedge \cdots \wedge p(n_0 + k) \rightarrow p(n_0 + k + 1) \qquad (5.11)$$

for $k \in \omega$. Define

$$q(k) := p(n_0) \wedge p(n_0 + 1) \wedge \cdots \wedge p(n_0 + k).$$

We proceed with the induction.

- Since $p(n_0)$ holds, we have $q(0)$.

- Assume $q(n)$ with $n \geq 0$. By definition of $q(n)$, we have $p(n_0)$ through $p(n_0 + n)$. Thus, $p(n_0 + n + 1)$ by (5.11) from which $q(n + 1)$ follows.

Therefore, $q(n)$ is true for all $n \in \omega$ (Theorem 5.4.1). Hence, $p(n)$ for all integers $n \geq n_0$. ■

Fibonacci Sequence

Leonardo of Pisa (known as Fibonacci) in his 1202 work *Liber abaci* posed a problem about how a certain population of rabbits increases with time (Fibonacci and Sigler 2002). Each rabbit that is at least 2 months old is considered an adult. It is a young rabbit if it is a month old. Otherwise, it is a baby. The rules that govern the population are as follows:

- No rabbits die.

- The population starts with a pair of adult rabbits.

- Each pair of adult rabbits will bear a new pair each month.

The population then grows according to the following table:

Month	Adult Pairs	Young Pairs	Baby Pairs
1	1	0	1
2	1	1	1
3	2	1	2
4	3	2	3
5	5	3	5
6	8	5	8

It appears that the number of adult (or baby) pairs at month n is given by the sequence,

$$1, 1, 2, 3, 5, 8, 13, 21, 34, \ldots .$$

This is known as the **Fibonacci sequence**, and each term of the sequence is called a **Fibonacci number**. Let F_n denote the nth term of the sequence. So

$$F_1 = 1, F_2 = 1, F_3 = 2, F_4 = 3, F_5 = 5, F_6 = 8, \ldots .$$

Each term of the sequence can be calculated recursively by

$$F_1 = 1,$$
$$F_2 = 1, \qquad\qquad (5.12)$$
$$F_n = F_{n-1} + F_{n-2} \ \ (n > 2).$$

Since we have only checked a few terms, we have not proved that F_n is equal to the number of adult pairs in the nth month. To show this, we use strong induction. Since the recursive definition starts by explicitly defining F_1 and F_2, the basis case for the induction will prove that the formula holds for $n = 1$ and $n = 2$.

- From the table, in each of the first 2 months, there is exactly one adult pair of rabbits. This coincides with $F_1 = 1$ and $F_2 = 1$ in (5.12).

- Let $n > 2$, and assume that F_k equals the number of adult pairs in the kth month for all $k \leq n$. Because of the third rule, the number of pairs of adults in any month is the same as the number of adult pairs in the previous month plus the number of baby pairs 2 months prior. Therefore,

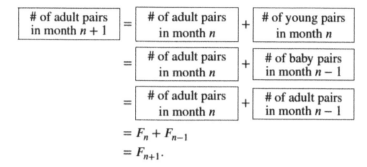

$$= F_n + F_{n-1}$$
$$= F_{n+1}.$$

It turns out that the Fibonacci sequence is closely related to another famous object of study in the history of mathematics. Letting $n \geq 1$, define

$$a_n = \frac{F_{n+1}}{F_n}.$$

The first seven terms of this sequence are

$$a_1 = 1/1 = 1,$$
$$a_2 = 2/1 = 2,$$
$$a_3 = 3/2 = 1.5,$$
$$a_4 = 5/3 \approx 1.667,$$
$$a_5 = 8/5 = 1.6,$$
$$a_6 = 13/8 = 1.625,$$
$$a_7 = 21/13 \approx 1.615.$$

This sequence has a limit that we call τ. To find this limit, notice that

$$\frac{F_{n+1}}{F_n} = \frac{F_n + F_{n-1}}{F_n} = 1 + \frac{F_{n-1}}{F_n}.$$

Because $a_{n-1} = F_n/F_{n-1}$ when $n > 1$,

$$a_n = 1 + \frac{1}{a_{n-1}},$$

and therefore,

$$a_n - 1 - \frac{1}{a_{n-1}} = 0.$$

Because

$$\lim_{n \to \infty} a_n = \lim_{n \to \infty} a_{n-1} = \tau,$$

we conclude that

$$\tau^2 - \tau - 1 = 0. \tag{5.13}$$

Therefore, $(1 \pm \sqrt{5})/2$ are the solutions to (5.13), but since $F_{n+1}/F_n > 0$, we take the positive value and find that

$$\tau = \frac{1 + \sqrt{5}}{2}.$$

The number τ is called the **golden ratio**. It was considered by the ancient Greeks to represent the ratio of the sides of the most beautiful rectangle.

■ **EXAMPLE 5.5.2**

Prove $F_n \leq \tau^{n-1}$ when $n \geq 2$ using strong induction.

- Since $F_2 = 1 < \tau^1 \approx 1.618$, the inequality holds for $n = 2$.

- Let $n \geq 3$, and assume that $F_k \leq \tau^{k-1}$ for all k such that $2 \leq k \leq n$. Because $\tau^{-1} = (\sqrt{5} - 1)/2$ and $\tau^{-2} = (3 - \sqrt{5})/2$,

$$\tau^{-1} + \tau^{-2} = 1.$$

Therefore, the induction hypothesis gives

$$F_{n+1} = F_n + F_{n-1} \leq \tau^{n-1} + \tau^{n-2} = \tau^n(\tau^{-1} + \tau^{-2}) = \tau^n.$$

Unique Factorization

Theorem 5.4.13 states that if a prime divides an integer, it divides one of the factors of the integer. It appears reasonable that any integer can then be written as a product that includes all of its prime divisors. For example, we can write $126 = 2 \cdot 3 \cdot 3 \cdot 7$, and this is essentially the only way in which we can write 126 as a product of primes. All of this is summarized in the next theorem. It is also known as the **fundamental theorem of arithmetic**. It is the reason the primes are important. They are the building blocks of the integers.

■ **THEOREM 5.5.3 [Unique Factorization]**

If $n > 1$, there exists a unique sequence of primes $p_0 \leq p_1 \leq \cdots \leq p_k$ $(k \in \omega)$ such that $n = p_0 p_1 \cdots p_k$.

PROOF
Prove existence with strong induction on n.

- When $n = 2$, we are done since 2 is prime.

- Assume that k can be written as the product of primes as described above for all k such that $2 \leq k < n$. If n is prime, we are done as in the basis case. So suppose n is composite. Then, there exist integers a and b such that $n = ab$ and $1 < a \leq b < n$. By the induction hypothesis, we can write

$$a = q_0 q_1 \cdots q_u$$

and

$$b = r_0 r_1 \cdots r_v,$$

where the q_i and r_j are primes. Now place these primes together in increasing order and relabel them as

$$p_0 \leq p_1 \leq \cdots \leq p_k$$

with $k = u + v$. Then, $n = p_0 p_1 \cdots p_k$ as desired.

For uniqueness, suppose that there are two sets of primes

$$p_0 \leq p_1 \leq \cdots \leq p_k \text{ and } q_0 \leq q_1 \leq \cdots \leq q_l$$

so that

$$n = p_0 p_1 \cdots p_k = q_0 q_1 \cdots q_l.$$

By canceling, if necessary, we can assume the sides have no common primes. If the cancellation yields $1 = 1$, the sets of primes are the same. In order to obtain a contradiction, assume that there is at least one prime remaining on the left-hand side. Suppose it is p_0. If the product on the right equals 1, then $p_0 \mid 1$, which is impossible. If there are primes remaining on the right, p_0 divides one of them by Lemma 5.4.12. This is also a contradiction, since the sides have no common prime factors because of the cancellation. Hence, the two sequences must be the same. ■

Unique Factorization allows us to make the following definition.

■ **DEFINITION 5.5.4**

Let $n \in \mathbb{Z}^+$. If $p_0, p_1, \ldots, p_{k-1}$ are distinct primes and $r_0, r_1, \ldots, r_{k-1}$ are natural numbers such that

$$n = p_0^{r_0} p_1^{r_1} \cdots p_{k-1}^{r_{k-1}},$$

then $p_0^{r_0} p_1^{r_1} \cdots p_{k-1}^{r_{k-1}}$ is called a **prime power decomposition** of n.

■ **EXAMPLE 5.5.5**

Consider the integer 360. It has $2^3 \cdot 3^2 \cdot 5^1$ as a prime power decomposition. If the exponents are limited to positive integers, the expression is unique. In this sense, we can say that $2^3 \cdot 3^2 \cdot 5^1$ is *the* prime power decomposition of 360. However, there are times when primes need to be included in the product that are not factors of the integer. By setting the exponent to zero, these primes can be included. For example, we can also write 360 as $2^3 \cdot 3^2 \cdot 5^1 \cdot 7^0$.

■ **EXAMPLE 5.5.6**

Suppose $n \in \mathbb{Z}$ such that $n > 1$. Use unique factorization (Theorem 5.5.3) to prove that n is a perfect square if and only if all powers in a prime power decomposition of n are even.

- Let n be a perfect square. This means that $n = k^2$ for some integer $k > 1$. Write a prime power decomposition of k,

$$k = p_0^{r_0} p_1^{r_1} \cdots p_{l-1}^{r_{l-1}}.$$

Therefore,

$$n = k^2 = p_0^{2r_0} p_1^{2r_1} \cdots p_{l-1}^{2r_{l-1}}.$$

- Assume all the powers are even in a prime power decomposition of n. Namely,

$$n = p_0^{r_0} p_1^{r_1} \cdots p_{l-1}^{r_{l-1}},$$

where there exists $u_i \in \mathbb{Z}$ so that $r_i = 2u_i$ for $i = 0, 1, \ldots, l - 1$. Thus,

$$n = p_0^{2u_0} p_1^{2u_1} \cdots p_{l-1}^{2u_{l-1}} = \left(p_0^{u_0} p_1^{u_1} \cdots p_{l-1}^{u_{l-1}} \right)^2,$$

a perfect square.

Exercises

1. Given each recursive definition, prove the formula for a_n holds for all positive integers n.
    (a) If $a_1 = -1$ and $a_n = -a_{n-1}$, then $a_n = (-1)^n$.
    (b) If $a_1 = 1$ and $a_n = 1/3 a_{n-1}$, then $a_n = (1/3)^{n-1}$.
    (c) If $a_1 = 0$, $a_2 = -6$, and $a_n = 5a_{n-1} - 6a_{n-2}$, then

$$a_n = 3 \cdot 2^n - 2 \cdot 3^n.$$

    (d) If $a_1 = 4$, $a_2 = 12$, and $a_n = 4a_{n-1} - 2a_{n-2}$, then

$$a_n = (2 + \sqrt{2})^n + (2 - \sqrt{2})^n.$$

    (e) If $a_1 = 1$, $a_2 = 5$, and $a_{n_1} = a_n + 2a_{n-1}$ for all $n > 2$, then

$$a_n = 2^n + (-1)^n.$$

    (f) If $a_1 = 3$, $a_2 = -3$, $a_3 = 9$, and $a_n = a_{n-1} + 4a_{n-2} - 4a_{n-3}$, then

$$a_n = 1 - (-2)^n.$$

    (g) If $a_1 = 3$, $a_2 = 10$, $a_3 = 21$, and $a_n = 3a_{n-1} - 3a_{n-2} + a_{n-3}$, then

$$a_n = n + 2n^2.$$

2. Let $g_1 = a$, $g_2 = b$, and $g_n = g_{n-1} + g_{n-2}$ for all $n > 2$. This sequence is called the **generalized Fibonacci sequence**. Show that $g_n = af_{n-2} + bf_{n-1}$ for all $n > 2$.

3. Let $n > 0$ be an integer. Prove.

   (a) $F_{n+2} > \tau^n$

   (b) $\displaystyle\sum_{i=1}^{n} F_i = F_{n+2} - 1$

4. Prove that Theorem 5.5.1 implies Theorem 5.4.1.

5. Let $\sigma = (1 - \sqrt{5})/2$ and demonstrate that $F_n = \dfrac{\tau^n - \sigma^n}{\sqrt{5}}$.

6. Let $n \geq 1$ and $a \in \mathbb{Z}$. Prove.

   (a) $a^{n+1} - 1 = (a + 1)(a^n - 1) - a(a^{n-1} - 1)$.

   (b) $a^n - 1 = (a - 1)(a^{n-1} + a^{n-2} + \cdots + a + 1)$.

7. For all $n \in \omega$, prove that 12 divides $n^4 - n^2$ (Definition 2.4.2).

8. Assume $e \mid a$ and $e \mid b$. Write prime power decompositions for a and b:

$$a = p_0^{r_0} p_1^{r_1} \cdots p_{k-1}^{r_{k-1}}$$

and

$$b = p_0^{s_0} p_1^{s_1} \cdots p_{k-1}^{s_{k-1}}.$$

Prove that there exist $t_0, t_1, \ldots t_{k-1} \in \omega$ such that

$$e = p_0^{t_0} p_1^{t_1} \cdots p_{k-1}^{t_{k-1}},$$

$t_i \leq r_i$, and $t_i \leq s_i$ for all $i = 0, 1, \ldots, k - 1$.

9. Prove that $a^3 \mid b^2$ implies $a \mid b$ for all $a, b \in \mathbb{Z}$.

10. Let $a \in \mathbb{Z}^+$. Let a have the property that for all primes p, if $p \mid a$, then $p^2 \mid a$. Prove that a is the product of a perfect square and a perfect cube.

11. Prove that $\gcd(F_n, F_{n+2}) = 1$ for all $n \in \mathbb{Z}^+$.

5.6   REAL NUMBERS

As $\mathbb{Z}$ is defined using ω and $\mathbb{Q}$ is defined using $\mathbb{Z}$, the set analog to $\mathbf{R}$ is defined using $\mathbb{Q}$. We start with a definition.

■ **DEFINITION 5.6.1**

Let $(A, \preccurlyeq)$ be a poset. The set B is an **initial segment** of A when $B \subseteq A$ and

$$\text{for all } a, b \in A, \text{ if } a \preccurlyeq b \text{ and } b \in B, \text{ then } a \in B. \tag{5.14}$$

An initial segment B of A is **proper** if $B \neq A$.

The condition (5.14) is called **downward closed**. Notice that for all $a \in \mathbf{R}$, both $(-\infty, a)$ and $(-\infty, a]$ are initial segments of $(\mathbf{R}, \leq)$. A poset is an initial segment of itself, but it is not proper.

■ DEFINITION 5.6.2

Let $(A, \preccurlyeq)$ be a poset with $b \in A$. Define

$$\mathrm{seg}_{\preccurlyeq}(A, b) = \{a \in A : a \prec b\}.$$

For example,

$$\mathrm{seg}_{\leq}(\mathbf{R}, 5) = (-\infty, 5)$$

in $(\mathbf{R}, \leq)$, and

$$\mathrm{seg}_{\leq}(\mathbb{Z}, 5) = \{\ldots, 0, 1, 2, 3, 4\}$$

in $(\mathbb{Z}, \leq)$. Both of these are proper initial segments. Notice that every initial segment of $\mathbb{Z}$ is of the form $\mathrm{seg}_{\leq}(\mathbb{Z}, n)$ for some $n \in \mathbb{Z}$.

Neither $\mathbb{Z}$ nor $\mathbb{Q}$ are well-ordered by $\leq$. If a poset is well-ordered, its initial segments have a particular form.

■ LEMMA 5.6.3

If $(A, \preccurlyeq)$ is a well-ordered set and B is a proper initial segment of A, there exists a unique $m \in A$ such that $B = \mathrm{seg}_{\preccurlyeq}(A, m)$.

PROOF

Suppose that $(A, \preccurlyeq)$ is well-ordered and $B \subseteq A$ is a proper initial segment. First, note that $A \setminus B$ is not empty since B is proper. Thus, $A \setminus B$ contains a least element m because $\preccurlyeq$ well-orders A.

- Let $a \in B$, which implies that $a \prec m$. Otherwise, m would be an element of B because B is downward closed. Hence, $a \in \mathrm{seg}_{\preccurlyeq}(A, m)$, which implies that $B \subseteq \mathrm{seg}_{\preccurlyeq}(A, m)$.

- Conversely, take $a \in \mathrm{seg}_{\preccurlyeq}(A, m)$, which means $a \prec m$. If $a \in A \setminus B$, then $m \preccurlyeq a$ because m is the least element of $A \setminus B$, so a must be an element of B by the trichotomy law (Theorem 4.3.21). Thus, $\mathrm{seg}_{\preccurlyeq}(A, B) \subseteq B$.

To prove uniqueness, let $m' \in A$ such that $m \neq m'$ and $B = \mathrm{seg}_{\preccurlyeq}(A, m')$. If $m \prec m'$, then $m \in \mathrm{seg}_{\preccurlyeq}(A, m')$, and if $m' \prec m$, then $m' \in B = \mathrm{seg}_{\preccurlyeq}(A, m)$. Both cases are impossible. ■

Dedekind Cuts

A basic property of $\mathbf{R}$ is that it is **complete**. This means that

every nonempty set of real numbers with an upper bound
has a real least upper bound

(Definition 4.3.12). For example, the set

$$A = \left\{1 - \frac{1}{n} : n \in \mathbf{Z}^+\right\}$$

is bounded from above, and its least upper bound is 1. Also,

$$B = \{3, 3.1, 3.14, 3.141, 3.1415, 3.14159, \dots\}$$

is bounded from above, and its least upper bound is π. Observe that both A and B are sets of rational numbers. The set A has a rational least upper bound, but B does not. This shows that the rational numbers are not complete. Intuitively, the picture is that of a number line. If **R** is graphed, there are no holes because of completeness, but if **Q** is graphed, there are holes. These holes represent the irrational numbers that when filled, complete the rational numbers resulting in the set of reals. We use this idea to construct a model of the real numbers from **Q** (Dedekind 1901).

■ DEFINITION 5.6.4

A set x of rational numbers is a **Dedekind cut** (or a **real number**) if x is a subset of $(\mathbb{Q}, \leq)$ such that

- x is nonempty,

- x is a proper initial segment of $\mathbb{Q}$,

- x does not have a greatest element.

Denote the set of Dedekind cuts by $\mathbb{R}$.

By Lemma 5.6.3, some Dedekind cuts are of the form $\text{seg}_<(\mathbb{Q}, a)$ for some $a \in \mathbb{Q}$. In this case, write $\mathbf{a} = \text{seg}_<(\mathbb{Q}, a)$. Therefore, $\mathbb{Q}$ can be embedded in $\mathbb{R}$ using the function $f : \mathbb{Q} \to \mathbb{R}$ defined by $f(a) = \mathbf{a}$ (Exercise 14). The elements of $\mathbb{R} \setminus \text{ran}(f)$ are the **irrational numbers**.

■ EXAMPLE 5.6.5

- The Dedekind cut that corresponds to the integer 7 is

$$7 = \text{seg}_\leq(\mathbb{Q}, 7).$$

Note that there is no gap between 7 and $\mathbb{Q} \setminus 7 = \{a \in \mathbb{Q} : a \geq 7\}$.

- The Dedekind cut x that corresponds to π includes

$$\{3, 3.1, 3.14, 3.141, 3.1415, 3.14159, \dots\}$$

as a subset. Notice that $\pi \notin x$ and $\pi \notin \mathbb{Q} \setminus x$ because π is not rational. This means that we imagine a gap between x and $\mathbb{Q} \setminus x$. This gap is where π is located.

Imagine that only the rational numbers have been placed on the number line (Section 5.3). For every Dedekind cut x, call the point on the number line where x and $\mathbb{Q} \setminus x$ meet a **cut**. Following our intuition, if the least upper bound of x is an element of $\mathbb{Q} \setminus x$, there is a point at the cut [Figure 5.1(a)]. This means that x represents a rational number, a point already on the number line. However, if the least upper bound of x is

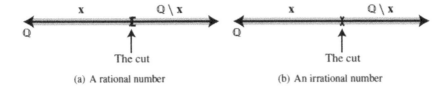

Figure 5.1   The cuts of two types of numbers.

not an element of $Q \setminus x$, there is no point at the cut [Figure 5.1(b)]. This means that x represents an irrational number. To obtain all real numbers, a point must be placed at each cut without a point, filling the entire number line. Therefore, the first step in showing that $\mathbb{R}$ is a suitable model for **R** is to prove that every set of Dedekind cuts with an upper bound must must have a least upper bound that is a Dedekind cut. That is, $\mathbb{R}$ must be shown to be complete. To accomplish this, we first define on order on $\mathbb{R}$.

■ **DEFINITION 5.6.6**

Let $x, y \in \mathbb{R}$. Define $x \leq y$ if and only if $x \subseteq y$, and define $x < y$ to mean $x \leq y$ and $x \neq y$.

For example, $3 \leq 4$ in $\mathbb{R}$ because $3 = \mathrm{seg}_<(Q, 3)$ and $4 = \mathrm{seg}_<(Q, 4)$. Because the order on Q is linear (Theorem 5.3.8), it is left to Exercise 5 to prove that we have defined a linear order on $\mathbb{R}$.

■ **THEOREM 5.6.7**

$(\mathbb{R}, \leq)$ is a linear order but it is not a well-order.

Now to prove that $\mathbb{R}$ is complete using the order of Definition 5.6.6.

■ **THEOREM 5.6.8**

Every nonempty subset of $\mathbb{R}$ with an upper bound has a real least upper bound.

PROOF

Let $\mathcal{F} \neq \emptyset$ and $\mathcal{F} \subseteq \mathbb{R}$. Let $m \in \mathbb{R}$ be an upper bound of $\mathcal{F}$. By Example 4.3.14, $\bigcup \mathcal{F}$ is the least upper bound of $\mathcal{F}$. We show that $\bigcup \mathcal{F} \in \mathbb{R}$.

- Take $x \in \mathcal{F}$. Since Dedekind cuts are nonempty, $x \neq \emptyset$, which implies that $\bigcup \mathcal{F} \neq \emptyset$.

- By hypothesis, $x \subseteq m$ for all $x \in \mathcal{F}$. Hence, $\bigcup \mathcal{F} \subseteq m$. Since $m \in \mathbb{R}$, we have that $m \neq Q$, so $\bigcup \mathcal{F} \subset Q$.

- Let $x \in \bigcup \mathcal{F}$ and $y \leq x$. Thus, $x \in a$ for some Dedekind cut $a \in \mathcal{F}$. Since a is downward closed, $y \in a$, so $y \in \bigcup \mathcal{F}$. Hence, with the previous part, $\bigcup \mathcal{F}$ is a proper initial segment of Q.

- Let $x \in \mathbf{a} \in \mathscr{F}$. Since $\mathbf{a}$ is a Dedekind cut, it has no greatest element. Thus, there exists $y \in \mathbf{a}$ such that $x < y$. Because $y \in \bigcup \mathscr{F}$, we see that $\bigcup \mathscr{F}$ has no greatest element. ∎

Since every nonempty bounded subset of real numbers has a least upper bound in $\mathbb{R}$, the set of Dedekind cuts does not have the same issue with gaps as $\mathbb{Q}$ does. For example, the least upper bound of the set of real numbers

$$\{2, 2.7, 2.71, 2.718, 2.7182, 2.71828, \dots \}$$

is the Dedekind cut that corresponds to $e \in \mathbb{R}$.

Arithmetic

As with the other sets of numbers that have been constructed using the axioms of **ZFC**, we now define addition and multiplication on $\mathbb{R}$. First, define the following Dedekind cuts:

- $\mathbf{0} = \text{seg}_{\le}(\mathbb{Q}, 0)$

- $\mathbf{1} = \text{seg}_{\le}(\mathbb{Q}, 1)$.

Let $\mathbf{x}, \mathbf{y} \in \mathbb{R}$. That

$$S = \{a + b : a \in \mathbf{x} \wedge b \in \mathbf{y}\}$$

is a Dedekind cut is Exercise 3. Assume that $\mathbf{0} \le \mathbf{x}$ and $\mathbf{0} \le \mathbf{y}$. We claim that the set

$$P_1 = \{ab : 0 \le a \in \mathbf{x} \wedge 0 \le b \in \mathbf{y}\} \cup \mathbf{0}$$

is a Dedekind cut.

- $P_1 \ne \varnothing$ because $\mathbf{0} \ne \varnothing$.

- Let $u \in \mathbb{Q} \setminus \mathbf{x}$ and $v \in \mathbb{Q} \setminus \mathbf{y}$. Let

$$m = \begin{cases} u & \text{if } u \ge v, \\ v & \text{if } v > u. \end{cases}$$

Then, $m^2 \notin P_1$, so $P_1 \ne \mathbb{Q}$.

- Let $0 \le a \in \mathbf{x}$ and $0 \le b \in \mathbf{y}$, and suppose that $w \in \mathbb{Q}$ such that $w < ab$. Hence, $wb^{-1} < a$, which implies that $wb^{-1} \in \mathbf{x}$ since $\mathbf{x}$ is downward closed. Therefore, $w \in P_1$ because

$$w = (wb^{-1})b.$$

- Again, let $0 \le a \in \mathbf{x}$ and $0 \le b \in \mathbf{y}$. Since $\mathbf{x}$ and $\mathbf{y}$ do not have greatest elements, there exists $u \in \mathbf{x}$ and $v \in \mathbf{y}$ such that $a < u$ and $b < v$. Then, by Exercise 5.3.15(b)

$$ab < uv \in P_1.$$

Furthermore, if $x < 0$ and $y < 0$, define

$$P_2 = \{ab : a \in x \wedge b \in y \wedge 0 \le -a \wedge 0 \le -b\} \cup 0, \tag{5.15}$$

if $x < 0$ and $0 \le y$, define

$$P_3 = \{ab : a \in x \wedge 0 \le b \in y\}, \tag{5.16}$$

or $0 \le x$ and $y < 0$, define

$$P_4 = \{ab : 0 \le a \in x \wedge b \in y\}, \tag{5.17}$$

Since P_1, P_2, P_3, P_4, and S are Dedekind cuts (Exercise 4), we can use them to define the two standard operations on $\mathbb{R}$.

■ **DEFINITION 5.6.9**

Let $x, y \in \mathbb{R}$. Define

$$x + y = \{a + b : a \in x \wedge b \in y\}$$

and

$$x \cdot y = \begin{cases} \{ab : 0 \le a \in x \wedge 0 \le b \in y\} \cup 0 & \text{if } 0 \le x \wedge 0 \le y, \\ \{ab : a \in x \wedge b \in y \wedge 0 \le -a \wedge 0 \le -b\} \cup 0 & \text{if } x < 0 \wedge y < 0, \\ \{ab : a \in x \wedge 0 \le b \in y\} & \text{if } x < 0 \wedge 0 \le y, \\ \{ab : 0 \le a \in x \wedge b \in y\} & \text{if } 0 \le x \wedge y < 0. \end{cases}$$

Since addition and multiplication on $\mathbb{Q}$ are associative and commutative, addition and multiplication are associative and commutative on $\mathbb{R}$.

■ **THEOREM 5.6.10**

Addition and multiplication of real numbers are associative and commutative.

PROOF
Let $x, y, z \in \mathbb{R}$. We prove that addition is associative, leaving the rest to Exercise 7.

$$\begin{aligned} x + (y + z) &= x + \{b + c : b \in y \wedge c \in z\} \\ &= \{a + v : a \in x \wedge v \in \{b + c : b \in y \wedge c \in z\}\} \\ &= \{a + (b + c) : a \in x \wedge (b \in y \wedge c \in z)\} \\ &= \{(a + b) + c : (a \in x \wedge b \in y) \wedge c \in z\} \\ &= \{u + c : u \in \{a + b : a \in x \wedge b \in y\} \wedge c \in z\} \\ &= \{a + b : a \in x \wedge b \in y\} + z \\ &= (x + y) + z. \ \blacksquare \end{aligned}$$

The Dedekind cuts **0** and **1** behave as expected. For example,

$$\mathbf{0} + \mathbf{4} = \{a + b : a \in \mathbf{0} \wedge b \in \mathbf{4}\} = \mathrm{seg}_{\leq}(\mathbb{Q}, 4) = \mathbf{4}$$

and

$$\mathbf{1} \cdot \mathbf{4} = \{ab : 0 \leq a \in \mathbf{1} \wedge 0 \leq b \in \mathbf{4}\} \cup \mathbf{0} = \mathrm{seg}_{\leq}(\mathbb{Q}, 4) = \mathbf{4}.$$

These equations suggest the following.

■ **THEOREM 5.6.11**

$\mathbb{R}$ has additive and multiplicative identities.

PROOF

Let $\mathbf{x} \in \mathbb{R}$. We first show that

$$\mathbf{x} = \{a + b : a \in \mathbf{x} \wedge b \in \mathbf{0}\} = \mathbf{x} + \mathbf{0}.$$

Take $u \in \mathbf{x}$. Since $\mathbf{x}$ has no greatest element, there exists $v \in \mathbf{x}$ such that $u < v$. Write $u = v + (u - v)$. Since $u - v < 0$, we have that $u \in \mathbf{x} + \mathbf{0}$. Conversely, let $b \in \mathbf{0}$. Since $b \in \mathbf{0}$ implies that $b < 0$, we have that $u + b < u$ (Exercise 5.3.16), and because $\mathbf{x}$ is downward closed, $u + b \in \mathbf{x}$.

We next show that

$$\mathbf{x} = \mathbf{x} \cdot \mathbf{1}.$$

We have two cases to consider.

- Let $\mathbf{0} \leq \mathbf{x}$. By Definition 5.6.9,

$$\mathbf{x} \cdot \mathbf{1} = \{a \cdot b : 0 \leq a \in \mathbf{x} \wedge 0 \leq b \in \mathbf{1}\} \cup \mathbf{0}.$$

  Let $0 \leq a \in \mathbf{x}$ and $0 \leq b < 1$. Then, $ab \leq a$ (Exercise 5.3.17), so $ab \in \mathbf{x}$ since $\mathbf{x}$ is downward closed. Conversely, take $a \in \mathbf{x}$. If $a < 0$, then $a \in \mathbf{0}$, and if $a = 0$, then $a = 0 \cdot 0$, so suppose $a > 0$. Since $\mathbf{x}$ has no greatest element, there exists $u \in \mathbf{x}$ such that $a < u$. This implies that $au^{-1} < 1$, so $a \in \mathbf{x} \cdot \mathbf{1}$ because $a = u(au^{-1})$.

- Let $\mathbf{x} < \mathbf{0}$ and proceed like in the previous case. ■

We leave the proof of the last result to Exercise 10.

■ **THEOREM 5.6.12**

- Every element of $\mathbb{R}$ has an additive inverse.

- Every nonzero element of $\mathbb{R}$ has a multiplicative inverse.

- The distributive law holds for $\mathbb{R}$.

Complex Numbers

The last set of numbers that we define are the complex numbers.

■ **DEFINITION 5.6.13**

Define $\mathbb{C} = \mathbb{R} \times \mathbb{R}$ to be the set of **complex numbers**. Denote $(a, b) \in \mathbb{C}$ by $a + bi$.

Observe that the standard embedding $f : \mathbb{R} \to \mathbb{C}$ defined by $f(x) = (x, 0)$ allows us to consider $\mathbb{R}$ as a subset of $\mathbb{C}$. We will not define an order on $\mathbb{C}$ but will define the standard two operations.

■ **DEFINITION 5.6.14**

Let $a + bi, c + di \in \mathbb{C}$.

- $(a + bi) + (c + di) = (a + c) + (b + d)i$.

- $(a + bi) \cdot (c + di) = (ac - bd) + (ad + bc)i$.

We leave the proof of the following to Exercise 11.

■ **THEOREM 5.6.15**

- $i^2 = -1 + 0i$.

- Addition and multiplication are associative and commutative.

- $\mathbb{C}$ has additive and multiplicative identities.

- Every element of $\mathbb{C}$ has an additive inverse.

- Every nonzero element of $\mathbb{C}$ has a multiplicative inverse.

- The distributive law holds in $\mathbb{C}$.

Exercises

1. Find the given initial segments in the indicate posets.
    - (a) $\text{seg}_|(\mathbb{Z}, 50)$ from Example 4.3.6
    - (b) $\text{seg}_{\preceq}(\{0, 1\}^*, 101010)$ from Example 4.3.7
    - (c) $\text{seg}_{\subseteq}(P(\mathbb{Z}), \{1, 2, 3, 4\})$ from Example 4.3.9

2. Let $(A, \preceq)$ be a poset. Assume that B and C are initial segments of A. Prove that $B \cap C$ is an initial segment of A. Is $B \cup C$ also an initial segment of A?

3. Let $x, y \in \mathbb{R}$. Prove that $S = \{a + b : a \in x \wedge b \in y\}$ is a Dedekind cut.

4. Prove that P_2 (5.15), P_3 (5.16), and P_4 (5.17) are Dedekind cuts.

5. Prove Theorem 5.6.7.

6. Show that every Dedekind cut has an upper bound in $\mathbb{R}$.

7. Finish the proof of Theorem 5.6.10.

8. Prove that 0 and 1 are Dedekind cuts.

9. Let $a, b \in \mathbb{R}$ and $ab = 0$. Prove that $a = 0$ or $b = 0$.

10. Prove Theorem 5.6.12.

11. Prove Theorem 5.6.15.

12. Prove that between any two real numbers is another real number.

13. Prove that between any two real numbers is a rational number.

14. Show that $\mathbb{Q}$ can be embedded in $\mathbb{R}$ by proving that $f : \mathbb{Q} \to \mathbb{R}$ defined by

$$f(a) = \text{seg}_\leq(\mathbb{Q}, a)$$

is an order isomorphism preserving $\leq$ with $\leq$. Furthermore, show that both ω and $\mathbb{Z}$ can be embedded into $\mathbb{R}$.

15. Let f be the function defined in Exercise 14. Show that $f(a+1)$ is an upper bound of $f(a)$ for all $a \in \mathbb{Q}$.

16. The absolute value function (Exercise 2.4.23) can be defined so that for all $x \in \mathbb{R}$, $|x| = x \cup -x$, where $-x$ refers to the additive inverse of x (Theorem 5.6.12). Let a be a positive real number. Prove the following for every $x, y \in \mathbb{R}$.
   (a) $|-x| = |x|$.
   (b) $|x^2| = |x|^2$.
   (c) $x \leq |x|$.
   (d) $|xy| = |x|\,|y|$.
   (e) $|x| < a$ if and only if $-a < x < a$.
   (f) $a < |x|$ if and only if $a < x$ or $x < -a$.

CHAPTER 6

ORDINALS AND CARDINALS

6.1 ORDINAL NUMBERS

In Chapter 5, we defined certain sets to represent collections of numbers. Despite being sets themselves, the elements of those sets were called numbers. We continue this association with sets as numbers but for a different purpose. While before we defined ω, $\mathbb{Z}$, $\mathbb{Q}$, $\mathbb{R}$, and $\mathbb{C}$ to represent $\mathbf{N}$, $\mathbf{Z}$, $\mathbf{Q}$, $\mathbf{R}$, and $\mathbf{C}$, the definitions of this chapter are intended to be a means by which all sets can be classified according to a particular criterion. Specifically, in the later part of the chapter, we will define sets for the purpose of identifying the size of a given set, and we begin the chapter by defining sets that are used to identify whether two well-ordered sets have the same order type (Definition 4.5.24). A crucial tool in this pursuit is the following generalization of Theorem 5.5.1 to well-ordered infinite sets.

■ **THEOREM 6.1.1 [Transfinite Induction 1]**

Let $(A, \preccurlyeq)$ be a well-ordered set. If $B \subseteq A$ and $\mathrm{seg}_{\preccurlyeq}(A, x) \subseteq B$ implies $x \in B$ for all $x \in A$, then $A = B$.

A First Course in Mathematical Logic and Set Theory, First Edition. Michael L. O'Leary.
© 2016 John Wiley & Sons, Inc. Published 2016 by John Wiley & Sons, Inc.

PROOF

To show that A is a subset of B, suppose that $A \setminus B$ is nonempty. Since A is well-ordered by $\preceq$, let m be the least element of $A \setminus B$. This implies that $\text{seg}_\preceq(A, m) \subseteq B$, so $m \in B$ by hypothesis, a contradiction. ∎

Note that transfinite induction restricted to ω is simply strong induction (Theorem 5.5.1). To see this, let the well-ordered set $(A, \preceq)$ of Theorem 6.1.1 be $(\omega, \leq)$. Define the set $B = \{k : p(k)\} \subseteq \omega$ for some formula $p(k)$. The conditional

$$\text{seg}_\leq(\omega, n) \subseteq B \to n \in B$$

implies $p(0)$ when $n = 0$ because $\text{seg}_\leq(\omega, 0) = \varnothing$ and implies

$$p(0) \wedge p(1) \wedge \cdots \wedge p(n-1) \to p(n)$$

when $n > 0$ because $\text{seg}_\leq(\omega, n) = n$.

Our first use of transfinite induction is the following lemma. It uses the terminology of Exercise 4.4.32 and is the first of a sequence of lemmas that will play a critical role.

◼ **LEMMA 6.1.2**

Let $(A, \preceq)$ be well-ordered. If $\varphi : A \to A$ is increasing, then $a \preceq \varphi(a)$ for all $a \in A$.

PROOF

Define $B = \{x \in A : x \preceq \varphi(x)\}$, where φ is an increasing function $A \to A$. Let $\text{seg}_\preceq(A, a) \subseteq B$. We note that a is the least element of $A \setminus \text{seg}_\preceq(A, a)$. Let $y \in \text{seg}_\preceq(A, a)$. This implies that $y \preceq \varphi(y) \prec \varphi(a)$ by definition of B and because $y \prec a$. Hence, $\varphi(a) \in A \setminus \text{seg}_\preceq(A, a)$. Thus, $a \preceq \varphi(a)$ and $A = B$ by transfinite induction (Theorem 6.1.1). ∎

◼ **LEMMA 6.1.3**

For all well-ordered sets $(A, \preceq)$ and $(A', \preceq')$, there exists at most one order isomorphism $\varphi : A \to A'$.

PROOF

Let $\varphi : A \to A'$ and $\psi : A \to A'$ be order isomorphisms. Since both φ^{-1} and ψ^{-1} are order isomorphisms $A' \to A$ (Theorem 4.5.26), $\psi^{-1} \circ \varphi$ and $\varphi^{-1} \circ \psi$ are order isomorphisms $A \to A$ (Theorem 4.5.27). We note that for every $b, c \in A$, if $b \prec c$, then $\varphi(b) \prec' \varphi(c)$ and then $\psi^{-1}(\varphi(b)) \prec \psi^{-1}(\varphi(c))$. This means that $\psi^{-1} \circ \varphi$ is increasing. A similar argument proves that $\varphi^{-1} \circ \psi$ is increasing. To show that $\varphi = \psi$, let $a \in A$. By Lemma 6.1.2,

$$a \preceq (\psi^{-1} \circ \varphi)(a)$$

and

$$a \preceq (\varphi^{-1} \circ \psi)(a).$$

Therefore, $\psi(a) \preceq' \varphi(a)$ and $\varphi(a) \preceq' \psi(a)$. Since $\preceq'$ is antisymmetric, we have that $\varphi(a) = \psi(a)$. ∎

■ **LEMMA 6.1.4**

No well-ordered set $(A, \preccurlyeq)$ is order isomorphic to any of its proper initial segments.

PROOF

Let $(A, \preccurlyeq)$ be a well-ordered set. Suppose that S is a proper initial segment of A. In order to obtain a contradiction, assume that $\varphi : A \rightarrow S$ is an order isomorphism. Take $a \in A \setminus S$. Since $\varphi(a) \in S$ and φ is increasing, we have that $a \preccurlyeq \varphi(a) \prec a$ by Lemma 6.1.2. ■

The next result follows from Lemma 6.1.4 (Exercise 1).

■ **LEMMA 6.1.5**

Distinct initial segments of a well-ordered set are not order isomorphic.

The lemmas lead to the following theorem.

■ **THEOREM 6.1.6**

If $(A, \preccurlyeq)$ and $(B, \preccurlyeq')$ are well-ordered sets, there exists an order isomorphism such that exactly one of the following holds.

- $A \cong B$.

- A is order isomorphic to a proper initial segment of B.

- B is order isomorphic to a proper initial segment of A.

PROOF

Let $(A, \preccurlyeq)$ and $(B, \preccurlyeq')$ be well-ordered sets. Appealing to Lemma 6.1.5, if $x \in A$, there is at most one $y \in B$ such that $\text{seg}_{\preccurlyeq}(A, x) \cong \text{seg}_{\preccurlyeq'}(B, y)$, so define the function

$$\varphi = \{(x, y) \in A \times B : \text{seg}_{\preccurlyeq}(A, x) \cong \text{seg}_{\preccurlyeq'}(B, y)\}.$$

We have a number of facts to prove.

- Let $y_1, y_2 \in \text{ran}(\varphi)$ such that $y_1 = y_2$. Take $x_1, x_2 \in A$ such that

$$\text{seg}_{\preccurlyeq}(A, x_1) \cong \text{seg}_{\preccurlyeq'}(B, y_1)$$

and

$$\text{seg}_{\preccurlyeq}(A, x_2) \cong \text{seg}_{\preccurlyeq'}(B, y_2).$$

Then, we have $\text{seg}_{\preccurlyeq}(A, x_1) \cong \text{seg}_{\preccurlyeq}(A, x_2)$, and $x_1 = x_2$ by Lemma 6.1.5. Therefore, φ is one-to-one.

- Take $x_1, x_2 \in \text{dom}(\varphi)$ and assume that $x_1 \preccurlyeq x_2$. This implies that

$$\text{seg}_{\preccurlyeq}(A, x_1) \subseteq \text{seg}_{\preccurlyeq}(A, x_2).$$

Then, by definition of φ, we have

$$\operatorname{seg}_\preccurlyeq(A, x_1) \cong \operatorname{seg}_{\preccurlyeq'}(B, \varphi(x_1))$$

and

$$\operatorname{seg}_\preccurlyeq(A, x_2) \cong \operatorname{seg}_{\preccurlyeq'}(B, \varphi(x_2)).$$

Hence, $\operatorname{seg}_{\preccurlyeq'}(B, \varphi(x_1))$ is order isomorphic to an initial segment S of $\operatorname{seg}_{\preccurlyeq'}(B, \varphi(x_2))$ (Exercise 17). If $S \neq \operatorname{seg}_{\preccurlyeq'}(B, \varphi(x_1))$, then B has two distinct isomorphic initial segments, contradicting Lemma 6.1.5. This implies that $\varphi(x_1) \prec' \varphi(x_2)$, so φ is order-preserving.

- Let $x_1, x_2 \in A$. Suppose that $x_1 \preccurlyeq x_2$ and $x_2 \in \operatorname{dom}(\varphi)$. This means that there exists $y_2 \in B$ such that

$$\operatorname{seg}_\preccurlyeq(A, x_2) \cong \operatorname{seg}_{\preccurlyeq'}(B, y_2).$$

If $x_1 = x_2$, then $x_1 \in \operatorname{dom}(\varphi)$, so assume that $x_1 \neq x_2$. Since $x_1 \prec x_2$, we have that $x_1 \in \operatorname{seg}_\preccurlyeq(A, x_2)$. Because φ is order-preserving,

$$\operatorname{seg}_\preccurlyeq(A, x_1) \cong \operatorname{seg}_{\preccurlyeq'}(B, y_1)$$

for some $y_1 \in \operatorname{seg}_{\preccurlyeq'}(B, y_2)$ (Exercise 17). Therefore, $(x_1, y_1) \in \varphi$, so $x_1 \in \operatorname{dom}(\varphi)$, proving that the domain of φ is an initial segment of A.

- That the range of φ is an initial segment of B is proved like the previous case.

If φ is a surjection and $\operatorname{dom}(\varphi) = A$, then φ is an order isomorphism $A \to B$, else $\varphi^{-1}[B]$ is a proper initial segment of A. If φ is not a surjection, $\varphi[A]$ is a proper initial segment of B. ∎

Ordinals

Theorem 6.1.6 is a sort of trichotomy law for well-ordered sets. Two well-ordered sets look alike, or one has a copy of itself in the other. This suggests that we should be able to choose certain well-ordered sets to serve as representatives of all the different types of well-ordered sets. No two of the chosen sets should be order isomorphic, but it should be the case that every well-ordered set is order isomorphic to exactly one of them. That is, we should be able to classify all of the well-ordered sets. This will be our immediate goal and is the purpose behind the next definition.

■ **DEFINITION 6.1.7**

The set α is an **ordinal number** (or simply an **ordinal**) if $(\alpha, \subseteq)$ is a well-ordered set and $\beta = \operatorname{seg}_\subseteq(\alpha, \beta)$ for all $\beta \in \alpha$. For ordinals, define

$$\operatorname{seg}(\alpha, \beta) = \operatorname{seg}_\subseteq(\alpha, \beta).$$

Definition 6.1.7 implies that ω and every natural number is an ordinal because they are well-ordered by $\subseteq$ and for all $n \in \omega \setminus \{0\}$,

$$n = \{0, 1, 2, \ldots, n-1\} = \text{seg}(\omega, n),$$

and for all $k \in n$,

$$k = \{0, 1, 2, \ldots, k-1\} = \text{seg}(n, k).$$

For example, $5 \in 7$ and

$$5 = \{0, 1, 2, 3, 4\} = \text{seg}(7, 5).$$

We now prove a sequence of basic results about ordinals. The first is similar to Theorem 5.2.8, so its proof is left to Exercise 5.

■ THEOREM 6.1.8

Ordinals are transitive sets.

■ THEOREM 6.1.9

The elements of ordinals are transitive sets.

PROOF

Let α be an ordinal and $\beta \in \alpha$. Take $\gamma \in \beta$ and $\delta \in \gamma$. Since α is transitive (Theorem 6.1.8), we have that $\gamma \in \alpha$. Therefore, $\gamma = \text{seg}(\alpha, \gamma)$ and $\beta = \text{seg}(\alpha, \beta)$, so

$$\delta \in \text{seg}(\alpha, \gamma) \subseteq \text{seg}(\alpha, \beta) = \beta,$$

which implies that β is transitive (Definition 5.2.7). ■

■ THEOREM 6.1.10

Every element of an ordinal is an ordinal.

PROOF

Let α be an ordinal and $\beta \in \alpha$. Notice that this implies that β is transitive (Theorem 6.1.9). Since $\beta \subseteq \alpha$, we have that $(\beta, \subseteq)$ is a well-ordered set by a subset axiom (5.1.8) and Theorem 4.3.26. Now take $\delta \in \beta$. Since $\delta \in \alpha$, we have that δ is transitive. Therefore, by Exercise 5.2.3,

$$\begin{aligned}
\delta &= \{\gamma : \gamma \in \delta\} \\
&= \{\gamma : \gamma \in \beta \wedge \gamma \in \delta\} \\
&\subseteq \{\gamma : \gamma \in \beta \wedge \gamma \subset \delta\} \\
&= \text{seg}(\beta, \delta) \\
&\subseteq \text{seg}(\alpha, \delta) \\
&= \delta.
\end{aligned}$$

From this, we conclude that $\delta = \text{seg}(\beta, \delta)$. ■

■ **THEOREM 6.1.11**

Let α and β be ordinals. Then, $\alpha \subset \beta$ if and only if $\alpha \in \beta$.

PROOF

If $\alpha \in \beta$, then $\alpha \subset \beta$ because β is transitive (Theorem 6.1.8 and Exercise 5.2.3). Conversely, suppose that $\alpha \subset \beta$. Let $\gamma \in \alpha$ and $\delta \subset \gamma$ with $\delta \in \beta$. Since β is an ordinal, $\delta = \text{seg}(\beta, \delta)$. Hence, $\delta = \text{seg}(\gamma, \delta)$, which implies that $\delta \in \gamma$ because γ is an ordinal by Theorem 6.1.10. Therefore, $\delta \in \alpha$ because α is transitive (Theorem 6.1.8). This shows that α is a proper initial segment of β with respect to $\subseteq$ (Definition 5.6.1). From this, it follows by Lemma 5.6.3 that $\alpha = \text{seg}(\beta, \zeta)$ for some $\zeta \in \beta$. Hence, $\alpha \in \beta$. ■

■ **THEOREM 6.1.12**

Every ordinal is well-ordered by $\in$.

The next theorem is an important part of the process of showing that the ordinals are the sets that classify all well-ordered sets according to their order types. It states that distinct ordinals are not order isomorphic with respect to $\subseteq$.

■ **THEOREM 6.1.13**

For all ordinals α and β, if $(\alpha, \subseteq) \cong (\beta, \subseteq)$, then $\alpha = \beta$.

PROOF

Let $\varphi : \alpha \to \beta$ be an order isomorphism preserving $\subseteq$. Define

$$A = \{\gamma \in \alpha : \varphi(\gamma) = \gamma\}.$$

Take $\delta \in \alpha$ and assume that $\text{seg}(\alpha, \delta) \subseteq A$. Then,

$$\begin{aligned}
\varphi(\delta) &= \text{seg}(\beta, \varphi(\delta)) \\
&= \varphi[\text{seg}(\alpha, \delta)] \\
&= \{\varphi(\gamma) : \gamma \in \alpha \wedge \gamma \subset \delta\} \\
&= \{\gamma : \gamma \subset \delta\} \\
&= \delta.
\end{aligned}$$

The first equality follows because $\varphi(\delta)$ is an ordinal in β, the second follows because φ is an order isomorphism, and the fourth equation follows by the assumption. Therefore, by transfinite induction (Theorem 6.1.1), $A = \alpha$, so φ is the identity map and $\alpha = \beta$. ■

Because of Theorem 6.1.13, we are able to prove that there is a trichotomy law for the ordinals with respect to $\subseteq$.

■ **THEOREM 6.1.14 [Trichotomy]**

For all ordinals α and β, exactly one of the following holds: $\alpha = \beta$, $\alpha \subset \beta$, or $\alpha \subset \beta$.

PROOF

Since $(\alpha, \subseteq)$ and $(\beta, \subseteq)$ are well-ordered, by Theorem 6.1.6, exactly one of the following holds.

- $\alpha \cong \beta$, which implies that $\alpha = \beta$ by Theorem 6.1.13.

- There exists $\delta \in \beta$ such that $\alpha \cong \text{seg}(\beta, \delta)$. Since $\text{seg}(\beta, \delta)$ is an ordinal (Theorem 6.1.10), $\alpha = \text{seg}(\beta, \delta)$, again by Theorem 6.1.13. Therefore, $\alpha \subset \beta$.

- There exists $\gamma \in \alpha$ such that $\beta \cong \text{seg}(\alpha, \gamma)$. As in the previous case, we have that $\beta \subset \alpha$. ∎

Because of Theorem 6.1.11, we can quickly conclude the following.

■ COROLLARY 6.1.15

For all ordinals α and β, exactly one of the following holds: $\alpha = \beta$, $\alpha \in \beta$, or $\alpha \in \beta$.

In addition to the ordinals having a trichotomy law, the least upper bound with respect to $\subseteq$ of a set of ordinals is also an ordinal (compare Example 4.3.14).

■ THEOREM 6.1.16

If $\mathscr{F}$ is a set of ordinals, $\bigcup \mathscr{F}$ is an ordinal.

PROOF

We show that $\bigcup \mathscr{F}$ satisfies the conditions of Definition 6.1.7. Since the elements of ordinals are ordinals, $\bigcup \mathscr{F}$ is a set of ordinals, and by Theorem 6.1.14, we see that $(\bigcup \mathscr{F}, \subseteq)$ is a linearly ordered set. Let $B \subseteq \bigcup \mathscr{F}$ and take $\alpha \in B$. We have two cases to consider.

- Suppose $\alpha \cap B = \varnothing$. Let $\beta \in B$. Then, $\beta \notin \alpha$, so by Theorem 6.1.11 and Theorem 6.1.14, $\alpha \subseteq \beta$. Hence, α is the least element of B.

- Let $\alpha \cap B$ be nonempty. Since α is an ordinal, there exists an ordinal δ that is the least element of $\alpha \cap B$ with respect to $\subseteq$. Let $\beta \in B$. If $\beta \subset \alpha$, then $\beta \in \alpha \cap B$, which implies that $\delta \subseteq \beta$. Also, if $\alpha \subseteq \beta$, then $\delta \subset \beta$. Since these are the only two options (Theorem 6.1.14), this implies that δ is the least element of B.

We conclude that $(\bigcup \mathscr{F}, \subseteq)$ is a well-ordered set.

Next, let $\beta \in \bigcup \mathscr{F}$. This means that there exists an ordinal $\alpha \in A$ such that $\beta \in \alpha$. Since $\text{seg}(\bigcup \mathscr{F}, \beta) \subseteq \beta$ by definition, take $\delta \in \beta$. Since α is transitive (Theorem 6.1.8), $\delta \in \alpha$. Therefore, $\delta \in \bigcup \mathscr{F}$, which implies that $\beta \subseteq \text{seg}(\bigcup \mathscr{F}, \beta)$. ∎

Classification

Let α be an ordinal number. We check the two conditions of Definition 6.1.7 to show that α^+ is an ordinal.

- Let B be a nonempty subset of $\alpha \cup \{\alpha\}$. If $B \cap \alpha \neq \varnothing$, then B has a least element with respect to $\subseteq$ since $(\alpha, \subseteq)$ is well-ordered. If $B = \{\alpha\}$, then α is the least element of B.

- Let $\beta \in \alpha \cup \{\alpha\}$. If $\beta \in \alpha$, then $\beta = \text{seg}(\alpha, \beta) = \text{seg}(\alpha^+, \beta)$ because α is an ordinal number. Otherwise, $\beta = \alpha = \text{seg}(\alpha^+, \alpha)$.

The ordinal α^+ is called a **successor ordinal** because it has a predecessor. For example, every positive natural number is a successor ordinal.

Now assume that $\alpha \neq \varnothing$ and α is an ordinal that is not a successor.

- Let $\delta \in \beta \in \alpha$. Since α is transitive (Theorem 6.1.8), $\delta \in \alpha$. Thus, we conclude that $\bigcup \{\beta : \beta \in \alpha\} \subseteq \alpha$.

- Now take $\delta \in \alpha$. This implies that δ is an ordinal (Theorem 6.1.10), so $\delta \subset \alpha$ by Theorem 6.1.11. Therefore, $\delta^+ \subseteq \alpha$, so $\delta^+ \subset \alpha$ since α is not a successor. Thus, $\delta^+ \in \alpha$, again by appealing to Theorem 6.1.11. Because $\delta \in \delta^+$, we have that $\alpha \subseteq \bigcup \{\beta : \beta \in \alpha\}$.

We conclude that for every nonempty ordinal α that is not a successor,

$$\alpha = \bigcup_{\beta \in \alpha} \beta.$$

Such an ordinal number is called a **limit ordinal**. For example, since every natural number is an ordinal, $\omega = \bigcup \{n : n \in \omega\}$ is a limit ordinal. Therefore, $\omega^+, \omega^{++}, \omega^{+++}, \ldots$ are also ordinals, but they are successors.

All of this proves the following.

■ THEOREM 6.1.17

A nonempty ordinal is either a successor or a limit ordinal.

Therefore, by Theorem 6.1.14 and Corollary 6.1.15, we can view the ordinals as sorted by $\subset$ giving

$$0 \subset 1 \subset 2 \subset \cdots \subset \omega \subset \omega^+ \subset \omega^{++} \subset \omega^{+++} \subset \cdots$$

and as sorted by $\in$ giving

$$0 \in 1 \in 2 \in \cdots \in \omega \in \omega^+ \in \omega^{++} \in \omega^{+++} \in \cdots.$$

Characterizing every ordinal as being equal to 0, a successor ordinal, or a limit ordinal allows us to restate Theorem 6.1.1. The form of the theorem generalizes Theorem 5.4.1 to infinite ordinals. Its proof is left to Exercise 8.

■ **THEOREM 6.1.18 [Transfinite Induction 2]**

If α is an ordinal and $A \subseteq \alpha$, then $A = \alpha$ if the following hold:

- $0 \in A$.

- If $\beta \in A$, then $\beta^+ \in A$.

- If β is a limit ordinal such that $\delta \in A$ for all $\delta \in \beta$, then $\beta \in A$.

We use this second form of transfinite induction to prove the sought-after classification theorem for well-ordered sets.

■ **THEOREM 6.1.19**

Let $(A, \preccurlyeq)$ be a well-ordered set. Then, $(A, \preccurlyeq) \cong (\alpha, \subseteq)$ for some ordinal α.

PROOF

Define
$$p(x, y) := y \text{ is an ordinal} \wedge \text{seg}_{\preccurlyeq}(A, x) \cong y.$$

Let $B = \{y : \exists x \, [x \in A \wedge p(x, y)]\}$. By Theorem 6.1.13, $p(x, y)$ defines a function, so by a replacement axiom (5.1.9), we conclude that B is a set. We have a number of items to prove.

Let $D \subseteq B$ and $D \neq \varnothing$. Let $C = \{a \in A : \exists \alpha [\alpha \in D \wedge p(a, \alpha)]\}$. Observe that C is not empty. Therefore, there exists a least element $m \in C$ with respect to $\preccurlyeq$. Take an ordinal $\delta_0 \in D$ such that
$$\text{seg}_{\preccurlyeq}(A, m) \cong \delta_0.$$

Let $\delta \in D$. This means that δ is an ordinal and
$$\text{seg}_{\preccurlyeq}(A, c) \cong \delta$$

for some $c \in C$. Since $m \preccurlyeq c$, we have that
$$\text{seg}_{\preccurlyeq}(A, m) \subseteq \text{seg}_{\preccurlyeq}(A, c).$$

Hence, δ_0 is isomorphic to a subset of δ, which implies that $\delta_0 \subseteq \delta$ (Theorem 6.1.13). We conclude that $(B, \subseteq)$ is a well-ordered set.

Let $E = \{\beta \in B : \text{seg}(B, \beta) = \beta\}$. Let $\text{seg}(B, \epsilon) \subseteq E$ for $\epsilon \in B$.

- First, suppose that $\epsilon = \gamma^+$ for some ordinal γ. Then, $\text{seg}(B, \gamma) = \gamma$, so
$$\text{seg}(B, \gamma) \cup \{\gamma\} = \gamma^+.$$

Also, $\gamma^+ \cong \text{seg}_{\preccurlyeq}(A, a)$ for some $a \in A$. Let m be the greatest element of $\text{seg}_{\preccurlyeq}(A, a)$ (Exercise 9). This implies that $\gamma \cong \text{seg}_{\preccurlyeq}(A, a) \setminus \{m\}$, so $\gamma \in B$. Hence,
$$\text{seg}(B, \gamma) \cup \{\gamma\} = \text{seg}(B, \gamma^+),$$

and we have $\epsilon \in E$.

- Second, let $\epsilon = \bigcup\{\gamma : \gamma \in \epsilon\}$. This means that $\mathrm{seg}(B, \gamma) = \gamma$ for all $\gamma \in \epsilon$. Therefore,

$$\mathrm{seg}(B, \epsilon) = \bigcup_{\gamma \in \epsilon} \mathrm{seg}(B, \gamma) = \bigcup_{\gamma \in \epsilon} \gamma = \epsilon,$$

and ϵ is again an element of E.

By transfinite induction (Theorem 6.1.18), $E = B$. This combined with $(B, \subseteq)$ being a well-ordered set means that B is an ordinal.

Define $\varphi : A \to B$ by $\varphi(x) = y \Leftrightarrow p(x, y)$. Since φ is an order isomorphism (Exercise 10), B is an ordinal that is order isomorphic to $(A, \preccurlyeq)$, and because of Theorem 6.1.13, it is the only one. ■

For any well-ordered set $(A, \leq)$, the unique ordinal α such that $A \cong \alpha$ guaranteed by Theorem 6.1.19 is called the **order type** of A. Compare this definition with Definition 4.5.24. For example, the order type of $(\{2n : n > 5 \wedge n \in \mathbb{Z}\}, \leq)$ is ω.

Burali-Forti and Hartogs

Suppose $\mathscr{A} = \{0, 4, 6, 9\}$. Then, $\bigcup \mathscr{A}$ equals the ordinal 9, which is the least upper bound of $\mathscr{A}$. Also, assume that $\mathscr{B} = \{5, 100, \omega\}$. Then, $\bigcup \mathscr{B} = \omega$. However, the least upper bound of $\mathscr{C} = \{n \in \omega : \exists k(k \in \omega \wedge n = 2k)\}$ is not an element of ω. Instead, the least upper bound of $\mathscr{C}$ is $\bigcup \mathscr{C} = \omega$. Moreover, notice that $\mathscr{A} \subseteq 10$, $\mathscr{B} \subseteq \omega^+$, and $\mathscr{C} \subseteq \omega^+$. We generalize this to the next theorem.

■ THEOREM 6.1.20

If $\mathscr{F}$ is a set of ordinals, there exists an ordinal α such that $\mathscr{F} \subseteq \alpha$.

PROOF

Take $\mathscr{F}$ to be a set of ordinals and let $\alpha \in \mathscr{F}$. Then, $\alpha \subseteq \bigcup \mathscr{F}$ and $\bigcup \mathscr{F}$ is an ordinal by Theorem 6.1.16. If $\alpha \subset \bigcup \mathscr{F}$, then $\alpha \in \bigcup \mathscr{F}$ by Theorem 6.1.11. If $\alpha = \bigcup \mathscr{F}$, then $\alpha \in \{\bigcup \mathscr{F}\}$. Thus, $\mathscr{F} \subseteq (\bigcup \mathscr{F})^+$. ■

Although every set of ordinals is a subset of an ordinal, there is no set of all ordinals, otherwise a contradiction would arise, as was first discovered by Cesare Burali-Forti (1897). This is why when we noted that $\subseteq$ gives the ordinals a linear order, we did not claim that $\subseteq$ is used to define a linearly ordered set containing all ordinals.

■ THEOREM 6.1.21 [Burali-Forti]

There is no set that has every ordinal as an element.

PROOF

Suppose $\mathscr{F} = \{\alpha : \alpha \text{ is an ordinal}\}$ is a set. This implies that $\bigcup \mathscr{F}$ is an ordinal by Theorem 6.1.16. However, for every $\alpha \in \mathscr{F}$, we have that $\alpha \in \alpha^+ \in A$, showing that $\mathscr{F} \subseteq \bigcup \mathscr{F}$. Since $\bigcup \mathscr{F} \in \mathscr{F}$, we also have $\bigcup \mathscr{F} \in \bigcup \mathscr{F}$, which contradicts Theorem 5.1.16. ■

The Burali-Forti theorem places a limit on what can be done with ordinals. One such example is a theorem of Friedrich Hartogs.

■ THEOREM 6.1.22 [Hartogs]

For every set A, there exists an ordinal α such that there are no injections of α into A.

PROOF

Let A be a set. Define

$$\mathcal{E} = \{\alpha : \alpha \text{ is an ordinal} \wedge \exists\psi(\psi \text{ is an injection } \alpha \rightarrow A)\}.$$

Notice that for every $\alpha \in \mathcal{E}$, there exists a bijection φ_α such that

$$\varphi_\alpha : \alpha \rightarrow B_\alpha$$

for some $B_\alpha \subseteq A$. Define a well-order $\preccurlyeq_\alpha$ on B_α by

$$\varphi_\alpha(\beta_1) \preccurlyeq_\alpha \varphi_\alpha(\beta_2) \text{ if and only if } \beta_1 \subseteq \beta_2 \text{ for all } \beta_1, \beta_2 \in \alpha.$$

Then, φ_α is an order isomorphism preserving $\subseteq$ with $\preccurlyeq_\alpha$. Next, define

$$\mathcal{F} = \{(B, \leq) : B \subseteq A \wedge \leq \text{ is a well-ordering of } B\}.$$

Since $\mathcal{F} \subseteq \mathbf{P}(A) \times \mathbf{P}(A \times A)$, we have that $\mathcal{F}$ is a set by the Power Set Axiom (5.1.7) and a Subset Axiom (5.1.8). Let

$$p(x, y) := x \in \mathcal{F} \wedge \exists\gamma(\gamma \text{ is an order isomorphism } x \rightarrow y$$
$$\text{preserving the order on } x \text{ with } \subseteq\}.$$

Suppose that $p((B, \leq), \alpha_1)$ and $p((B, \leq), \alpha_2)$. By Theorem 6.1.13, we have that $\alpha_1 = \alpha_2$, so $p(x, y)$ defines a function with domain $\mathcal{F}$. Moreover, $\mathcal{E}$ is a subset of the range of this function because $p((B_\alpha, \preccurlyeq_\alpha), \alpha)$ due to φ_α. Therefore, $\mathcal{E}$ is a set by a replacement axiom (5.1.9) and a subset axiom, and $\mathcal{E}$ cannot contain all ordinals by the Burali-Forti theorem (6.1.21). ■

Transfinite Recursion

Theorem 6.1.19 only applies to well-ordered sets, so, for example, it does not apply to $(\mathbb{Z}, \leq)$ or $(\mathbb{R}, \leq)$. However, if we change the order on $\mathbb{Z}$ from the standard $\leq$ to $\preccurlyeq$ defined so as to put $\mathbb{Z}$ into this order,

$$0, 1, -1, 2, -2, 3, -3, \ldots,$$

then $(\mathbb{Z}, \preccurlyeq)$ is a well-ordered set of order type ω. That this can be done even with sets like $\mathbb{R}$ is due to a theorem first proved by Zermelo, which is often called the **well-ordering theorem**. Its proof requires some preliminary work.

Let A be a set and α an ordinal. By Definition 4.4.13, $^\alpha A$ is the set of all functions $\alpha \rightarrow A$. Along these lines, define

$$^{<\alpha}A = \{\varphi : \exists\beta(\beta \in \alpha \wedge \varphi \text{ is a function } \beta \rightarrow A)\}.$$

For example, $f, g \in {}^{<5}\mathbb{Z}$, where

$$f = \{(0, 1), (1, 2), (2, 3), (3, 4)\}$$

and

$$g = \{(0, -4), (1, 14)\}.$$

Also, $f, g \in {}^{<\omega}\mathbb{Z}$, but the identity function on ω is not an element of ${}^{<\omega}\mathbb{Z}$ because its domain is ω. We should also note that for any set A,

$$^{<\varnothing}A = \varnothing.$$

We use this notation in the following generalization of recursion to infinite ordinals.

■ THEOREM 6.1.23 [Transfinite Recursion]

Let α be an ordinal. For every function $\psi : {}^{<\alpha}A \rightarrow A$, there exists a unique function $\varphi : \alpha \rightarrow A$ such that for every $\beta \in \alpha$,

$$\varphi(\beta) = \psi(\varphi \restriction \beta).$$

PROOF

To prove uniqueness, in addition to φ, let φ' be a function $\alpha \rightarrow A$ such that for all $\beta \in \alpha$,

$$\varphi'(\beta) = \psi(\varphi' \restriction \beta).$$

Define $B = \{\beta \in \alpha : \varphi(\beta) = \varphi'(\beta)\}$. We use transfinite induction (Theorem 6.1.1) to show that $B = \alpha$. Suppose $\mathrm{seg}(\alpha, \delta) \subseteq B$ with $\delta \in \alpha$. That is,

$$\forall \beta[\beta \in \delta \rightarrow \varphi(\beta) = \varphi'(\beta)].$$

This implies that $\varphi \restriction \delta = \varphi' \restriction \delta$. Therefore,

$$\varphi(\delta) = \psi(\varphi \restriction \delta) = \psi(\varphi' \restriction \delta) = \varphi'(\delta),$$

so $\delta \in B$, and we conclude that $\varphi = \varphi'$.

We prove existence indirectly. Suppose that α is the least ordinal (Exercise 2) such that

there exists a function $\psi_0 : {}^{<\alpha}A \rightarrow A$ such that
for every $\varphi : \alpha \rightarrow A$, there exists $\beta \in \alpha$
such that $\varphi(\beta) \neq \psi_0(\varphi \restriction \beta)$.

Since the theorem is trivially true for $\alpha = 0$, we have two cases to consider.

- Let $\alpha = \delta^+$ for some ordinal δ. By minimality of α, we have a function $\varphi_\delta : \delta \rightarrow A$ such that for all $\beta \in \delta$,

$$\varphi_\delta(\beta) = \psi_0(\varphi_\delta \restriction \beta).$$

Extend φ_δ to $\bar{\varphi}_\delta : \alpha \to A$ by defining $\bar{\varphi}_\delta(\delta) = \psi_0(\varphi_\delta)$, so

$$\bar{\varphi}_\delta(\delta) = \psi_0(\bar{\varphi}_\delta \restriction \delta).$$

This contradicts the minimality of α.

- Let α be a limit ordinal. For each $\delta \in \alpha$, there exists a unique $\varphi_\delta : \delta \to A$ such that

$$\varphi_\delta(\delta) = \psi(\varphi_\delta \restriction \delta).$$

Notice that $\delta \in \gamma$ implies that φ_γ is an extension of φ_δ, otherwise $\varphi_\gamma \restriction \delta$ would have the property

$$(\varphi_\gamma \restriction \delta)(\beta) = \psi_0([\varphi_\gamma \restriction \delta] \restriction \beta)$$

for all $\beta \in \delta$ yet $\varphi_\delta \neq \varphi_\gamma \restriction \delta$. This contradicts the uniqueness of φ_δ. Therefore, $\{\varphi_\delta : \delta \in \alpha\}$ is a chain, so, as in Exercise 4.4.13, define the function $\varphi : \alpha \to A$ by

$$\varphi = \bigcup_{\delta \in \alpha} \varphi_\delta.$$

To check that φ is the function given by the theorem, take $\beta \in \alpha$. Since α is a limit ordinal, $\beta^+ \in \alpha$ and

$$\varphi(\beta) = \varphi_{\beta^+}(\beta) = \psi_0(\varphi_{\beta^+} \restriction \beta) = \psi_0(\varphi \restriction \beta),$$

again contradicting the minimality of α. ■

Theorem 6.1.23 has a corollary that can be viewed as an extension of Theorem 5.2.14. Its proof is left to Exercise 18.

■ COROLLARY 6.1.24

Let A be a set and $a \in A$. For every ordinal α, if ψ is a function $A \to A$, there exists a unique function $\varphi : \alpha \to A$ such that

- $\varphi(0) = a$,

- $\varphi(\beta^+) = \psi(\varphi(\alpha))$ for all $\beta \in \alpha$,

- $\varphi(\gamma) = \bigcup\{\varphi(\beta) : \beta \in \gamma\}$ for all limit ordinals $\gamma \in \alpha$.

We are now ready to prove that every set can be well-ordered. The following theorem is equivalent to the axiom of choice (Exercise 20).

■ THEOREM 6.1.25 [Zermelo]

For any set A, there exists a relation R on A such that (A, R) is a well-ordered set.

PROOF

Take a set A and let $\xi : P(A) \to A$ be a choice function (Corollary 5.1.11). By Theorem 6.1.22, there is an ordinal α such that no injection $\alpha \to A$ exists. Hence, we have $\varphi : A \to \alpha$ that is one-to-one. Let $B \subseteq A$. Since every element of α is an ordinal, there exists an ordinal $\delta \subseteq \alpha$ such that $\varphi[B] = \delta$ (Theorem 6.1.16). Define

$$\psi_B = \varphi \restriction B.$$

Then, $\psi_B^{-1} \in {}^{<\alpha}A$. Let

$$P = \{\psi_B^{-1} : B \in P(A)\},$$

and define

$$h_1 : P \to P(A)$$

by $h_1(\psi_B^{-1}) = B$. Also, define

$$h_2 : P \to A$$

by $h_2 = \xi \circ h_1$. Since $P \subseteq {}^{<\alpha}A$, extend h_2 to some

$$h : {}^{<\alpha}A \to A$$

such that $h \restriction P = h_2$. By transfinite recursion (Theorem 6.1.23), there exists a function $f : \alpha \to A$ such that for all $\beta \in \alpha$,

$$f(\beta) = h(f \restriction \beta).$$

Define Φ so that for all $\beta \in \alpha$,

$$\Phi(\beta) = \begin{cases} h(A \setminus f[\beta]) & \text{if } A \setminus f[\beta] \neq \varnothing, \\ A & \text{if } A \setminus f[\beta] = \varnothing. \end{cases}$$

Notice that $A \notin A$ by Theorem 5.1.16. Let β_0 be the least ordinal such that $\Phi(\beta_0) = A$. Then, $\Phi \restriction \beta_0$ is a bijection $\beta_0 \to A$ [Exercise 19(a)]. Lastly, define the relation R on A by

$$a \, R \, b \text{ if and only if } \Phi^{-1}(a) \subseteq \Phi^{-1}(b),$$

for all $a, b \in A$. Since $\subseteq$ is a well-order on β_0, R is a well-order on A [Exercise 19(b)]. ∎

Exercises

1. Prove Lemma 6.1.5.

2. Does $\text{seg}(\bigcup A, \beta) = \bigcup_{\alpha \in A} \text{seg}(\alpha, \beta)$ for all sets of ordinals A with $\beta \in A$? Explain.

3. Explain why $\{0, 2, 3, 4, 5\}$ is not an ordinal.

4. Prove that $\varnothing$ is an element of every ordinal.

5. Prove that an ordinal is a transitive set (Theorem 6.1.8).

6. Let A be a set of ordinals. Prove.
   (a) $\bigcap A$ is an ordinal.
   (b) A has a least element with respect to $\subseteq$.

7. Let B be a nonempty subset of the ordinal α. Prove that there exists $\beta \in B$ such that β and B are disjoint.

8. Prove Theorem 6.1.18.

9. From the proof of Theorem 6.1.19, prove that $\text{seg}_{\preccurlyeq}(A, a)$ has a greatest element.

10. Prove that A is order isomorphic to B in the proof of Theorem 6.1.19.

11. The proof of Theorem 6.1.19 contains many isomorphisms without explicitly identifying the isomorphism. Find these functions and prove that they are order isomorphisms.

12. Let R be a well-ordering on A and suppose that A has no greatest element. Show the the order type of (A, R) is a limit ordinal.

13. Find a transitive set that is not an ordinal.

14. Theorem 6.1.21 comes from the **Burali-Forti paradox**. Like Russell's paradox (page 225), it arises when any formula is allowed to define a set. In this case, suppose that $A = \{\alpha : \alpha$ is an ordinal$\}$ and assume that A is a set. Prove that A is an ordinal that must include all ordinals as its elements.

15. Prove that there exists a function F such that $F(n)$ is the nth Fibonacci number.

16. Prove that for every function $h : {}^{<\omega}A \to A$, there is a unique function $f : \omega \to A$ such that for all $n \in \omega$, $f(n) = h(f \restriction n)$.

17. Let $(A, \preccurlyeq)$ and $(B, \preccurlyeq')$ be well-ordered sets and $\varphi : A \to B$ be an order-preserving surjection. Prove that for every $a \in A$, there exists $b \in B$ such that

$$\varphi[\text{seg}_{\preccurlyeq}(A, a)] = \text{seg}_{\preccurlyeq'}(B, b).$$

18. Prove Corollary 6.1.24.

19. Prove the following from the proof of Zermelo's theorem (6.1.25).
   (a) $\Phi \restriction \beta_0$ is a bijection.
   (b) R is a well-order on A.

20. Prove that Theorem 6.1.25 implies Axiom 5.1.10.

21. Prove that Zorn's lemma (5.1.13) implies Theorem 6.1.25 without using the axiom of choice.

6.2 EQUINUMEROSITY

How can we determine whether two sets are of the same size? One possibility is to count their elements. What happens, however, if the sets are infinite? We need another method. Suppose $A = \{12, 47, 84\}$ and $B = \{17, 101, 200\}$. We can see that these two sets are the same size without counting. Define a function $f : A \to B$ so that $f(12) = 17$, $f(47) = 101$, and $f(84) = 200$. This function is a bijection. Since each element is paired with exactly one element of the opposite set, A and B must be the same size. This is the motivation behind our first definition.

■ **DEFINITION 6.2.1**

The sets A and B are **equinumerous** (written as $A \approx B$) if there exists a bijection $\varphi : A \to B$. If A and B are not equinumerous, write $A \not\approx B$.

■ **EXAMPLE 6.2.2**

Take $n \in \mathbb{Z}$ such that $n \neq 0$ and define

$$n\mathbb{Z} = \{nk : k \in \mathbb{Z}\}.$$

We prove that $\mathbb{Z} \approx n\mathbb{Z}$. To show this, we must find a bijection $f : \mathbb{Z} \to n\mathbb{Z}$. Define $f(k) = nk$.

- Assume $x_1, x_2 \in \mathbb{Z}$, and let $f(x_1) = f(x_2)$. Then $nx_1 = nx_2$, which yields $x_1 = x_2$ since $n \neq 0$. Thus, f is one-to-one.

- Let $y \in n\mathbb{Z}$. This means that $y = nk$ for some $k \in \mathbb{Z}$, so $y = f(k)$. This shows that f is onto and, hence, a bijection.

■ **EXAMPLE 6.2.3**

To see $\mathbb{Z}^+ \approx \mathbb{Z}$, define a one-to-one correspondence so that each even integer is paired with a nonnegative integer and every odd integer is paired with a negative integer (Figure 6.1). Let $g : \mathbb{Z}^+ \to \mathbb{Z}$ be defined by

$$g(n) = \begin{cases} k - 1 & \text{if } n = 2k \text{ for some } k \in \mathbb{Z}^+, \\ -k & \text{if } n = 2k - 1 \text{ for some } k \in \mathbb{Z}^+. \end{cases}$$

Notice that $g(4) = 1$ since $4 = 2(2)$, and $g(5) = -3$ because $5 = 2(3) - 1$. This function is a bijection (Exercise 12).

Equinumerosity plays a role similar to that of equality of integers. This is seen in the next theorem. In fact, the theorem resembles Definition 4.2.4. Despite this, it does not demonstrate the existence of an equivalence relation. This is because an equivalence relation is a relation on a set, so, to define an equivalence relation, the next result would require a set of all sets, contradicting Corollary 5.1.17.

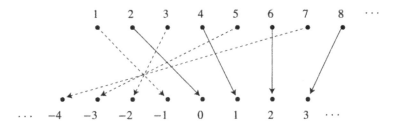

Figure 6.1 $\mathbb{Z} \approx \mathbb{Z}^+$.

■ THEOREM 6.2.4

Let A, B, and C be sets.

- $A \approx A$. (Reflexive)

- If $A \approx B$, then $B \approx A$. (Symmetric)

- If $A \approx B$ and $B \approx C$, then $A \approx C$. (Transitive)

PROOF

- $A \approx A$ since the identity map is a bijection.

- Assume $A \approx B$. Then, there exists a bijection $\varphi : A \to B$. Therefore, φ^{-1} is a bijection. Hence, $B \approx A$.

- By Theorem 4.5.23, the composition of two bijections is a bijection. Therefore, $A \approx B$ and $B \approx C$ implies $A \approx C$. ■

The symmetric property allows us to conclude that $n\mathbb{Z} \approx \mathbb{Z}$ and $\mathbb{Z} \approx \mathbb{Z}^+$ from Examples 6.2.2 and 6.2.3. The transitivity part of Theorem 6.2.4 allows us to conclude from this that $n\mathbb{Z} \approx \mathbb{Z}^+$.

■ EXAMPLE 6.2.5

Show $(0, 1) \approx \mathbb{R}$. We do this in two parts. First, let $f : (0, 1) \to (-\pi/2, \pi/2)$ be defined by $f(x) = \pi x - \pi/2$. This function is a one-to-one correspondence since its graph is a nonvertical, nonhorizontal line (Exercise 4.5.4). Second, define $g : (-\pi/2, \pi/2) \to \mathbb{R}$ to be the function $g(x) = \tan x$. From trigonometry, we know that tangent is a bijection on $(-\pi/2, \pi/2)$. Hence,

$$(0, 1) \approx (-\pi/2, \pi/2)$$

and

$$(-\pi/2, \pi/2) \approx \mathbb{R}.$$

Therefore, $(0, 1) \approx \mathbb{R}$ by Theorem 6.2.4.

Order

If $\approx$ resembles equality, the following resembles the $\leq$ relation.

■ DEFINITION 6.2.6

The set B **dominates** the set A (written as $A \preceq B$) if there exists an injection $\varphi : A \to B$. If B does not dominate A, write $A \npreceq B$. Furthermore, define $A \prec B$ to mean $A \preceq B$ but $A \napprox B$.

■ EXAMPLE 6.2.7

If $A \subseteq B$, then $A \preceq B$. This is proved using the inclusion map (Exercise 4.4.26). For instance, let $\imath : \mathbb{Z}^+ \to \mathbb{Z}$ be the inclusion map and $f : \mathbb{Z} \to \mathbb{R}$ be defined as

$$f(n) = \mathrm{seg}_<(\mathbb{Q}, n).$$

Then, $\mathbb{Z}^+ \preceq \mathbb{R}$ because $f \circ \imath$ is an injection. Similarly, $\mathbb{Q} \preceq C$. However, $A \subset B$ does not imply $A \napprox B$. As an example, $n\mathbb{Z} \approx \mathbb{Z}$, but $n\mathbb{Z} \subset \mathbb{Z}$ when $n \neq \pm 1$.

Another method used to prove that $A \preceq B$ is to find a surjection $B \to A$. Consider the sets $A = \{1, 2\}$ and $B = \{3, 4, 5\}$. Define $f : B \to A$ to be the surjection given by $f(3) = 1$, $f(4) = 2$, and $f(5) = 2$. This is the inverse of the relation R in Figure 4.1. To show that B dominates A, we must find an injection $A \to B$. To do this, modify f^{-1} by deleting $(2, 4)$ and call the resulting function g. Observe that $g(1) = 3$ and $g(2) = 5$, which is an injection, so $A \preceq B$.

■ THEOREM 6.2.8

If there exists a surjection $\varphi : A \to B$, then $B \preceq A$.

PROOF
Let $\varphi : A \to B$ be onto. Define a relation $R \subseteq B \times A$ by

$$R = \{(b, a) : \varphi(a) = b\}.$$

Since φ is onto,
$$\mathrm{dom}(R) = \mathrm{ran}(\varphi) = B.$$

Corollary 5.1.11 yields a function f so that $\mathrm{dom}(f) = \mathrm{dom}(R)$ and $f \subseteq R$. We claim that f is one-to-one. Indeed, let $b_1, b_2 \in B$. Assume that we have $f(b_1) = f(b_2)$. Let $a_1 = f(b_1)$ and $a_2 = f(b_2)$ where $a_1, a_2 \in A$. This means $a_1 = a_2$. Also, $\varphi(a_1) = b_1$ and $\varphi(a_2) = b_2$ because $f \subseteq R$. Since φ is a function, $b_1 = b_2$. ■

■ EXAMPLE 6.2.9

Let R be an equivalence relation on a set A. The map $\varphi : A \to A/R$ defined by $\varphi(a) = [a]_R$ is a surjection. Therefore, $A/R \preceq A$.

■ **EXAMPLE 6.2.10**

We know that $\mathbb{Z}^+ \preceq \mathbb{R}$ by Example 6.2.7. We can also prove this by using the function $f : \mathbb{R} \to \mathbb{Z}^+$ defined by $f(x) = |\|x\|| + 1$ and appealing to Theorem 6.2.8.

The next theorem states that $\preceq$ closely resembles an antisymmetric relation (Definition 4.3.1). Cantor was the first to publish a statement of it (1888). He proved it using the axiom of choice, but it was later shown that it can be proved in **ZF**. It was proved independently by Ernst Schröder and Felix Bernstein around 1890. The proof given here follows that of Julius König (1906).

■ **THEOREM 6.2.11 [Cantor–Schröder–Bernstein]**

If $A \preceq B$ and $B \preceq A$, then $A \approx B$.

PROOF

Let $f : A \to B$ and $g : B \to A$ be injections. To prove that A is equinumerous to B, we define a one-to-one correspondence $h : A \to B$. To do this, we recursively define two sequences of sets by first letting

$$C_0 = A \setminus \operatorname{ran}(g)$$

and

$$D_0 = f[C_0].$$

Then, for $n \in \omega$,

$$C_{n+1} = g[D_n]$$

and

$$D_n = f[C_n].$$

This is illustrated in Figure 6.2. Note that both $\{C_n : n \in \omega\}$ and $\{D_n : n \in \omega\}$ are pairwise disjoint because f and g are one-to-one (Exercise 11). Define h by

$$h = f \restriction \left(\bigcup_{n \in \omega} C_n \right) \cup g^{-1} \restriction \left(\operatorname{ran}(g) \setminus \bigcup_{n \in \omega} C_n \right).$$

We show that h is the desired function.

- Let $x_1, x_2 \in A$ such that $x_1 \neq x_2$. Since both f and g are one-to-one, we only need to check the case when $x_1 \in C_k$ for some $k \in \omega$ and $x_2 \notin C_n$ for all $n \in \omega$. Then, $f(x_1) \in D_k$ but $g^{-1}(x_2) \notin D_k$, so $f(x_1) \neq g^{-1}(x_2)$. That is, $h(x_1) \neq h(x_2)$.

- Take $y \in B$. If $y \in D_k$ for some $k \in \omega$, then $y = f(x)$ for some $x \in C_k$. That is, $y = h(x)$. Now suppose $y \notin \bigcup_{n \in \omega} D_n$. Clearly, $g(y) \notin C_0$. If $g(y) \in C_k$ for some $k > 0$, then $y \in D_{k-1}$, a contradiction. Hence, $g(y) = x$ for some $x \in \operatorname{ran}(g) \setminus \bigcup_{n \in \omega} C_n$. This implies that we have $h(x) = g^{-1}(x) = y$. ■

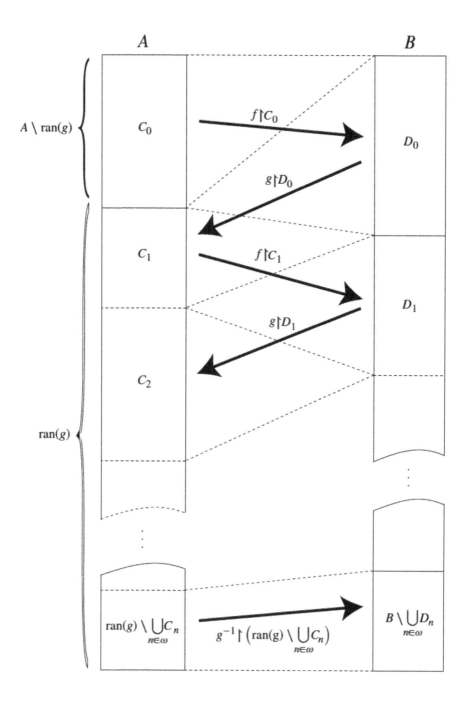

Figure 6.2 The Cantor–Schröder–Bernstein theorem.

■ **EXAMPLE 6.2.12**

Because $(0, 1) \subseteq [0, 1]$, we have that $(0, 1) \preceq [0, 1]$, and because the function
$f : [0, 1] \to (0, 1)$ defined by

$$f(x) = \frac{1}{2}x + \frac{1}{4}$$

is an injection, we have that $[0, 1] \preceq (0, 1)$, so we conclude by the Cantor–
Schröder–Bernstein theorem (6.2.11) that $(0, 1) \approx [0, 1]$.

Diagonalization

The strict inequality $A \prec B$ is sometimes difficult to prove because we must show that
there does not exist a bijection from A onto B. The next method was developed by
Cantor (1891) to accomplish this for infinite sets. It is called **diagonalization**.

Let M be the set of all functions $f : \omega \to \{m, w\}$, where $m \neq w$. To show that
$\omega \prec M$, we prove two facts:

- There exists an injection $\omega \to M$.

- There is no one-to-one correspondence between ω and M.

Cantor's method does both of these at once. Let $\varphi : \omega \to M$ be a function. Writing
the functions of the range of φ as infinite tuples, let

$$f_i = (a_{i0}, a_{i1}, a_{i2}, \ldots, a_{ij}, \ldots),$$

where $a_{ij} \in \{m, w\}$ for all $i, j \in \omega$. For example,

$$\varphi(4)(3) = f_4(3) = a_{4,3}.$$

Now, write the functions in order:

$$f_0 = (\boldsymbol{a_{00}}, a_{01}, a_{02}, \ldots, a_{0j}, \ldots),$$
$$f_1 = (a_{10}, \boldsymbol{a_{11}}, a_{12}, \ldots, a_{1j}, \ldots),$$
$$f_2 = (a_{20}, a_{21}, \boldsymbol{a_{22}}, \ldots, a_{2j}, \ldots),$$
$$\vdots$$
$$f_i = (a_{i0}, a_{i1}, a_{i2}, \ldots, \boldsymbol{a_{ii}}, \ldots, a_{ij}, \ldots),$$
$$\vdots$$

From this, a function $f \in M$ that is not in the list can be found by identifying the
elements on the diagonal and defining $f(n)$ to be the opposite of a_{nn}. In other words,
define for all $i \in \omega$,

$$b_i = \begin{cases} m & \text{if } a_{ii} = w, \\ w & \text{if } a_{ii} = m, \end{cases}$$

and $f(n) = b_n$ is an element of M not in the list because for all $n \in \omega$,

$$f(n) \neq a_{nn}.$$

Since the function φ mapping ω to M is arbitrary, there are injections $\omega \to M$ but none of them are onto. Therefore,

$$\omega \prec M.$$

Furthermore, note that the elements of $[0, 1]$ can be uniquely represented as binary numbers of the form

$$0 . a_0 \, a_1 \, a_2 \, \ldots \, a_i \, \ldots,$$

where $a_i \in \{0, 1\}$ for each $i \in \omega$. For example

$$1 = 0.1111111 \ldots$$

and

$$1/2 = 0.1000000 \ldots.$$

Therefore,

$$M \approx [0, 1].$$

Hence, we can conclude like Cantor that since $[0, 1] \approx \mathbb{R}$ (Examples 6.2.5 and 6.2.12),

$$\omega \prec \mathbb{R}.$$

Cantor's diagonalization argument can be generalized, but we first need a definition. Let A be a set and $B \subseteq A$. The function

$$\chi_B : A \to \{0, 1\}$$

is called a **characteristic function** and is defined by

$$\chi_B(a) = \begin{cases} 1 & \text{if } a \in B, \\ 0 & \text{if } a \notin B. \end{cases}$$

For example, if $A = \mathbb{Z}$ and $B = \{0, 1, 3, 5\}$, then $\chi_B(1) = 1$ but $\chi_B(2) = 0$. Moreover, for every set A,

$$^A 2 = \{\chi_B : B \subseteq A\}.$$

The characteristic function plays an important role in the proof of the next theorem.

■ **THEOREM 6.2.13**

If A is a set, $A \prec {}^A 2$.

PROOF
Since the function $\psi : A \to {}^A2$ defined by

$$\psi(a) = \chi_{\{a\}}$$

is an injection, ${}^A2$ dominates A. To show that A is not equinumerous with ${}^A2$, we show that there is no surjection $A \to {}^A2$. Let $\varphi : A \to {}^A2$ be a function, and for all $a \in A$, write $\varphi(a) = \chi_{B_a}$ for some $B_a \subseteq A$. Define χ so that

$$\chi(a) = \begin{cases} 1 & \text{if } \chi_{B_a}(a) = 0, \\ 0 & \text{if } \chi_{B_a}(a) = 1. \end{cases}$$

Therefore, $\chi \notin \operatorname{ran}(\varphi)$ because $\chi_{B_a}(a) \neq \chi(a)$ for all $a \in A$. However, $\chi \in {}^A2$. To prove this, define

$$B = \{a \in A : \chi_{B_a}(a) = 0\}.$$

We conclude that $\chi(a) = \chi_B(a)$ because if $\chi_{B_a}(a) = 0$, then $a \in B$, so $\chi(a) = 1$ and $\chi_B(a) = 1$, but if $\chi_{B_a}(a) = 1$, then $a \notin B$, so $\chi(a) = 0$ and $\chi_B(a) = 0$. Therefore, $\chi = \chi_B$, and φ is not onto. ∎

By Exercise 14,

$$P(A) \approx {}^A2.$$

This result combined with Theorem 6.2.13 quickly yield the following.

■ COROLLARY 6.2.14

If A is a set, $A \prec P(A)$.

From the theorem, we conclude that there exists a sequence of sets

$$\omega \prec P(\omega) \prec P(P(\omega)) \prec P(P(P(\omega))) \prec \cdots.$$

Thus, there are larger and larger magnitudes of infinity.

Exercises

1. Given $\chi_B : \mathbb{Z} \to \{0, 1\}$ with $B = \{2, 5, 19, 23\}$, find
   (a) $\chi_B(1)$
   (b) $\chi_B(2)$
   (c) $\chi_B(-10)$
   (d) $\chi_B(19)$

2. Find a surjection $\varphi : \omega \times \omega \to \omega$ showing that $\omega \preceq \omega \times \omega$.

3. Prove the given equations.
   (a) $\chi_{A \cup B} = \chi_A + \chi_B - \chi_A \chi_B$
   (b) $\chi_{A \cap B} = \chi_A \chi_B$

4. Prove that there exists a bijection between the given pairs of sets.

(a) $[0, \pi]$ and $[-1, 1]$

(b) $[-\pi/2, \pi/2]$ and $[-1, 1]$

(c) $(0, \infty)$ and $\mathbb{R}$

(d) ω and $\mathbb{Z}$

(e) $\mathbb{Z}^+$ and $\mathbb{Z}^-$

(f) $\{(x, 0) : x \in \mathbb{R}\}$ and $\mathbb{R}$

(g) $\mathbb{Z}$ and $\mathbb{Z} \times \mathbb{Z}$

(h) $\{(x, y) \in \mathbb{R} \times \mathbb{R} : y = 2x + 4\}$ and $\mathbb{R}$

5. Let $a < b$ and $c < d$, where $a, b, c, d \in \mathbb{R}$. Prove.

(a) $(a, b) \approx (c, d)$

(b) $[a, b] \approx [c, d]$

(c) $(a, b) \approx [c, d]$

(d) $(a, b) \approx (c, d]$

6. Prove $\mathbb{R} \approx \mathbb{C}$.

7. Let A, B, C, and D be nonempty sets. Prove.

(a) $\mathbb{N} \preceq \mathbb{Z}^-$

(b) $A \cap B \preceq \mathbf{P}(A)$

(c) $A \preceq A \times B$

(d) $[0, 2] \preceq [5, 7]$

(e) $A \cap B \preceq A$

(f) $^A B \preceq A \times B$

(g) $A \times \{0\} \preceq (A \cup B) \times \{1\}$

(h) $A \times B \times C \preceq A \times B \times C \times D$

(i) $A \setminus B \preceq C \times A$

8. Prove.

(a) If $A \preceq B$ and $B \approx C$, then $A \preceq C$.

(b) If $A \approx B$ and $B \preceq C$, then $A \preceq C$.

(c) If $A \preceq B$ and $B \preceq C$, then $A \preceq C$.

(d) If $A \preceq B$ and $C \preceq D$, then $^A C \preceq {}^B D$.

(e) If $A \approx B$, then $\mathbf{P}(A) \approx \mathbf{P}(B)$.

(f) If $A \approx B$ and $C \approx D$, then $A \times C \approx B \times D$.

(g) If $A \approx B$, $a \in A$, and $b \in B$, then $A \setminus \{a\} \approx B \setminus \{b\}$.

(h) If $A \setminus B \approx B \setminus A$, then $A \approx B$.

(i) If $A \subseteq B$ and $A \approx A \cup C$, then $B \approx A \cup B \cup C$.

(j) If $C \subseteq A$, $B \subseteq D$, and $A \cup B \approx B$, then $C \cup D \approx D$.

9. Given the function $\varphi : A \to B$, prove that $\varphi \approx A$ and $\varphi \preceq \mathrm{ran}(\varphi)$.

10. Without appealing to the Cantor–Schröder–Bernstein theorem (6.2.11), prove that $(0, 1) \approx [0, 1]$.

11. Prove that the sets $\{C_n : n \in \omega\}$ and $\{D_n : n \in \omega\}$ from the proof of Theorem 6.2.11 are pairwise disjoint.

12. Show that g is a bijection, where $g : \mathbb{Z}^+ \to \mathbb{Z}$ is defined by

$$g(n) = \begin{cases} k - 1 & \text{if } n = 2k \text{ for some } k \in \omega, \\ -k & \text{if } n = 2k + 1 \text{ for some } k \in \omega. \end{cases}$$

13. Let A be an infinite set and $\{a_1, a_2, \dots\}$ be a set of distinct elements from A. Prove that $A \to A \setminus \{a_1\}$ is a bijection, where

$$\varphi(x) = \begin{cases} a_{n+1} & \text{if } x = a_n, \\ x & \text{otherwise.} \end{cases}$$

14. For any set A, prove that $P(A) \approx {}^A 2$.

15. Prove that if $f : A \to B$ is a surjection, there exists a function $g : B \to A$ such that $f \circ g = I_B$.

16. Use the power set to prove that there is no set of all sets.

6.3 CARDINAL NUMBERS

Let A be a set. Define

$$B = \{\beta : \beta \text{ is an ordinal} \land \beta \approx A\}.$$

By Zermelo's theorem (6.1.25), A can be well-ordered, so by Theorem 6.1.19, A is order isomorphic to some ordinal, so B is nonempty. Moreover, B is subset of an ordinal (Theorem 6.1.20). Therefore, $(B, \subseteq)$ has a least element, which has the property that it is not equinumerous to any of its elements. This allows us to define the second of our new types of number (page 283). This type will be used to denote the size of a set.

■ DEFINITION 6.3.1

An ordinal κ is a **cardinal number** (or simply a **cardinal**) if $\alpha \not\approx \kappa$ for every $\alpha \in \kappa$.

Observe that every infinite cardinal is a limit ordinal. This is because $\alpha \approx \alpha^+$ for every infinite ordinal α. However, a limit ordinal might not be a cardinal.

Let κ and λ be cardinals. Suppose that $A \approx \kappa$ and $A \approx \lambda$. By Theorem 6.2.4, we have that $\kappa \approx \lambda$. If $\kappa \in \lambda$ or $\lambda \in \kappa$, this would contradict the definition of a cardinal number. Therefore, $\kappa = \lambda$ (Corollary 6.1.15), and we conclude that every set is equinumerous to exactly one cardinal.

■ **DEFINITION 6.3.2**

The **cardinality** of a set A is denoted by $|A|$ and defined as the unique cardinal equinumerous to A.

Observe that the cardinality of a cardinal κ is κ.

■ **EXAMPLE 6.3.3**

Let $\mathscr{A}$ be a set of cardinals. By Theorem 6.1.16, we know that $\bigcup \mathscr{A}$ is an ordinal. Now we show that it is also a cardinal. Suppose that α is an ordinal such that $\alpha \approx \bigcup \mathscr{A}$. By Definition 6.3.1, we must show that $\bigcup \mathscr{A} \subseteq \alpha$. Suppose there exists an ordinal $\beta \in \bigcup \mathscr{A}$ such that $\beta \notin \alpha$. This means that there exists a cardinal $\kappa \in \mathscr{A}$ such that $\beta \in \kappa$. This is impossible because

$$\alpha \leq \beta < \kappa \leq \bigcup \mathscr{A} \approx \alpha.$$

Finite Sets

Intuitively, we know what a finite set is. Both of the sets $A = \{0, 2, 3, 5, 8, 10\}$ and $B = \{n \in \mathbb{Z} : (n-1)(n+3) = 0\}$ are examples because we can count all of their elements and find that there is only one ordinal equinumerous to A and only one ordinal equinumerous to B. That is, $|A| = 6$ and $|B| = 2$. This suggests the following definition.

■ **DEFINITION 6.3.4**

For every set A, if there exists $n \in \omega$ such that $A \approx n$, then A is **finite**. If A is not finite, it is **infinite**.

As we will see, finite sets are fundamentally different from infinite sets. There are properties that finite sets have in addition to the number of their elements that infinite sets do not have. Let us consider some of those properties of finite sets.

■ **LEMMA 6.3.5**

Let n be a positive natural number. If $y \in n$, then $n \setminus \{y\} \approx n^-$.

PROOF
We proceed by mathematical induction.

- When $n = 1$, it must be the case that $y = 0$, so $n \setminus \{y\} = 0 = 1^-$.

- Take $y \in n + 1$. If $y = n$, then $(n+1) \setminus \{n\} = n$, so suppose that $y < n$. By induction, there exists a bijection $g : n \setminus \{y\} \to n^-$. Then, define $f : (n+1) \setminus \{y\} \to n$ by $f(m) = g(m)$ for all $m < n$ and $f(n) = n$. The function f is a bijection (Exercise 3). ■

Lemma 6.3.5 is used to prove a characteristic property of finite cardinals.

■ THEOREM 6.3.6

No natural number is equinumerous to a proper subset of itself.

PROOF

Let $n \in \omega$ be minimal such that there exists $A \subset n$ and $n \approx A$. Since $\varnothing$ has no proper subsets and the only proper subset of 1 is 0, we can assume that $n \geq 2$. Let $f : A \to n$ be a bijection and $x \in A$ and $y \in n \setminus A$ such that $f(x) = y$. We check the following.

- Let $a \in A \setminus \{x\}$. If $f(a) = y$, then we contradict the hypothesis that f is one-to-one because $f(x) = y$ and $x \neq a$. Thus, $f(a) \in n \setminus \{y\}$.

- Take $b \in n \setminus \{y\}$. Since f is a surjection, there exists $a \in A$ such that $f(a) = b$. If $a = x$, then $\{b, y\} \subseteq [x]_f$ (Definition 4.2.7), which is impossible because f is a function. Thus, $a \in A \setminus \{x\}$, and we conclude that $f \restriction (A \setminus \{x\})$ is onto $n \setminus \{y\}$.

- Since the restriction of a one-to-one function is one-to-one, $f \restriction (A \setminus \{x\})$ is one-to-one.

Hence, $f_0 = f \restriction (A \setminus \{x\})$ is a bijection with range $n \setminus \{y\}$. We have two cases to consider.

- Suppose $n^- \notin A$. This implies that $A \setminus \{x\} \subseteq n^-$. Since $x \in A$, we have that $x \neq n^-$, so $x \in n^-$. Thus, $A \setminus \{x\} \subset n^-$. By Lemma 6.3.5, there exists a bijection $g : n \setminus \{y\} \to n^-$, so we have $A \setminus \{x\} \approx n^-$ because $g \circ f_0$ is a bijection, contradicting the minimality of n.

- Assume $n^- \in A$. Define $A' = A \setminus \{n^-\} \cup \{y\}$. Since $y \notin A$, $A' \approx A$, which implies that $A' \approx n$. Replace A with A' in the previous argument and use $f \restriction (A' \setminus \{y\})$ to contradict the minimality of n. ■

■ COROLLARY 6.3.7

Every finite set is equinumerous to exactly one natural number.

PROOF

Let A be finite. This means that $A \approx n$ for some $n \in \omega$. Let $m \in \omega$ also have the property that $A \approx m$. This implies that $n \approx m$. Hence, by Theorem 6.3.6, we conclude that $n = m$ because $n \subseteq m$ or $m \subseteq n$. ■

There are many results that follow directly from Theorem 6.3.6. The following six corollaries are among them.

■ COROLLARY 6.3.8 [Pigeonhole Principle]

Let A and B be finite sets with $B \prec A$. There is no one-to-one function $A \to B$.

PROOF

There exists unique $m, n \in \omega$ such that $A \approx m$ and $B \approx n$ by Corollary 6.3.7. Assume that $f : A \rightarrow B$ is one-to-one. Then,

$$m \approx A \preceq B \approx n,$$

so m is equinumerous to a subset of n. This implies that $m \subseteq n$. However, $n \subset m$ because $B \prec A$, which contradicts Theorem 6.1.14. ∎

■ **COROLLARY 6.3.9**

No finite set is equinumerous to a proper subset of itself.

■ **COROLLARY 6.3.10**

A set equinumerous to a proper subset of itself is infinite.

Because $f : \omega \rightarrow \omega \setminus \{0\}$ defined by $f(n) = n + 1$ is a bijection, we have the next result by Corollary 6.3.10.

■ **COROLLARY 6.3.11**

ω is infinite.

The proofs of the last two corollaries are left to Exercise 5.

■ **COROLLARY 6.3.12**

If A is a proper subset of a natural number n, there exists $m < n$ such that $A \approx m$.

■ **COROLLARY 6.3.13**

Let $A \subseteq B$. If B is finite, A is finite, and if A is infinite, B is infinite.

Countable Sets

Since ω is the first infinite ordinal, it is also a cardinal. Therefore,

$$|A| = \omega \text{ if and only if } A \approx \omega.$$

As sets go, finite sets and those equinumerous with ω are small, so we classify them together using the next definition.

■ **DEFINITION 6.3.14**

A set A is **countable** if $A \preceq \omega$.

Sometimes countable sets are called **discrete** or **denumerable**. For example, the bijection $f : \mathbb{Z}^+ \rightarrow \omega$ defined by $f(n) = n^-$ shows that $\mathbb{Z}^+$ is countable. Moreover, a nonempty finite set is countable and can be written as

$$\{a_0, a_1, \ldots, a_{n-1}\},$$

for some positive integer n. A countably infinite set can be written as

$$\{a_0, a_1, a_2, \ldots \},$$

where there are infinitely many distinct elements of the set.

■ **EXAMPLE 6.3.15**

The set of rational numbers is a countable set. To prove this, define a bijection $f : \omega \to \mathbb{Q}$ by first mapping the even natural numbers to the nonnegative rational numbers. The function is defined along the path indicated in Figure 6.3. When a rational number that previously has been used is encountered, it is skipped. To complete the definition, associate the odd naturals with the negative rational numbers using a path as in the diagram. This function is a bijection, so we conclude that $\mathbb{Q}$ is countable.

We have defined countability in terms of bijections. Now let us identify a condition for countability using surjections.

■ **THEOREM 6.3.16**

A set A is countable if and only if there exists a function from ω onto A.

PROOF

If $\varphi : \omega \to A$ is a surjection, by Theorem 6.2.8, $A \preceq \omega$. Conversely, suppose A is countable. We have two cases to check.

- Suppose $A \approx \omega$. Then, there is a surjection from the set of natural numbers to A.

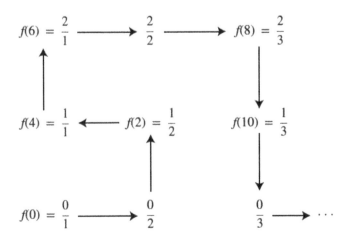

Figure 6.3 The rational numbers are countable.

- Now let $A \approx n$ for some $n \in \omega$. If $A = \varnothing$, then $\varnothing$ is a surjection $\omega \to A$. Thus, assume $A \neq \varnothing$. This means that we can write

$$A = \{a_0, a_1, \ldots, a_{n-1}\}.$$

Define $\varphi : \omega \to A$ by

$$\varphi(i) = \begin{cases} a_i & \text{if } i = 0, 1, \ldots, n-1, \\ a_{n-1} & \text{otherwise.} \end{cases}$$

This function is certainly onto. ∎

If A_i is countable for all $i = 0, 1, \ldots, n-1$, then $A_0 \times A_1 \times \cdots \times A_{n-1}$ is countable (Exercise 15). In particular, $\omega \times \omega$ and $\mathbb{Z} \times \mathbb{Z}$ are countable. We also have the next theorem.

■ THEOREM 6.3.17

The union of a countable family of countable sets is countable.

PROOF

Let $\{A_\alpha : \alpha \in I\}$ be a family of countable sets with I countable. Since we have that $\bigcup \varnothing = \varnothing$ is countable (Example 3.4.12), we can assume that I is nonempty. For each $\alpha \in I$, there exists a surjection in $^\omega(A_\alpha)$ by Theorem 6.3.16. Therefore, by Corollary 5.1.11, there exists

$$f : I \to {}^\omega(A_\alpha)$$

such that $f(\alpha)$ is a surjection $\omega \to A_\alpha$ for all $\alpha \in I$. Because I is countable, we have a surjection $g : \omega \to I$, and since $\omega \times \omega$ is countable, we have another surjection $h : \omega \to \omega \times \omega$. We now define

$$\psi : \omega \times \omega \to \bigcup_{\alpha \in I} A_\alpha$$

by $\psi(m, n) = f(g(m))(n)$ and let

$$\varphi : \omega \to \bigcup_{\alpha \in I} A_\alpha$$

be defined by $\varphi = \psi \circ h$. To check that φ is onto, let $a \in A_\alpha$, some $\alpha \in I$. Since g is onto, there exists $i \in \omega$ so that $g(i) = \alpha$. Furthermore, since $f(\alpha)$ is onto, we have $j \in \omega$ such that $f(\alpha)(j) = a$, and since h is onto, there exists $k \in \omega$ so that $h(k) = (i, j)$. Therefore,

$$\varphi(k) = (\psi \circ h)(k) = \psi(i, j) = f(g(i))(j) = f(\alpha)(j) = a. ∎$$

For example, $\bigcup\{n\mathbb{Z} : n \in \omega\}$ and $\bigcup\{\mathbb{Q} \times \{n\} : n \in \omega\}$ are countable.

Alephs

Cantor denoted the first infinite cardinal ω by $\aleph_0$. The symbol $\aleph$ (**aleph**) is the first letter of the Hebrew alphabet. The next magnitude of infinity is $\aleph_1$, which seems to exist by Theorem 6.2.13. This continues and gives an increasing sequence of infinite cardinals, and since natural numbers must be less than any infinite cardinal, we have

$$0 < 1 < 2 < \cdots < \aleph_0 < \aleph_1 < \aleph_2 < \cdots .$$

For instance, $4 < \aleph_1$, $\aleph_0 \leq \aleph_0$, and $\aleph_3 < \aleph_7$.

■ **EXAMPLE 6.3.18**

Although he was unable to prove it, Cantor suspected that $\aleph_1 = |\mathbb{R}|$. This conjecture is called the **continuum hypothesis** (CH). However, it is possible that $\aleph_1 < |\mathbb{R}|$. It is also possible that $\aleph_1 = |\mathbb{R}|$. Cantor was unable to prove CH because it is **undecidable** assuming the axioms of **ZFC**. In other words, it is an ST-sentence that can be neither proved nor disproved from **ZFC**.

Now to define the other alephs. Pick an ordinal α. Define the function h by

$$h(g) = \text{least infinite cardinal not in } \text{ran}(g),$$

where g is a function with $\text{dom}(g) \in \alpha$. For example, if $\alpha = 5$ and

$$g = \{(0, \kappa_0), (1, \kappa_1), (2, \kappa_2), (3, \kappa_3), (4, \kappa_4)\},$$

then $h(g)$ is the least infinite cardinal not in $\{\kappa_0, \kappa_1, \kappa_2, \kappa_3, \kappa_4\}$. By Transfinite Recursion (6.1.23), there exists a unique function f with domain α and

$$f(\beta) = \text{least infinite cardinal not in } \text{ran}(f \restriction \beta)$$

for all $\beta \in \alpha$. Define

$$\aleph_\beta = f(\beta).$$

Since $h(0) = \omega$, we have that $\aleph_0 = \omega$. Moreover, the definition of f implies that

$$\aleph_1 = \text{least infinite cardinal not in } \{\aleph_0\},$$
$$\aleph_2 = \text{least infinite cardinal not in } \{\aleph_0, \aleph_1\},$$
$$\aleph_3 = \text{least infinite cardinal not in } \{\aleph_0, \aleph_1, \aleph_2\},$$
$$\vdots$$
$$\aleph_\omega = \text{least infinite cardinal not in } \{\aleph_n : n \in \omega\},$$
$$\vdots$$
$$\aleph_\alpha = \text{least infinite cardinal not in } \{\aleph_\beta : \beta \in \alpha\}$$
$$\vdots$$

The question at this point is whether the alephs name all of the infinite cardinals. The next theorem answers the question.

■ THEOREM 6.3.19

For every infinite cardinal κ, there exists an ordinal α such that $\kappa = \aleph_\alpha$.

PROOF

Suppose $\kappa \neq \aleph_\alpha$ for every ordinal α. We can assume that κ is the minimal such cardinal. Thus, for all cardinals $\lambda \in \kappa$, there exists an ordinal β_λ such that $\lambda = \aleph_{\beta_\lambda}$. Therefore, the least infinite cardinal not an element of

$$\{\aleph_{\beta_\lambda} : \lambda \in \kappa \wedge \lambda \text{ is a cardinal}\}$$

is the next aleph, but this is κ. ■

■ EXAMPLE 6.3.20

The **generalized continuum hypothesis** (GCH) states that for every ordinal α,

$$\aleph_{\alpha^+} = |P(\aleph_\alpha)|.$$

When $\alpha = 0$, GCH implies that

$$\aleph_1 = |P(\aleph_0)| = |\mathbb{R}|,$$

which is CH (Example 6.3.18). Like CH, GCH is undecidable in **ZFC**.

Like the ordinals, the cardinals can be divided into two classes.

■ DEFINITION 6.3.21

Let κ be a nonzero cardinal number. If $\kappa \in \omega$ or there exists an ordinal α such that $\kappa = \aleph_{\alpha^+}$, then κ is a **successor cardinal**. Otherwise, κ is a **limit cardinal**.

For example, the positive natural numbers and $\aleph_1$ are successor cardinals, while $\aleph_0$ and $\aleph_\omega$ are limit cardinals. Notice that if κ is a limit cardinal,

$$\kappa = \bigcup\{\lambda : \lambda \in \kappa \wedge \lambda \text{ is a cardinal}\}.$$

Exercises

1. Prove that $|\alpha| = |\alpha^+|$ for every ordinal α.

2. Let $p(x)$ be a formula. Prove that if $p(\alpha)$ is false for some ordinal α, then there exists a least ordinal β such that $p(\beta)$ is false.

3. Prove that the function f in the proof of Lemma 6.3.5 is a bijection.

4. Show that the following attempted generalization of Lemma 6.3.5 is false: Let α be an ordinal and $a \in A$. If $|A| = \aleph_{\alpha^+}$, then $|A \setminus \{a\}| = \aleph_\alpha$.

5. Demonstrate Corollaries 6.3.12 and 6.3.13.

6. Let A and B be finite sets. Prove the following.

(a) $|A \cup B| = |A| + |B| - |A \cap B|$
(b) $|A \cap B| = |A| - |A \setminus B|$
(c) $|A \times B| = |A| \cdot |B|$

7. Prove that the intersection and union of finite sets is finite is finite.

8. Prove that every finite set has a choice function without using the axiom of choice.

9. Let R and R^{-1} be well-orderings of a set A. Prove that A is finite.

10. Show that $\mathbb{Z}$ is countable.

11. Let A be infinite. Find infinite sets B and C such that $A = B \cup C$.

12. If B is countable, prove that $|A \times B| = |A|$.

13. Let A and B be sets and A is countable. Prove that B is countable when $A \approx B$.

14. Let A and B be countable sets. Show that the given sets are countable.
    (a) $A \cup B$
    (b) $A \cap B$
    (c) $A \times B$
    (d) $A \setminus B$

15. Let $A_0, A_1, \ldots, A_{n-1}$ be countable sets. Prove that the given sets are countable.
    (a) $A_0 \cup A_1 \cup \cdots \cup A_{n-1}$
    (b) $A_0 \cap A_1 \cap \cdots \cap A_{n-1}$
    (c) $A_0 \times A_1 \times \cdots \times A_{n-1}$

16. Prove.
    (a) If $A \cup B$ is countable, A and B are countable.
    (b) If A is countable, ${}^2\omega \approx {}^A\omega$

17. Let $\mathscr{F}$ be a set of cardinals. Prove that $\bigcup \mathscr{F}$ is a cardinal.

18. A real number is **algebraic** if it is a root of a nonzero polynomial with integer coefficients. A real number that is not algebraic is **transcendental**. Prove that the set of algebraic numbers is countable and the set of transcendental numbers is uncountable.

19. Take a set A and define $B = \{\alpha : \alpha \text{ is an ordinal} \wedge \alpha \leq A\}$. [See Hartogs' theorem (6.1.22).] Prove the following.
    (a) B is a cardinal.
    (b) $|A| < B$.
    (c) B is the least cardinal such that $|A| < B$.

20. Assuming GCH, find $|P(P(P(P(P(\mathbb{R})))))|$.

21. Prove that for all ordinals α and β, if $\alpha \in \beta$, then $\aleph_\alpha \in \aleph_\beta$.

22. Recursively define the following function using $\beth$ (**beth**), the second letter of the Hebrew alphabet. Let α be an ordinal.

$$\beth_0 = \aleph_0,$$
$$\beth_{\alpha^+} = 2^{\beth_\alpha},$$
$$\beth_\alpha = \bigcup_{\beta \in \alpha} \beth_\beta \text{ if } \alpha \text{ is a limit.}$$

(a) Use $\beth$ to restate GCH.

(b) Use transfinite recursion (Corollary 6.1.24) to prove that $\beth$ defines a function.

6.4   ARITHMETIC

Since every natural number is both an ordinal and a cardinal, we want to extend the operations on ω to all of the ordinals and all of the cardinals. Since the purpose of the ordinals is to characterize well-ordered sets but the purpose of the cardinals is to count, we expect the two extensions to be different.

Ordinals

Definitions 5.2.15 and 5.2.18 define what it means to add and multiply finite ordinals. When generalizing these two definitions to the infinite ordinals, we must take care because addition and multiplication should be binary operations, but if we defined these operations on all ordinals, their domains would not be sets by the Burali-Forti Theroem (6.1.21), resulting in the operations not being sets. Therefore, we choose an ordinal and define addition and multiplication on it. Since $1 + 1 \notin 2$, the ordinal must be a limit ordinal.

■ **DEFINITION 6.4.1**

Let ζ be a limit ordinal. For all $\alpha, \beta \in \zeta$,

- $\alpha + 0 = \alpha$,

- $\alpha + \beta^+ = (\alpha + \beta)^+$,

- $\alpha + \beta = \bigcup \{\alpha + \delta : \delta \in \beta\}$ if β is a limit ordinal.

As with addition of natural numbers (Definition 5.2.15), to prove that Definition 6.4.1 gives a binary operation, let $\psi : \zeta \to \zeta$ be the successor function. By transfinite recursion (Corollary 6.1.24), there exist unique functions $\varphi_\alpha : \alpha \to \zeta$ for all $\alpha \in \zeta$ such that

- $\varphi_\alpha(0) = \alpha$

- $\varphi_\alpha(\beta^+) = \psi(\varphi_\alpha(\beta)) = \varphi_\alpha(\beta)^+$

- for all limit ordinals $\beta \in \zeta$,

$$\varphi_\alpha(\beta) = \bigcup_{\delta \in \beta} \varphi_\alpha(\delta).$$

Define $A : \zeta \times \zeta \to \zeta$ by $A(\alpha, \beta) = \varphi_\alpha(\beta)$. By the uniqueness of each φ_α, the binary operation A is the function of Definition 6.4.1, so

$$\alpha + \beta = A(\alpha, \beta).$$

Furthermore, take ζ' to be a limit ordinal such that $\zeta \subseteq \zeta'$ and define $\psi' : \zeta' \to \zeta'$ to be the successor function. As above, there are unique functions φ'_α for all $\alpha \in \zeta'$ with the same properties as φ_α. Notice that

$$\varphi_\alpha = \varphi'_\alpha \restriction \zeta.$$

Otherwise, $\varphi'_\alpha \restriction \zeta$ would have the same properties as φ_α yet be a different function, contradicting the uniqueness given by transfinite recursion. Next, define the binary operation $A' : \zeta' \times \zeta' \to \zeta'$ by $A'(\alpha, \beta) = \varphi'_\alpha(\beta)$. Therefore,

$$A = A' \restriction (\zeta \times \zeta).$$

This implies that although addition is not defined as a binary operation on all ordinals, we can add any two ordinals and obtain the same sum independent of the ordinal on which the addition is defined.

Consider $m \in \omega$. Since ω is a limit ordinal,

$$m + \omega = \bigcup \{m + n : n \in \omega\} = \omega.$$

However,

$$\omega + 1 = \omega + 0^+ = (\omega + 0)^+ = \omega^+ = \omega \cup \{\omega\},$$

and

$$\omega + 2 = \omega + 1^+ = (\omega + 1)^+ = (\omega \cup \{\omega\})^+ = \omega \cup \{\omega\} \cup \{\omega \cup \{\omega\}\} = \omega^{++}.$$

Therefore, addition of infinite ordinals is not commutative. Moreover, an order isomorphism can be defined between $\omega + n$ and $(\{0\} \times \omega) \cup (\{1\} \times n)$ ordered lexicographically (Exercise 4.3.16) as

0	1	2	...	n	...	ω	ω^+	ω^{++}	...
$\updownarrow$	$\updownarrow$	$\updownarrow$		$\updownarrow$		$\updownarrow$	$\updownarrow$	$\updownarrow$	
$(0,0)$	$(0,1)$	$(0,2)$	...	$(0,n)$	...	$(1,0)$	$(1,1)$	$(1,2)$	...

This means that $\omega + n$ looks like ω followed by a copy of n. Generalizing, the ordinal $\omega + \omega$ looks like ω followed by a copy of ω. In particular,

$$\omega + \omega = \bigcup_{n \in \omega} \omega + n,$$

which means that the proof its existence requires a replacement axiom (5.1.9). All of this suggests the next result.

■ **LEMMA 6.4.2**

Let ζ be a limit ordinal and $\alpha, \beta \in \zeta$. If $x \in \alpha + \beta$, then $x \in \alpha$ or $x \in \alpha + b$ for some $b \in \beta$.

PROOF
Define

$$A = \{z \in \zeta : \forall x (x \in \alpha + z \to x \in \alpha \vee \exists g \in z[x \in \alpha + g])\}.$$

Clearly, $0 \in A$, so take $\gamma \in \zeta$ such that $\text{seg}(\zeta, \gamma) \subseteq A$.

- Let $\gamma = \delta^+$ for some ordinal δ. Take $x \in \alpha + \delta^+$, which implies that $x \in (\alpha + \delta)^+$. This means that $x \in \alpha + \delta$ or $x = \alpha + \delta$. If $x \in \alpha + \delta$, then we are done. If $x = \alpha + \delta$, then $x \in (\alpha + \delta)^+ = \alpha + \delta^+$.

- Let γ be a limit ordinal. Take $x \in \alpha + \gamma$. This means that $x \in \alpha + \delta$ for some $\delta \in \gamma$. Therefore, $x \in \alpha$ or $x \in \alpha + d$ for some $d \in \delta \subset \gamma$. ■

Using Lemma 6.4.2, we can prove the next useful result.

■ **LEMMA 6.4.3**

Let ζ be a limit ordinal. If $\alpha, \beta \in \zeta$, then $\alpha + \beta = \alpha \cup \{\alpha + b : b \in \beta\}$.

PROOF
Define

$$A = \{z \in \zeta : \alpha + z = \alpha \cup \{\alpha + g : g \in z\}\}.$$

We have that $0 \in A$, so assume $\text{seg}(\zeta, \gamma) \subseteq A$, where $\gamma \in \zeta$.

- Suppose $\gamma = \delta^+$. Then,

$$\begin{aligned} \alpha \cup \{\alpha + d : d \in \delta^+\} &= \alpha \cup \{\alpha + d : d \in \delta\} \cup \{\alpha + \delta\} \\ &= (\alpha + \delta) \cup \{\alpha + \delta\} \\ &= (\alpha + \delta)^+ \\ &= \alpha + \delta^+. \end{aligned}$$

- Let γ be a limit ordinal. Take $x \in \alpha + \gamma$. By Lemma 6.4.2, we have $x \in \alpha$ or $x \in \alpha + g$ for some $g \in \gamma$. If the former, we are done, so suppose the latter. In this case, the assumption gives

$$\alpha + g = \alpha \cup \{\alpha + g' : g' \in g\}.$$

Thus, $x = \alpha + g'$ for some $g' \in g \in \gamma$. Conversely, if $x \in \alpha$, then $x \in \alpha + 0$, so $x \in \alpha + \gamma$. For the other case, let $x \in \{\alpha + g : g \in \gamma\}$. This implies that $x \in (\alpha + g)^+ = \alpha + g^+$ for some $g \in \gamma$. Since $g^+ \in \gamma$,

$$x \in \bigcup \{\alpha + g' : g' \in \gamma\} = \alpha + \gamma. \blacksquare$$

Although ordinal addition is not commutative, it does have other familiar properties as noted in the next result.

■ **THEOREM 6.4.4**

Addition of ordinals is associative, and 0 is the additive identity.

PROOF
Let ζ be a limit ordinal. Define

$$A = \{\beta \in \zeta : 0 + \beta = \beta\}.$$

Suppose γ is an ordinal such that $\operatorname{seg}(\lambda, \gamma) \subseteq A$. We have two cases to check.

- Let $\gamma = \delta^+$ for some ordinal δ. Then,

$$0 + \gamma = 0 + \delta^+ = (0 + \delta)^+ = \delta^+ = \gamma.$$

- Let γ be a limit ordinal. We then have

$$0 + \gamma = \bigcup_{\delta \in \gamma}(0 + \delta) = \bigcup_{\delta \in \gamma}\delta = \gamma.$$

In both cases, $\gamma \in A$, so by transfinite induction, $A = \zeta$. Since we can also prove that

$$\zeta = \{\beta \in \zeta : \beta + 0 = \beta\},$$

we conclude that 0 is the additive identity.

To prove that ordinal addition is associative, we proceed by transfinite induction. Let $\alpha, \beta, \delta \in \zeta$. Define

$$B = \{z \in \zeta : \alpha + (\beta + z) = (\alpha + \beta) + z\}.$$

Assume that $\operatorname{seg}(\zeta, \gamma) \subseteq B$. Then, $\gamma \in B$ because by Lemma 6.4.3, we have

$$
\begin{aligned}
\alpha + (\beta + \gamma) &= \alpha \cup \bigcup\{\alpha + x : x \in \beta + \gamma\} \\
&= \alpha \cup \bigcup\{\alpha + x : x \in \beta \vee \exists g(g \in \gamma \to x = \beta + g)\} \\
&= \alpha \cup \bigcup(\{\alpha + x : x \in \beta\} \cup \{\alpha + (\beta + g) : g \in \gamma\}) \\
&= \alpha \cup \bigcup\{\alpha + x : x \in \beta\} \cup \bigcup\{(\alpha + \beta) + g : g \in \gamma\} \\
&= (\alpha \cup \beta) \cup \bigcup\{\alpha + \beta + g : g \in \gamma\} \\
&= (\alpha + \beta) + \gamma.
\end{aligned}
$$

Note that Exercise 3.4.28(e) is used on the fourth equality. ■

■ **DEFINITION 6.4.5**

Let ζ be a limit ordinal. For all $\alpha, \beta \in \zeta$,

- $\alpha \cdot 0 = 0$,

- $\alpha \cdot \beta^+ = \alpha \cdot \beta + \alpha$,

- $\alpha \cdot \beta = \bigcup\{\alpha \cdot \delta : \delta \in \beta\}$ if β is a limit ordinal.

As with ordinal addition, ordinal multiplication is well-defined by transfinite recursion (Exercise 3). Also, as with addition of ordinals, certain expected properties hold, while others do not. The next two results are the analog of Lemma 6.4.3 and Theorem 6.4.4. Their proofs are Exercise 4.

■ LEMMA 6.4.6

If α and β are ordinals, $\alpha \cdot \beta = \{\alpha \cdot b + a : b \in \beta \wedge a \in \alpha\}$.

■ THEOREM 6.4.7

Multiplication of ordinals is associative, and 1 is the multiplicative identity.

Observe that

$$0 \cdot \omega = \bigcup\{0 \cdot n : n \in \omega\} = 0.$$

Also, $\omega \cdot 1 = 1 \cdot \omega = \omega$ (Theorem 6.4.7), so multiplication on the right by a natural number behaves as we would expect in that

$$\omega \cdot 2 = \omega \cdot 1 + \omega = \omega + \omega \tag{6.1}$$

and

$$\omega \cdot 3 = \omega \cdot 2 + \omega = \omega + \omega + \omega,$$

but

$$2 \cdot \omega = \bigcup\{2 \cdot n : n \in \omega\} = \omega$$

and

$$3 \cdot \omega = \bigcup\{3 \cdot n : n \in \omega\} = \omega.$$

Hence, multiplication of ordinals is not commutative. Because of this, it is not surprising that there are issues with the distributive law. For ordinals, there is a left distributive law but not a right distributive law (Exercise 9).

■ THEOREM 6.4.8 [Left Distributive Law]

$\alpha \cdot (\beta + \delta) = \alpha \cdot \beta + \alpha \cdot \delta$ for all ordinals α, β, and δ.

Since addition of ordinals is an operation on a limit ordinal ζ, we know that for all ordinals $\alpha, \beta, \delta \in \zeta$,

$$\alpha = \beta \Rightarrow \alpha + \delta = \beta + \delta$$

and

$$\alpha = \beta \Rightarrow \alpha \cdot \delta = \beta \cdot \delta.$$

The next result gives information regarding how ordinal multiplication behaves with an inequality.

■ **THEOREM 6.4.9**

Let α, β, and δ be ordinals.

- If $\alpha \subset \beta$, then $\delta + \alpha \subset \delta + \beta$.

- If $\alpha \subseteq \beta$, then $\alpha + \delta \subseteq \beta + \delta$.

- If $\alpha \subset \beta$ and $\delta \neq 0$, then $\delta \cdot \alpha \subset \delta \cdot \beta$.

- If $\alpha \subseteq \beta$, then $\alpha \cdot \delta \subseteq \beta \cdot \delta$.

PROOF

We prove the third part, leaving the others to Exercise 7. Let $\alpha \subset \beta$ and $\delta \neq 0$. Then, by Lemma 6.4.6,

$$\delta \cdot \alpha = \{\delta \cdot a + d : a \in \alpha \wedge d \in \delta\} \subset \{\delta \cdot b + d : b \in \beta \wedge d \in \delta\} = \delta \cdot \beta. \blacksquare$$

Finally, we define exponentiation so that it generalizes exponentiation on ω (Exercise 5.2.17). It is a binary operation (Exercise 3).

■ **DEFINITION 6.4.10**

Let ζ be a limit ordinal and α, $\beta \in \zeta$.

- $\alpha^0 = 1$

- $\alpha^{\beta^+} = \alpha^\beta \cdot \alpha$

- $\alpha^\beta = \bigcup\{\alpha^\delta : \delta \in \beta\}$ if β is a limit ordinal.

For example,

$$\omega^1 = \omega^{0^+} = \omega^0 \cdot \omega = 1 \cdot \omega = \omega,$$

and

$$\omega^2 = \omega^{1^+} = \omega^1 \cdot \omega = \omega \cdot \omega,$$

so raising an ordinal to a natural number appears to behave as expected. Also,

$$1^\omega = \bigcup\{1^n : n \in \omega\} = \bigcup\{1\} = 1$$

and

$$2^\omega = \bigcup\{2^n : n \in \omega\} = \omega.$$

We leave the proof of the following properties of ordinal exponentiation to Exercise 11.

■ **THEOREM 6.4.11**

Let α, β, δ be ordinals.

- $\alpha^{\beta+\delta} = \alpha^{\beta} \cdot \alpha^{\delta}$.
- $(\alpha^{\beta})^{\delta} = \alpha^{\beta\,\delta}$.

Cardinals

Even though the finite cardinals are the same sets as the finite ordinals and every infinite cardinal is a limit ordinal, the arithmetic defined on the cardinals will only apply when the given sets are viewed as cardinals. The definitions for addition, multiplication, and exponentiation for cardinals are not given recursively.

■ **DEFINITION 6.4.12**

Let κ and λ be cardinals.

- $\kappa + \lambda = |(\kappa \times \{0\}) \cup (\lambda \times \{1\})|$
- $\kappa \cdot \lambda = |\kappa \times \lambda|$
- $\kappa^{\lambda} = |{}^{\lambda}\kappa|$.

Since ordinal arithmetic was simply a generalization of the arithmetic of natural numbers, that the addition worked in the finite case was not checked. Here this is not the case, so let us check

$$2 + 3 = |(2 \times \{0\}) \cup (3 \times \{1\})| = |\{(0,0), (1,0), (0,1), (1,1), (2,1)\}| = 5.$$

Also,

$$3 + 2 = |(3 \times \{0\}) \cup (2 \times \{1\})| = |\{(0,0), (1,0), (2,0), (0,1), (1,1)\}| = 5.$$

This suggests that cardinal addition is commutative. This and other basic results are given in the next theorem. Some details of the proof are left to Exercise 14.

■ **THEOREM 6.4.13**

Addition of cardinals is associative and commutative, and 0 is the additive identity.

PROOF

Let κ, λ, and μ be cardinals. Addition is associative because

$$(\kappa \times \{0\}) \cup [([\lambda \times \{0\}] \cup [\mu \times \{1\}]) \times \{1\}]$$

is equinumerous to

$$[([\kappa \times \{0\}] \cup [\lambda \times \{1\}]) \times \{0\}] \cup (\mu \times \{1\}),$$

it is commutative because

$$(\kappa \times \{0\}) \cup (\lambda \times \{1\}) \approx (\lambda \times \{1\}) \cup (\kappa \times \{0\}),$$

and 0 is the additive identity because

$$(\kappa \times \{0\}) \cup (0 \times \{1\}) = (\kappa \times \{0\}). \blacksquare$$

Now let us multiply

$$2 \cdot 3 = |\{0, 1\} \times \{0, 1, 2\}| = |\{(0,0), (0, 1), (0, 2), (1,0), (1, 1), (1, 2)\}| = 6$$

and

$$3 \cdot 2 = |\{0, 1, 2\} \times \{0, 1\}| = |\{(0,0), (0, 1), (1,0), (1, 1), (2,0), (2, 1)\}| = 6.$$

As with cardinal addition, it seems that cardinal multiplication is commutative. This and other results are stated in the next theorem. Its proof is left to Exercise 15.

■ **THEOREM 6.4.14**

Multiplication of cardinals is associative and commutative, and 1 is the multiplicative identity.

As with ordinal arithmetic (Theorem 6.4.8), cardinal arithmetic has a left distribution law, but since cardinal multiplication is commutative, cardinal arithmetic also has a right distribution law.

■ **THEOREM 6.4.15 [Distributive Law]**

Let κ, λ, and μ be cardinals.

- $\kappa \cdot (\lambda + \mu) = \kappa \cdot \lambda + \kappa \cdot \mu$.
- $(\kappa + \lambda) \cdot \mu = \kappa \cdot \mu + \lambda \cdot \mu$.

PROOF
The left distribution law holds because

$$\kappa \times ([\lambda \times \{0\}] \cup [\mu \times \{1\}]) \approx ([\kappa \times \lambda] \times \{0\}) \cup ([\kappa \times \mu] \times \{1\}).$$

The remaining details of the proof are left to Exercise 16. ■

The last of the operations of Definition 6.4.12 is exponentiation. Let κ be a cardinal. Observe that since there is exactly one function $0 \to \kappa$ (Exercise 4.4.16),

$$\kappa^0 = |{}^0\kappa| = 1$$

and if $\kappa \neq 0$,

$$0^\kappa = |{}^\kappa 0| = 0$$

because there are no functions $\kappa \to 0$, and by Theorem 6.2.13,

$$\kappa < 2^\kappa.$$

In addition, cardinal exponentiation follows other expected rules.

■ **THEOREM 6.4.16**

Let κ, λ, and μ be cardinals.

- $\kappa^{\lambda+\mu} = \kappa^\lambda \cdot \kappa^\mu$.

- $(\kappa \cdot \lambda)^\mu = \kappa^\mu \cdot \lambda^\mu$.

- $(\kappa^\lambda)^\mu = \kappa^{\lambda \cdot \mu}$.

PROOF

We prove the last part and leave the rest to Exercise 17. Define

$$\varphi : {}^\mu\left({}^\lambda\kappa\right) \to {}^{\lambda\times\mu}\kappa$$

such that for all $\psi \in {}^\mu\left({}^\lambda\kappa\right)$, $\alpha \in \lambda$ and $\beta \in \mu$,

$$\varphi(\psi)(\alpha, \beta) = \psi(\alpha)(\beta).$$

We claim that φ is a bijection.

- Let $\psi_1, \psi_2 \in {}^\mu\left({}^\lambda\kappa\right)$. Assume that $\varphi(\psi_1) = \varphi(\psi_2)$. Take $\alpha \in \lambda$ and $\beta \in \mu$. Then,

$$\psi_1(\alpha)(\beta) = \varphi(\psi_1)(\alpha, \beta) = \varphi(\psi_2)(\alpha, \beta) = \psi_2(\alpha)(\beta).$$

  Therefore, $\psi_1 = \psi_2$, and φ is one-to-one.

- Let $\psi \in {}^{\lambda\times\mu}\kappa$. For $\alpha \in \lambda$ and $\beta \in \mu$, define $\psi'(\alpha)(\beta) = \psi(\alpha, \beta)$. This implies that $\varphi(\psi') = \psi$, so φ is onto. ■

Since every infinite cardinal number can be represented by an $\aleph$, let us determine how to calculate using this notation. We begin with a lemma.

■ **LEMMA 6.4.17**

If $n \in \omega$ and κ an infinite cardinal, $n + \kappa = n \cdot \kappa = \kappa$.

PROOF

Let n be a natural number. Define $\varphi : (n \times \{0\}) \cup (\kappa \times \{1\}) \to \kappa$ by

$$\varphi(i, 0) = i \text{ for all } i < n,$$
$$\varphi(\alpha, 1) = n + \alpha \text{ for all } \alpha \in \kappa.$$

For example, if $n = 5$, then $\varphi(4, 0) = 4$, $\varphi(0, 1) = 5$, $\varphi(6, 1) = 11$, $\varphi(\omega, 1) = \omega$, and $\varphi(\omega + 1, 1) = \omega + 1$. Therefore, $n + \kappa = \kappa$ because φ is a bijection. That $n \cdot \kappa = \kappa$ is left to Exercise 18. ■

Lemma 6.4.17 allows us to compute with alephs.

■ **THEOREM 6.4.18**

Let α and β be ordinals and $n \in \omega$.

- $n + \aleph_\alpha = n \cdot \aleph_\alpha = \aleph_\alpha$.

- $\aleph_\alpha + \aleph_\beta = \aleph_\alpha \cdot \aleph_\beta = \begin{cases} \aleph_\alpha & \text{if } \alpha \geq \beta, \\ \aleph_\beta & \text{otherwise.} \end{cases}$

PROOF

The first part follows by Lemma 6.4.17. To prove the addition equation from the second part, let α and β be ordinals. Without loss of generality assume that $\alpha \subseteq \beta$. By Definition 6.4.12,

$$\aleph_\alpha + \aleph_\beta = |(\aleph_\alpha \times \{0\}) \cup (\aleph_\beta \times \{1\})|.$$

Since $\aleph_\beta = |\aleph_\beta \times \{1\}|$,

$$\aleph_\beta \leq |(\aleph_\alpha \times \{0\}) \cup (\aleph_\beta \times \{1\})|.$$

Furthermore, because $\aleph_\alpha \subseteq \aleph_\beta$,

$$\begin{aligned} |(\aleph_\alpha \times \{0\}) \cup (\aleph_\beta \times \{1\})| &\leq |(\aleph_\beta \times \{0\}) \cup (\aleph_\beta \times \{1\})| \\ &= |\aleph_\beta \times \{0, 1\}| \\ &= \aleph_\beta. \end{aligned}$$

Because of Lemma 6.4.17, the last equality holds since $\aleph_\beta$ is infinite and $\{0, 1\}$ is finite. Hence,

$$\aleph_\alpha + \aleph_\beta \approx \aleph_\beta$$

by the Cantor–Schröder–Bernstein theorem (6.2.11). Since both $\aleph_\alpha + \aleph_\beta$ and $\aleph_\beta$ are cardinals, $\aleph_\alpha + \aleph_\beta = \aleph_\beta$. ■

For example, $\aleph_5 + \aleph_9 = \aleph_5 \cdot \aleph_9 = \aleph_9$. More generally, we quickly have the following corollary by Theorem 6.3.19.

■ **COROLLARY 6.4.19**

For every infinite cardinal κ, both $\kappa + \kappa = \kappa$ and $\kappa \cdot \kappa = \kappa$.

Exercises

1. Let $\mathscr{A} = \{A_0, A_1, \ldots, A_{n-1}\}$ be a pairwise disjoint family of sets. Assuming that the sets are distinct, prove that the cardinality of $\bigcup \mathscr{A}$ is equal to the sum

$$|A_0| + |A_1| + \cdots + |A_{n-1}|.$$

2. Let α and β be ordinals. Let $\varphi : \alpha \to \beta$ be a function such that $\varphi(\delta) \in \varphi(\gamma)$ for all $\delta \in \gamma \in \alpha$ (Compare Lemma 6.1.2). Prove the following.

(a) $\alpha \subseteq \beta$.

(b) $\delta \subseteq \varphi(\delta)$ for all $\delta \in \alpha$.

3. Let ζ be a limit ordinal. Use transfinite recursion to prove that ordinal multiplication (Definition 6.4.5) and ordinal exponentiation (Definition 6.4.10) are binary operations on ζ.

4. Prove Lemma 6.4.6 and Theorem 6.4.7.

5. For every ordinal α, prove that $0 \cdot \alpha = 0$.

6. Prove that for all $n \in \omega$, $n + \omega = n \cdot \omega$ as ordinals. Can this be generalized to $\alpha + \beta = \alpha \cdot \beta$ for ordinals $\alpha \in \beta$ with β being infinite? If so, is the $\alpha \in \beta$ required?

7. Prove the remaining parts of Theorem 6.4.9.

8. Find ordinals α, β, and δ such that the following properties hold.

(a) $\alpha \subset \beta$ but $\beta + \delta \subseteq \alpha + \delta$.

(b) $\alpha \subset \beta$ but $\beta \cdot \delta \subseteq \alpha \cdot \delta$.

9. Let α, β, and δ be ordinals.

(a) Prove that $\alpha \cdot (\beta + \delta) = \alpha \cdot \beta + \alpha \cdot \delta$.

(b) Show that it might be the case that $(\alpha + \beta) \cdot \delta \neq \alpha \cdot \delta + \beta \cdot \delta$.

(c) For which ordinals does the right distribution law hold?

10. Let α, β, and δ be ordinals. Prove the following.

(a) $\alpha + \beta \in \alpha \cdot \delta$ if and only if $\beta \in \delta$.

(b) $\alpha + \beta = \alpha + \delta$ if and only if $\beta = \delta$.

(c) If $\alpha + \delta \in \beta + \delta$, then $\alpha \in \beta$.

(d) $\alpha \cdot \beta \in \alpha \cdot \delta$ if and only if $\beta \in \delta$ and $\alpha \neq 0$.

(e) If $\alpha \cdot \beta = \alpha \cdot \delta$, then $\beta = \delta$ or $\alpha = 0$.

11. Prove Theorem 6.4.11.

12. Let α, β, and δ be ordinals. Prove the following.

(a) $\alpha^\beta \in \alpha^\delta$ if and only if $\beta \in \alpha$ and $1 \in \alpha$.

(b) If $\alpha \in \beta$, then $\alpha^\delta \subseteq \beta^\delta$.

(c) If $\alpha^\delta \in \beta^\delta$, then $\alpha \in \beta$.

(d) If $1 \in \alpha$, then $\beta \subseteq \alpha^\beta$.

(e) If $\alpha \in \beta$, there exists a unique ordinal γ such that $\alpha + \gamma = \beta$.

13. Prove that the ordinal $\omega + \omega$ is not a cardinal.

14. Provide the details for the proof of Theorem 6.4.13.

15. Prove Theorem 6.4.14.

16. Provide the details to the proof of Theorem 6.4.15.

17. Prove the remaining parts of Theorem 6.4.16.

18. Prove that for any natural number n and infinite cardinal κ, $n \cdot \kappa = \kappa$.

19. Prove that if κ is inifinte, $(\kappa^+)^\kappa = 2^\kappa$.

20. Is there a cancellation law for ordinals or for cardinals?

21. Prove that $\kappa + \lambda = \kappa \cdot \lambda = \lambda$, given that κ is a countable cardinal and λ is an infinite cardinal.

22. Generalize Exercise 21 by showing that if κ and λ are cardinals with λ infinite such that $\kappa \leq \lambda$, then $\kappa + \lambda = \kappa \cdot \lambda = \lambda$.

23. Let κ and λ be cardinals with $\aleph_0 \leq \lambda$. Show that if $2 \leq \kappa < \lambda$, then $\kappa^\lambda = 2^\lambda$.

24. Prove for all ordinals α that $|\alpha| < \aleph_\alpha$.

25. For all ordinals α, define **Hartogs' function** by

$$\Gamma(\alpha) = \{\beta : \beta \text{ is an ordinal} \wedge \beta \leq \alpha\}.$$

Prove that $\Gamma(\alpha)$ is an ordinal and $\alpha < \Gamma(\alpha)$ for all ordinals α.

26. Using Exercise 25, define an **initial number** ω_α as follows.

$$\omega_0 = \omega,$$
$$\omega_{\gamma+} = \Gamma(\omega_\gamma),$$
$$\omega_\gamma = \bigcup_{\delta \in \gamma} \omega_\delta \text{ if } \gamma \text{ is a limit ordinal.}$$

Prove that initial numbers are limit ordinals and ω_1 is the first uncountable ordinal.

27. Prove that there is no greatest initial number.

28. For all ordinals α, show that $\aleph_\alpha = |\omega_\alpha|$. Can we write $\aleph_\alpha = \omega_\alpha$?

29. For all countable ordinals α, show that $2^{\aleph_\alpha} = \aleph_{\alpha+}$ implies that $2^{\aleph_{\omega_1}} = \aleph_{\omega_1+1}$.

6.5 LARGE CARDINALS

Since every cardinal is a limit ordinal, every cardinal κ can be written in the form

$$\kappa = \bigcup \{\alpha : \alpha \in \kappa\}.$$

In particular, for the limit cardinal $\aleph_{\omega+\omega}$, we have that

$$\aleph_{\omega+\omega} = \bigcup \{\alpha : \alpha \in \aleph_{\omega+\omega}\}. \tag{6.2}$$

Notice that (6.2) is the union of a set with $\aleph_{\omega+\omega}$ elements. However, we also have

$$\aleph_{\omega+\omega} = \bigcup \{\aleph_\alpha : \alpha \in \omega + \omega\}, \tag{6.3}$$

and

$$\aleph_{\omega+\omega} = \bigcup \{\aleph_{\omega+n} : n \in \omega\}. \tag{6.4}$$

Both (6.3) and (6.4) are unions of sets with $\aleph_0$ elements. The next definition is introduced to handle these differences. Since infinite cardinals are limit ordinals, the definition is given for limit ordinals.

■ **DEFINITION 6.5.1**

The **cofinality** of a limit ordinal α is denoted by $\mathrm{cf}(\alpha)$ and defined as the least cardinal λ such that there exists $\mathscr{F} \subseteq \alpha$ with $|\mathscr{F}| = \lambda$ and $\alpha = \bigcup \mathscr{F}$.

Observe that $\mathrm{cf}(\alpha) \leq |\alpha|$ because $\alpha = \bigcup \alpha$. There can be other sets B such that $\alpha = \bigcup B$, and they might have different cardinalities. Any set of ordinals B with this property is said to be **cofinal** in α. Moreover, we can write $B = \{\beta_\delta : \delta \in \kappa\}$ for some cardinal κ, so

$$\alpha = \bigcup_{\delta \in \kappa} \beta_\delta,$$

and when $\kappa = \mathrm{cf}(\alpha)$,

$$\alpha = \bigcup_{\delta \in \mathrm{cf}(\alpha)} \beta_\delta.$$

■ **EXAMPLE 6.5.2**

Since the finite union of a finite set is finite, the cofinality of any infinite set must be infinite. Therefore, because

$$\omega = \bigcup_{n \in \omega} n,$$

we see that $\mathrm{cf}(\omega) = \aleph_0$. Also, since

$$\aleph_\omega = \bigcup_{n \in \omega} \aleph_n,$$

we conclude that $\mathrm{cf}(\aleph_\omega) = \aleph_0$ and $\{\aleph_n : n \in \omega\}$ is cofinal in $\aleph_\omega$. However, $\mathrm{cf}(\aleph_1) = \aleph_1$ because the countable union of countable sets is countable (Theorem 6.3.17).

Regular and Singular Cardinals

Since infinite cardinals are limit ordinals, we can classify the cardinals based on their cofinalities. We make the following definition.

■ **DEFINITION 6.5.3**

A cardinal κ is **regular** if $\kappa = \mathrm{cf}(\kappa)$, else it is **singular**.

Notice that Example 6.5.2 shows that $\aleph_0$ and $\aleph_1$ are regular but $\aleph_\omega$ is singular. This implies a direction to follow to characterize the cardinal numbers. We begin with the successors.

■ **THEOREM 6.5.4**

Successor cardinals are regular.

PROOF

Let α be an ordinal and let $\mathscr{F} \subseteq \aleph_{\alpha+1}$ such that $\aleph_{\alpha+1} = \bigcup \mathscr{F}$. This implies that $|\beta| \leq \aleph_\alpha$ for all $\beta \in \mathscr{F}$. Thus,

$$\bigcup \mathscr{F} \leq |\mathscr{F}| \cdot \aleph_\alpha.$$

By Theorem 6.4.18, we conclude that $\aleph_{\alpha+1} \leq |\mathscr{F}|$, so $\mathrm{cf}(\aleph_{\alpha+1}) = \aleph_{\alpha+1}$ by the Cantor–Schröder–Bernstein theorem (6.2.11). ∎

Since $\aleph_0$ is regular, Theorem 6.5.4 tells us where to find the singular cardinals.

■ **COROLLARY 6.5.5**

A singular cardinal is an uncountable limit cardinal.

Because we have not proved the converse of Corollary 6.5.5, we investigate the cofinality of certain limit cardinals in an attempt to determine which limit cardinals are singular.

■ **THEOREM 6.5.6**

If α is a limit ordinal, $\mathrm{cf}(\aleph_\alpha) = \mathrm{cf}(\alpha)$.

PROOF

The proof uses the Cantor–Schröder–Bernstein theorem (6.2.11). Let α be a limit ordinal.

- We first show that $\mathrm{cf}(\aleph_\alpha) \leq \mathrm{cf}(\alpha)$. Let A be a cofinal subset of α such that $|A| = \mathrm{cf}(\alpha)$. Notice that if $\beta \in A$, then $\aleph_\beta \in \aleph_\alpha$. On the other hand, take $\delta \in \aleph_\alpha$, which implies that $|\delta| < \aleph_\alpha$. Therefore, there exists $\gamma \in \alpha$ such that $|\delta| = \aleph_\gamma$. Since A is cofinal in α, there exists $\zeta \in A$ such that $\delta \in \zeta$. Hence, $\delta \in \aleph_\zeta$, which implies that $\delta \in \bigcup\{\aleph_\beta : \beta \in A\}$. Therefore,

$$\aleph_\alpha = \bigcup\{\aleph_\beta : \beta \in A\},$$

from which follows,

$$\mathrm{cf}(\aleph_\alpha) \leq |A| = \mathrm{cf}(\alpha).$$

- We now show $\mathrm{cf}(\alpha) \leq \mathrm{cf}(\aleph_\alpha)$. Let $A \subseteq \aleph_\alpha$ so that $\aleph_\alpha = \bigcup A$ and $|A| = \mathrm{cf}(\aleph_\alpha)$. Define

$$\mathscr{F} = \{\delta \in \alpha : \exists \gamma (\gamma \in A \wedge |\gamma| = \aleph_\delta)\}.$$

Then, $\beta = \bigcup \mathscr{F}$ is an ordinal by Theorem 6.1.16. For all $\zeta \in A$, we have that $\zeta \in \aleph_{\beta+1}$ because $|\zeta| \leq \aleph_\beta$. Hence,

$$\bigcup A \subseteq \aleph_\beta,$$

from which follows that $\alpha \in \beta$, which means that $\alpha \subseteq \bigcup \mathscr{F}$. Therefore, since the elements of $\mathscr{F}$ are ordinals of α, we have that $\alpha = \bigcup \mathscr{F}$, so

$$\mathrm{cf}(\alpha) \leq |\mathscr{F}| = |A| = \mathrm{cf}(\aleph_\alpha). \quad ∎$$

Theorem 6.5.6 confirms the result of Example 6.5.2 because

$$cf(\aleph_\omega) = cf(\omega) = \aleph_0. \tag{6.5}$$

Also,

$$cf(\aleph_{\omega+\omega}) = cf(\omega + \omega) = \aleph_0, \tag{6.6}$$

so $\aleph_{\omega+\omega}$ is singular. Observe that by (6.5) and (6.6),

$$cf(cf(\aleph_\omega)) = cf(cf(\omega)) = cf(\aleph_0) = \aleph_0$$

and

$$cf(cf(\aleph_{\omega+\omega})) = cf(cf(\omega + \omega)) = cf(\aleph_0) = \aleph_0.$$

The next result generalizes this and proves that $cf(cf(\alpha)) = cf(\alpha)$ for every limit ordinal α.

■ **THEOREM 6.5.7**

For any limit ordinal α, $cf(\alpha)$ is a regular cardinal.

PROOF

Let α be a limit ordinal and write $\alpha = \bigcup \{A_\gamma : \gamma \in cf(\alpha)\}$. For every ordinal $\gamma \in cf(\alpha)$, define $\alpha_\gamma = \bigcup_{\delta \in \gamma} A_\delta$. Then, $\{\alpha_\gamma : \gamma \in cf(\alpha)\}$ is a chain of ordinals (Theorem 6.1.16) and

$$\alpha = \bigcup \{\alpha_\gamma : \gamma \in cf(\alpha)\}.$$

Now write

$$cf(\alpha) = \bigcup \{\beta_\gamma : \gamma \in cf(cf(\alpha))\}.$$

Define

$$B = \{\alpha_{\beta_\gamma} : \gamma \in cf(cf(\alpha))\}.$$

Let $\zeta \in \alpha$. This implies that $\zeta \in \alpha_{\gamma_0}$ for some $\gamma_0 \in cf(\alpha)$. Then, $\gamma_0 \in \beta_{\gamma_1}$ for some $\gamma_1 \in cf(cf(\alpha))$. Hence,

$$\zeta \in \alpha_{\gamma_0} \subseteq \alpha_{\beta_{\gamma_1}} \in B,$$

so $\zeta \in \bigcup B$. Therefore, $\alpha = \bigcup B$, and this implies that $cf(\alpha) \leq cf(cf(\alpha))$. Since the opposite inequality always holds, by the Cantor–Schröder–Bernstein theorem (6.2.11), $cf(\alpha) = cf(cf(\alpha))$, which means that $cf(\alpha)$ is regular. ■

Although the Continuum Hypothesis cannot be proved, it is possible to discover some information about the value of $2^{\aleph_0}$. Notice how its proof resembles Cantor's diagonalization (page 303). It is due to König (1905).

■ **THEOREM 6.5.8 [König]**

If κ is an infinite cardinal and $cf(\kappa) \leq \lambda$, then $\kappa \prec \kappa^\lambda$.

PROOF

Suppose that κ is am infinite cardinal number and $cf(\kappa) = \lambda$. Write

$$\kappa = \bigcup \{\delta_\alpha : \alpha \in \lambda\}.$$

Let $\mathscr{F} = \{f_\beta : \beta \in \kappa\}$ be a subset of $^\lambda\kappa$. Define $g : \lambda \to \kappa$ such that

$$g(\alpha) = \text{least element of } \kappa \setminus \{f_\beta(\delta_\alpha) : \beta \in \delta_\alpha\}.$$

For any $\alpha \in \lambda$,

$$g(\alpha) \neq f_\beta(\delta_\alpha) \text{ for all } \beta \in \delta_\alpha.$$

Therefore,

$$g \neq f_\beta \text{ for all } \beta \in \delta_\alpha. \tag{6.7}$$

Since (6.7) is true for all $\alpha \in \lambda$ and $\{\delta_\alpha : \alpha \in \lambda\}$ is cofinal in κ, we conclude that $g \neq f_\beta$ for all $\beta \in \kappa$. Therefore, $g \notin \mathscr{F}$, so $\kappa \prec \kappa^\lambda$. Note that the same argument leads to this conclusion if $cf(\kappa) \prec \lambda$ (Exercise 4). ∎

■ COROLLARY 6.5.9

Let κ be an infinite cardinal. Then, $\kappa \prec cf(2^\kappa)$.

PROOF

Suppose that $cf(2^\kappa) \leq \kappa$. By König's theorem (6.5.8), Theorem 6.4.16, and Corollary 6.4.19,

$$2^\kappa \prec (2^\kappa)^{cf(2^\kappa)} \leq (2^\kappa)^\kappa = 2^{\kappa \cdot \kappa} = 2^\kappa. ∎$$

By Corollary 6.5.9,

$$\aleph_0 \prec cf(2^{\aleph_0}),$$

but by Example 6.5.2, we know that $cf(\aleph_\omega) = \aleph_0$, so

$$cf(\aleph_\omega) \prec cf(2^{\aleph_0}).$$

Hence, even though we cannot prove what the cardinality of $2^{\aleph_0}$ is, we do know that it is not $\aleph_\omega$.

Inaccessible Cardinals

As we have noted, $\aleph_0$ is both a regular and a limit cardinal. Are there any others with this property?

■ DEFINITION 6.5.10

A regular limit cardinal that is uncountable is called **weakly inaccessible**.

It is not possible using the axioms of **ZFC** to prove the existence of a weakly inaccessible cardinal. Here is another class of cardinals "beyond" the weakly inaccessible cardinals.

■ **DEFINITION 6.5.11**

The cardinal κ is **strongly inaccessible** if it is an uncountable regular cardinal such that $2^\lambda < \kappa$ for all $\lambda < \kappa$.

Since every strongly inaccessible cardinal is weakly inaccessible (Exercise 1), it is not possible to prove from the axioms of **ZFC** that a strongly inaccessible cardinal exists. However, it is apparent that assuming GCH, κ is a weakly inaccessible cardinal if and only if κ is a strongly inaccessible cardinal (Exercise 10). Cardinal numbers such as these are known as **large cardinals** because assumptions beyond the axioms of **ZFC** are required to "reach" them.

Exercises

1. Prove that every strongly inaccessible cardinal is weakly inaccessible.

2. Let $n \in \omega$. Show that $\mathrm{cf}(\aleph_n) = \aleph_n$.

3. For all limit ordinals α, β, and δ, show that if α is cofinal in β and β is cofinal in δ, then α is cofinal in δ.

4. Rewrite the proof of König's theorem (6.5.8) assuming that $\mathrm{cf}(\kappa) < \lambda$.

5. Let α and β be limit ordinals. Prove that $\mathrm{cf}(\alpha) = \mathrm{cf}(\alpha)$ if and only if $(\alpha, \subseteq)$ and $(\beta, \subseteq)$ have order isomorphic cofinal subsets.

6. Let α be a countable limit ordinal. Show that $\mathrm{cf}(\alpha) = \aleph_0$.

7. Let κ and λ be cardinals such that κ is infinite and $2 \le \lambda$. Show the following.
   (a) $\kappa < \mathrm{cf}(\lambda^\kappa)$.
   (b) $\kappa < \kappa^{\mathrm{cf}(\kappa)}$.

8. Assume GCH and let α and β be ordinals. Prove.
   (a) If $\aleph_\beta < \mathrm{cf}(\aleph_\alpha)$, then $\aleph_\alpha^{\aleph_\beta} = \aleph_\alpha$.
   (b) If $\mathrm{cf}(\aleph_\alpha) \le \aleph_\beta < \aleph_\alpha$, then $\aleph_\alpha^{\aleph_\beta} = \aleph_{\alpha+}$.

9. Let α be a limit ordinal. Show that $\mathrm{cf}(\beth_\alpha) = \mathrm{cf}(\alpha)$. (See Exercise 6.3.22.)

10. Prove that GCH implies that a cardinal is weakly inaccessible if and only if it is strongly inaccessible.

11. Let κ be a cardinal. Prove the following biconditionals.
    (a) κ is weakly inaccessible if and only if κ is regular and $\aleph_\kappa = \kappa$.
    (b) κ is strongly inaccessible if and only if κ is regular and $\beth_\kappa = \kappa$.

CHAPTER 7

MODELS

7.1 FIRST-ORDER SEMANTICS

We now return to logic. In Section 1.5, we proved that propositional logic is both sound and complete (Theorems 1.5.9 and 1.5.15). We now do the same for first-order logic. We have an added complication in that this logic involves formulas with variables. Sometimes the variables are all bound resulting in a sentence (Definition 2.2.14), but other times the formula will have free occurrences. We need additional machinery to handle this. Throughout this chapter, let A be a first-order alphabet and S its set of theory symbols. We start with the fundamental definition (compare Definition 4.1.1).

■ **DEFINITION 7.1.1**

The pair $\mathfrak{A} = (A, \mathfrak{a})$ is an **S-structure** if $A \neq \varnothing$ and $\mathfrak{a}$ is a function with domain S such that

- $\mathfrak{a}(c)$ is an element of A for every constant $c \in S$,

- $\mathfrak{a}(R)$ is an n-ary relation on A for every n-ary relation symbol $R \in S$,

- $\mathfrak{a}(f)$ is an n-ary function on A for every n-ary function symbol $f \in S$.

A First Course in Mathematical Logic and Set Theory, First Edition. Michael L. O'Leary.
© 2016 John Wiley & Sons, Inc. Published 2016 by John Wiley & Sons, Inc.

The set A is the **domain** of $\mathfrak{A}$ and is denoted by $\mathrm{dom}(\mathfrak{A})$. The domain of the function $\mathfrak{a}$ is the **signature** of the structure. If $S = \{s_0, s_1, s_2, \dots\}$, we often write the structure as

$$(A, \mathfrak{a}(s_0), \mathfrak{a}(s_1), \mathfrak{a}(s_2), \dots)$$

or

$$(A, s_0^{\mathfrak{A}}, s_1^{\mathfrak{A}}, s_2^{\mathfrak{A}}, \dots)$$

The font used to identify a structure and its function is the traditional one. It is called **Fraktur** and can be found in the appendix.

The purpose of the function $\mathfrak{a}$ is to associate a symbol with a particular object in the domain of the structure. For this reason, if s is an element of the signature of the structure, the object $\mathfrak{a}(s)$ is called the **meaning** of s and the symbol s is the **name** of $\mathfrak{a}(s)$.

■ EXAMPLE 7.1.2

We are familiar with the constant, function, and relation symbols of NT (Example 2.1.4). We define an NT-structure $\mathfrak{A}$ with domain ω. To do this, specify the function $\mathfrak{a}$:

$$\mathfrak{a}(0) = \varnothing,$$
$$\mathfrak{a}(1) = \{\varnothing\},$$
$$\mathfrak{a}(+) = \{((m, n), m + n) : m, n \in \omega\},$$
$$\mathfrak{a}(\cdot) = \{((m, n), m \cdot n) : m, n \in \omega\}.$$

Notice that $\varnothing$ is the meaning of 0 and 1 is the name of $\{\varnothing\}$. Also, observe that $\mathfrak{a}(+)$ and $\mathfrak{a}(\cdot)$ are the addition and multiplication functions on ω (Definitions 5.2.15 and 5.2.18). These operations are usually represented by $+$ and $\cdot$, but these symbols already appear in NT. Therefore, for the structure $\mathfrak{A}$,

$$0^{\mathfrak{A}} = \mathfrak{a}(0),$$
$$1^{\mathfrak{A}} = \mathfrak{a}(1),$$
$$+^{\mathfrak{A}} = \mathfrak{a}(+),$$
$$\cdot^{\mathfrak{A}} = \mathfrak{a}(\cdot),$$

so $\mathfrak{A} = (\omega, \mathfrak{a})$ is an NT-structure with signature $\{0, 1, +, \cdot\}$ that can be written as

$$(\omega, \varnothing, \{\varnothing\}, \{((m, n), m + n) : m, n \in \omega\}, \{((m, n), m \cdot n) : m, n \in \omega\})$$

or, more compactly,

$$(\omega, 0^{\mathfrak{A}}, 1^{\mathfrak{A}}, +^{\mathfrak{A}}, \cdot^{\mathfrak{A}}).$$

■ **EXAMPLE 7.1.3**

When the symbols of a structure's signature are not needed to represent the meanings of the symbols, the notation of Example 7.1.2 is not needed. This is common for GR-structures. For example, if $\mathfrak{b}(e) = 0$ and $\mathfrak{b}(\circ) = +$, then

$$(\mathbb{Z}, \mathfrak{b}) = (\mathbb{Z}, 0, +)$$

is a GR-structure with signature $\{e, \circ\}$ (Example 2.1.5), and if $\mathfrak{c}(e) = 1$ and $\mathfrak{c}(\circ) = \cdot$, then

$$(\mathbb{R} \setminus \{0\}, \mathfrak{c}) = (\mathbb{R} \setminus \{0\}, 1, \cdot)$$

is also a GR-structure with signature $\{e, \circ\}$.

Satisfaction

The purpose of a structure is to serve as a universe for a given language. Recall that the terms of a language represent objects and the formulas of a language describe the properties of objects (Figure 2.3). With this in mind, we now make sense of the terms and formulas within structures. The first step in doing this is to define how to give meaning to terms. This is done so that the terms represent elements of the domain of a structure. The second step is to develop a method by which it can be determined which formulas hold true in a structure and which do not. We begin by defining the interpretation of terms.

■ **DEFINITION 7.1.4**

Let $\mathfrak{A} = (A, \mathfrak{a})$ be an S-structure. Define an **S-interpretation** of $\mathfrak{A}$ to be a function $I : \text{TERMS}(A) \to A$ that has the following properties:

- If x is a variable symbol, then $I(x) \in A$, thus assigning a value to x.

- If c is a constant symbol, $I(c) = \mathfrak{a}(c)$.

- If f is a function symbol and $t_0, t_1, \ldots, t_{n-1}$ are S-terms,

$$I(f(t_0, t_1, \ldots, t_{n-1})) = \mathfrak{a}(f)(I(t_0), I(t_1), \ldots, I(t_{n-1})).$$

Definition 7.1.4 is an example of a definition by **induction on terms**. This means that first the definition was made for variable and constant symbols, and then assuming that it was made for terms in general, the definition is made for functions applied to terms. Induction on terms simply follows Definition 2.1.7. A proof done by induction on terms is one that uses the same process to prove a result about terms.

■ **EXAMPLE 7.1.5**

Let $(\omega, 0^{\mathfrak{A}}, 1^{\mathfrak{A}}, +^{\mathfrak{A}}, \cdot^{\mathfrak{A}})$ be the NT-structure from Example 7.1.2. Let I be an NT-interpretation such that

$$I(x) = 5,$$
$$I(y) = 7,$$
$$I(0) = 0^{\mathfrak{A}},$$
$$I(1) = 1^{\mathfrak{A}}.$$

Then, $I(x + y) = 12$ is the interpretation of $x + y$ because

$$I(x + y) = \mathfrak{a}(+)(I(x), I(y)) = \mathfrak{a}(+)(5, 7) = 5 +^{\mathfrak{A}} 7,$$

and $I(0 \cdot 1) = 0^{\mathfrak{A}}$ because

$$I(0 \cdot 1) = \mathfrak{a}(\cdot)(I(0), I(1)) = \mathfrak{a}(\cdot)(0, 1) = 0^{\mathfrak{A}} \cdot^{\mathfrak{A}} 1^{\mathfrak{A}}.$$

We are now ready for the main definition. It describes what it means for a formula to be interpreted as true. The definition, which is foundational to model theory, is generally attributed to Alfred Tarski. This contribution is found in two papers, "Der wahrheitsbegriff in den formalisierten sprachen" (1935) and "Arithmetical extensions of relational systems" with Robert Vaught (1957).

■ DEFINITION 7.1.6

Let $\mathfrak{A} = (A, \mathfrak{a})$ be an S-structure and I be an S-interpretation of $\mathfrak{A}$. Assume that p and q are S-formulas and $t_0, t_1, \dots, t_{n-1}$ are S-terms. Define $\vDash$ as follows:

- $\mathfrak{A} \vDash t_0 = t_1 \; [I] \Leftrightarrow I(t_0) = I(t_1)$.

- $\mathfrak{A} \vDash R(t_0, t_1, \dots, t_{n-1}) \; [I] \Leftrightarrow (I(t_0), I(t_1), \dots, I(t_{n-1})) \in \mathfrak{a}(R)$.

- $\mathfrak{A} \vDash \neg p \; [I] \Leftrightarrow (\text{not } \mathfrak{A} \vDash p \; [I])$.

- $\mathfrak{A} \vDash p \to q \; [I] \Leftrightarrow (\text{if } \mathfrak{A} \vDash p \; [I] \text{ then } \mathfrak{A} \vDash q \; [I])$.

- $\mathfrak{A} \vDash \exists x p \; [I] \Leftrightarrow (\mathfrak{A} \vDash p \; [I_x^a] \text{ for some } a \in A)$, where for every $u \in A$, the function I_x^u is the S-interpretation of $\mathfrak{A}$ such that if y is a variable symbol,

$$I_x^u(y) = \begin{cases} u & \text{if } y = x, \\ I(y) & \text{if } y \neq x, \end{cases}$$

and if c is a constant symbol, $I_x^a(c) = I(c)$.

Definition 7.1.6 is an example of a definition by **induction on formulas**. This means that first the definition was made for the basic formulas involving equality and relation symbols, and then assuming that it was made for formulas in general, the definition is made for formulas written using connectives or quantifiers. Induction on formulas simply follows Definition 2.1.9. A proof done by induction on formulas is one that uses the same process to prove a result about terms.

Definition 7.1.6 can be extended to the other connectives and the existential quantifier using the next theorem.

■ **THEOREM 7.1.7**

Let $\mathfrak{A} = (A, \mathfrak{a})$ be an S-structure and I be an S-interpretation of $\mathfrak{A}$. Assume that p and q are S-formulas.

- $\mathfrak{A} \models p \wedge q\, [I] \Leftrightarrow (\mathfrak{A} \models p\, [I]$ and $\mathfrak{A} \models q\, [I])$.

- $\mathfrak{A} \models p \vee q\, [I] \Leftrightarrow (\mathfrak{A} \models p\, [I]$ or $\mathfrak{A} \models q\, [I])$.

- $\mathfrak{A} \models p\, [I] \leftrightarrow q \Leftrightarrow (\mathfrak{A} \models p\, [I]$ if and only if $\mathfrak{A} \models q\, [I])$.

- $\mathfrak{A} \models \forall x p\, [I] \Leftrightarrow (\mathfrak{A} \models p\, [I_x^a]$ for all $a \in A)$.

PROOF
Since $p \vee q \Leftrightarrow \neg p \to q$ (page 59), we have the following:

$$\mathfrak{A} \models p \vee q\, [I] \Leftrightarrow \mathfrak{A} \models \neg p \to q\, [I]$$
$$\Leftrightarrow \text{if } \mathfrak{A} \models \neg p\, [I], \text{then } \mathfrak{A} \models q\, [I]$$
$$\Leftrightarrow \text{if not } \mathfrak{A} \models p\, [I], \text{then } \mathfrak{A} \models q\, [I]$$
$$\Leftrightarrow \mathfrak{A} \models p\, [I] \text{ or } \mathfrak{A} \models q\, [I].$$

Since $\forall x p \Leftrightarrow \neg \exists x \neg p$ by QN, we have the following:

$$\mathfrak{A} \models \forall x p\, [I] \Leftrightarrow \mathfrak{A} \models \neg \exists x \neg p\, [I]$$
$$\Leftrightarrow \text{not } \mathfrak{A} \models \exists x \neg p\, [I]$$
$$\Leftrightarrow \text{not } (\mathfrak{A} \models \neg p\, [I_x^a] \text{ for some } a \in A)$$
$$\Leftrightarrow \text{not } (\text{not } \mathfrak{A} \models p\, [I_x^a] \text{ for some } a \in A)$$
$$\Leftrightarrow \mathfrak{A} \models p\, [I_x^a] \text{ for all } a \in A.$$

The remaining parts are left to Exercise 1. ■

■ **EXAMPLE 7.1.8**

Let the ST-structure $\mathfrak{A}$ be defined as $(\omega, \in)$. Since ST has only relation symbols, $\mathfrak{A}$ is called a **purely relational** structure. Although we usually make a notational distinction between a name and its meaning, we do not do this with the $\in$ symbol. Likewise, the equality symbol $=$ can be used in a formula and during any interpretation of the formula. To see how this works, define

$$p := \forall x \forall y \forall z (x \in y \wedge y \in z \to x \in z).$$

We show that $\mathfrak{A} \models p$ for any ST-interpretation I. To determine what needs to be done to accomplish this, we work backwards using Definition 7.1.6 and Theorem 7.1.7. Let I be an ST-interpretation and assume that

$$\mathfrak{A} \models \forall x \forall y \forall z (x \in y \wedge y \in z \to x \in z)\, [I].$$

This implies that

$$\mathfrak{A} \vDash x \in y \wedge y \in z \to x \in z \; [((I_x^a)_y^b)_z^c] \text{ for all } a, b, c \in \omega.$$

How this formula is interpreted is still based on the order of connectives found in Definition 1.1.5. Therefore, since the conjunction has precedence, we apply Definition 7.1.6 to find that

$$\text{for all } a, b, c \in \omega, \text{ if } \mathfrak{A} \vDash x \in y \wedge y \in z \; [((I_x^a)_y^b)_z^c],$$
$$\text{then } \mathfrak{A} \vDash x \in z \; [((I_x^a)_y^b)_z^c].$$

Hence, by Theorem 7.1.7,

$$\text{for all } a, b, c \in \omega, \text{ if } \mathfrak{A} \vDash x \in y \; [((I_x^a)_y^b)_z^c] \text{ and } \mathfrak{A} \vDash y \in z \; [((I_x^a)_y^b)_z^c],$$
$$\text{then } \mathfrak{A} \vDash x \in z \; [((I_x^a)_y^b)_z^c].$$

We now apply the interpretation using Definition 7.1.4 to find that

$$\text{for all } a, b, c \in \omega, \text{ if } ((I_x^a)_y^b)_z^c(x) \in ((I_x^a)_y^b)_z^c(y) \text{ and } ((I_x^a)_y^b)_z^c(y) \in ((I_x^a)_y^b)_z^c(z),$$
$$\text{then } ((I_x^a)_y^b)_z^c(x) \in ((I_x^a)_y^b)_z^c(z).$$

Therefore,

$$\text{for all } a, b, c \in \omega, \text{ if } a \in b \text{ and } b \in c, \text{ then } a \in c,$$

which follows by Theorem 5.2.8. This means that we can back through the steps to prove that $\mathfrak{A} \vDash p \; [I]$.

Typically, formulas cannot be understood as true or false on their own. They have to be examined against a given universe. The structure is that universe, and the examining is done by the interpretation. For this reason, the structure-interpretation pair forms the basis for our work in first-order logic, so we give it a name.

■ DEFINITION 7.1.9

Let I be an S-interpretation of the S-structure $\mathfrak{A}$. Let p be an S-formula. The pair $(\mathfrak{A}, I)$ is called an **S-model**. Additionally, if $\mathfrak{A} \vDash p \; [I]$, then $(\mathfrak{A}, I)$ is a **model** of p and **satisfies** p. If $(\mathfrak{A}, I)$ is not a model of p, write $\mathfrak{A} \nvDash p \; [I]$.

■ EXAMPLE 7.1.10

Let $\mathfrak{A}$ be the NT-structure with domain ω and signature $\{0, 1, +, \cdot\}$ from Example 7.1.2. Let I be an NT-interpretation of $\mathfrak{A}$ such that

$$I(x) = 0^{\mathfrak{A}}, \; I(x_i) = i +^{\mathfrak{A}} 1^{\mathfrak{A}}, \text{ and } I(1) = 1^{\mathfrak{A}}.$$

- $(\mathfrak{A}, I)$ is a model of $x_4 + x_7 = x_7 + x_4$ because by Theorem 5.2.17,

$$\begin{aligned}
I(x_4 + x_7) &= I(x_4) +^{\mathfrak{A}} I(x_7) \\
&= 5 +^{\mathfrak{A}} 8 \\
&= 8 +^{\mathfrak{A}} 5 \\
&= I(x_7) +^{\mathfrak{A}} I(x_4) \\
&= I(x_7 + x_4).
\end{aligned}$$

- $(\mathfrak{A}, I)$ is a model of $\forall x(x = x_2 \to x + 1 = x_2 + 1)$. To see this, take $n \in \omega$ and assume that $\mathfrak{A} \vDash x = x_2 [I_x^n]$. This implies that $n = I(x_2)$. Hence,

$$n +^{\mathfrak{A}} 0^{\mathfrak{A}} = I(x_2) +^{\mathfrak{A}} 0^{\mathfrak{A}},$$
$$n +^{\mathfrak{A}} I(1) = I(x_2) +^{\mathfrak{A}} I(1),$$
$$I_x^n(x) +^{\mathfrak{A}} I_x^n(1) = I_x^n(x_2) +^{\mathfrak{A}} I_x^n(1),$$
$$I_x^n(x + 1) = I_x^n(x_2 + 1).$$

Therefore, $\mathfrak{A} \vDash x + 1 = x_2 + 1 [I_x^n]$. We conclude that for all $n \in \omega$,

$$\text{if } \mathfrak{A} \vDash x = x_2 [I_x^n], \text{ then } \mathfrak{A} \vDash x + 1 = x_2 + 1 [I_x^n],$$

so for all $n \in \omega$,

$$\mathfrak{A} \vDash x = x_2 \to x + 1 = x_2 + 1 [I_x^n].$$

This implies that

$$\mathfrak{A} \vDash \forall x(x = x_2 \to x + 1 = x_2 + 1) [I].$$

Definition 7.1.9 can be generalized to sets of formulas.

■ **DEFINITION 7.1.11**

If $(\mathfrak{A}, I)$ is an S-model and $\mathcal{F}$ is a set of S-formulas,

$$\mathfrak{A} \vDash \mathcal{F} [I] \text{ if and only if } \mathfrak{A} \vDash p [I] \text{ for all } p \in \mathcal{F}.$$

If $\mathfrak{A} \vDash \mathcal{F} [I]$, then $(\mathfrak{A}, I)$ is a **model** of $\mathcal{F}$ and **satisfies** $\mathcal{F}$.

For example, using the NT-model of Example 7.1.10, we see that

$$\mathfrak{A} \vDash \{x_4 + x_7 = x_7 + x_4, \forall x(x = x_2 \to x + 1 = x_2 + 1)\} [I], \qquad (7.1)$$

so $(\mathfrak{A}, I)$ is a model of $\{x_4 + x_7 = x_7 + x_4, \forall x(x = x_2 \to x + 1 = x_2 + 1)\}$. Here is a more involved example.

■ **EXAMPLE 7.1.12**

Let $\mathfrak{B}$ be the GR-structure $(\mathbb{Z}, \mathfrak{b})$ so that $\mathfrak{b}$ is defined by $\mathfrak{b}(e) = 0$ and $\mathfrak{b}(\mathrm{o}) = +$. Let the interpretation J have the property that

$$J(x_i) = \begin{cases} i/2 & \text{if } i \text{ is even,} \\ (i+1)/2 & \text{if } i \text{ is odd.} \end{cases}$$

- $\mathfrak{B} \vDash \exists x(x_7 \circ x = e) [J]$. This is because

$$\mathfrak{B} \vDash x_7 \circ x = e [J_x^{-4}],$$

and this holds since $4 + (-4) = 0$.

- $\mathfrak{B} \models \forall x \exists y (x \circ y = e) \, [J]$. To prove this, take $n \in \mathbb{Z}$. Then,

$$\mathfrak{B} \models \exists y (x \circ y = e) \, [J_x^n]$$

because

$$\mathfrak{B} \models x \circ y = e \, [(J_x^n)_y^{-n}].$$

This last satisfaction holds because $-n \in \mathbb{Z}$ and $n + -n = 0$.

- The previous two satisfactions demonstrate that

$$\mathfrak{B} \models \{\exists x (x_7 \circ x = e), \forall x \exists y (x \circ y = e)\} \, [J]. \tag{7.2}$$

The previous work with models leads to the following definition.

■ **DEFINITION 7.1.13**

The S-formula p is **S-satisfiable** (denoted by $\mathrm{Sat}_S p$) if there is an S-structure $\mathfrak{A}$ and S-interpretation I such that $(\mathfrak{A}, I)$ is a model for p. The set of S-formulas $\mathscr{F}$ is **S-satisfiable** (denoted by $\mathrm{Sat}_S \mathscr{F}$) if there is a model for $\mathscr{F}$.

By (7.1), we have that

$$\mathrm{Sat}_{\mathbf{NT}} \{x_4 + x_7 = x_7 + x_4, \forall x[(x + x_5) + x_8 = x + (x_5 + x_8)]\},$$

and by (7.2), we have that $\mathrm{Sat}_{\mathbf{GR}} \{\exists x (x_7 \circ x = e), \forall x \exists y (x \circ y = e)\}$.

Groups

As noted in Example 2.1.5, the GR symbols are intended for the study of a set with a single binary operation defined on it. The basic example is $\mathbb{Z}$ with addition. Although other properties will be added later, the language developed for this is designed to handle just the basic properties of this pair. Namely, the operation should be associative, the set should have an identity, and every element should have an inverse. This means that the axioms that will define this theory will be GR-sentences, and we need only three.

■ **AXIOMS 7.1.14 [Group]**

- **G1.** $\forall x \forall y \forall z \, [x \circ (y \circ z) = (x \circ y) \circ z]$.

- **G2.** $\forall x (e \circ x = x \land x \circ e = x)$.

- **G3.** $\forall x \exists y (x \circ y = e \land y \circ x = e)$.

■ **EXAMPLE 7.1.15**

Define a GR-structure $\mathfrak{G} = (G, \mathfrak{g})$ by letting the domain $G = \{0\}$, $\mathfrak{g}(e) = 0$, and $\mathfrak{g}(\circ) = +$. The binary operation $+$ is addition of integers and $0 \in \mathbb{Z}$ (Section 5.3). Let I be an interpretation of $\mathfrak{G}$. We show that $\mathfrak{G} \models \{G1, G2, G3\}\ [I]$.

- Let $a, b, c \in G$. Since 0 is the only element of G, observe that by Definition 7.1.4,

$$((I_x^a)_y^b)_z^c(x \circ [y \circ z]) = ((I_x^a)_y^b)_z^c(x) + ((I_x^a)_y^b)_z^c(y \circ z) \tag{7.3}$$

$$= ((I_x^a)_y^b)_z^c(x) + [((I_x^a)_y^b)_z^c(y) + ((I_x^a)_y^b)_z^c(z)] \tag{7.4}$$

$$= 0 + (0 + 0) \tag{7.5}$$

$$= 0 + 0 + 0 \tag{7.6}$$

$$= [((I_x^a)_y^b)_z^c(x) + ((I_x^a)_y^b)_z^c(y)] + ((I_x^a)_y^b)_z^c(z) \tag{7.7}$$

$$= ((I_x^a)_y^b)_z^c(x \circ y) + ((I_x^a)_y^b)_z^c(z) \tag{7.8}$$

$$= ((I_x^a)_y^b)_z^c([x \circ y] \circ z). \tag{7.9}$$

We see that (7.6) follows from (7.5) by Theorem 5.3.5. Therefore, by Theorem 7.1.7,

$$\mathfrak{G} \models x \circ (y \circ z) = (x \circ y) \circ z\ [((I_x^a)_y^b)_z^c]\ \text{for all } a, b, c \in G,$$

which implies that

$$\mathfrak{G} \models \forall z[x \circ (y \circ z) = (x \circ y) \circ z]\ [(I_x^a)_y^b]\ \text{for all } a, b \in G.$$

Therefore,

$$\mathfrak{G} \models \forall y \forall z[x \circ (y \circ z) = (x \circ y) \circ z]\ [I_x^a]\ \text{for all } a \in G,$$

so

$$\mathfrak{G} \models \forall x \forall y \forall z[x \circ (y \circ z) = (x \circ y) \circ z]\ [I].$$

- Let $a \in G$. Again, by Definition 7.1.4,

$$I_x^a(e \circ x) = I_x^a(e) + I_x^a(x) = 0 + 0 = 0 = I_x^a(x),$$

so by Definition 7.1.6,

$$\mathfrak{G} \models e \circ x = x\ [I_x^a].$$

Also,

$$I_x^a(x \circ e) = I_x^a(x) + I_x^a(e) = 0 + 0 = 0 = I_x^a(x),$$

so

$$\mathfrak{G} \models x \circ e = x\ [I_x^a].$$

Therefore, by Theorem 7.1.7,

$$\mathfrak{G} \models e \circ x = x \wedge x \circ e = x \; [I_x^a].$$

Since a was arbitrarily chosen,

$$\mathfrak{G} \models \forall x (e \circ x = x \wedge x \circ e = x) \; [I].$$

- Take $a \in G$. Because

$$(I_x^a)_y^0(x \circ y) = (I_x^a)_y^0(x) + (I_x^a)_y^0(y) = 0 + 0 = 0 = (I_x^a)_y^0(e),$$

we have

$$\mathfrak{G} \models x \circ y = e \; [(I_x^a)_y^0].$$

Similarly,

$$\mathfrak{G} \models y \circ x = e \; [(I_x^a)_y^0],$$

so

$$\mathfrak{G} \models x \circ y = e \wedge y \circ x = e \; [(I_x^a)_y^0].$$

Therefore, since $0 \in G$,

there exists $b \in G$ such that $\mathfrak{G} \models x \circ y = e \wedge y \circ x = e \; [(I_x^a)_y^b],$

and since a was arbitrarily chosen, we have that

for all $a \in G$, there exists $b \in G$ such that
$$\mathfrak{G} \models x \circ y = e \wedge y \circ x = e \; [(I_x^a)_y^b].$$

Then, by Definition 7.1.6, we conclude that

for all $a \in G$, $\mathfrak{G} \models \exists y (x \circ y = e \wedge y \circ x = e) \; [I_x^a],$

so by Theorem 7.1.7,

$$\mathfrak{G} \models \forall x \exists y (x \circ y = e \wedge y \circ x = e) \; [I].$$

Based on Example 7.1.15, we conclude that $\{G1, G2, G3\}$ is GR-satisfiable (Definition 7.1.13). That is, there exists a model in which **G1**, **G2**, and **G3** are interpreted as true. The name of the model was first used by Évariste Galois in the early 1830s.

■ **DEFINITION 7.1.16**

A GR-structure that models the group axioms is called a **group**.

A group with a commutative binary operation, one that satisfies

$$\forall x \forall y (x \circ y = y \circ x),$$

is known as an **abelian group.** It is named after the Norwegian mathematician Niels Abel. Using the GR-structure $\mathfrak{G}$ and interpretation I of Example 7.1.15, the set of GR-sentences $\{G1, G2, G3, \forall x \forall y (x \circ y = y \circ x)\}$ is shown to be GR-satisfiable.

■ **EXAMPLE 7.1.17**

Using definitions of Examples 4.2.6 and 4.2.9 for $\mathbb{Z}$, define

$$\mathbb{Z}_n = \{[a]_n : a \in \mathbb{Z}\},$$

and on this set, specify $+$ by

$$[a]_n + [b]_n = [a+b]_n.$$

As we have seen, the meaning of the symbol $+$ is determined by context. The $+$ on the left is the new definition, but the $+$ on the right is standard addition (Definition 5.3.4). With this definition, a generalization of Example 4.4.30 shows that $+$ is a binary operation. We check that $\mathfrak{G} = (\mathbb{Z}_n, [0]_n, +)$ is a group. Let I be a GR-interpretation of $\mathfrak{G}$ such that $I(e) = [0]_n$. We have three axioms to check.

- Let $[a]_n, [b]_n, [c]_n \in \mathbb{Z}_n$, where $a, b, c \in \mathbb{Z}$. Observe that by Definition 7.1.4,

$$((I_x^{[a]_n})_y^{[b]_n})_z^{[c]_n}(x \circ [y \circ z])$$

$$= ((I_x^{[a]_n})_y^{[b]_n})_z^{[c]_n}(x) + ((I_x^{[a]_n})_y^{[b]_n})_z^{[c]_n}(y \circ z) \qquad (7.10)$$

$$= ((I_x^{[a]_n})_y^{[b]_n})_z^{[c]_n}(x) + [((I_x^{[a]_n})_y^{[b]_n})_z^{[c]_n}(y) + ((I_x^{[a]_n})_y^{[b]_n})_z^{[c]_n}(z)] \qquad (7.11)$$

$$= [a]_n + ([b]_n + [c]_n)$$

$$= [a]_n + ([b+c]_n)$$

$$= [a + (b+c)]_n \qquad (7.12)$$

$$= [a + b + c]_n \qquad (7.13)$$

$$= [a+b]_n + [c]_n$$

$$= ([a]_n + [b]_n) + [c]_n$$

$$= [((I_x^{[a]_n})_y^{[b]_n})_z^{[c]_n}(x) + ((I_x^{[a]_n})_y^{[b]_n})_z^{[c]_n}(y)] + ((I_x^{[a]_n})_y^{[b]_n})_z^{[c]_n}(z) \qquad (7.14)$$

$$= ((I_x^{[a]_n})_y^{[b]_n})_z^{[c]_n}(x \circ y) + ((I_x^{[a]_n})_y^{[b]_n})_z^{[c]_n}(z) \qquad (7.15)$$

$$= ((I_x^{[a]_n})_y^{[b]_n})_z^{[c]_n}([x \circ y] \circ z). \qquad (7.16)$$

Therefore,

$$((I_x^{[a]_n})_y^{[b]_n})_z^{[c]_n}(x \circ [y \circ z]) = ((I_x^{[a]_n})_y^{[b]_n})_z^{[c]_n}([x \circ y] \circ z)$$
$$\text{for all } [a]_n, [b]_n, [c]_n \in \mathbb{Z}_n,$$

which implies that

$$\mathfrak{G} \models \forall x \forall y \forall z [x \circ (y \circ z) = (x \circ y) \circ z] \; [I].$$

Notice that (7.13) follows from (7.12) by Theorem 5.3.5. More importantly, notice that (7.10), (7.11), (7.14), (7.15), and (7.16) mimic (7.3), (7.4), (7.7), (7.8), and (7.9) of Example 7.1.15. The other equalities are

*	e	a	b	c
e	e	a	b	c
a	a	e	c	b
b	b	c	e	a
c	c	b	a	e

Figure 7.1   The Klein-4 group.

specific to this example. We conclude that in order to prove that $\mathfrak{G}$ is a model of **G1**, we only need to show the work specific to the structure $\mathfrak{G}$. We use this to prove $\mathfrak{G} \vDash \{$**G2, G3**$\}$ $[I]$.

- Let $a \in \mathbb{Z}$. Then,

$$[0]_n + [a]_n = [0 + a]_n = [a]_n,$$

and

$$[a]_n + [0]_n = [a + 0]_n = [a]_n.$$

Therefore, $\mathfrak{G} \vDash$ **G2** $[I]$.

- Let $b \in \mathbb{Z}$. Then,

$$[b]_n + [-b]_n = [b + (-b)]_n = [0]_n$$

and

$$[-b]_n + [b]_n = [(-b) + b]_n = [0]_n.$$

Hence, $\mathfrak{G} \vDash$ **G3** $[I]$.

Therefore, $\mathfrak{G}$ is a group. Moreover, because for all $a, b \in \mathbb{Z}$,

$$[a]_n + [b]_n = [a + b]_n = [b + a]_n = [b]_n + [a]_n,$$

$\mathfrak{G}$ is an abelian group.

■ EXAMPLE 7.1.18

The structures $(\omega^+, \varnothing, +)$, $(\mathbb{Z}, 1, \cdot)$, and $(\mathbb{R}, 1, \cdot)$ are not groups, but $(\mathbb{Z}, 0, +)$, $(\mathbb{Q}, 0, +)$, $(\mathbb{R}, 0, +)$, $(\mathbb{Q} \setminus \{0\}, 1, \cdot)$, and $(\mathbb{R} \setminus \{0\}, 1, \cdot)$ are. These are examples of infinite groups. When the group's set is finite, we use the term **order** to refer to its cardinality. The group $\mathfrak{G} = (\{\epsilon\}, \epsilon, *)$, where $* = \{((\epsilon, \epsilon), \epsilon)\}$, is the only group of order 1 (Example 7.1.15). This means that any other group with one element, such as $(\{0\}, 0, +)$ or $(\{1\}, 1, \cdot)$, has the same structure as $\mathfrak{G}$. We say that these three groups are **isomorphic**. Any two groups of order 2 will be isomorphic, and any groups of order 3 will also be isomorphic. There are essentially two groups of order 4, one being the **Klein-4** group (Figure 7.1) and the other being the group of Example 7.1.17 when $n = 4$.

■ **EXAMPLE 7.1.19**

All of the examples of groups given so far have been abelian. Here is one that is not. For all $n \in \mathbb{Z}^+$, define $M_n(\mathbb{R})$ as the set of $n \times n$ matrices with real entries. In other words, each matrix has n rows, n columns, and looks like

$$\begin{bmatrix} a_{1,1} & \cdots & a_{1,n} \\ \vdots & \ddots & \vdots \\ a_{m,1} & \cdots & a_{m,n} \end{bmatrix},$$

where $a_{i,j} \in \mathbb{R}$ for $i = 1, \ldots, n$ and $j = 1, \ldots, n$. As an example,

$$\begin{bmatrix} 1 & 2 & 3 \\ 4 & 5 & 6 \\ 7 & 8 & 9 \end{bmatrix} \in M_3(\mathbb{R}).$$

Define **matrix multiplication** for 2×2 matrices by

$$\begin{bmatrix} a_{1,1} & a_{1,2} \\ a_{2,1} & a_{2,2} \end{bmatrix} \cdot \begin{bmatrix} b_{1,1} & b_{1,2} \\ b_{2,1} & b_{2,2} \end{bmatrix} = \begin{bmatrix} a_{1,1}b_{1,1} + a_{1,2}b_{2,1} & a_{1,1}b_{1,2} + a_{1,2}b_{2,2} \\ a_{2,1}b_{1,1} + a_{2,2}b_{2,1} & a_{2,1}b_{1,2} + a_{2,2}b_{2,2} \end{bmatrix}.$$

For example,

$$\begin{bmatrix} 1 & 2 \\ 3 & 4 \end{bmatrix} \cdot \begin{bmatrix} 0 & -1 \\ 2 & 1 \end{bmatrix} = \begin{bmatrix} 4 & 1 \\ 8 & 1 \end{bmatrix}. \tag{7.17}$$

This multiplication is not commutative because

$$\begin{bmatrix} 0 & -1 \\ 2 & 1 \end{bmatrix} \cdot \begin{bmatrix} 1 & 2 \\ 3 & 4 \end{bmatrix} = \begin{bmatrix} -3 & -4 \\ 5 & 8 \end{bmatrix}, \tag{7.18}$$

but it is associative [Exercise 12(a)]. The **identity matrix** for $M_2(\mathbb{R})$ is

$$I_2 = \begin{bmatrix} 1 & 0 \\ 0 & 1 \end{bmatrix}.$$

Notice that I_2 is the multiplicative identity. If $A \in M_n(\mathbb{R})$, then A is **invertible** if there exists $B \in M_n(\mathbb{R})$ such that $AB = BA = I_n$. For $n = 2$,

$$\begin{bmatrix} 2 & 1 \\ 0 & -3 \end{bmatrix}$$

is invertible, but

$$\begin{bmatrix} 1 & 0 \\ 5 & 0 \end{bmatrix}$$

is not. All of this can be generalized to any $n \times n$ matrix. Finally, define

$$M_n^*(\mathbb{R}) = \{A \in M_n(\mathbb{R}) : A \text{ is invertible}\}.$$

Let $GL(n, \mathbb{R})$ denote the group $(M_n^*(\mathbb{R}), I_n, \cdot)$. This is called the **general linear group** of degree n.

Consequence

We now generalize the notion of logical implication (Definition 1.2.2) to the theory of models.

■ **DEFINITION 7.1.20**

Let $\mathscr{F}$ be a set of S-formulas. An S-formula p is an **S-consequence** of $\mathscr{F}$ (written as $\mathscr{F} \vDash p$) when $\mathfrak{A} \vDash \mathscr{F} [I]$ implies $\mathfrak{A} \vDash p [I]$ for any S-model $(\mathfrak{A}, I)$. If $\{p\} \vDash q$, simply write $p \vDash q$.

Definition 7.1.20 implies that if the S-formula q is not an S-consequence of $\mathscr{F}$, there exists an S-structure $\mathfrak{A}$ with interpretation I such that $\mathfrak{A} \vDash \mathscr{F} [I]$ but $\mathfrak{A} \nvDash q [I]$. For example, define

$$p := \forall x \forall y (x + y = y + x).$$

We know that $\mathrm{GL}(n, \mathbb{R})$ is a group, but under any GR-interpretation I of $\mathrm{GL}(n, \mathbb{R})$,

$$\mathrm{GL}(n, \mathbb{R}) \nvDash p [I]$$

because of (7.17) and (7.18). Therefore, p is not an S-consequence of the group axioms (7.1.14). In other words, not all groups are abelian groups.

■ **EXAMPLE 7.1.21**

Suppose that

$$\mathscr{F} = \{\forall x \forall y (x + y = y + x), \forall x \exists y (x + y = 0)\}.$$

Let $(\mathfrak{A}, I)$ be an NT-model of $\mathscr{F}$. Let $A = \mathrm{dom}(\mathfrak{A})$. Then,

$$\mathfrak{A} \vDash \forall x \exists y (x + y = 0) [I],$$

so by Theorem 7.1.7,

$$\text{for all } u \in A, \; \mathfrak{A} \vDash \exists y (x + y = 0) [I_x^u],$$

which by Definition 7.1.6 implies that

$$\text{there exists } v \in A \text{ such that for all } u \in A, \; \mathfrak{A} \vDash x + y = 0 \; [(I_x^u)_y^v].$$

Hence, for arbitrary a (UI) and particular b (EI) in A,

$$(I_x^a)_y^b(+)((I_x^a)_y^b(x), (I_x^a)_y^b(y)) = (I_x^a)_y^b(0).$$

However, since

$$\mathfrak{A} \vDash \forall x \forall y (x + y = y + x) [I],$$

we find that

$$(I_x^a)_y^b(+)((I_x^a)_y^b(x), (I_x^a)_y^b(y)) = (I_x^a)_y^b(+)((I_x^a)_y^b(y), (I_x^a)_y^b(x)),$$

so,

$$(I_x^a)_y^b(+)((I_x^a)_y^b(y), (I_x^a)_y^b(x)) = (I_x^a)_y^b(0).$$

Therefore, by EG and UG,

there exists $v \in A$ such that for all $u \in A$, $I(+)(I(v), I(u)) = I(0)$,

and we can reverse the steps above to find that

$$\mathfrak{A} \models \exists x \forall y(y + x = 0) \, [I],$$

so we conclude that

$$\mathcal{F} \models \exists x \forall y(y + x = 0).$$

We say that an S-sentence p is **valid** if $\varnothing \models p$ and write $\models p$. This means that if an S-sentence p is valid, every S-structure is a model of p since every S-structure is a model of the empty set using any S-interpretation (Exercise 7). For example, $\forall x(x = x)$ and $P \vee \neg P$ are valid.

We now connect the notions of consequence and satisfaction.

■ THEOREM 7.1.22

Let $\mathcal{F}$ be a set of S-formulas and p be an S-formula. Then,

$$\mathcal{F} \models p \text{ if and only if not } \text{Sat}_S \, \mathcal{F} \cup \{\neg p\}.$$

PROOF

The following are equivalent:

- $\mathcal{F} \models p$.

- For every S-model $(\mathfrak{A}, I)$, if $\mathfrak{A} \models \mathcal{F} \, [I]$, then $\mathfrak{A} \models p \, [I]$.

- There does not exist an S-model $(\mathfrak{A}, I)$ so that $\mathfrak{A} \models \mathcal{F} \, [I]$ and $\mathfrak{A} \not\models p \, [I]$.

- There does not exist an S-model $(\mathfrak{A}, I)$ so that $\mathfrak{A} \models \mathcal{F} \, [I]$ and $\mathfrak{A} \models \neg p \, [I]$.

- There does not exist an S-model $(\mathfrak{A}, I)$ such that $\mathfrak{A} \models \mathcal{F} \cup \{\neg p\} \, [I]$.

- Not $\text{Sat}_S \, \mathcal{F} \cup \{\neg p\}$. ■

■ EXAMPLE 7.1.23

Since Zorn's lemma was proved from the axioms of **ZFC** (Theorem 5.1.13), as in Example 7.1.17, we can use the work specific to the proof of Zorn's lemma to conclude that **ZFC** $\models$ Zorn's lemma, so by Theorem 7.1.22, there is no ST-model that satisfies **ZFC** and the negation of Zorn's lemma.

■ **EXAMPLE 7.1.24**

The S-formula p is valid if $\neg p$ is not S-satisfiable. To see this, suppose that p is not valid. This means that there is an S-model $(\mathfrak{A}, I)$ so that $\mathfrak{A} \not\models p\ [I]$. Hence, by Definition 7.1.6, $\mathfrak{A} \models \neg p\ [I]$. Therefore, $\text{Sat}\{\neg p\}$. That the converse is true is Exercise 19.

Compare the next definition with Definition 1.3.1.

■ **DEFINITION 7.1.25**

Let p and q be S-formulas. Then, p is **logically equivalent** to q means $p \models q$ if and only if $q \models p$.

Notice Definition 7.1.25 implies that the formulas p and q are logically equivalent if and only if $\models (p \leftrightarrow q)$. For example, by De Morgan's law, $\neg(p \wedge q)$ is logically equivalent to $\neg p \vee \neg q$, and by QN, we conclude that $\neg \forall x p(x)$ is logically equivalent to $\exists x \neg p(x)$.

Coincidence

Let $\mathfrak{A} = (\mathbb{Z}, \mathfrak{a})$ and $\mathfrak{B} = (\mathbb{Z}, \mathfrak{b})$ be GR-structures such that

$$\mathfrak{a}(e) = \mathfrak{b}(e) = 0,$$
$$\mathfrak{a}(\circ) = \mathfrak{b}(\circ) = +.$$

Let I be an GR-interpretation of $\mathfrak{A}$ and J be a GR-interpretation of $\mathfrak{B}$ such that

$$I(x) = J(x) = 3.$$

Other assignments of these functions are not identified. Consider the following deduction:

$$-3 + 3 = 0.$$
$$n + 3 = 0 \text{ for some } n \in \mathbb{Z}.$$
$$I_y^n(y) + I_y^n(x) = I_y^n(e) \text{ for some } n \in \mathbb{Z}.$$
$$I_y^n(y \circ x) = I_y^n(e) \text{ for some } n \in \mathbb{Z}.$$
$$\mathfrak{A} \models y \circ x = e\ [I_y^n] \text{ for some } n \in \mathbb{Z}.$$

Therefore,

$$\mathfrak{A} \models \exists y(y \circ x = e)\ [I].$$

By replacing I with J and $\mathfrak{A}$ with $\mathfrak{B}$ in the deduction, we conclude that

$$\mathfrak{B} \models \exists y(y \circ x = e)\ [J].$$

Since I and J agree on their interpretations of e, $\circ$, and x, it is not surprising that they should agree on their interpretation of any $\{e, \circ\}$-formula with x as its only free variable. The generalization of this to terms and formulas is the next two results.

■ **LEMMA 7.1.26 [Coincidence for Terms]**

Let S and T be sets of theory symbols. Let $\mathfrak{A} = (A, \mathfrak{a})$ be an S-structure and $\mathfrak{B} = (B, \mathfrak{b})$ be a T-structure such that $A = B$. Let I be an interpretation of $\mathfrak{A}$ and J be an interpretation of $\mathfrak{B}$. If $I \upharpoonright \mathsf{VAR} = J \upharpoonright \mathsf{VAR}$ and $\mathfrak{a}(u) = \mathfrak{b}(u)$ for all $u \in \mathsf{S} \cap \mathsf{T}$, then $I(t) = J(t)$ for every $(\mathsf{S} \cap \mathsf{T})$-term t.

PROOF

By induction on $(\mathsf{S} \cap \mathsf{T})$-terms.

- Let x be a variable symbol. Then, $I(x) = J(x)$ by hypothesis.

- Let c be a constant symbol in $\mathsf{S} \cap \mathsf{T}$. We have

$$I(c) = \mathfrak{a}(c) = \mathfrak{b}(c) = J(c).$$

- Suppose $I(t_i) = J(t_i)$ for all $(\mathsf{S} \cap \mathsf{T})$-terms t_i with $i = 0, 1, \dots, n-1$. Then,

$$\begin{aligned}
I(f(t_0, t_1, \dots, t_{n-1})) &= \mathfrak{a}(f)(I(t_0), I(t_1), \dots, I(t_{n-1})) \\
&= \mathfrak{a}(f)(J(t_0), J(t_1), \dots, J(t_{n-1})) \\
&= \mathfrak{b}(f)(J(t_0), J(t_1), \dots, J(t_{n-1})) \\
&= J(f(t_0, t_1, \dots, t_{n-1})). \blacksquare
\end{aligned}$$

■ **LEMMA 7.1.27 [Coincidence for Formulas]**

Let S and T be sets of theory symbols. Let $\mathfrak{A} = (A, \mathfrak{a})$ be an S-structure and $\mathfrak{B} = (B, \mathfrak{b})$ be a T-structure such that $A = B$. Let I be an interpretation of $\mathfrak{A}$ and J be an interpretation of $\mathfrak{B}$. If $I \upharpoonright \mathsf{VAR} = J \upharpoonright \mathsf{VAR}$ and $\mathfrak{a}(u) = \mathfrak{b}(u)$ for every $u \in \mathsf{S} \cap \mathsf{T}$, then $\mathfrak{A} \vDash p \, [I]$ if and only if $\mathfrak{B} \vDash p \, [J]$ for all $(\mathsf{S} \cap \mathsf{T})$-formulas p.

PROOF

By induction on $(\mathsf{S} \cap \mathsf{T})$-formulas.

- Let t_0 and t_1 be $(\mathsf{S} \cap \mathsf{T})$-terms. Then, by Lemma 7.1.26,

$$\mathfrak{A} \vDash t_0 = t_1 \, [I] \Leftrightarrow I(t_0) = I(t_1) \Leftrightarrow J(t_0) = J(t_1) \Leftrightarrow \mathfrak{B} \vDash t_0 = t_1 \, [J].$$

- Let $t_0, t_1, \dots, t_{n-1}$ be $(\mathsf{S} \cap \mathsf{T})$-terms and R a relation symbol of $\mathsf{S} \cap \mathsf{T}$. Then, by Lemma 7.1.26,

$$\begin{aligned}
\mathfrak{A} \vDash R(t_0, t_1, \dots, t_{n-1}) \, [I] &\Leftrightarrow (I(t_0), I(t_1), \dots, I(t_{n-1})) \in \mathfrak{a}(R) \\
&\Leftrightarrow (J(t_0), J(t_1), \dots, J(t_{n-1})) \in \mathfrak{a}(R) \\
&\Leftrightarrow (J(t_0), J(t_1), \dots, J(t_{n-1})) \in \mathfrak{b}(R) \\
&\Leftrightarrow \mathfrak{B} \vDash R(t_0, t_1, \dots, t_{n-1}) \, [J].
\end{aligned}$$

Now let p be an $(\mathsf{S} \cap \mathsf{T})$-formula.

- $\mathfrak{A} \vDash \neg p \, [I] \Leftrightarrow \mathfrak{A} \nvDash p \, [I] \Leftrightarrow \mathfrak{B} \nvDash p \, [J] \Leftrightarrow \mathfrak{B} \vDash \neg p \, [J]$.

- Assume that $\mathfrak{A} \models p\ [I]$ implies $\mathfrak{A} \models q\ [I]$. Also, suppose that $\mathfrak{B} \models p\ [J]$. Then, $\mathfrak{A} \models p\ [I]$ by induction, so $\mathfrak{A} \models q\ [I]$. Thus, $\mathfrak{B} \models q\ [J]$ by induction. The converse is proved similarly, so we have

$$
\begin{aligned}
\mathfrak{A} \models p \to q\ [I] &\Leftrightarrow \text{if } \mathfrak{A} \models p\ [I] \text{ then } \mathfrak{A} \models q\ [I] \\
&\Leftrightarrow \text{if } \mathfrak{B} \models p\ [J] \text{ then } \mathfrak{B} \models q\ [J] \\
&\Leftrightarrow \mathfrak{B} \models p \to q\ [J].
\end{aligned}
$$

- Note that for all $b \in A$,

$$
I_x^b \upharpoonright \mathsf{VAR} = J_x^b \upharpoonright \mathsf{VAR}
$$

because

$$
I_x^b(x) = u = J_x^b(x)
$$

and if $y \neq x$,

$$
I_x^b(y) = I(y) = J(y) = J_x^b(y).
$$

Therefore, by induction and since $A = B$,

$$
\begin{aligned}
\mathfrak{A} \models \exists x p\ [I] &\Leftrightarrow \mathfrak{A} \models p\ [I_x^a] \text{ for some } a \in A \\
&\Leftrightarrow \mathfrak{A} \models p\ [J_x^a] \text{ for some } a \in B \\
&\Leftrightarrow \mathfrak{A} \models \exists x p\ [J]. \ \blacksquare
\end{aligned}
$$

■ EXAMPLE 7.1.28

Define the sets of theory symbols $S = \{0, 1, +, \cdot, \leq\}$ and $T = \{0, 1, +, *, \geq\}$. Let $\mathfrak{A} = (\omega, \mathfrak{a})$ be an S-structure and $\mathfrak{B} = (\omega, \mathfrak{b})$ be a T-structure where

$$
\begin{aligned}
\mathfrak{a}(0) &= \mathfrak{b}(0) = \varnothing, \\
\mathfrak{a}(1) &= \mathfrak{b}(1) = \{\varnothing\}, \\
\mathfrak{a}(+) &= \mathfrak{b}(+).
\end{aligned}
$$

Notice, for example, that under the right interpretation, $\mathfrak{A}$ could be a model of $\forall x(x \cdot 1 = x)$ since it is an S-formula, but it does not make sense for $\mathfrak{B}$ to be a model of the same sentence because $\forall x(x \cdot 1 = x)$ is not a T-formula. Now, let I be an S-interpretation of $\mathfrak{A}$ and J be a T-interpretation of $\mathfrak{B}$ such that they agree on all variable symbols. Since we have the hypotheses of Lemma 7.1.27 satisfied, let us confirm the lemma. Consider the $(S \cap T)$-formula $\forall x(x + 0 = x)$. Assume

$$
\mathfrak{A} \models \forall x(x + 0 = x)\ [I],
$$

so

$$
\mathfrak{A} \models (x + 0 = x)\ [I_x^n] \text{ for all } n \in \omega.
$$

This implies that

$$\mathfrak{a}(+)(I_x^n(x), I_x^n(0)) = I_x^n(x) \text{ for all } n \in \omega.$$

Since I and J agree on all variable symbols, $\mathfrak{a}(0) = \mathfrak{b}(0)$, and $\mathfrak{a}(+) = \mathfrak{b}(+)$,

$$\mathfrak{b}(+)(J_x^n(x), J_x^n(0)) = J_x^n(x) \text{ for all } n \in \omega.$$

Therefore,

$$\mathfrak{B} \vDash (n + 0 = a) \, [J_x^n] \text{ for all } n \in \omega,$$

which gives

$$\mathfrak{B} \vDash \forall x(x + 0 = x) \, [J].$$

The purpose of the coincidence lemmas (7.1.26 and 7.1.27) is to minimize the use of interpretation functions, especially when modeling sentences.

■ **LEMMA 7.1.29**

Let $\mathfrak{A}$ be an S-structure. If p is an S-sentence,

$$\mathfrak{A} \vDash p \, [I] \text{ if and only if } \mathfrak{A} \vDash p \, [J]$$

for all S-interpretations I and J of $\mathfrak{A}$.

PROOF
Suppose that I and J are S-interpretations of $\mathfrak{A}$. Let $\mathfrak{A} \vDash p \, [I]$. Since p has no free variables, $\mathfrak{A} \vDash p \, [J]$ by the proof of Lemma 7.1.27. ■

Lemma 7.1.29 implies that any interpretation will do when modeling sentences. Therefore, we make the next definition.

■ **DEFINITION 7.1.30**

For any S-sentence p and S-structure $\mathfrak{A} = (A, \mathfrak{a})$, write $\mathfrak{A} \vDash p$ if $\mathfrak{A} \vDash p \, [I]$ for all S-interpretations I of $\mathfrak{A}$.

Lemma 7.1.29 can be used to quickly prove the next result.

■ **THEOREM 7.1.31**

For any S-structure $\mathfrak{A}$ and S-sentence p,

$$\mathfrak{A} \vDash p \text{ if and only if } \mathfrak{A} \vDash p \, [I] \text{ for some S-interpretation } I \text{ of } \mathfrak{A}.$$

Therefore, letting $\mathfrak{B}$ be the GR-structure of Example 7.1.12, by (7.2), we have that

$$\mathfrak{B} \vDash \forall x \exists y(x \circ y = 0).$$

The coincidence lemmas (7.1.26 and 7.1.27) also minimize the use of the sets of theory symbols. Consider the following.

■ **THEOREM 7.1.32**

Let $S \subseteq T$ be theory symbol sets. If $\mathcal{F}$ is a set of S-formulas, $\mathcal{F}$ is S-satisfiable if and only if $\mathcal{F}$ is T-satisfiable.

PROOF

Let $\mathcal{F}$ be a set of of S-formulas. First, suppose that $(A, a) \vDash \mathcal{F} [I]$, where $\text{dom}(a) = S$ and $\text{dom}(I) = \text{TERMS}(S)$. Let a' be an extension of a to T and I' be an extension of I such that $\text{dom}(I') = \text{TERMS}(T)$. Notice that this implies that a and a' agree on $S = S \cap T$. Therefore, $(A, a') \vDash \mathcal{F} [I']$ by Lemma 7.1.27.

Conversely, assume that $(A, a) \vDash \mathcal{F} [I]$ such that both $\text{dom}(a) = T$ and $\text{dom}(I) = \text{TERMS}(T)$. Let $a' = a \restriction S$ and $I' = I \cap \text{TERMS}(S)$. This implies that $(A, a') \vDash \mathcal{F} [I']$ by Lemma 7.1.27. ■

There is terminology to name the relationship between the structures found in the proof of Theorem 7.1.32. Let the theory symbols S be a subset of the theory symbols T. Let $\mathfrak{A} = (A, a)$ be an S-structure and $\mathfrak{A}'(A', a')$ be an T-structure. If $A = A'$ and $a = a' \restriction S$, we call $\mathfrak{A}'$ an **expansion** of $\mathfrak{A}$ and $\mathfrak{A}$ a **reduct** of $\mathfrak{A}'$. Hence, in the first part of the proof, we started with a structure and then moved to an expansion, and in the second part, we started with a structure and then moved to a reduct.

Theorems 7.1.31 and 7.1.32 motivate the next two definitions.

■ **DEFINITION 7.1.33**

Let $S \subseteq T$ be sets of theory symbols. An S-formula p is **satisfiable** (denoted by Sat p) if there exists an T-structure that is a model for p. The set of S-formulas $\mathcal{F}$ is **satisfiable** (denoted by Sat $\mathcal{F}$) if there exists an T-structure that is a model for $\mathcal{F}$.

■ **DEFINITION 7.1.34**

Let $S \subseteq T$ be sets of theory symbols. Assume that $\mathcal{T}$ is a set of S-sentences. An S-sentence p is a **consequence** of $\mathcal{T}$ (denoted by $\mathcal{T} \vDash p$) when $\mathfrak{A} \vDash \mathcal{T}$ implies $\mathfrak{A} \vDash p$ for every T-structure $\mathfrak{A}$.

■ **EXAMPLE 7.1.35**

The group axioms state that in a group there is an identity and there are inverses. Based on what we know about the integers, we should be able to prove more about these elements. For example, we expect that in a group, both

there is exactly one identity

and

every element has a unique inverse

are true. The uniqueness of the identity is left to Exercise 20. To show the uniqueness of inverses, let $\mathfrak{G} = (G, e, \circ)$ be a group, and take $a \in G$. Suppose that

$a', a'' \in G$ and are inverses of a. Then,

$$a' = a' \circ e = a' \circ (a \circ a'') = (a' \circ a) \circ a'' = e \circ a'' = a''.$$

Therefore, the uniqueness of inverses is a consequence (Definition 7.1.34) of the group axioms.

Rings

Consider the equation $2x + 1 = 0$. The exact steps needed to find its solution are

$$(2x + 1) + -1 = 0 + -1,$$
$$2x + (1 + -1) = 0 + -1,$$
$$2x + 0 = 0 + -1,$$
$$2x = -1,$$
$$1/2(2x) = 1/2(-1),$$
$$(1/2 \cdot 2)x = -1/2,$$
$$1x = -1/2,$$
$$x = -1/2.$$

Now examine the steps. There are two operations, addition and multiplication. We used inverses and identities. The associative law was also used. When studying these steps, we realize that they cannot be performed within $(\mathbb{Z}, 0, +)$ even though the initial equation had only integer coefficients. This means that the group idea needs to be expanded. This is done by including two symbols to represent addition and multiplication. Since these two operations can have their own identities, replace e with $\bigcirc$ to represent the additive identity. The ideas behind the group axioms are then extended using RI-sentences.

■ **AXIOMS 7.1.36 [Ring]**

- **R1.** $\forall x \forall y \forall z \, [x \oplus (y \oplus z) = (x \oplus y) \oplus z]$
  $\forall x \forall y \forall z \, [x \otimes (y \otimes z) = (x \otimes y) \otimes z]$

- **R2.** $\forall x \forall y (x \oplus y = y \oplus x)$

- **R3.** $\forall x (0 \oplus x = x)$

- **R4.** $\forall x \exists y (x \oplus y = \bigcirc)$

- **R5.** $\forall x \forall y \forall z \, [x \otimes (y \oplus z) = x \otimes y \oplus x \otimes z]$
  $\forall x \forall y \forall z \, [(x \oplus y) \otimes z = x \otimes z \oplus y \otimes z]$

■ **DEFINITION 7.1.37**

An RI-structure $\mathfrak{R} = (R, 0, +, \cdot)$ that models the ring axioms is called a **ring**. If there exists an multiplicative identity in R, then $\mathfrak{R}$ is a ring with **unity**.

The additive inverse of a is $-a$, and the multiplicative inverse of a is a^{-1} assuming that $a \neq 0$. We usually write $a - b$ instead of $a + (-b)$. Notice that if $\mathfrak{R} = (R, 0, +, \cdot)$ is a ring, its reduct $(R, 0, +)$ is a group. Also, letting $+$ and $\cdot$ denote addition and multiplication on $\mathbb{Z}$,

$$3 = (\mathbb{Z}, 0, +, \cdot)$$

is a ring with unity. Also, $\mathbb{Q}$, $\mathbb{R}$, and $\mathbb{C}$ are the domains of rings with unity using the typical operations of addition and multiplication.

■ **EXAMPLE 7.1.38**

Axioms 7.1.36 do not require that the ring multiplication be commutative. Let $\mathfrak{R} = (R, 0, +, \cdot)$. Then, $\mathfrak{R}$ is a **commutative ring**. if

$$\mathfrak{R} \vDash \forall x \forall y (x \otimes y = y \otimes x).$$

- Let $+$ and $\cdot$ denote standard addition and multiplication on $\mathbb{Z}$. Let $n \in \mathbb{Z}$. We conclude that $\mathfrak{S} = (n\mathbb{Z}, 0, +, \cdot)$ is a commutative ring. It is without unity if $n \neq \pm 1$.

- Take $[a]_n$, $[b]_n \in \mathbb{Z}_n$ and define $+$ as in Example 7.1.17 and multiplication defined by

$$[a]_n \cdot [b]_n = [ab]_n.$$

  Then, $\mathfrak{T} = (\mathbb{Z}_n, [0]_n, +, \cdot)$ is a commutative ring (Exercise 29).

Axioms 7.1.36 also do not state that when the additive identity is multiplied by any element of the ring, the result is the additive identity. It is not among the axioms because it can be proved. Take $a \in R$. By **R3**, $0 + 0 = 0$, so by **R5**,

$$0 \cdot a = (0 + 0) \cdot a = 0 \cdot a + 0 \cdot a.$$

By **R4** and since $+$ is a binary operation,

$$0 \cdot a + -(0 \cdot a) = (0 \cdot a + 0 \cdot a) + -(0 \cdot a).$$

Because of **R1**,

$$0 \cdot a + -(0 \cdot a) = 0 \cdot a + [0 \cdot a + -(0 \cdot a)].$$

Hence, $0 = 0 \cdot a + 0$, which implies that $0 = 0 \cdot a$. Therefore,

$$\mathfrak{R} \vDash \forall x (\bigcirc = \bigcirc \otimes x),$$

and $\forall x (\bigcirc = \bigcirc \otimes x)$ is a consequence (Definition 7.1.34) of the ring axioms.

■ **EXAMPLE 7.1.39**

Let $n \in \mathbb{Z}^+$, define **matrix addition** on $M_n(\mathbb{R})$ entrywise. For instance,

$$\begin{bmatrix} 1 & 2 & 1 \\ 3 & 4 & -4 \\ 5 & 6 & 0 \end{bmatrix} + \begin{bmatrix} 1 & 0 & 8 \\ 2 & -5 & 0 \\ 0 & -2 & 3 \end{bmatrix} = \begin{bmatrix} 2 & 2 & 9 \\ 5 & -1 & -4 \\ 5 & 4 & 3 \end{bmatrix}.$$

Let $\mathbf{0}_n$ be the **zero matrix**. It is the $n \times n$ matrix with all of its entries equal to 0. As in Example 7.1.19, let $\cdot$ represent matrix multiplication and I_n the identity matrix. Prove that

$$\mathfrak{M}_n(\mathbb{R}) = (M_n(\mathbb{R}), \mathbf{0}_n, +, \cdot)$$

is a ring.

- To see that matrix addition is associative, we rely on the fact that standard addition of real numbers is associative. Take three matrices from $M_2(\mathbb{R})$ and add:

$$\begin{aligned} &\begin{bmatrix} a_{1,1} & a_{1,2} \\ a_{2,1} & a_{2,2} \end{bmatrix} + \left(\begin{bmatrix} b_{1,1} & b_{1,2} \\ b_{2,1} & b_{2,2} \end{bmatrix} + \begin{bmatrix} c_{1,1} & c_{1,2} \\ c_{2,1} & c_{2,2} \end{bmatrix} \right) \\ &= \begin{bmatrix} a_{1,1} & a_{1,2} \\ a_{2,1} & a_{2,2} \end{bmatrix} + \begin{bmatrix} b_{1,1} + c_{1,1} & b_{1,2} + c_{1,2} \\ b_{2,1} + c_{2,1} & b_{2,2} + c_{2,2} \end{bmatrix} \\ &= \begin{bmatrix} a_{1,1} + (b_{1,1} + c_{1,1}) & a_{1,2} + (b_{1,2} + c_{1,2}) \\ a_{2,1} + (b_{2,1} + c_{2,1}) & a_{2,2} + (b_{2,2} + c_{2,2}) \end{bmatrix} \\ &= \begin{bmatrix} (a_{1,1} + b_{1,1}) + c_{1,1} & (a_{1,2} + b_{1,2}) + c_{1,2} \\ (a_{2,1} + b_{2,1}) + c_{2,1} & (a_{2,2} + b_{2,2}) + c_{2,2} \end{bmatrix} \\ &= \begin{bmatrix} a_{1,1} + b_{1,1} & a_{1,2} + b_{1,2} \\ a_{2,1} + b_{2,1} & a_{2,2} + b_{2,2} \end{bmatrix} + \begin{bmatrix} c_{1,1} & c_{1,2} \\ c_{2,1} & c_{2,2} \end{bmatrix} \\ &= \left(\begin{bmatrix} a_{1,1} & a_{1,2} \\ a_{2,1} & a_{2,2} \end{bmatrix} + \begin{bmatrix} b_{1,1} & b_{1,2} \\ b_{2,1} & b_{2,2} \end{bmatrix} \right) + \begin{bmatrix} c_{1,1} & c_{1,2} \\ c_{2,1} & c_{2,2} \end{bmatrix}. \end{aligned}$$

- Since

$$\begin{bmatrix} 0 & 0 \\ 0 & 0 \end{bmatrix} + \begin{bmatrix} a_{1,1} & a_{1,2} \\ a_{2,1} & a_{2,2} \end{bmatrix} = \begin{bmatrix} 0 + a_{1,1} & 0 + a_{1,2} \\ 0 + a_{2,1} & 0 + a_{2,2} \end{bmatrix}$$
$$= \begin{bmatrix} a_{1,1} & a_{1,2} \\ a_{2,1} & a_{2,2} \end{bmatrix},$$

the zero matrix is the additive identity.

- To prove that every element of $M_2(\mathbb{R})$ has an additive inverse, take

$$A = \begin{bmatrix} a_{1,1} & a_{1,2} \\ a_{2,1} & a_{2,2} \end{bmatrix} \in M_2(\mathbb{R}).$$

Then,

$$-A = \begin{bmatrix} -a_{1,1} & -a_{1,2} \\ -a_{2,1} & -a_{2,2} \end{bmatrix}$$

because $A + (-A) = 0_2$. Generalizing, conclude that

$$\mathfrak{M}_n(\mathbb{R}) \models \{G1, G2, G3\},$$

making the GR-structure $(M_n(\mathbb{R}), 0_n, +)$ a group.

- Matrix addition is commutative because addition on $\mathbb{R}$ is commutative. Therefore, $(M_n(\mathbb{R}), 0_n, +)$ is an abelian group.

- Matrix multiplication is associative.

- Lastly, to show that the operations are distributive, we must show for all $A, B, C \in M_2(\mathbb{R})$,
$$A(B + C) = AB + AC$$
and
$$(A + B)C = AC + BC.$$

Therefore, $\mathfrak{M}_n(\mathbb{R}) \models \{R1, R2, R3, R4, R5\}$, so $\mathfrak{M}_n(\mathbb{R})$ is a ring.

- Since I_n is the multiplicative identity, $\mathfrak{R}$ is a ring with unity, and since matrix multiplication is not commutative, $\mathfrak{R}$ is a noncommutative ring. This proves that $\forall x \forall y (x \otimes y = y \otimes x)$ is not a consequence of the ring axioms.

■ **EXAMPLE 7.1.40**

If R is the domain of a ring, $a, b \in R \setminus \{0\}$ are **zero divisors** of the ring means that $a \cdot b = 0$. Defining addition and multiplication coordinatewise (Exercise 18), the ring $(\mathbb{Z} \times \mathbb{Z}, (0,0), +, \cdot)$ has zero divisors such as

$$(1,0) \cdot (0,1) = (0,0).$$

Other examples can be found in $M_2(\mathbb{R})$ where

$$\begin{bmatrix} 1 & 1 \\ 0 & 0 \end{bmatrix} \cdot \begin{bmatrix} 1 & 1 \\ -1 & -1 \end{bmatrix} = \begin{bmatrix} 0 & 0 \\ 0 & 0 \end{bmatrix}.$$

However,

$$\begin{bmatrix} 1 & 1 \\ -1 & -1 \end{bmatrix} \cdot \begin{bmatrix} 1 & 1 \\ 0 & 0 \end{bmatrix} = \begin{bmatrix} 1 & 1 \\ -1 & -1 \end{bmatrix},$$

showing that an element can be a **left zero divisor** but not a **right zero divisor**. This situation is common for rings where multiplication is not commutative. We do, however, have many rings that do not have zero divisors. An **integral domain** is a commutative ring with unity that does not have zero divisors. The rings $(\mathbb{Z}, 0, +, \cdot)$, $(\mathbb{Q}, 0, +, \cdot)$, $(\mathbb{R}, 0, +, \cdot)$, and $(\mathbb{C}, 0, +, \cdot)$ are integral domains.

The equation $2x + 1 = 0$ is written with elements of $\mathbb{Z}$ and the operations of regular addition and multiplication. Although $\mathbb{Z}$ has no zero divisors, there is no integer that is a solution to this equation. To solve the equation, we need the existence of multiplicative inverses. Let $(R, 0, +, \cdot)$ be a ring with unity. If $u \in R$ has the property that there exists $v \in R$ such that $u \cdot v = v \cdot u = 1$, then u is called a **unit**. Notice that units are multiplicative inverses of each other. With this terminology, we make the next definition.

■ **DEFINITION 7.1.41**

Let $(R, 0, +, \cdot)$ be a ring with unity.

- If all nonzero elements of R are units, R is called a **division ring** or sometimes a **skew field**.

- A commutative division ring is called a **field**.

The reason that the equation on page 353 can be solved the way it was is that $\mathbb{R}$ with addition and multiplication form a field.

■ **EXAMPLE 7.1.42**

While $\mathbb{Z}$ is not a field with standard addition and multiplication, $\mathbb{Q}$, $\mathbb{R}$, and $\mathbb{C}$ are. A more interesting structure is $(\mathbb{Z}_p, [0]_p, +, \cdot)$ when p is a prime. To prove that it is a field, let $[a]_p \in \mathbb{Z}_p$ so that $[a]_p \neq [0]_p$. We must find an element of $\mathbb{Z}_p$ so that when it is multiplied with $[a]_p$ the result is $[1]_p$. Since $[a]_p \neq [0]_p$, p does not divide a. Hence, p and a are relatively prime, so there are integers u and v such that $ua + vp = 1$. We are then able to calculate:

$$[u]_p \cdot [a]_p = [ua]_p$$
$$= [1 - vp]_p$$
$$= [1]_p + [-vp]_p$$
$$= [1]_p + [0]_p$$
$$= [1]_p .$$

■ **EXAMPLE 7.1.43**

Let $\mathfrak{R}$ be a division ring and take u and v to be elements of the domain of $\mathfrak{R}$. Let 1 be unity. Assume $uv = 0$ and $u \neq 0$. Then, u^{-1} exists, and we can calculate

$$u^{-1}(uv) = u^{-1}0,$$
$$(u^{-1}u)v = 0,$$
$$1v = 0,$$
$$v = 0.$$

Therefore, $\mathfrak{R}$ has no zero divisors.

Exercises

1. Prove the remaining parts of Theorem 7.1.7.

2. Let $\preccurlyeq$ be a linear order on a nonempty set A. Let R be a binary relation symbol. Define the $\{R\}$-structure $\mathfrak{A} = (A, \preccurlyeq)$. Let I be an S-interpretation of $\mathfrak{A}$ such that $I(R) = \preccurlyeq$. Prove the following.
   (a) $\mathfrak{A} \vDash \exists x(xRx)\,[I]$.
   (b) $\mathfrak{A} \vDash \forall x \forall y(xRy \vee yRx)\,[I]$.
   (c) $\mathfrak{A} \vDash \forall x \forall y(xRy \wedge yRx \rightarrow x = y)\,[I]$.

3. Let $\mathfrak{B}$ be the NT-structure $(\mathbb{Z}, 0^{\mathfrak{B}}, 1^{\mathfrak{B}}, +^{\mathfrak{B}}, \cdot^{\mathfrak{B}})$, where $0^{\mathfrak{B}}$ and $1^{\mathfrak{B}}$ are the numbers 0 and 1 in $\mathbb{Z}$ while $+^{\mathfrak{B}}$ and $\cdot^{\mathfrak{B}}$ are the standard operations of addition and multiplication of integers. Let I be a NT-interpretation such that $I(x) = 2$ and $I(y) = -2$. Prove the following.
   (a) $\mathfrak{B} \vDash x + y = 0\,[I]$.
   (b) $\mathfrak{B} \vDash \exists x([x + 1] + 1 = 0)\,[I]$.
   (c) $\mathfrak{B} \vDash \exists x \forall y(x \cdot y = y)\,[I]$.
   (d) $\mathfrak{B} \vDash \forall x \forall y \forall z(\neg z = 0 \wedge x \cdot z = y \cdot z \rightarrow x = y)\,[I]$.

4. Show that $\mathfrak{A} \vDash \forall x[(x + x_5) + x_8 = x + (x_5 + x_8)]$, where $\mathfrak{A}$ is the NT-structure of Example 7.1.10.

5. Find a set of theory symbols S, an S-structure $\mathfrak{A}$, and an S-interpretation I such that $\mathfrak{A} \vDash p\,[I]$ for each given formula p.
   (a) $x + y = ([1 + 1] + 1) + 1$
   (b) $x/y + z = 10$
   (c) $\exists x \exists y(x < y \wedge x + 1 = y)$
   (d) $\forall x \forall y \forall z(xRy \wedge yRz \rightarrow zRx)$
   (e) $\forall x \forall y(x \cdot y = 0 \rightarrow x = 0 \vee y = 0)$

6. For each formula in Exercise 5, find a model $(\mathfrak{A}, I)$ such that $\mathfrak{A} \nvDash p\,[I]$ for each given formula p.

7. Prove that every S-structure is a model of the empty set.

8. Let A be a set. Is $(\mathbf{P}(A), \varnothing, \cap)$ a group? Explain.

9. Explain why $(\mathbb{Z}^+, 0, +)$, $(\mathbb{Z}, 1, \cdot)$, and $(\mathbb{R}, 1, \cdot)$ are not groups, where the operations are the standard ones.

10. Suppose that $*$ is an operation on $\mathbb{Z}$ defined by $x * y = x + y + 2$.
    (a) Identify the identity e and the inverses with respect to $*$.
    (b) Prove that $(\mathbb{Z}, e, *)$ is a group, where e is the identity found in Exercise 10(a).
    (c) Solve $8 * x = 10$.

11. Let 0 represent the zero function $\mathbb{R} \rightarrow \mathbb{R}$ and $+$ be function addition. That is, For all $x \in \mathbb{R}$,

$$(f + g)(x) = f(x) + g(x).$$

(a) Prove that $(^R\mathbb{R}, 0, +)$ is a group.

(b) Is $^R\mathbb{R}$ the domain of a group where the binary operation is function division? If so, what is its identity?

(c) Is $^R\mathbb{R}$ the domain of a group where the binary operation is composition? If so, what is identity?

12. Let n be a positive integer.

   (a) Prove that matrix multiplication is associative.

   (b) Solve the equation in $M_2(\mathbb{R})$:

$$\begin{bmatrix} 1 & 4 \\ -3 & 0 \end{bmatrix} + \begin{bmatrix} a & b \\ c & d \end{bmatrix} = \begin{bmatrix} -3 & 8 \\ 0 & -6 \end{bmatrix}.$$

   (c) Show that $M_n(\mathbb{R})$ is not the domain of a group under matrix multiplication.

13. Let $(G, \epsilon, *)$ and $(G', \epsilon', *')$ be two groups. For all $a, b \in G$ and $a', b' \in G'$ define $(a, a') \cdot (b, b') = (a * b, a' *' b')$.

   (a) Confirm that $\cdot$ is a binary operation on $G \times G'$.

   (b) Show that $(G \times G', (\epsilon, \epsilon'), \cdot)$ is a group. Prove that it is abelian if and only if both of the given groups are abelian.

14. Let n be an integer. Prove that $(n\mathbb{Z}, 0, +, \cdot)$ is a commutative ring.

15. Why is

$$\left\{ \begin{bmatrix} a & b \\ c & d \end{bmatrix} : a, b, c, d, \in \mathbb{Z}^+ \right\}$$

not the domain of a ring under the standard matrix operations?

16. Prove that the set

$$\left\{ \begin{bmatrix} a & 0 \\ 0 & b \end{bmatrix} : a, b \in \mathbb{R} \right\}$$

is the domain of a ring with the standard matrix operations.

17. Both $+$ (function addition) and $\circ$ (composition) are binary operations on $^R\mathbb{R}$, but $(^R\mathbb{R}, 0, +, \circ)$ is not a ring. Identify which ring axioms fail.

18. Let $(R, 0, +, \cdot)$ and $(R', 0', +', \cdot')$ be rings. Define addition and multiplication on $R \times R'$ so that for all $(a, b), (c, d) \in R \times R'$,

$$(a, b) + (c, d) = (a + c, b +' d),$$

and

$$(a, b) \cdot (c, d) = (a \cdot c, b \cdot' d).$$

Prove that $(R \times R', (0, 0), +, \cdot)$ is a ring.

19. Prove the converse of Example 7.1.24.

20. Prove that

$$\{G1, G2, G3\} \models \forall x \forall y [\forall z (x \circ z = z \wedge z \circ x = x)$$
$$\wedge \forall z (y \circ z = z \wedge z \circ y = z) \rightarrow x = y].$$

21. Let $\mathfrak{G} = (G, \epsilon, *)$ be a group so that

$$\mathfrak{G} \models \forall x \forall y[(a \circ b) \circ (a \circ b) = a \circ a \circ b \circ b].$$

Show that $\mathfrak{G}$ is abelian.

22. Prove that $\forall x(O \otimes x = O)$ is a consequence of the ring axioms.

23. Let $-$ be a unary function symbol. Define $\text{RI}' = \text{RI} \cup \{-\}$. Show that the given sentences are consequences of the ring axioms and $\forall x(-x \oplus x = O)$.
    (a) $\forall x \forall y[-(x \otimes y) = -x \otimes y \wedge -x \otimes y = x \otimes -y]$
    (b) $\forall x \forall y(-a \otimes -b = a \otimes b)$
    (c) $\forall x \forall[-(a \oplus b) = -a \oplus -b]$
    (d) $-O = O$

24. This exercise uses the notation of Exercise 23. Let $\mathfrak{R}$ be a ring with unity. Let $\mathfrak{R}'$ be the expansion of $\mathfrak{R}$ to $\text{RI}' = \text{RI} \cup \{-\}$. Assume that for all $r \in \text{dom}(\mathfrak{R}')$,

$$-^{\mathfrak{R}}(r) \oplus^{\mathfrak{R}} r = O^{\mathfrak{R}}.$$

Prove that $\mathfrak{R}' \models \forall x[\forall y(x \otimes y = y \wedge y \otimes x = y) \rightarrow \forall z(-z = -x \otimes z)]$.

25. Let p and q be S-formulas. Prove that the given S-sentences are valid.
    (a) $p \vee \neg p$
    (b) $p \rightarrow q \leftrightarrow \neg p \vee q$
    (c) $\exists x(p \vee q) \leftrightarrow \exists x p \vee \exists x q$
    (d) $\forall x(p \wedge q) \leftrightarrow \forall x p \wedge \forall x q$

26. Let R be a binary relation symbols and f be a binary function symbol. Show that the given sentences are satisfiable.
    (a) $\exists x(x = x)$
    (b) $\exists x \exists y \exists z(\neg x = y \wedge \neg x = z \wedge \neg y = z)$
    (c) $\exists x \forall y(Rxy \vee x = y)$
    (d) $\forall x \forall y(f xy = 1)$
    (e) $\forall x \forall y[Rxy \rightarrow \exists z(Rxz \wedge Rzy)]$

27. Let S and T be sets of theory symbols such that $S \subseteq T$. Let $\mathfrak{A}$ be a reduct to S of the T-structure $\mathfrak{B}$. Prove that $\mathfrak{A} \models p$ if and only if $\mathfrak{B} \models p$ for all S-sentences p.

28. Suppose that $p_0, p_1, \ldots, p_{n-1}$ are S-sentences. For every S-structure $\mathfrak{A}$, prove that $\mathfrak{A} \models p_0 \wedge p_1 \wedge \cdots \wedge p_{n-1}$ if and only if $\mathfrak{A} \models p_i$ for all $i = 0, 1, \ldots, n-1$.

29. Answer the following about $(\mathbb{Z}_n, [0]_n, +, \cdot)$:
    (a) Prove that addition and multiplication of congruence classes is well-defined.
    (b) Show that the additive identity is $[0]_n$.
    (c) For all $a \in \mathbb{Z}$, show that $-[a]_n = [n - a]_n$.
    (d) Show that $[1]_n$ is the multiplicative identity.
    (e) Prove that $(\mathbb{Z}_n, [0]_n, +, \cdot)$ is a commutative ring.
    (f) Prove that the ring contains zero divisors when n is not prime.

30. Prove that $\forall x \forall y \forall z(x \oplus y = x \oplus z \rightarrow y = z)$ is a consequence of the ring axioms.

31. Let $\mathfrak{R}$ be an integral domain. Prove the following.
    (a) $\mathfrak{R} \models \forall x \forall y(x \otimes y = O \rightarrow x = O \vee y = 0)$.
    (b) $\mathfrak{R} \models \forall x \forall y \forall z(x \otimes y = x \otimes z \wedge x \neq O \rightarrow y = z)$.

32. Suppose that $\mathfrak{R} = (R, 0, +, \cdot)$ is a commutative ring with unity. Show that if $\mathfrak{R} \models \forall x \forall y \exists z(x \otimes z \oplus y = O)$, then $\mathfrak{R}$ a field.

33. Is $(\{0\}, 0, +, \cdot)$ a field? Explain.

7.2 SUBSTRUCTURES

When looking for examples of groups, the GR-structure $(\mathbb{Z}, 0, +)$ is often the first to come to mind. The benefit of this example is that not only are we familiar with the integers but it has the property that many of its subsets also form groups. Let $n \in \mathbb{Z}$. Addition on $\mathbb{Z}$ restricted to $n\mathbb{Z} \times n\mathbb{Z}$ is an associative binary operation on $n\mathbb{Z}$, every element of $n\mathbb{Z}$ has an additive inverse in $n\mathbb{Z}$, and $0 \in n\mathbb{Z}$, so the GR-structure $(n\mathbb{Z}, 0, +)$ is a group. Since $n \neq \pm 1$ implies that $n\mathbb{Z} \subset \mathbb{Z}$, there are infinitely many different examples of GR-structures, all within $(\mathbb{Z}, 0, +)$. We generalize this idea to arbitrary structures.

■ **DEFINITION 7.2.1**

If $\mathfrak{A} = (A, a)$ and $\mathfrak{B} = (B, b)$ are S-structures, $\mathfrak{A}$ is a **substructure** of $\mathfrak{B}$ (written as $\mathfrak{A} \subseteq \mathfrak{B}$) means that $A \subseteq B$ and the following properties hold.

- $a(c) = b(c)$ for all constant symbols c.

- $a(R) = b(R) \cap A^n$ for every n-ary relation symbol R.

- $a(f) = b(f) \restriction A^n$ for every n-ary function symbol f.

If $\mathfrak{A}$ is a substructure of $\mathfrak{B}$, then $\mathfrak{B}$ is an **extension** of $\mathfrak{A}$.

Note the difference between a substructure and a reduct and between an extension and an expansion (page 352). For all $n \in \mathbb{Z}$, the group $(n\mathbb{Z}, 0, +)$ is a substructure of $(\mathbb{Z}, 0, +)$, and $(\mathbb{Z}, 0, +)$ is an extension of $(n\mathbb{Z}, 0, +)$. Here both structures have the same set of theory symbols, and the domain of one is a subset of the other. However, $(\mathbb{Z}, 0, +)$ is a reduct of $(\mathbb{Z}, 0, 1, +, \cdot)$, and $(\mathbb{Z}, 0, 1, +, \cdot)$ is an expansion of $(\mathbb{Z}, 0, +)$. In this case, the domains are the same, but the theory symbol set of the one is a subset of the theory symbol set of the other.

■ **EXAMPLE 7.2.2**

Let R be a binary relation symbol. Let $\mathfrak{A} = ([0, 1], a)$ and $\mathfrak{B} = ([0, 2], b)$ be $\{R\}$-structures such that $a(R)$ and $b(R)$ are both standard less-than. That is,

$$a(R) = \{(x, y) \in \mathbb{R} \times \mathbb{R} : 0 \leq x < y \leq 1\}$$

and

$$\mathfrak{b}(R) = \{(x, y) \in \mathbb{R} \times \mathbb{R} : 0 \leq x < y \leq 2\}.$$

We conclude that $\mathfrak{A} \subseteq \mathfrak{B}$ because of the following:

- $[0, 1] \subseteq [0, 2]$.

- There are no constant symbols.

- $\mathfrak{a}(R) = \mathfrak{b}(R) \cap ([0, 1] \times [0, 1])$.

- There are no function symbols.

■ EXAMPLE 7.2.3

Let $n \in \mathbb{Z}$. Define the NT-structure $\mathfrak{B} = (\mathbb{Z}, \mathfrak{b})$, where $\mathfrak{b}(0)$ is the additive identity of $\mathbb{Z}$, $\mathfrak{b}(1)$ is the multiplicative identity of $\mathbb{Z}$, $\mathfrak{b}(+)$ is standard addition on $\mathbb{Z}$, and $\mathfrak{b}(\cdot)$ is standard multiplication on $\mathbb{Z}$. Let $\mathfrak{A}_n = (n\mathbb{Z}, \mathfrak{a})$ such that

$$\mathfrak{a}(0) = \mathfrak{b}(0),$$
$$\mathfrak{a}(1) = \mathfrak{b}(1),$$
$$\mathfrak{a}(+) = \mathfrak{b}(+) \restriction (n\mathbb{Z} \times n\mathbb{Z}),$$
$$\mathfrak{a}(\cdot) = \mathfrak{b}(\cdot) \restriction (n\mathbb{Z} \times n\mathbb{Z}).$$

Then, $\mathfrak{A}$ is a substructure of $\mathfrak{B}$.

In particular, Example 7.2.3 gives

$$\mathfrak{A}_8 \subseteq \mathfrak{A}_4 \subseteq \mathfrak{A}_2,$$

which implies that $\mathfrak{A}_8 \subseteq \mathfrak{A}_2$. This is a special case of the next theorem.

■ THEOREM 7.2.4

Let $\mathfrak{A}$, $\mathfrak{B}$, and $\mathfrak{C}$ be S-structures.

- $\mathfrak{A} \subseteq \mathfrak{A}$.

- If $\mathfrak{A} \subseteq \mathfrak{B}$ and $\mathfrak{B} \subseteq \mathfrak{C}$, then $\mathfrak{A} \subseteq \mathfrak{C}$.

PROOF

That $\mathfrak{A}$ is a substructure of itself is clear, so suppose that $\mathfrak{A}$ is a substructure of $\mathfrak{B}$ and $\mathfrak{B}$ is a substructure of $\mathfrak{C}$. Write $\mathfrak{A} = (A, \mathfrak{a})$, $\mathfrak{B} = (B, \mathfrak{b})$, and $\mathfrak{C} = (C, \mathfrak{c})$. Then, for all constant symbols c,

$$\mathfrak{a}(c) = \mathfrak{b}(c) = \mathfrak{c}(c).$$

Since $\mathfrak{A} \subseteq \mathfrak{B}$, $\mathfrak{a}(R) = \mathfrak{b}(R) \cap A^n$, and since $\mathfrak{B} \subseteq \mathfrak{C}$, $\mathfrak{b}(R) = \mathfrak{c}(R) \cap B^n$ for every n-ary relation symbol R, so

$$\mathfrak{a}(R) = \mathfrak{c}(R) \cap B^n \cap A^n = \mathfrak{c}(R) \cap A^n.$$

Also, $\mathfrak{a}(f) = \mathfrak{b}(f) \restriction A^n$ and $\mathfrak{b}(f) = \mathfrak{c}(f) \restriction B^n$ for all n-ary function symbols f, so

$$\mathfrak{a}(f) = (\mathfrak{c}(f) \restriction B^n) \restriction A^n = \mathfrak{c}(f) \restriction A^n.$$

Therefore, $\mathfrak{A}$ is a substructure of $\mathfrak{C}$. ■

Subgroups

Let a be an element of a group $(G, \epsilon, *)$. For all positive integers n, define a^n to be the result of operating a with itself n times. That is,

$$a^1 = a, a^2 = a * a, a^3 = a * a * a, \dots$$

and

$$a^m * a^n = a^{m+n}.$$

Further, define $a^0 = e$ and a^{-1} to be the inverse of a. With this notation, we observe that

$$(a * b)^{-1} = b^{-1} * a^{-1}$$

and

$$a^{-n} = (a^n)^{-1} = (a^{-1})^n.$$

We then gather all of these elements into a set,

$$\langle a \rangle = \{a^n : n \in \mathbb{Z}\},$$

and define the following.

■ DEFINITION 7.2.5

A group $\mathfrak{G}$ is **cyclic** if there exists $a \in \text{dom}(\mathfrak{G})$ such that $\text{dom}(\mathfrak{G}) = \langle a \rangle$. The element a is called a **generator** of $\mathfrak{G}$.

For example, $(\mathbb{Z}, 0, +)$ is a cyclic group. Both 1 and -1 are generators. However, $\mathbb{Q}$ and $\mathbb{R}$ paired with addition do not form cyclic groups. As for finite groups, each $\mathbb{Z}_n$ is cyclic, generated by $[1]_n$, but the Klein-4 group (Example 7.1.18) is not cyclic because $a^2 = e$ for all a in the group.

An element a of a group might not generate the entire group, but since $e \in \langle a \rangle$ and both a^n and a^{-n} are elements of $\langle a \rangle$, the set generated by a forms a group using the operation from $\mathfrak{G}$.

■ DEFINITION 7.2.6

A substructure $\mathfrak{H}$ of a group $\mathfrak{G}$ that is a group is called a **subgroup** of $\mathfrak{G}$.

Every group with at least two elements has at least two subgroups, itself (the **improper subgroup**) and the subgroup with domain $\{\epsilon\}$ (the **trivial subgroup**). A group that has at most these two subgroups is called **simple**. For example, $(\mathbb{Z}_2, 0, +)$ and $(\mathbb{Z}_3, 0, +)$ form simple groups, but $(\mathbb{Z}_4, 0, +)$ does not because it has a subgroup with domain $\{[0]_4, [2]_4\}$. Other examples of nonsimple groups are $(\mathbb{R}, 0, +)$ because $(\mathbb{Z}, 0, +)$

is one of its subgroups and $(\langle 2 \rangle, 0, +)$ because $(\langle 6 \rangle, 0, +)$ is one of its subgroups. These subgroups that are not improper are called **proper**.

It is tempting to define a subgroup simply as a substructure of a group, but this would not work if the subgroup is to be a group. For example, viewing ω as a subset of $\mathbb{Z}$ via (5.8) allows $(\omega, 0, +)$ to be a GR-substructure of $(\mathbb{Z}, 0, +)$, but

$$(\omega, 0, +) \models \{\mathbf{G1}, \mathbf{G2}\}$$

yet

$$(\omega, 0, +) \not\models \mathbf{G3}.$$

This example suggests the following.

■ THEOREM 7.2.7

A substructure $\mathfrak{H}$ of a group $\mathfrak{G}$ is a subgroup of $\mathfrak{G}$ if and only if $\mathfrak{H} \models \mathbf{G3}$.

PROOF

Write $\mathfrak{G} = (G, \mathfrak{g})$ and $\mathfrak{H} = (H, \mathfrak{h})$ and let $\mathfrak{H} \subseteq \mathfrak{G}$. If $\mathfrak{H}$ is a subgroup, then $\mathfrak{H} \models \mathbf{G3}$. To prove the converse, assume $\mathfrak{H} \models \mathbf{G3}$.

- Let $x, y, z \in H$. Since $h(\circ) = g(\circ) \restriction (H \times H)$,

$$\begin{aligned}
\mathfrak{h}(\circ)(x, \mathfrak{h}(\circ)(y, z)) &= \mathfrak{g}(\circ)(x, \mathfrak{g}(\circ)(y, z)) \\
&= \mathfrak{g}(\circ)(\mathfrak{g}(\circ)(x, y), z) \\
&= \mathfrak{h}(\circ)(\mathfrak{h}(\circ)(x, y), z).
\end{aligned}$$

The second equality holds because the interpretation of $\circ$ in $\mathfrak{G}$ is associative.

- Let $x \in H$. Because $\mathfrak{h}(e) = \mathfrak{g}(e)$,

$$\mathfrak{h}(\circ)(\mathfrak{h}(e), x) = \mathfrak{g}(\circ)(\mathfrak{g}(e), x) = x$$

and

$$\mathfrak{h}(\circ)(x, \mathfrak{h}(e)) = \mathfrak{g}(\circ)(x, \mathfrak{g}(e)) = x.$$

Therefore, $\mathfrak{H}$ is a group and, thus, a subgroup of $\mathfrak{G}$. ■

The standard way to show that a subset of a group forms a subgroup is not to show directly that the set satisfies the three group axioms or to appeal to Theorem 7.2.7. Instead, what is typically done in algebra is to check that the conditions of the next theorem are satisfied by the set.

■ THEOREM 7.2.8

If $\mathfrak{G} = (G, \epsilon, *)$ is a group and $H \subseteq G$, there exists a subgroup of $\mathfrak{G}$ with domain H if

- H is closed under $*$,
- $\epsilon \in H$,
- $a^{-1} \in H$ for all $a \in H$.

PROOF

Suppose that the three hypotheses of the theorem hold.

- Let $a, b \in H$. By the first hypothesis, $a * b \in H$. Therefore, $* \restriction (H \times H)$ is a binary operation on H.

- Since $*$ is associative on G, the restriction of $*$ to H must be associative.

- The second hypothesis gives H an identity element.

- Every element of H has its inverse in H by the third hypothesis.

Therefore, $(H, \epsilon, * \restriction [H \times H])$ is a group. Since $H \subseteq \mathrm{dom}(\mathfrak{G})$, we conclude that $(H, \epsilon, * \restriction [H \times H])$ is a subgroup of $\mathfrak{G}$. ∎

■ EXAMPLE 7.2.9

To illustrate the theorem, take a group $\mathfrak{G} = (G, \epsilon, *)$ and a family of subgroups $(H_i, \epsilon, * \restriction [H_i \times H_i])$ for all $i \in I$. Although the union of subgroups might not be a subgroup (Exercise 9), we can show that

$$\left(\bigcap_{i \in I} H_i, \epsilon, * \restriction \left[\bigcap_{i \in I} H_i \times \bigcap_{i \in I} H_i \right] \right)$$

is a subgroup of $\mathfrak{G}$.

- By Exercise 3.4.22(b), $\bigcap_{i \in I} H_i \subseteq G$.

- Let $a, b \in \bigcap_{i \in I} H_i$. This means that $a, b \in H_i$ for all $i \in I$. Since each H_i is closed under the operation of $\mathfrak{G}$, $a * b \in H_i$ for all $i \in I$. Hence,

$$a * b \in \bigcap_{i \in I} H_i.$$

- Since $\epsilon \in H_i$ for every $i \in I$, we must have $\epsilon \in \bigcap_{i \in I} H_i$.

- Take a to be an element of $\bigcap_{i \in I} H_i$. Then, $a^{-1} \in H_i$ for all $i \in I$, so $a^{-1} \in \bigcap_{i \in I} H_i$.

Now we return to cyclic groups.

■ THEOREM 7.2.10

A subgroup of a cyclic group is cyclic.

PROOF

Let $\mathfrak{G} = (G, \epsilon, *)$ be a cyclic group with generator a. Let $\mathfrak{H} = (H, \epsilon, *)$ be a subgroup of $\mathfrak{G}$. If $\mathfrak{H}$ is the trivial subgroup, the subgroup is cyclic with generator ϵ. So suppose that $\mathfrak{H}$ is not the trivial subgroup. Because $\mathfrak{G}$ is cyclic, there exists a least natural number $n > 0$ such that $a^n \in H$ (Theorem 5.2.13). Suppose that a^n is not a generator of $\mathfrak{H}$. This means that there exists $m \in \omega$ with $m > n$ such that $a^m \in H$ but $a^m \notin \langle a^n \rangle$. This combined with the division algorithm (Theorem 4.3.31) yields unique natural numbers q and r such that $m = nq + r$ and $0 < r < n$. Therefore,

$$a^m = a^{nq+r} = a^{nq} * a^r,$$

and from this, we conclude that

$$a^r = a^{-nq} * a^m.$$

Since $a^{-nq}, a^m \in H$, a^r is an element of H. This contradicts the minimality of n because $r < n$. Thus, a^n is a generator of $\mathfrak{H}$. ∎

Subrings

Some of the examples of rings had domains that were subsets of other rings. For example, $n\mathbb{Z}$ is a subset of $\mathbb{Z}$, and $\mathbb{Q}$ is a subset of $\mathbb{R}$. Generalizing leads to the next definition.

■ DEFINITION 7.2.11

A substructure $\mathfrak{S}$ of a ring $\mathfrak{R}$ that is a ring is called a **subring** of $\mathfrak{R}$.

A subring of $\mathfrak{R}$ such that its domain is a proper subset of the domain of $\mathfrak{R}$ is called a **proper subring**. The ring itself is called the **improper subring**. The subring with domain $\{0\}$ is the **trivial subring**.

■ EXAMPLE 7.2.12

- $(\{[0]_9, [3]_9, [6]_9\}, [0]_9, +, \cdot)$ is a subring of $(\mathbb{Z}_9, [0]_9, +, \cdot)$.

- $(\mathbb{Z}, 0, +, \cdot)$ is a subring of $(\mathbb{R}, 0, +, \cdot)$.

- $(M_2(\mathbb{R}), 0_2, +, \cdot)$ is a subring of $(M_2(\mathbb{C}), 0_2, +, \cdot)$.

As with subgroups, a substructure of a ring is not necessarily a subring, but we do have results similar to those for groups found in Theorems 7.2.8 and 7.2.7. They are stated without proof since they follow quickly from Definition 7.2.11.

■ THEOREM 7.2.13

A substructure $\mathfrak{S}$ of a ring $\mathfrak{R}$ is a subring of $\mathfrak{R}$ if and only if $\mathfrak{S} \vDash$ **R4**.

We follow the convention that if $\mathfrak{R}$ represents an arbitrary ring, $\mathfrak{R} = (R, 0, +, \cdot)$ and if $\mathfrak{R}'$ also represents an arbitrary ring, $\mathfrak{R}' = (R', 0', +', \cdot')$. This will help us with our notation.

■ **THEOREM 7.2.14**

If $\mathfrak{R}$ is a ring and $S \subseteq R$, there exists a subring of $\mathfrak{R}$ with domain S if

- S is closed under $+$ and $\cdot$,

- $0 \in S$,

- $-a \in S$ for all $a \in R$.

The subring found while proving Theorem 7.2.14 is $(S, 0, + \restriction [S \times S], \cdot \restriction [S \times S])$.

■ **EXAMPLE 7.2.15**

We use Theorem 7.2.14 to show that $\mathfrak{S} = (S, \mathbf{0}_2, + \restriction [S \times S], \cdot \restriction [S \times S])$ is a subring of $\mathfrak{M}_2(\mathbb{R})$ (Example 7.1.39), where

$$S = \left\{ \begin{bmatrix} a & 0 \\ 0 & b \end{bmatrix} : a, b \in \mathbb{R} \right\}.$$

- Let $a, b, a', b' \in \mathbb{R}$, and assume that

$$A = \begin{bmatrix} a & 0 \\ 0 & b \end{bmatrix} \text{ and } B = \begin{bmatrix} a' & 0 \\ 0 & b' \end{bmatrix}.$$

Then,

$$A + B = \begin{bmatrix} a + a' & 0 \\ 0 & b + b' \end{bmatrix}$$

and

$$AB = \begin{bmatrix} aa' & 0 \\ 0 & bb' \end{bmatrix}.$$

These are elements of S.

- Clearly, the zero matrix is in S. (Let $a = b = 0$.)

- Take $a, b \in \mathbb{R}$ and write

$$A = \begin{bmatrix} a & 0 \\ 0 & b \end{bmatrix}.$$

Hence,

$$-A = \begin{bmatrix} -a & 0 \\ 0 & -b \end{bmatrix}$$

is an element of S.

■ **EXAMPLE 7.2.16**

Let $\mathfrak{S}$ and $\mathfrak{T}$ be subrings of a ring $\mathfrak{R}$. Let S be the domain of $\mathfrak{S}$ and T be the domain of $\mathfrak{T}$. Check the conditions of Theorem 7.2.14 to show that there exists a subring of $\mathfrak{R}$ with domain $S \cap T$.

- To prove closure, let $x, y \in S \cap T$. This means that $x+y \in S$ and $x+y \in T$. Hence, $x + y \in S \cap T$. Similarly, $xy \in S \cap T$.

- Since $0 \in S$ and $0 \in T$, $0 \in S \cap T$.

- Suppose $x \in S \cap T$. Then $x \in S$ and $x \in T$. Since these are subrings, $-x \in S$ and $-x \in T$. Thus, $-x \in S \cap T$.

Ideals

The subring $\mathfrak{S}$ of the ring $\mathfrak{R}$ in Example 7.2.15 lacks a property that is often desirable to have in a subring. Observe that

$$\begin{bmatrix} 1 & 0 \\ 0 & 1 \end{bmatrix} \cdot \begin{bmatrix} 1 & 2 \\ 3 & 4 \end{bmatrix} = \begin{bmatrix} 1 & 2 \\ 3 & 4 \end{bmatrix} \notin S,$$

so in general it is false that $AB \in S$ and $BA \in S$ for all $A \in S$ and $B \in M_2(\mathbb{R})$.

■ **DEFINITION 7.2.17**

Let $\mathfrak{R}$ be a ring with domain R and $\mathfrak{I}$ be a subring of $\mathfrak{R}$ with domain I.

- If $ra \in I$ and $ar \in I$ for all $r \in R$ and $a \in I$, then $\mathfrak{I}$ is an **ideal** of $\mathfrak{R}$.

- If $ra \in I$ for all $r \in R$ and $a \in I$, then $\mathfrak{I}$ is a **left ideal**.

- If $ar \in I$ for all $r \in R$ and $a \in I$, then $\mathfrak{I}$ is a **right ideal**.

A ring $\mathfrak{R}$ is an ideal of itself, called the **improper** ideal of $\mathfrak{R}$. All other ideals of $\mathfrak{R}$ are **proper**, including the ideal formed by $\{0\}$. Furthermore, in a commutative ring, there is no difference between a left and right ideal. However, if the ring is not commutative, a left ideal might not be a right ideal.

■ **EXAMPLE 7.2.18**

Define

$$I = \left\{ \begin{bmatrix} a & 0 \\ b & 0 \end{bmatrix} : a, b \in \mathbb{R} \right\}.$$

Using matrix multiplication,

$$\begin{bmatrix} x & y \\ z & w \end{bmatrix} \cdot \begin{bmatrix} a & 0 \\ b & 0 \end{bmatrix} = \begin{bmatrix} xa + yb & 0 \\ za + wb & 0 \end{bmatrix} \in I,$$

but

$$\begin{bmatrix} 1 & 0 \\ 1 & 0 \end{bmatrix} \cdot \begin{bmatrix} 1 & 1 \\ 1 & 1 \end{bmatrix} = \begin{bmatrix} 1 & 1 \\ 1 & 1 \end{bmatrix} \notin I.$$

Therefore, the subring $(I, 0_2, + \upharpoonright [I \times I], \cdot \upharpoonright [I \times I])$ of $\mathfrak{M}_2(\mathbb{R})$ is a left ideal but not a right ideal.

■ EXAMPLE 7.2.19

Let $I = \{[0]_4, [2]_4\}$. Then, $\mathfrak{I} = (I, [0]_4, + \upharpoonright [I \times I], \cdot \upharpoonright [I \times I])$ is a subring of the ring $\mathfrak{R} = (\mathbb{Z}_4[0]_4, +, \cdot)$. It also forms an ideal. To see this, check the calculations:

$[0]_4 \cdot [0]_4 = [0]_4$	$[0]_4 \cdot [0]_4 = [0]_4,$
$[1]_4 \cdot [0]_4 = [0]_4$	$[0]_4 \cdot [1]_4 = [0]_4,$
$[2]_4 \cdot [0]_4 = [0]_4$	$[0]_4 \cdot [2]_4 = [0]_4,$
$[3]_4 \cdot [0]_4 = [0]_4$	$[0]_4 \cdot [3]_4 = [0]_4,$
$[0]_4 \cdot [2]_4 = [0]_4$	$[2]_4 \cdot [0]_4 = [0]_4,$
$[1]_4 \cdot [2]_4 = [2]_4$	$[2]_4 \cdot [1]_4 = [2]_4,$
$[2]_4 \cdot [2]_4 = [0]_4$	$[2]_4 \cdot [2]_4 = [0]_4,$
$[3]_4 \cdot [2]_4 = [2]_4$	$[2]_4 \cdot [3]_4 = [2]_4.$

When we multiply any element of $\mathbb{Z}_4$ by an element of I on either side, the result is an element of I.

■ EXAMPLE 7.2.20

Let $\mathfrak{R}$ be a ring. It is left to Exercise 18 to show that the ring

$$\mathfrak{S} = (R \times \{0\}, (0,0), +', \cdot')$$

is a subring of

$$\mathfrak{T} = (R \times R, (0,0), +, \cdot)$$

with $+$ and $\cdot$ defined coordinatewise and $+'$ and $\cdot'$ being the restrictions of $+$ and $\cdot$ to $R \times \{0\}$. To show that $\mathfrak{S}$ is an ideal of $\mathfrak{T}$, let $(r, s) \in R \times R$ and $(a, 0) \in R \times \{0\}$. We then calculate

$$(r, s) \cdot (a, 0) = (ra, 0) \in R \times \{0\}$$

and

$$(a, 0) \cdot (r, s) = (ar, 0) \in R \times \{0\}.$$

■ EXAMPLE 7.2.21

Let $\mathfrak{R}$ be a ring. Let S and T be subsets of R such that

$$\mathfrak{S} = (S, 0, + \upharpoonright [S \times S], \cdot \upharpoonright [S \times S])$$

and
$$\mathfrak{T} = (T, 0, + \upharpoonright [T \times T], \cdot \upharpoonright [T \times T])$$

are ideals of $\mathfrak{R}$. Define
$$S + T = \{s + t : s \in S \text{ and } t \in T\}.$$

We prove that $S + T$ is the domain of an ideal of $\mathfrak{R}$.

- Take $x, y \in S + T$. This means that $x = s + t$ and $y = s' + t'$ for some $s, s' \in S$ and $t, t' \in T$. Then,

$$
\begin{aligned}
x + y &= (s + t) + (s' + t') \\
&= s + (t + s') + t' \\
&= s + (s' + t) + t' \\
&= (s + s') + (t + t').
\end{aligned}
$$

Thus, $x + y \in S + T$ since $s + s' \in S$ and $t + t' \in T$. Also,
$$xy = (s + t)(s' + t') = (s + t)s' + (s + t)t'.$$

Since $s + t \in R$ and $\mathfrak{S}$ is an ideal, $(s + t)s' \in S$. Likewise, $(s + t)t'$ is an element of T. Hence, $xy \in S + T$.

- We know that $0 \in S + T$ because $0 = 0 + 0$ and $0 \in S$ and $0 \in T$.

- Let $s \in S$ and $t \in T$. Then, since $-s \in S$ and $-t \in t$,
$$-(s + t) = -s + -t \in S + T.$$

- Let $r \in R$, $s \in S$, and $t \in T$. Since $rs \in S$ and $rt \in T$,
$$r(s + t) = rs + rt \in S + T,$$

and since $sr \in S$ and $tr \in T$,
$$(s + t)r = sr + tr \in S + T.$$

Cyclic subgroups (Definition 7.2.5) are generated by a single element. The corresponding notion in rings is the following definition.

■ DEFINITION 7.2.22

Let $\mathfrak{R}$ be a ring. For every $a \in R$, define
$$\langle a \rangle = \{ra : r \in R\}.$$

If $\mathfrak{I}$ is a left ideal of $\mathfrak{R}$ such that $\mathrm{dom}(\mathfrak{I}) = \langle a \rangle$ for some $a \in R$, then $\mathfrak{I}$ is called a **principal ideal left ideal** and a is a **generator** of $\mathfrak{I}$. If $\mathfrak{R}$ is commutative, $\mathfrak{I}$ is an ideal of $\mathfrak{R}$ called a **principal ideal**.

The set $n\mathbb{Z}$ is the domain of an ideal of the ring $(\mathbb{Z}, 0, +, \cdot)$. It is a principal ideal because $n\mathbb{Z} = \langle n \rangle$. (See Exercises 16 and 21.) In fact, every element of a ring generates a left ideal of the ring.

■ **THEOREM 7.2.23**

For every ring $\mathfrak{R}$ and $a \in R$, the ring $\mathfrak{I} = (\langle a \rangle, 0, + \upharpoonright [\langle a \rangle \times \langle a \rangle], \cdot \upharpoonright [\langle a \rangle \times \langle a \rangle])$ is a left ideal of $\mathfrak{R}$.

PROOF
First, show that $\mathfrak{I}$ is a subring.

- Let $r, s \in R$. Then, because $r + s$ and $(ra)s$ are elements of dom(R),

$$ra + sa = (r + s)a \in \langle a \rangle$$

and

$$(ra)(sa) = [(ra)s] a \in \langle a \rangle.$$

- Since $0a = 0$ [Exercise 7.1.22], we have that $0 \in \langle a \rangle$.

- If $r \in R$, then $(-r)a \in \langle a \rangle$ and $-ra + ra = (-r + r)a = 0a = 0$.

To prove that $\mathfrak{I}$ is a left ideal of $\mathfrak{R}$, take $r, s \in R$. Then, $r(sa) \in \langle a \rangle$ because $rs \in R$ and $r(sa) = (rs)a$. ■

Notice that a principal left ideal might not be a two-sided ideal. Example 7.2.18 combined with the next example illustrates this fact.

■ **EXAMPLE 7.2.24**

Let $\mathfrak{I} = (I, 0_2, + \upharpoonright [I \times I], \cdot \upharpoonright [I \times I])$ be a left ideal of $\mathfrak{M}_2(\mathbb{R})$, where

$$I = \left\{ \begin{bmatrix} a & 0 \\ b & 0 \end{bmatrix} : a, b \in \mathbb{R} \right\}.$$

By Theorem 7.2.23, it is a principal left ideal of $\mathfrak{M}_2(\mathbb{R})$ because

$$I = \left\langle \begin{bmatrix} 1 & 0 \\ 0 & 0 \end{bmatrix} \right\rangle.$$

To prove this, it suffices to take $a, b \in \mathbb{R}$ and observe that

$$\begin{bmatrix} a & 0 \\ b & 0 \end{bmatrix} = \begin{bmatrix} a & 0 \\ b & 0 \end{bmatrix} \cdot \begin{bmatrix} 1 & 0 \\ 0 & 0 \end{bmatrix}$$

implies

$$\begin{bmatrix} a & 0 \\ b & 0 \end{bmatrix} \in \left\langle \begin{bmatrix} 1 & 0 \\ 0 & 0 \end{bmatrix} \right\rangle.$$

■ **EXAMPLE 7.2.25**

Every ideal in the ring $3 = (\mathbb{Z}, 0, +, \cdot)$ is principal. To see this, let $\mathfrak{I}$ be an ideal of 3. We must find an element of $\mathbb{Z}$ that generates $I = \mathrm{dom}(\mathfrak{I})$. We have two cases to consider:

- If $I = \{0\}$, then $I = \langle 0 \rangle$.

- Suppose $I \neq \{0\}$. This means that $I \cap \mathbb{Z}^+ \neq \varnothing$. By Theorem 5.2.13 and Exercise 5.3.1, I must contain a minimal positive integer. Call it m. We claim that $I = \langle m \rangle$. It is clear that $\langle m \rangle \subseteq I$, so take $a \in I$ and divide it by m. The division algorithm (Theorem 4.3.31) gives $q, r \in \omega$ so that

$$a = mq + r$$

  with $0 \leq r < m$. Then, $r \in I$ because $r = a - mq$ and $a, mq \in I$. If $r > 0$, then we have a contradiction of the fact that m is the smallest positive integer in I. Hence, $r = 0$ and $a = mq$. This means that $a \in \langle m \rangle$.

3 is an example of a **principal ideal domain**, an integral domain in which every ideal is principal.

Exercises

1. Let R be a binary relation symbol. Define the $\{R\}$-structures $\mathfrak{A} = (\mathbb{Q}, <_\mathbb{Q})$ and $\mathfrak{B} = (\mathbb{R}, <_\mathbb{R})$, where $<_\mathbb{Q}$ refers to standard less-than on $\mathbb{Q}$ and $<_\mathbb{R}$ refers to standard less-than on $\mathbb{R}$. Prove that $\mathfrak{A}$ is a substructure of $\mathfrak{B}$.

2. Let $\mathfrak{A} = (A, \mathfrak{a})$, $\mathfrak{B} = (B, \mathfrak{b})$, and $\mathfrak{C} = (C, \mathfrak{c})$ be S-structures such that $\mathfrak{B} \subseteq \mathfrak{A}$ and $\mathfrak{C} \subseteq \mathfrak{A}$. Define $\mathfrak{D} = (D, \mathfrak{d})$ such that $D = B \cap C$, $\mathfrak{d}(c) = \mathfrak{a}(c)$ for all constant symbols $c \in S$, $\mathfrak{d}(R) = \mathfrak{b}(R) \cap \mathfrak{c}(R)$ for all relation symbols $R \in S$, and $\mathfrak{d}(f) = \mathfrak{b}(f) \cap \mathfrak{c}(f)$ for all function symbols $f \in S$. Prove that $\mathfrak{D} \subseteq \mathfrak{A}$.

3. Let κ be a cardinal. Define $\mathfrak{A}_\alpha = (A_\alpha, \mathfrak{a}_\alpha)$ to be an S-structure for all $\alpha \in \kappa$. The family $\{\mathfrak{A}_\gamma : \gamma \in \kappa\}$ is called a **chain** of S-structures if for all $\alpha \in \beta \in \kappa$, we have that $\mathfrak{A}_\alpha \subseteq \mathfrak{A}_\beta$. Define the S-structure $\bigcup_{\gamma \in \kappa} \mathfrak{A}_\gamma = (\bigcup_{\gamma \in \kappa} A_\gamma, \mathfrak{a})$ so that for every relation symbol $R \in S$,

$$\mathfrak{a}(R) = \bigcup_{\gamma \in \kappa} \mathfrak{a}_\alpha(R),$$

and for every function symbol $f \in S$,

$$\mathfrak{a}(f) = \bigcup_{\gamma \in \kappa} \mathfrak{a}_\alpha(f).$$

Prove the following.
  (a) $\{\mathfrak{a}_\gamma(R) : \gamma \in \kappa\}$ is a chain with respect to $\subseteq$ for all relation symbols $R \in S$.
  (b) $\bigcup_{\gamma \in \kappa} \mathfrak{a}_\gamma(f)$ is a function.
  (c) $\mathfrak{A}_\alpha \subseteq \bigcup_{\gamma \in \kappa} \mathfrak{A}_\gamma$ for all $\alpha \in \kappa$.

4. Let $S \subseteq T$ be sets of subject symbols. Let $\mathfrak{A}$ and $\mathfrak{B}$ be T-structures. Prove that if $\mathfrak{A} \subseteq \mathfrak{B}$, then the reduct of $\mathfrak{A}$ to S is a substructure of the reduct of $\mathfrak{B}$ to S.

5. Find S-substructures of the given S-structures. If possible, find a S-sentence that is true in the given structure but not true in the substructure.

    (a) $(\mathbb{Z} \cup P(\mathbb{Z}), \in)$, $S = ST$

    (b) $(\omega, \varnothing, \{\varnothing\}, +, \cdot)$, $S = NT$

    (c) $(C, 0, +)$, $S = GR$

    (d) $(\mathbb{R}, 0, +, \cdot, <)$, $S = OF$

6. Let $\mathfrak{G}$ be a group with domain G. Find all subgroups of $\mathfrak{G}$ given the following and assuming the standard operations for each set.

    (a) G is the Klein-4 group

    (b) $G = \mathbb{Z}_5$

    (c) $G = \mathbb{Z}_8$

    (d) $G = \mathbb{Z}_2 \times \mathbb{Z}_3$

    (e) $G = \mathbb{Z}_2 \times \mathbb{Z}_6$

7. Show that $\{f \in {}^{\mathbb{R}}\mathbb{R} : f(0) = 0\}$ is the domain of a subgroup of the group $({}^{\mathbb{R}}\mathbb{R}, 0, +)$.

8. Define $H = \{A \in M_n(\mathbb{R}) : a_{1,1} + a_{2,2} + \cdots + a_{n,n} = 0\}$. Prove that $(H, 0_n, +)$ is a subgroup of the group $(M_n(\mathbb{R}), 0_n, +)$.

9. Let $\{\mathfrak{H}_i : i \in I\}$ be a family of subgroups of the group $\mathfrak{G}$. Give an example to show that $\bigcup_{i \in I} \mathfrak{H}_i$ is not necessarily the domain of a subgroup of $\mathfrak{G}$.

10. Let $(H, \epsilon, * \restriction [H \times H])$ and $(K, \epsilon, * \restriction [K \times K])$ be two subgroups of an abelian group $\mathfrak{G} = (G, \epsilon, *)$. Define

$$HK = \{a * b : a \in H \wedge b \in K\}.$$

Prove that $(HK, \epsilon, * \restriction [HK \times HK])$ is a subgroup of $\mathfrak{G}$.

11. For any group $\mathfrak{G} = (G, \epsilon, *)$, let $S \subseteq G$ and define

$$H = \{a \in G : a * b = b * a \text{ for all } b \in S\}.$$

Prove that $(H, \epsilon, * \restriction [H \times H])$ is a subgroup of $\mathfrak{G}$.

12. Demonstrate that simple groups are cyclic.

13. Prove Theorem 7.2.14.

14. Prove that the subrings of a ring with unity are rings with unity.

15. Let $\mathfrak{R}$ be a ring with domain R. Find all ideals of $\mathfrak{R}$ given the following and assuming the standard operations for each set.

    (a) $R = \mathbb{Z}_2$

    (b) $R = \mathbb{Z}_6$

    (c) $R = \mathbb{Z}_7$

(d) $R = \mathbb{Z}_{12}$

16. Let $n \in \mathbb{Z}$. Prove that $n\mathbb{Z}$ is the domain of an ideal of the ring $(\mathbb{Z}, 0, +, \cdot)$.

17. Let $\mathfrak{R} = (\mathbb{Z} \times \mathbb{Z}, (0, 0), +, \cdot)$.
    (a) Prove that $\{(2m, 2n) : m, n \in \mathbb{Z}\}$ is the domain of an ideal of $\mathfrak{R}$.
    (b) Prove that $\{(2n, 2n) : n \in \mathbb{Z}\}$ is not the domain of an ideal of $\mathfrak{R}$.
    (c) Prove that $\mathbb{Z} \times 3\mathbb{Z}$ is the domain of a principal ideal of $\mathfrak{R}$.
    (d) Is $\{(2m, 2n) : m, n \in \mathbb{Z}\}$ the domain of a principal ideal of $\mathfrak{R}$?

18. Prove the $\mathfrak{S}$ is a subring of $\mathfrak{T}$ in Example 7.2.20.

19. Let $\mathfrak{R}$ be a ring with unity and $\mathfrak{I}$ an ideal of $\mathfrak{R}$. Let R be the domain of $\mathfrak{R}$ and I be the domain of $\mathfrak{I}$. Prove that if u is a unit and $u \in I$, then $I = R$.

20. Prove that a field has no proper, nontrivial ideals.

21. Prove that $n\mathbb{Z}$ is the domain of a principal ideal of the ring $(\mathbb{Z}, 0, +, \cdot)$ for any integer n.

22. Show that for any ideal $\mathfrak{I}$, if $a \in \mathrm{dom}(\mathfrak{I})$, then $\langle a \rangle \subseteq \mathrm{dom}(\mathfrak{I})$.

23. Take a ring $\mathfrak{R}$ and let $a \in \mathrm{dom}(\mathfrak{R})$. Show $\langle a \rangle$ is the domain of a left ideal of $\mathfrak{R}$ but not necessarily a right ideal if $\mathfrak{R}$ is not commutative.

24. Let u be a unit of a ring $\mathfrak{R}$. Show for all $a \in \mathrm{dom}(\mathfrak{R})$, $\langle a \rangle = \langle ua \rangle$.

25. An ideal $\mathfrak{P} = (P, 0, +, \cdot)$ of a commutative ring $\mathfrak{R} = (R, 0, +, \cdot)$ is **prime** means for all $a, b \in R$, if $ab \in P$, then $a \in P$ or $b \in P$. Let $p \in \mathbb{Z}^+$ be a prime number (Example 2.4.18). Prove that $(p\mathbb{Z}, 0, + \restriction p\mathbb{Z}, \cdot \restriction p\mathbb{Z})$ is a prime ideal of $(\mathbb{Z}, 0, +, \cdot)$.

26. Prove that the trivial subring of an integral domain is a prime ideal.

27. Let $\mathfrak{R}$ be a commutative ring and $\mathfrak{M}$ a proper ideal of $\mathfrak{R}$. If no proper ideal of $\mathfrak{R}$ has $\mathfrak{M}$ as a proper ideal, $\mathfrak{M}$ is a **maximal ideal** of $\mathfrak{R}$. Assume that p is a prime number. Prove that $p\mathbb{Z}$ is the domain of a maximal ideal of $(\mathbb{Z}, 0, +, \cdot)$.

28. Prove that the trivial ideal is the maximal ideal of a field.

29. Let $p \in \mathbb{Z}$ be prime. Prove that $\{(pa, b) : a, b \in \mathbb{Z}\}$ is the domain of a maximal ideal of the ring with domain $\mathbb{Z} \times \mathbb{Z}$ (Example 7.2.20).

30. Use Zorn's lemma (Theorem 5.1.13) to prove that every commutative ring with unity has a maximal ideal.

7.3 HOMOMORPHISMS

The function $f : [0, \infty) \to (-\infty, 0]$ defined by $f(x) = -x$ preserves $\leq$ with $\geq$ (Example 4.5.25). Define the $\{\preccurlyeq\}$-structures $\mathfrak{A} = ([0, \infty), \mathfrak{a})$ and $\mathfrak{B} = ((-\infty, 0], \mathfrak{b})$, where $\mathfrak{a}(\preccurlyeq) = \leq$ and $\mathfrak{b}(\preccurlyeq) = \geq$. When $x_0, x_1 \in [0, \infty)$, observe that

$$(x_0, x_1) \in \mathfrak{a}(\preccurlyeq) \text{ if and only if } (f(x_0), f(x_1)) \in \mathfrak{b}(\preccurlyeq).$$

because

$$x_0 \leq x_1 \text{ if and only if } -x_0 \geq -x_1.$$

Therefore, we know that $\preceq$ behaves in $\mathfrak{B}$ as it behaves in $\mathfrak{A}$ because of f. We generalize this notion to all S-structures in the next definition.

■ **DEFINITION 7.3.1**

Let $\mathfrak{A} = (A, \mathfrak{a})$ and $\mathfrak{B} = (B, \mathfrak{b})$ be S-structures. A function $\varphi : A \rightarrow B$ is a **homomorphism** $\mathfrak{A} \rightarrow \mathfrak{B}$ if it **preserves the structure** of $\mathfrak{A}$ in $\mathfrak{B}$. This means that φ satisfies the following conditions.

- $\varphi(\mathfrak{a}(c)) = \mathfrak{b}(c)$ for all constant symbols $c \in$ S.

- If R is an n-ary relation symbol in S, then for all $a_0, a_1, \dots, a_{n-1} \in A$,

$$(a_0, a_1, \dots, a_{n-1}) \in \mathfrak{a}(R)$$

  if and only if

$$(\varphi(a_0), \varphi(a_1), \dots, \varphi(a_{n-1})) \in \mathfrak{b}(R).$$

- If f is an n-ary function symbol in S, for all $a_0, a_1, \dots, a_{n-1} \in A$,

$$\varphi(\mathfrak{a}(f)(a_0, a_1, \dots, a_{n-1})) = \mathfrak{b}(f)(\varphi(a_0), \varphi(a_1), \dots, \varphi(a_{n-1})).$$

When $\varphi : A \rightarrow B$ is a homomorphism, write $\varphi : \mathfrak{A} \rightarrow \mathfrak{B}$.

There are many important examples of homomorphisms that can be found in algebra. For instance, let $\mathfrak{R}$ and $\mathfrak{R}'$ be rings and $\varphi : \mathfrak{R} \rightarrow \mathfrak{R}'$ be a homomorphism. By Definition 7.3.1, this means that

$$\varphi(0) = 0'.$$

Because $\oplus$ and $\otimes$ are the only function symbols in RI, for all $a, b \in R$,

$$\varphi(a + b) = \varphi(a) +' \varphi(b)$$

and

$$\varphi(a \cdot b) = \varphi(a) \cdot' \varphi(b).$$

This together with a similar analysis of homomorphisms between groups motivates the next definition.

■ **DEFINITION 7.3.2**

- A **group homomorphism** is a homomorphism $\mathfrak{G} \rightarrow \mathfrak{G}'$, where $\mathfrak{G}$ and $\mathfrak{G}'$ are groups.

- A **ring homomorphism** is an homomorphism $\mathfrak{R} \rightarrow \mathfrak{R}'$, where $\mathfrak{R}$ and $\mathfrak{R}'$ are rings.

Throughout this section the focus will be on ring homomorphisms.

■ **EXAMPLE 7.3.3**

Let $\mathfrak{R}$ and $\mathfrak{R}'$ be rings. The function $\psi : R \to R'$ so that $\psi(r) = 0'$ for all $r \in R$ is a ring homomorphism. This function is called a **zero map**.

■ **EXAMPLE 7.3.4**

Define the function $\varphi : \mathbb{Z} \to \mathbb{Z} \times \mathbb{Z}$ by $\varphi(n) = (n, 0)$. Assume that $+$ and $\cdot$ are the standard operations on $\mathbb{Z}$ while $+'$ and $\cdot'$ are the coordinatewise operations on $\mathbb{Z} \times \mathbb{Z}$. Let $m, n \in \mathbb{Z}$. Then,

- $\varphi(0) = (0, 0)$

- $\varphi(m + n) = (m + n, 0) = (m, 0) +' (n, 0) = \varphi(m) +' \varphi(n)$

- $\varphi(mn) = (mn, 0) = (m, 0) \cdot' (n, 0) = \varphi(m) \cdot' \varphi(n).$

Therefore, φ is a ring homomorphism $(\mathbb{Z}, 0, +, \cdot) \to (\mathbb{Z} \times \mathbb{Z}, (0, 0), +', \cdot')$.

The homomorphism of Example 7.3.4 provides a good opportunity to clarify what a ring homomorphism does. Take the integers 1 and 4. Adding them together in $\mathbb{Z}$ yields 5. The images of these integers under φ are $\varphi(1) = (1, 0)$, $\varphi(4) = (4, 0)$, and $\varphi(5) = (5, 0)$. Observe that $(1, 0) + (4, 0) = (5, 0)$ in $\mathbb{Z} \times \mathbb{Z}$, illustrating that the ring homomorphism φ preserves the addition structure of $\mathbb{Z}$ in $\mathbb{Z} \times \mathbb{Z}$. That is, with respect to addition, both sets behave the same way. Multiplication also has the property. For example, when 3 is multiplied with 6 the result is 18 in $\mathbb{Z}$, and their images yield $\varphi(3)\varphi(6) = \varphi(18)$ in $\mathbb{Z} \times \mathbb{Z}$. This is illustrated in Figure 7.2.

Although ring homomorphisms will always preserve the additive identity by definition, this is not a condition that needs to be checked.

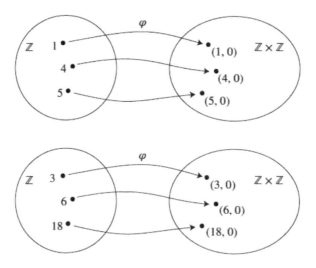

Figure 7.2   The ring homomorphism $\varphi : \mathbb{Z} \to \mathbb{Z} \times \mathbb{Z}$ defined by $\varphi(n) = (n, 0)$.

■ **THEOREM 7.3.5**

Let $\mathfrak{R}$ and $\mathfrak{R}'$ be rings. Then, $\varphi : R \to R'$ is a ring homomorphism if and only if for all $x, y \in R$,

- $\varphi(x + y) = \varphi(x) +' \varphi(y)$, and

- $\varphi(x \cdot y) = \varphi(x) \cdot' \varphi(y)$.

PROOF
Since sufficiency is clear, assume that φ preserves addition and multiplication. Then,
$$\varphi(0) +' \varphi(0) = \varphi(0 + 0) = \varphi(0).$$
Adding $-\varphi(0)$ to both sides yields $\varphi(0) = 0'$. ■

In addition to preserving operations and the additive identity, ring homomorphisms also preserve inverses. Remember, if $\varphi : R \to R'$ is a function and $a \in R$, then $-a$ is the additive inverse of a in R and $-\varphi(a)$ is the additive inverse of $\varphi(a)$ inside of R'.

■ **THEOREM 7.3.6**

Let $\mathfrak{R}$ and $\mathfrak{R}'$ be rings. If $\varphi : \mathfrak{R} \to \mathfrak{R}'$ is a ring homomorphism, for all $a \in R$, $\varphi(-a) = -\varphi(a)$.

PROOF
If $\varphi : \mathfrak{R} \to \mathfrak{R}'$ is a ring homomorphism, then for all $x \in R$,
$$\varphi(x) +' \varphi(-x) = \varphi(x + -x) = \varphi(0) = 0'. ■$$

■ **EXAMPLE 7.3.7**

Check Theorem 7.3.6 using the function φ as defined in the Example 7.3.4. Since 5 and -5 are additive inverses in $\mathbb{Z}$, we have that
$$(5, 0) + (-5, 0) = (0, 0),$$
so $\varphi(5)$ and $\varphi(-5)$ are additive inverses in $\mathbb{Z} \times \mathbb{Z}$.

Ring homomorphisms also preserve subrings and ideals.

■ **THEOREM 7.3.8**

Let $\mathfrak{R}$ and $\mathfrak{R}'$ be rings and $\varphi : \mathfrak{R} \to \mathfrak{R}'$ be a ring homomorphism. Let $\mathfrak{I}$ be an ideal of $\mathfrak{R}$ with domain I and $\mathfrak{I}'$ be an ideal of $\mathfrak{R}'$ with domain I'.

- If φ is onto, $(\varphi[I], 0', +' \restriction \varphi[I], \cdot' \restriction \varphi[I])$ is an ideal of $\mathfrak{R}'$.

- $\mathfrak{I} = (\varphi^{-1}[I'], 0, + \restriction \varphi^{-1}[I'], \cdot \restriction \varphi^{-1}[I'])$ is an ideal of $\mathfrak{R}$.

PROOF

The first part is left to Exercise 6. To prove the second, first prove that $\mathfrak{J}$ is a subring of $\mathfrak{R}$.

- To show closure, let $x_1, x_2 \in \varphi^{-1}[I']$. This means that $\varphi(x_1)$ and $\varphi(x_2)$ are elements of I'. Hence,

$$\varphi(x_1 + x_2) = \varphi(x_1) +' \varphi(x_2) \in I'$$

  and

$$\varphi(x_1 \cdot x_2) = \varphi(x_1) \cdot' \varphi(x_2) \in I'.$$

  Therefore, $x_1 + x_2, x_1 \cdot x_2 \in \varphi^{-1}[I']$.

- Since $0' \in I'$ and $\varphi(0) = 0'$, it follows that $0 \in \varphi^{-1}[I']$.

- Let $x \in \varphi^{-1}[I']$. Then, $\varphi(x) \in I'$, and we find that

$$\varphi(-x) = -\varphi(x) \in I'.$$

  Thus, $-x \in \varphi^{-1}[I']$.

To see that $\mathfrak{J}$ is an ideal of $\mathfrak{R}$, take $r \in R$ and $a \in \varphi^{-1}[I']$. This means that $\varphi(a) \in I'$. Then,

$$\varphi(ra) = \varphi(r)\varphi(a) \in I'$$

since $\mathfrak{J}'$ is an ideal. Thus, $ra \in \varphi^{-1}[I']$. Similarly, $ar \in \varphi^{-1}[I']$. ∎

■ **EXAMPLE 7.3.9**

The function $\varphi : \mathbb{Z} \to \mathbb{Z}/6\mathbb{Z}$ defined by $\varphi(n) = n + 6\mathbb{Z}$ is an onto homomorphism. (See Exercise 7.) The image of $2\mathbb{Z}$ under this map is

$$\varphi[2\mathbb{Z}] = \{n + 6\mathbb{Z} : n \in 2\mathbb{Z}\} = \{0 + 6\mathbb{Z}, 2 + 6\mathbb{Z}, 4 + 6\mathbb{Z}\}.$$

The pre-image of $I = \{0 + 6\mathbb{Z}, 3 + 6\mathbb{Z}\}$ is

$$\varphi^{-1}[I] = \{n \in \mathbb{Z} : \exists k \in \mathbb{Z} (n = 6k \vee n = 3 + 6k)\} = 3\mathbb{Z}.$$

Notice that both $\varphi[2\mathbb{Z}]$ and $\varphi^{-1}[I]$ are domains of ideals of their respective rings.

Take a ring $\mathfrak{R}$ and let R be its domain such that $|R| \geq 2$. Let 0 be the additive identity of $\mathfrak{R}$. We specify two functions. First, define $\pi : R \times R \times R \to R \times R$ by $\pi(x, y, z) = (x, y)$. This is a function similar to that found in Example 4.5.16. Notice that $\text{ran}(\pi) = R \times R$, which means that π is onto, and

$$A = \{(x, y, z) : \pi(x, y, z) = (0, 0)\} = \{(0, 0, z) : z \in R\}.$$

Because $|A| > 1$, we conclude that π is not one-to-one. Second, define the function $\psi : R \times R \to R \times R \times R$ so that $\psi(x, y) = (x, y, 0)$. Compare ψ with Example 4.5.7.

This function is not onto because $\text{ran}(\psi) = R \times R \times \{0\}$, but it does appear to be one-to-one because

$$B = \{(x, y) : \psi(x, y) = (0, 0, 0)\} = \{(0, 0)\}.$$

The sets A and B that contain the elements of the domain that are mapped to the identity of the range appear important to determining whether a function is one-to-one. For this reason, such sets are named.

■ DEFINITION 7.3.10

Let $\mathfrak{R}$ and $\mathfrak{R}'$ be rings. Let $\varphi : R \to R'$ be a function. The **kernel** of φ is

$$\ker(\varphi) = \{x \in R : \varphi(x) = 0'\}.$$

Notice that Definition 7.3.10 implies that $\ker(\varphi) = \varphi^{-1}[\{0'\}]$. Therefore, $\ker(\varphi)$ is an ideal of $\mathfrak{R}$ (Theorem 7.3.8). Similarly, because $\text{ran}(\varphi) = \varphi[R]$, we conclude that $\text{ran}(\varphi)$ is an ideal of $\mathfrak{R}'$.

■ EXAMPLE 7.3.11

Let $\varphi : \mathbb{Z} \to \mathbb{Z}_5$ defined by $\varphi(n) = [n]_5$.

- To find the kernel, assume $\varphi(n) = [0]_5$. By the assumption, $[n]_5 = [0]_5$. So, $n \in [0]_5$, which means $5 \mid n$. Hence,

$$\ker(\varphi) \subseteq 5\mathbb{Z}.$$

  Since the steps are reversible, $\ker(\varphi) = 5\mathbb{Z}$.

- Because $\text{ran}(\varphi) = \{[n]_5 : n \in \mathbb{Z}\}$ encompasses all congruence classes modulo 5, φ is onto and $\text{ran}(\varphi) = \mathbb{Z}_5$.

We now prove that the kernel does provide a test for whether a ring homomorphism is an injection. Since any ring homomorphism $\varphi : \mathfrak{R} \to \mathfrak{R}'$ maps the additive identity of $\mathfrak{R}$ to the additive identity of $\mathfrak{R}'$, it is always the case that $0 \in \ker(\varphi)$. Therefore, to show that $\ker(\varphi) = \{0\}$, we only need to prove $\ker(\varphi) \subseteq \{0\}$.

■ THEOREM 7.3.12

Let $\mathfrak{R}$ and $\mathfrak{R}'$ be rings such that 0 is the additive identity of $\mathfrak{R}$. Suppose that the function $\varphi : \mathfrak{R} \to \mathfrak{R}'$ is a ring homomorphism. Then, φ is one-to-one if and only if $\ker(\varphi) = \{0\}$.

PROOF

- Suppose that φ is one-to-one. Take $x \in \ker(\varphi)$. This means that $\varphi(x) = 0'$, where $0'$ is the additive identity of $\mathfrak{R}'$. Since φ is an injection, $x = 0$. This implies that $\ker(\varphi) = \{0\}$.

- Now let $\ker(\varphi) = \{0\}$ and assume $\varphi(x_1) = \varphi(x_2)$, where $x_1, x_2 \in \text{dom}(\mathfrak{R})$. We then have $\varphi(x_1) - \varphi(x_2) = 0$. Since φ is a homomorphism, we have

$$\varphi(x_1 - x_2) = \varphi(x_1) - \varphi(x_2)$$

by Theorem 7.3.6. Hence, $x_1 - x_2 \in \ker(\varphi)$, which means $x_1 - x_2 = 0$ by hypothesis. Therefore, $x_1 = x_2$. ■

A similar proof can be used to demonstrate that same result for group homomorphisms.

■ **THEOREM 7.3.13**

Let $\mathfrak{G}$ and $\mathfrak{G}'$ be groups such that ϵ is the additive identity of $\mathfrak{G}$. Suppose that the function $\varphi : \mathfrak{G} \to \mathfrak{G}'$ is a group homomorphism. Then, φ is one-to-one if and only if $\ker(\varphi) = \{\epsilon\}$.

Isomorphisms

Define $\psi : \mathbb{Z}_5 \times \mathbb{Z}_6 \to \mathbb{Z}_6 \times \mathbb{Z}_5$ by $\psi([a]_5, [b]_6) = ([b]_6, [a]_5)$. Let $+$ be coordinatewise addition on $\mathbb{Z}_5 \times \mathbb{Z}_6$ and $+'$ be coordinatewise addition on $\mathbb{Z}_6 \times \mathbb{Z}_5$. We prove that this function preserves addition and is a bijection.

- Let $a, b, c, d \in \mathbb{Z}$. Then,

$$\begin{aligned}
\psi(([a]_5, [b]_6) + ([c]_5, [d]_6)) &= \psi([a]_5 + [c]_5, [b]_6 + [d]_6) \\
&= \psi([a + c]_5, [b + d]_6) \\
&= ([b + d]_6, [a + c]_5) \\
&= ([b]_6 + [d]_6, [a]_5 + [c]_5) \\
&= ([b]_6, [a]_5) +' ([d]_6, [c]_5) \\
&= \psi([a]_5, [b]_6) +' \psi([c]_5, [d]_6).
\end{aligned}$$

- Let $([a]_5, [b]_6) \in \ker(\psi)$. In other words, $([b]_6, [a]_5) = ([0]_6, [0]_5)$. Hence, $5 \mid a$ and $6 \mid b$, which implies that $[a]_5 = [0]_5$ and $[b]_6 = [0]_6$ (Example 4.2.6). Thus, ψ is an injection by Theorem 7.3.13.

- To see that ψ is onto, take $([c]_6, [d]_5) \in \mathbb{Z}_6 \times \mathbb{Z}_5$. Then,

$$\psi([d]_5, [c]_6) = ([c]_6, [d]_5).$$

The function ψ generalizes to the next definition. Note the similarity between this definition and that of an order isomorphism (Definition 4.5.24).

■ **DEFINITION 7.3.14**

Let $\mathfrak{A}$ and $\mathfrak{B}$ be S-structures. An **isomorphism** $\mathfrak{A} \to \mathfrak{B}$ is a homomorphism $\mathfrak{A} \to \mathfrak{B}$ that is a bijection. An isomorphism $\mathfrak{A} \to \mathfrak{A}$ is called an **automorphism**. If there is an isomorphism $\varphi : \mathfrak{A} \to \mathfrak{B}$, the S-structures are **isomorphic** and we write $\mathfrak{A} \cong \mathfrak{B}$.

If $\mathfrak{A} = (\mathbb{Z}_5 \times \mathbb{Z}_6, ([0]_5, [0]_6), +)$ and $\mathfrak{B} = (\mathbb{Z}_6 \times \mathbb{Z}_5, ([0]_6, [0]_5), +')$, then ψ shows that $\mathfrak{A} \cong \mathfrak{B}$.

Like order isomorphisms (Example 4.5.28), there are three basic isomorphism results. The proof of this is left to Exercise 8.

■ **THEOREM 7.3.15**

Let $\mathfrak{A}$, $\mathfrak{B}$, and $\mathfrak{C}$ be S-structures.

- $\mathfrak{A} \cong \mathfrak{A}$.

- If $\mathfrak{A} \cong \mathfrak{B}$, then $\mathfrak{B} \cong \mathfrak{A}$.

- If $\mathfrak{A} \cong \mathfrak{B}$ and $\mathfrak{B} \cong \mathfrak{C}$, then $\mathfrak{A} \cong \mathfrak{C}$.

■ **EXAMPLE 7.3.16**

Let $S = \{0, <, \cdot\}$. Let $\mathfrak{A} = (^2\omega, \mathfrak{a})$, where

$$\mathfrak{a}(0)(0) = 0 \text{ and } \mathfrak{a}(0)(1) = 0,$$

and for all $f, g \in {}^2\omega$,

$$(f, g) \in \mathfrak{a}(<) \text{ if and only if } f(0) < g(0) \vee [f(0) = g(0) \wedge f(1) < g(1)],$$

that is, $<$ is interpreted as a lexicographical order (compare Exercise 4.3.16), and

$$\mathfrak{a}(\cdot) \text{ is function multiplication.}$$

This means that $\mathfrak{a}(\cdot)(f, g)(x) = f(x)g(x)$. Also, let $\mathfrak{B} = (\omega \times \omega \times \{0\}, \mathfrak{b})$, where

$$\mathfrak{b}(0) = (0, 0, 0),$$

and for all $m, n, m', n' \in \omega$,

$$((m, n, 0), (m', n', 0)) \in \mathfrak{b}(<) \text{ if and only if } n < n' \vee [n = n' \wedge m < m'],$$

and

$$\mathfrak{b}(\cdot) \text{ is coordinate-wise multiplication.}$$

That is,

$$\mathfrak{b}(\cdot)((m, n, 0), (m', n', 0)) = (mm', nn', 0).$$

Show that $\mathfrak{A} \cong \mathfrak{B}$ by showing that $\varphi : \mathfrak{A} \to \mathfrak{B}$ defined by

$$\varphi(\psi) = (\psi(1), \psi(0), 0)$$

is an S-isomorphism.

- $\varphi(\mathfrak{a}(0)) = (\mathfrak{a}(0)(1), \mathfrak{a}(0)(0), 0) = (0, 0, 0) = \mathfrak{b}(0)$.

- Let $f, g \in {}^2\omega$. Then,

$$(f, g) \in \mathfrak{a}(<) \Leftrightarrow f(0) < g(0) \vee [f(0) = g(0) \wedge f(1) < g(1)]$$
$$\Leftrightarrow ((f(1), f(0), 0), (g(1), g(0), 0)) \in \mathfrak{b}(<)$$
$$\Leftrightarrow (\varphi(f), \varphi(g)) \in \mathfrak{b}(<).$$

- Let $f, g \in {}^2\omega$. Then,

$$\varphi(\mathfrak{a}(\cdot)(f, g)) = \varphi(\{(0, f(0)g(0)), (1, f(1)g(1))\})$$
$$= (f(1)g(1), f(0)g(0), 0)$$
$$= \mathfrak{b}(\cdot)((f(1), f(0), 0), (g(1), g(0), 0))$$
$$= \mathfrak{b}(\cdot)(\varphi(f), \varphi(g)).$$

Therefore, φ is an homomorphism. That φ is a bijection is Exercise 9. Hence, $\mathfrak{A} \cong \mathfrak{B}$, and by Theorem 7.3.15, we have that $\mathfrak{B} \cong \mathfrak{A}$.

Because there is a bijection between $\mathbb{Z}_5 \times \mathbb{Z}_6$ and $\mathbb{Z}_6 \times \mathbb{Z}_5$, they have the same cardinality. Since the bijection that is typically chosen is also a homomorphism, the algebraic structure of the two rings are the same. Putting this together, we conclude that the two rings "look" the same as rings. The only difference is in the labeling of their elements. To formalize this notion, we make the next definition.

■ DEFINITION 7.3.17

- A group homomorphism that is a bijection is a **group isomorphism**.

- A ring homomorphism that is a bijection is a **ring isomorphism**.

We say that two rings R and R' are **isomorphic** if there is an isomorphism $\psi : R \to R'$. If two rings are isomorphic, write $R \cong R'$. For example, we saw that

$$\mathbb{Z}_5 \times \mathbb{Z}_6 \cong \mathbb{Z}_6 \times \mathbb{Z}_5.$$

The next example illustrates what we mean by "looking" the same.

■ EXAMPLE 7.3.18

Suppose that $\varphi : \mathfrak{R} \to \mathfrak{R}'$ is a ring isomorphism.

- If $\mathfrak{R}$ is an integral domain, $\mathfrak{R}'$ is an integral domain.

- If $\mathfrak{R}$ is a division ring, $\mathfrak{R}'$ is a division ring.

- If $\mathfrak{R}$ is a field, $\mathfrak{R}'$ is a field.

The proofs of the first two are left to Exercise 13. For example, suppose that $\mathfrak{R}$ is a field. We show that $\mathfrak{R}'$ is also a field.

- Let 1 be unity from $\mathfrak{R}$. Then, $\varphi(1)$ is unity from $\mathfrak{R}'$ (Exercise 14).

- Let $y_0, y_1 \in R'$. Since φ is onto, there exists $x_0, x_1 \in R$ so that $\varphi(x_0) = y_0$ and $\varphi(x_1) = y_1$. Hence,

$$
\begin{aligned}
y_0 \cdot' y_1 &= \varphi(x_0) \cdot' \varphi(x_1) \\
&= \varphi(x_0 \cdot x_1) \\
&= \varphi(x_1 \cdot x_0) \\
&= \varphi(x_1) \cdot' \varphi(x_0) \\
&= y_1 \cdot' y_0.
\end{aligned}
$$

- Let $y_0 \in R' \setminus \{0'\}$. There exists $x_0 \in R$ such that $\varphi(x_0) = y_0$. Since φ is one-to-one, $x_0 \neq 0$. Thus, because $\mathfrak{R}$ is a field, there exists $x_1 \in R \setminus \{0\}$ such that $x_0 \cdot x_1 = 1$, so

$$y_0 \cdot' \varphi(x_1) = \varphi(x_0) \cdot' \varphi(x_1) = \varphi(x_0 \cdot x_1) = \varphi(1).$$

The result of the next example is known as the **fundamental homomorphism theorem**.

■ **EXAMPLE 7.3.19**

Let $\mathfrak{R}$ and $\mathfrak{R}'$ be rings. Let $\varphi : \mathfrak{R} \to \mathfrak{R}'$ be a surjective ring homomorphism. Define $\psi : R/\ker(\varphi) \to R'$ by

$$\psi(a + \ker(\varphi)) = \varphi(a).$$

To prove that ψ is well-defined, let $a, b \in R$ such that $a + \ker(\varphi) = b + \ker(\varphi)$. This implies that $a - b \in \ker(\varphi)$, so

$$0 = \varphi(a - b) = \varphi(a) -' \varphi(b).$$

That is, $\varphi(a) = \varphi(b)$. Now to show that ψ is a ring homomorphism.

- Take $a, b \in R$. Then,

$$
\begin{aligned}
\psi(a + b + \ker(\varphi)) &= \varphi(a + b) \\
&= \varphi(a) +' \varphi(b) \\
&= \psi(a + \ker(\varphi)) +' \psi(b + \ker(\varphi)),
\end{aligned}
$$

and

$$
\begin{aligned}
\psi(a \cdot b + \ker(\varphi)) &= \varphi(a \cdot b) \\
&= \varphi(a) \cdot' \varphi(b) \\
&= \psi(a + \ker(\varphi)) \cdot' \psi(b + \ker(\varphi)).
\end{aligned}
$$

- To prove that ψ is onto, take $b \in R'$. Since φ is onto, there exists $a \in R$ such that $\varphi(a) = b$. Therefore,

$$\psi(a + \ker(\varphi)) = \varphi(a) = b.$$

- To prove that ψ is one-to-one, let $a, b \in R$ and assume that $\psi(a) = \psi(b)$. Since ψ is a ring homomorphism,

$$\psi(a - b) = \psi(a) -' \psi(b) = 0.$$

Hence, $a - b \in \ker(\varphi)$, which implies that $a + \ker(\varphi) = b + \ker(\varphi)$. ∎

Elementary Equivalence

Let A be a first-order language with theory symbols S. Suppose that $\mathfrak{A} = (A, \mathfrak{a})$ and $\mathfrak{B} = (B, \mathfrak{b})$ are S-structures and I is an interpretation of $\mathfrak{A}$. Let $\varphi : \mathfrak{A} \to \mathfrak{B}$ be an isomorphism. Define

$$I_\varphi : \text{TERMS(A)} \to B$$

by $I_\varphi(t) = (\varphi \circ I)(t)$ for all $t \in \text{TERMS(A)}$. We confirm that I_φ is an interpretation of $\mathfrak{B}$.

- If x is a variable symbol, $I_\varphi(x) = (\varphi \circ I)(x) \in B$.

- Let c be a constant symbol. Then, $I_\varphi(c) = (\varphi \circ I)(c) = \varphi(\mathfrak{a}(c)) = \mathfrak{b}(c)$.

- Let f be an n-ary function symbol and $t_0, t_1, \ldots, t_{n-1} \in \text{TERMS(A)}$. Then,

$$\begin{aligned}
I_\varphi(f(t_0, t_1, \ldots, t_{n-1})) &= (\varphi \circ I)(f(t_0, t_1, \ldots, t_{n-1})) \\
&= \varphi(\mathfrak{a}(f)(I(t_0), I(t_1), \ldots, I(t_{n-1}))) \\
&= \mathfrak{b}(f)(\varphi(I(t_0)), \varphi(I(t_1)), \ldots, \varphi(I(t_{n-1}))) \\
&= \mathfrak{b}(f)(I_\varphi(t_0), I_\varphi(t_1), \ldots, I_\varphi(t_{n-1})).
\end{aligned}$$

An example will clarify the use of the interpretation I_φ.

■ EXAMPLE 7.3.20

Let $\mathfrak{G}_0 = (\mathbb{Z}, 0, +)$ and $\mathfrak{G}_1 = (2\mathbb{Z}, 0, + \restriction 2\mathbb{Z})$. Assume that I is an interpretation of $\mathfrak{G}_0$. Define the group isomorphism $f : \mathbb{Z} \to 2\mathbb{Z}$ by $f(n) = 2n$. We consider the GR-formula $y \circ x = e$.

$$\begin{aligned}
\mathfrak{A} \vDash \exists x (y \circ x = e) [I] &\Leftrightarrow \mathfrak{A} \vDash (y \circ x = e) [I_x^a] \text{ for some } a \in \mathbb{Z} \\
&\Leftrightarrow I_x^a(y \circ x) = I_x^a(e) \text{ for some } a \in \mathbb{Z} \\
&\Leftrightarrow f(I_x^a(y \circ x)) = f(I_x^a(e)) \text{ for some } a \in \mathbb{Z} \\
&\Leftrightarrow f(I_x^a(y)) + f(I_x^a(x)) = f(I_x^a(e)) \text{ for some } a \in \mathbb{Z} \\
&\Leftrightarrow f(I_x^a(y)) + f(a) = f(I_x^a(e)) \text{ for some } a \in \mathbb{Z} \\
&\Leftrightarrow f(I_x^b(y)) + b = f(I_x^b(e)) \text{ for some } b \in 2\mathbb{Z} \\
&\Leftrightarrow f(I_x^b(y)) + f(I_x^b(x)) = f(I_x^b(e)) \text{ for some } b \in 2\mathbb{Z} \\
&\Leftrightarrow f(I_x^b(y \circ x)) = f(I_x^b(e)) \text{ for some } b \in 2\mathbb{Z} \\
&\Leftrightarrow \mathfrak{B} \vDash y \circ x = e [(I_f)_x^b] \text{ for some } b \in 2\mathbb{Z} \\
&\Leftrightarrow \mathfrak{B} \vDash \exists x (y \circ x = e) [I_f].
\end{aligned}$$

Using f, we see that the interpretation of $\mathfrak{A}$ gives rise to an interpretation of $\mathfrak{B}$. For example, notice that if $I(y) = 3$, then $f(I(y)) = 6$. Hence, in $\mathfrak{A}$,

$$\exists x(y \circ x = e) \text{ is interpreted as } 3 + x = 0 \text{ for some } x \in A,$$

and the witness of $\exists x(y \circ x = e)$ is -3. In $\mathfrak{B}$,

$$\exists x(y \circ x = e) \text{ is interpreted as } 6 + x = 0 \text{ for some } x \in B$$

and the witness of $\exists x(y \circ x = e)$ is -6.

■ LEMMA 7.3.21

Let $\mathfrak{A} = (A, \mathfrak{a})$ and $\mathfrak{B} = (B, \mathfrak{b})$ be S-structures. Assume that $\varphi : \mathfrak{A} \to \mathfrak{B}$ is an isomorphism. Let I be an S-interpretation of $\mathfrak{A}$. Then, $(I_x^a)_\varphi = (I_\varphi)_x^{\varphi(a)}$ for all $a \in A$.

PROOF

We proceed by induction on terms, relying on Definitions 7.1.4 and 7.3.1.

- Let x and y be variable symbols. Then,

$$(I_x^a)_\varphi(x) = \varphi(I_x^a(x)) = \varphi(a) = (I_\varphi)_x^{\varphi(a)}(x),$$

and if $y \neq x$,

$$(I_x^a)_\varphi(y) = \varphi(I_x^a(y)) = \varphi(I(y)) = (I_\varphi)_x^{\varphi(a)}(y).$$

- If c is a constant symbol, then

$$(I_x^a)_\varphi(c) = \varphi(I_x^a(c)) = \varphi(I(c)) = (I_\varphi)_x^{\varphi(a)}(c).$$

- Let f be an n-ary function symbol and $t_0, t_1, \ldots, t_{n-1}$ be S-terms. Take $a \in A$. Then,

$$
\begin{aligned}
(I_x^a)_\varphi & (f(t_0, t_1, \ldots, t_{n-1})) \\
&= \varphi(I_x^a(f(t_0, t_1, \ldots, t_{n-1}))) \\
&= \varphi(\mathfrak{a}(f)(I_x^a(t_0), I_x^a(t_1), \ldots, I_x^a(t_{n-1}))) \\
&= \mathfrak{b}(f)(\varphi(I_x^a(t_0)), \varphi(I_x^a(t_1)), \ldots, \varphi(I_x^a(t_{n-1}))) \\
&= \mathfrak{b}(f)((I_x^a)_\varphi(t_0), (I_x^a)_\varphi(t_1), \ldots, (I_x^a)_\varphi(t_{n-1})) \\
&= \mathfrak{b}(f)(((I_\varphi)_x^{\varphi(a)})(t_0), ((I_\varphi)_x^{\varphi(a)})(t_1), \ldots, ((I_\varphi)_x^{\varphi(a)})(t_{n-1})) \\
&= (I_\varphi)_x^{\varphi(a)}(f(t_0, t_1, \ldots, t_{n-1})).
\end{aligned}
$$

The fifth equality follows by induction. ■

If $\varphi : \mathfrak{A} \to \mathfrak{B}$ is an isomorphism, the structures look the same. Thus, formulas should be interpreted in $\mathfrak{B}$ essentially in the same way that they are interpreted in $\mathfrak{A}$. In other words, given an interpretation I of $\mathfrak{A}$, there should be an interpretation of $\mathfrak{B}$ that is like I. This interpretation is I_φ.

■ **LEMMA 7.3.22**

Let $\mathfrak{A}$ and $\mathfrak{B}$ be S-structures with isomorphism $\varphi : \mathfrak{A} \to \mathfrak{B}$. Assume that I is an S-interpretation of $\mathfrak{A}$. Then, for every S-formula p, $\mathfrak{A} \vDash p\ [I]$ if and only if $\mathfrak{B} \vDash p\ [I_\varphi]$.

PROOF

We proceed by induction on formulas, relying on Theorem 7.1.7.

- Suppose that t_0 and t_1 are S-terms. Since φ is one-to-one,

$$\begin{aligned}
\mathfrak{A} \vDash t_0 = t_1\ [I] &\Leftrightarrow I(t_0) = I(t_1) \\
&\Leftrightarrow \varphi(I(t_0)) = \varphi(I(t_1)) \\
&\Leftrightarrow I_\varphi(t_0) = I_\varphi(t_1) \\
&\Leftrightarrow \mathfrak{B} \vDash t_0 = t_1\ [I_\varphi].
\end{aligned}$$

- Let R be an n-ary relation symbol and $t_0, t_1, \ldots, t_{n-1}$ be S-terms. Since φ is an isomorphism,

$$\begin{aligned}
\mathfrak{A} \vDash R(t_0, t_1, \ldots, t_{n-1})\ [I] \\
\Leftrightarrow (I(t_0), I(t_1), \ldots, I(t_{n-1})) \in I(R) \\
\Leftrightarrow (\varphi(I(t_0)), \varphi(I(t_1)), \ldots, \varphi(I(t_{n-1}))) \in \varphi(I(R)) \\
\Leftrightarrow (I_\varphi(t_0), I_\varphi(t_1), \ldots, I_\varphi(t_{n-1})) \in I_\varphi(R) \\
\Leftrightarrow \mathfrak{B} \vDash R(t_0, t_1, \ldots, t_{n-1})\ [I_\varphi].
\end{aligned}$$

- Assume that q is an S-formula. Then,

$$\mathfrak{A} \vDash \neg q\ [I] \Leftrightarrow \mathfrak{A} \nvDash q\ [I] \Leftrightarrow \mathfrak{B} \nvDash q\ [I_\varphi] \Leftrightarrow \mathfrak{B} \vDash \neg q\ [I_\varphi],$$

where the middle equivalence holds by induction.

- Let q and r be S-formulas. Suppose that $\mathfrak{A} \vDash q \to r\ [I]$. This means that $\mathfrak{A} \vDash q\ [I]$ implies $\mathfrak{A} \vDash r\ [I]$. Assume $\mathfrak{B} \vDash q\ [I_\varphi]$. By induction, we have that $\mathfrak{A} \vDash q\ [I]$, whence $\mathfrak{A} \vDash r\ [I]$. Again, by induction, $\mathfrak{B} \vDash r\ [I_\varphi]$. Therefore, $\mathfrak{B} \vDash q \to r\ [I_\varphi]$. The converse is proved similarly.

- Let q be an S-formula. Then, by Lemma 7.3.21 and since φ is an isomorphism,

$$\begin{aligned}
\mathfrak{A} \vDash \exists x q\ [I] &\Leftrightarrow \mathfrak{A} \vDash q\ [I_x^a] \text{ for some } a \in A \\
&\Leftrightarrow \mathfrak{B} \vDash q\ [(I_x^a)_\varphi] \text{ for some } a \in A \\
&\Leftrightarrow \mathfrak{B} \vDash q\ [(I_\varphi)_x^{\varphi(a)}] \text{ for some } a \in A \\
&\Leftrightarrow \mathfrak{B} \vDash q\ [(I_\varphi)_x^b] \text{ for some } b \in B \\
&\Leftrightarrow \mathfrak{B} \vDash \exists x q\ [I_\varphi]. \ ■
\end{aligned}$$

Let $\mathfrak{G}_1 = (2\mathbb{Z}, 0, +)$ and $\mathfrak{G}_2 = (3\mathbb{Z}, 0, +)$. These isomorphic groups are infinite and cyclic, each with exactly two generators. The group $\mathfrak{G}_1$ is generated by 2 and -2, and $\mathfrak{G}_2$ is generated by 3 and -3. When Lemma 7.3.22 is restricted to sentences, we conclude that isomorphic structures model the same sentences. Hence, the GR-sentences satisfied by $\mathfrak{G}_1$ are exactly those GR-sentences satisfied by $\mathfrak{G}_2$. For example,

$$\mathfrak{G}_1 \vDash \forall x \forall y (x \circ y = y \circ x),$$

and

$$\mathfrak{G}_2 \vDash \forall x \forall y (x \circ y = y \circ x).$$

Also,

$$\mathfrak{G}_1 \vDash \exists x \forall y \exists z (y = x \circ z),$$

and

$$\mathfrak{G}_2 \vDash \exists x \forall y \exists z (y = x \circ z).$$

Therefore, we make the next definition and follow it immediately with the theorem that follows from Lemma 7.3.22 and Theorem 7.1.31.

■ DEFINITION 7.3.23

The S-structures $\mathfrak{A}$ and $\mathfrak{B}$ are **elementary equivalent** (denoted by $\mathfrak{A} \equiv \mathfrak{B}$) if for all S-sentences p, $\mathfrak{A} \vDash p$ if and only if $\mathfrak{B} \vDash p$.

■ THEOREM 7.3.24

For all S-structures $\mathfrak{A}$ and $\mathfrak{B}$, if $\mathfrak{A} \cong \mathfrak{B}$, then $\mathfrak{A} \equiv \mathfrak{B}$.

Theorem 7.3.24 implies that if structures are isomorphic, there is no first-order sentence that can be used to distinguish between the two. That is, there is no sentence p such that $\mathfrak{A} \vDash p$ but $\mathfrak{B} \nvDash p$ when $\mathfrak{A} \cong \mathfrak{B}$. Conversely, if there is a sentence that is satisfied by one structure but not the other, the structures cannot be isomorphic.

■ EXAMPLE 7.3.25

Let $\mathfrak{A}$ and $\mathfrak{B}$ be ST-structures such that $\mathfrak{A} = (\omega^+, \in)$ and $\mathfrak{B} = (\omega, \in)$. Observe that

$$\mathfrak{A} \vDash \exists x \forall y (y \in x \vee y = x)$$

but

$$\mathfrak{B} \nvDash \exists x \forall y (y \in x \vee y = x).$$

Therefore, by Theorem 7.3.24, we have that $\mathfrak{A}$ is not isomorphic to $\mathfrak{B}$.

■ EXAMPLE 7.3.26

Let $\mathfrak{A} = (A, \mathfrak{a})$ and $\mathfrak{B} = (B, \mathfrak{b})$ be $\{R\}$-structures, where R is a binary relation symbol. Let $A = \{a_0, a_1\}$ with $a_0 \neq a_1$. Assume that the structures are not isomorphic yet $\mathfrak{A} \equiv \mathfrak{B}$. Define

$$p := \exists x (x = x),$$

$$q := \forall x \exists y(x \neq y),$$

and

$$r := \forall x \forall y \forall z(x \neq y \wedge x \neq z \rightarrow y = z).$$

Then, $\mathfrak{A} \models p$ because A is nonempty, $\mathfrak{A} \models q$ because A has at least two elements, and $\mathfrak{A} \models r$ because A has at most two elements. Hence,

$$\mathfrak{A} \models p \wedge q \wedge r.$$

Since $\mathfrak{A} \equiv \mathfrak{B}$, we have that $\mathfrak{B} \models p \wedge q \wedge r$, so $|B| = 2$. Write $B = \{b_0, b_1\}$ with $b_0 \neq b_1$. This implies that there are only two bijections $A \rightarrow B$. Call them φ_0 and φ_1, and write

$$\varphi_0 = \{(a_0, b_0), (a_1, b_1)\}$$

and

$$\varphi_1 = \{(a_0, b_1), (a_1, b_0)\}.$$

By assumption, neither of these functions are $\{R\}$-homomorphisms, so they fail to preserve R. Suppose that $(a_0, a_1) \in \mathfrak{a}(R)$ is the particular element that causes the second condition of Definition 7.3.1 to fail. Therefore,

$$\mathfrak{A} \models \exists x \exists y[x \neq y \wedge R(x, y)].$$

Since φ_0 does not preserve R, $(b_0, b_1) \notin \mathfrak{b}(R)$, and since φ_1 does not preserve R, $(b_1, b_0) \notin \mathfrak{b}(R)$. Hence,

$$\mathfrak{B} \not\models \exists x \exists y[x \neq y \wedge R(x, y)],$$

which is impossible because $\mathfrak{A} \equiv \mathfrak{B}$, so φ_0 or φ_1 must be an $\{R\}$-isomorphism. The argument generalizes for all structures with a domain of cardinality 2 (Exercise 17) and then to all structures with a finite domain (Exercise 18). Therefore, the converse of Theorem 7.3.24 holds, provided that the domains of the structures are finite.

Elementary Substructures

Let $\mathfrak{A} = (A, \mathfrak{a})$ and $\mathfrak{B} = (B, \mathfrak{b})$ be S-structures such that $\mathfrak{A} \subseteq \mathfrak{B}$. Let I be an interpretation of $\mathfrak{A}$. By Definition 7.1.4, we have the following:

- For all variable symbols x, $I(x) \in A \subseteq B$.

- For all constant symbols c, $I(c) \in A \subseteq B$.

- For all n-ary function symbols f and S-terms $t_0, t_1, \ldots, t_{n-1}$, because $I(t_k) \in A$ for $k = 0, 1, \ldots, n-1$ and $\mathfrak{a}(f) = \mathfrak{b}(f) \upharpoonright A^n$,

$$I(f(t_0, t_1, \ldots, t_{n-1})) = \mathfrak{a}(f)(I(t_0), I(t_1), \ldots, I(t_{n-1}))$$
$$= \mathfrak{b}(f)(I(t_0), I(t_1), \ldots, I(t_{n-1})).$$

This implies that I is also an interpretation of $\mathfrak{B}$. Moreover, when p is a quantifier-free S-formula, the interpretation of p in $\mathfrak{B}$ requires no element of $B \setminus A$. This implies the next theorem. Its proof is left to Exercise 21.

■ **THEOREM 7.3.27**

Let $\mathfrak{A}$ and $\mathfrak{B}$ be S-structures such that $\mathfrak{A} \subseteq \mathfrak{B}$. For all S-interpretations I of $\mathfrak{A}$ and all quantifier-free S-formulas p, $\mathfrak{A} \models p[I]$ if and only if $\mathfrak{B} \models p[I]$.

Next, consider $\mathfrak{B} \models \exists x p[I]$. This means that $\mathfrak{B} \models p[I_x^b]$ for some $b \in B$. If b is also in A, then $\mathfrak{A} \models p[I_x^b]$, but this conclusion is not guaranteed if $b \in B \setminus A$. For example, let $\mathfrak{A} = (\mathbb{Z}, <)$ and $\mathfrak{B} = (\mathbb{Q}, <)$ be $\{R\}$-structures, where R is a binary relation symbol. Since $\mathbb{Z} \subseteq \mathbb{Q}$, we have that $\mathfrak{A}$ is a substructure of $\mathfrak{B}$. However, because of differences between $\mathbb{Q}$ and $\mathbb{Z}$,

$$\mathfrak{B} \models \forall x_1 \forall x_2 \exists y(x_1 < y < x_2)$$

but

$$\mathfrak{A} \not\models \forall x_1 \forall x_2 \exists y(x_1 < y < x_2).$$

Hence, in order for $\mathfrak{A} \models p[I]$ if and only if $\mathfrak{B} \models p[I]$ for even quantified formulas p, a stronger condition is required.

■ **DEFINITION 7.3.28**

$\mathfrak{A}$ is an **elementary substructure** of $\mathfrak{B}$ (written as $\mathfrak{A} \preceq \mathfrak{B}$) if $\mathfrak{A} \subseteq \mathfrak{B}$ and for all interpretations I of $\mathfrak{A}$ and S-formulas p,

$$\mathfrak{A} \models p[I] \text{ if and only if } \mathfrak{B} \models p[I].$$

If $\mathfrak{A} \preceq \mathfrak{B}$, then $\mathfrak{B}$ is an **elementary extension** of $\mathfrak{A}$.

Suppose that $\mathfrak{A}$ and $\mathfrak{B}$ are S-structures such that $\mathfrak{A}$ is an elementary substructure of $\mathfrak{B}$. Let p be an S-sentence. Since p is also an S-formula, $\mathfrak{A} \models p$ if and only if $\mathfrak{B} \models p$ by Definition 7.3.28. Therefore, $\mathfrak{A}$ and $\mathfrak{B}$ are elementary equivalent, proving the next result.

■ **THEOREM 7.3.29**

Let $\mathfrak{A}$ and $\mathfrak{B}$ be S-structures. If $\mathfrak{A} \preceq \mathfrak{B}$, then $\mathfrak{A} \equiv \mathfrak{B}$.

However, the converse of Theorem 7.3.29 does not hold. It is possible for $\mathfrak{A} \subseteq \mathfrak{B}$ and $\mathfrak{A} \equiv \mathfrak{B}$ yet $\mathfrak{A}$ not be an elementary substructure of $\mathfrak{B}$. If this is to be the case, there must be a formula p with at least one free variable that is satisfied by one of the structures but not the other.

■ **EXAMPLE 7.3.30**

Let $\mathfrak{A} = ([0, 1], \mathfrak{a})$ and $\mathfrak{B} = ([0, 2], \mathfrak{b})$ be the $\{R\}$-structures of Example 7.2.2, where $\mathfrak{a}(R)$ is standard $<$ on $[0, 1]$ and $\mathfrak{b}(R)$ is standard $<$ on $[0, 2]$. Recall that $\mathfrak{A}$ is a substructure of $\mathfrak{B}$. Since f defined by $f(x) = 2x$ is an $\{R\}$-isomorphism $[0, 1] \to [0, 2]$, we conclude that $\mathfrak{A} \equiv \mathfrak{B}$ (Theorem 7.3.24). However, define the formula

$$p(x) := \exists y R(x, y)$$

and let I be an interpretation such that $I(x) = 1$. Then,

$$\mathfrak{A} \nvDash p(x) \ [I]$$

and

$$\mathfrak{B} \vDash p(x) \ [I].$$

Therefore, $\mathfrak{A}$ is not an elementary substructure of $\mathfrak{B}$.

Because of Theorem 7.3.27, proving that one structure is an elementary substructure of another reduces to a check of one particular condition. This condition is found in the next theorem, which is due to Tarski and Vaught (1957).

■ **THEOREM 7.3.31 [Tarski–Vaught]**

Let $\mathfrak{A}$ and $\mathfrak{B}$ be S-structures such that $\mathfrak{A} \subseteq \mathfrak{B}$. Let A be the domain of $\mathfrak{A}$. The following are equivalent.

- $\mathfrak{A}$ is an elementary substructure of $\mathfrak{B}$.

- For every S-formula p and S-interpretation I of $\mathfrak{A}$,

$$\text{if } \mathfrak{B} \vDash \exists x p \ [I] \text{ then } \mathfrak{B} \vDash p \ [I_x^a] \text{ for some } a \in A. \qquad (7.19)$$

PROOF

Suppose that $\mathfrak{A} \preceq \mathfrak{B}$. Let p be an S-formula and I an S-interpretation of $\mathfrak{A}$. Let $\mathfrak{B} \vDash \exists x p \ [I]$. By hypothesis, we have that $\mathfrak{A} \vDash \exists x p \ [I]$. Thus, $\mathfrak{A} \vDash p \ [I_x^a]$ for some $a \in A$, so again by hypothesis, $\mathfrak{B} \vDash p \ [I_x^a]$ for some $a \in A$.

Conversely, let p be an S-formula and I an S-interpretation of $\mathfrak{A}$. We proceed by induction on formulas to show that $\mathfrak{A} \preceq \mathfrak{B}$.

- If p is $t_0 = t_1$ or $R(t_0, t_1, \ldots, t_{n-1})$ for S-terms $t_0, t_1, \ldots, t_{n-1}$, then p has no quantifiers, and the result follows by Theorem 7.3.27.

- Let p be $\neg q$. Then,

$$\begin{aligned}
\mathfrak{A} \vDash p \ [I] &\Leftrightarrow \mathfrak{A} \vDash \neg q \ [I] \\
&\Leftrightarrow \mathfrak{A} \nvDash q \ [I] \\
&\Leftrightarrow \mathfrak{B} \nvDash q \ [I] \\
&\Leftrightarrow \mathfrak{B} \vDash \neg q \ [I] \\
&\Leftrightarrow \mathfrak{B} \vDash p \ [I].
\end{aligned}$$

- Suppose p is $q \rightarrow r$. Assume that $\mathfrak{A} \vDash q \rightarrow r \ [I]$ and let $\mathfrak{B} \vDash q \ [I]$. Then, $\mathfrak{A} \vDash q \ [I]$ by induction, so $\mathfrak{A} \vDash r \ [I]$. Again, by induction, $\mathfrak{B} \vDash r \ [I]$. Therefore, $\mathfrak{B} \vDash r \rightarrow q \ [I]$. The converse is proved similarly.

- Now let p be $\exists x q$. First, because $\mathfrak{A} \subseteq \mathfrak{B}$,

$$\begin{aligned}
\mathfrak{A} \vDash \exists x q \ [I] &\Rightarrow \mathfrak{A} \vDash q \ [I_x^a] \text{ for some } a \in A \\
&\Rightarrow \mathfrak{B} \vDash q \ [I_x^a] \text{ for some } a \in B \\
&\Rightarrow \mathfrak{B} \vDash \exists x q \ [I].
\end{aligned}$$

Second, by (7.19),

$$\mathfrak{B} \models \exists x q \ [I] \Rightarrow \mathfrak{B} \models q \ [I_x^a] \text{ for some } a \in A$$
$$\Rightarrow \mathfrak{A} \models q \ [I_x^a] \text{ for some } a \in A$$
$$\Rightarrow \mathfrak{A} \models \exists x q \ [I]. \blacksquare$$

■ **EXAMPLE 7.3.32**

Prove that $\mathfrak{A} = (\mathbb{Q}, \leq)$ is an elementary substructure of $\mathfrak{B} = (\mathbb{R}, \leq)$. To do this, follow the test given by Theorem 7.3.31. Let p be an S-formula and I be an S-interpretation of $\mathfrak{A}$. Assume that $\mathfrak{B} \models \exists x p \ [I]$. This implies that

$$\mathfrak{B} \models p \ [I_x^b] \text{ for some } b \in \mathbb{R}.$$

Take $a_0, a_1, a \in \mathbb{Q}$ such that $b, a \in (a_0, a_1)$ and define the functions

$$f_0 : (-\infty, a_0] \rightarrow (-\infty, a_0],$$
$$f_1 : (a_0, a) \rightarrow (a_0, b),$$
$$f_2 : [b, a_1) \rightarrow [a, a_1),$$
$$f_3 : [a_1, \infty) \rightarrow [a_1, \infty),$$

so that f_0 is the identity on $(-\infty, a_0]$, f_3 is the identity on $[a_1, \infty)$, and f_1 and f_2 are order isomorphisms (Exercise 22). Let $f = f_0 \cup f_1 \cup f_2 \cup f_3$. Then, f is an order isomorphism $\mathbb{R} \rightarrow \mathbb{R}$ such that $f(b) = a$, $f(a_0) = f(a_0)$, and $f(a_1) = f(a_1)$. Hence, f is an automorphism (Exercise 23). Therefore, by Lemmas 7.3.21 and 7.3.22,

$$\mathfrak{B} \models p \ [I_x^b] \Leftrightarrow \mathfrak{B} \models p \ [(I_x^b)_f] \Leftrightarrow \mathfrak{B} \models p \ [(I_f)_x^{f(b)}],$$

which implies that

$$\mathfrak{B} \models p \ [I_x^b] \text{ for some } b \in \mathbb{Q}.$$

Thus, $\mathfrak{A} \preceq \mathfrak{B}$, which also implies that $\mathfrak{A} \equiv \mathfrak{B}$ (Theorem 7.3.29).

Notice that this example implies that there is no first-order sentence that can distinguish between $(\mathbb{Q}, <)$ and $(\mathbb{R}, <)$ even though they are not isomorphic. Also, this example shows that the converse of Theorem 7.3.24 is false if the domains of the structures are infinite.

Certainly, for all S-structures $\mathfrak{A}$, $\mathfrak{B}$, and $\mathfrak{C}$, we have the following by Exercise 19:

- $\mathfrak{A} \equiv \mathfrak{A}$.

- If $\mathfrak{A} \equiv \mathfrak{B}$, then $\mathfrak{B} \equiv \mathfrak{A}$.

- If $\mathfrak{A} \equiv \mathfrak{B}$ and $\mathfrak{B} \equiv \mathfrak{C}$, then $\mathfrak{A} \equiv \mathfrak{C}$.

Similar to this and Theorem 7.2.4, we have the following result for elementary substructures.

■ **THEOREM 7.3.33**

Let $\mathfrak{A}$, $\mathfrak{B}$, and $\mathfrak{C}$ be S-structures.

- $\mathfrak{A} \leq \mathfrak{A}$.

- If $\mathfrak{A} \leq \mathfrak{B}$ and $\mathfrak{B} \leq \mathfrak{C}$, then $\mathfrak{A} \leq \mathfrak{C}$.

- If $\mathfrak{A} \leq \mathfrak{C}$, $\mathfrak{B} \leq \mathfrak{C}$, and $\mathfrak{A} \subseteq \mathfrak{B}$, then $\mathfrak{A} \leq \mathfrak{B}$.

PROOF

We prove the second part and leave the others to Exercise 25. Let $\mathfrak{A} \leq \mathfrak{B}$ and $\mathfrak{B} \leq \mathfrak{C}$. This first means that $\mathfrak{A} \subseteq \mathfrak{B}$ and $\mathfrak{B} \subseteq \mathfrak{C}$. Thus, by Theorem 7.2.4, $\mathfrak{A} \subseteq \mathfrak{C}$. Let I be an interpretation of $\mathfrak{A}$. The proof is completed for quantifier-free formulas using Theorem 7.3.27 (Exercise 24), so let p be an S-formula with a quantifier.

- Suppose $\mathfrak{A} \vDash \exists x p\ [I]$. This implies that $\mathfrak{A} \vDash p\ [I_x^a]$ for some $a \in A$. Then, $\mathfrak{B} \vDash p\ [I_x^a]$ from which it follows that $\mathfrak{C} \vDash p\ [I_x^a]$. Since $A \subseteq C$, we conclude that $\mathfrak{C} \vDash \exists x p\ [I]$.

- Conversely, let $\mathfrak{C} \vDash \exists x p\ [I]$. By Theorem 7.3.31, there exists $b \in B$ such that $\mathfrak{C} \vDash p\ [I_x^b]$. By hypothesis, $\mathfrak{B} \vDash p\ [I_x^b]$, so $\mathfrak{B} \vDash \exists x p\ [I]$. Again, by Theorem 7.3.31, there exists $a \in A$ such that $\mathfrak{B} \vDash p\ [I_x^a]$. From this follows that $\mathfrak{A} \vDash p\ [I_x^a]$, whence $\mathfrak{A} \vDash \exists x p\ [I]$. ■

Exercises

1. Let $\mathfrak{A}$, $\mathfrak{B}$, and $\mathfrak{C}$ be S-structures such that $\mathfrak{A} \subseteq \mathfrak{B}$. Let $\varphi : \mathfrak{B} \to \mathfrak{C}$ be a homomorphism. Prove that $\varphi[\mathrm{dom}(\mathfrak{A})]$ is the domain of a substructure of $\mathfrak{C}$.

2. Prove that the following are group homomorphisms. Assume in each instance that $\circ$ is interpreted as the standard addition on the set.
    (a) $\varphi : \mathbb{Z} \times \mathbb{Z} \to \mathbb{Z}$ where $\varphi(a, b) = a$
    (b) $\varphi : \mathbb{Z}_{12} \to \mathbb{Z}_6$ where $\varphi([a]_{12}) = [a]_6$
    (c) $\psi : \mathbb{Z} \times \mathbb{Z} \to M_{2,2}(\mathbb{R})$ where

$$\psi(a, b) = \begin{bmatrix} a & 0 \\ 0 & b \end{bmatrix}$$

3. Prove that the zero map is a group homomorphism.

4. Let $\mathfrak{Z} = (\mathbb{Z}, +)$ and $\mathfrak{Z}_n = (\mathbb{Z}_n, +)$ with $n \in \mathbb{Z}^+$. Prove that $\varphi : \mathbb{Z} \to \mathbb{Z}_n$ defined by $\varphi(x) = [x]_n$ is a ring homomorphism $\mathfrak{Z} \to \mathfrak{Z}_n$.

5. Let $\mathfrak{C} = (\mathbb{C}, 0 + 0i, +, \cdot)$. Define $\psi : \mathbb{C} \to \mathbb{C}$ by $\psi(a + bi) = a - bi$ for all $a, b \in \mathbb{R}$. Prove the following.
    (a) ψ is a ring homomorphism $\mathfrak{C} \to \mathfrak{C}$.
    (b) ψ is one-to-one and onto.

6. Prove the first part of Theorem 7.3.8.

7. Let $\mathfrak{R} = (R, 0, +, \cdot)$ be a ring and $\mathfrak{I}$ be an ideal of $\mathfrak{R}$ with domain I. Let the function $\varphi : R \to R/I$ be defined as $\varphi(a) = a + I$ for all $a \in R$.
   (a) Prove that φ is a homomorphism.
   (b) Show that $\ker(\varphi) = I$.
   (c) Prove that φ is onto.

8. Prove Theorem 7.3.15.

9. Prove that the function φ of Example 7.3.16 is a bijection.

10. Let $\mathfrak{A}$ and $\mathfrak{B}$ be S-structures. Write $\mathfrak{A} \precsim \mathfrak{B}$ if there exists a one-to-one homomorphism $\varphi : \mathfrak{A} \to \mathfrak{B}$. Prove the following.
   (a) $\mathfrak{A} \precsim \mathfrak{A}$
   (b) If $\mathfrak{A} \precsim \mathfrak{B}$ and $\mathfrak{B} \precsim \mathfrak{C}$, then $\mathfrak{A} \precsim \mathfrak{C}$.
   (c) It is possible for $\mathfrak{A} \precsim \mathfrak{B}$ and $\mathfrak{B} \precsim \mathfrak{C}$ yet $\mathfrak{A}$ and $\mathfrak{B}$ are different S-structures.

11. Let $\mathfrak{A}$ and $\mathfrak{A}'$ be $\{R\}$-structures with equal domains that are finite, where R is a binary relation symbol. Assume that

$$\mathfrak{A} \vDash X \text{ has a least element with respect to } R$$

and

$$\mathfrak{A}' \vDash X \text{ has a least element with respect to } R$$

for every $X \subseteq \text{dom}(\mathfrak{A})$. Prove that $\mathfrak{A} \cong \mathfrak{A}'$.

12. Assume that S is finite and that the domain of the S-structure $\mathfrak{A}$ is finite. Find a S-sentence p such that $\mathfrak{B} \vDash p$ if and only if $\mathfrak{B} \cong \mathfrak{A}$, for all S-structures $\mathfrak{B}$.

13. Suppose that $\varphi : \mathfrak{R} \to \mathfrak{R}'$ is a ring isomorphism as in Example 7.3.18. Prove.
   (a) If $\mathfrak{R}$ is an integral domain, $\mathfrak{R}'$ is an integral domain.
   (b) If $\mathfrak{R}$ is a division ring, $\mathfrak{R}'$ is a division ring.

14. Let $\varphi : \mathfrak{R} \to \mathfrak{R}'$ be a ring isomorphism. Prove that 1 is unity from $\mathfrak{R}$ if and only if $\varphi(1)$ is unity from $\mathfrak{R}'$.

15. Prove or show false the given elementary equivalences.
   (a) $(\mathbb{R}, 0, +) \equiv (\mathbb{C}, 0, +)$.
   (b) $(\mathbb{Z}, 0, +) \equiv (\mathbb{Q}, 0, +)$.
   (c) $(\mathbb{Z}, \leq) \equiv (\mathbb{Q}, \leq)$.
   (d) $(\omega, \leq) \equiv (\mathbb{Z}^+, \leq)$.
   (e) $(\omega, \leq) \equiv (\mathbb{Z}^-, \leq)$.

16. Let A and B be sets and assume that $S = \varnothing$. Prove that $(A) \equiv (B)$.

17. Generalize the argument of Example 7.3.26 to prove that $\mathfrak{A} \equiv \mathfrak{B}$ implies that $\mathfrak{A} \cong \mathfrak{B}$ if the cardinality of the domain of $\mathfrak{A}$ equals 2.

18. Generalize the argument of Example 7.3.26 to prove that $\mathfrak{A} \equiv \mathfrak{B}$ implies that $\mathfrak{A} \cong \mathfrak{B}$ if the domain of $\mathfrak{A}$ is finite.

19. Let $\mathfrak{A}$, $\mathfrak{B}$, and $\mathfrak{C}$ be S-structures. Prove the following.
    (a) $\mathfrak{A} \equiv \mathfrak{A}$.
    (b) If $\mathfrak{A} \equiv \mathfrak{B}$, then $\mathfrak{B} \equiv \mathfrak{A}$.
    (c) If $\mathfrak{A} \equiv \mathfrak{B}$ and $\mathfrak{B} \equiv \mathfrak{C}$, then $\mathfrak{A} \equiv \mathfrak{C}$.

20. Let $\mathfrak{A}$ and $\mathfrak{B}$ be elementary equivalent S-structures. Prove that there exists an S-structure $\mathfrak{C}$ such that $\mathfrak{A} \le \mathfrak{C}$ and $\mathfrak{B} \le \mathfrak{C}$.

21. Prove Theorem 7.3.27.

22. Find the order isomorphisms f_1 and f_2 in Example 7.3.32.

23. Prove that the function f from Example 7.3.32 is a $\{\le\}$-isomorphism.

24. Assume that for all S-structures $\mathfrak{A}$, $\mathfrak{B}$, and $\mathfrak{C}$, if $\mathfrak{A} \le \mathfrak{B}$ and $\mathfrak{B} \le \mathfrak{C}$, then $\mathfrak{A} \vDash p$ if and only if $\mathfrak{C} \vDash p$ for all quantifier-free p.

25. Prove the first and third parts of Theorem 7.3.33.

26. Let $\mathscr{F} = \{\mathfrak{A}_\gamma : \gamma \in \kappa\}$ be a chain of S-structures for some cardinal κ (Exercise 7.2.3). Assume that $\mathfrak{A}_\alpha \le \mathfrak{A}_\beta$ when $\alpha \in \beta \in \kappa$, making $\mathscr{F}$ an **elementary chain**. Prove that $\mathfrak{A}_\alpha \le \bigcup_{\gamma \in \kappa} \mathfrak{A}_\gamma$ for all $\alpha \in \kappa$.

27. Find a set of subject symbols S and a chain of S-structures $\{\mathfrak{A}_n : n \in \omega\}$ such that $\mathfrak{A}_i \equiv \mathfrak{A}_j$ for all $i, j \in \omega$ but $\mathfrak{A}_0 \not\equiv \bigcup_{n \in \omega} \mathfrak{A}_n$.

7.4 THE THREE PROPERTIES REVISITED

This section is an extension of Section 1.5 to first-order logic. First, we define what it means to be consistent with the ultimate goal to show that any consistent system has a model. Next, we show that first-order logic is sound in that every sentence that can be proved is true, provided we have the correct understanding of what it means *to prove* and what it means *to be true*. Lastly, we show that first-order logic is complete in that every true sentence can be proved.

Consistency

Consider the following propositions representing Euclid's axioms for his geometry (Euclid 1925, I Postulates).

- **Eu1.** A line can be drawn through two distinct points.

- **Eu2.** A line segment can be drawn between two distinct points.

- **Eu3.** A circle can be drawn given any center and radius.

- **Eu4.** All right angles are congruent to each another.

- **Eu5.** If a line falling on two straight lines make the interior angles on the same side less than two right angles, the two lines intersect on the side on which the angles are less than two right angles.

There are three sets of proposition here. First, is the list itself,

$$\mathscr{E} = \{\mathbf{Eu1}, \mathbf{Eu2}, \mathbf{Eu3}, \mathbf{Eu4}, \mathbf{Eu5}\}.$$

Second, is the set of consequences (Definition 7.1.20) of $\mathscr{E}$,

$$\mathscr{E}_1 = \{p : \mathscr{E} \vDash p\}.$$

Third, is the set of propositions provable (Definition 1.2.13) from $\mathscr{E}$,

$$\mathscr{E}_2 = \{p : \mathscr{E} \vdash p\}.$$

For Euclidean geometry, $\mathscr{E}_1 = \mathscr{E}_2$. This means that both $\mathscr{E}_1$ and $\mathscr{E}_2$ can be said to describe all of the propositions that are in the geometry. It is the job of the geometer to discover what those propositions are.

We want to similarly analyze first-order logic. Given the rules of logic and a set of theory symbols, we want the set of consequences to be equal to the set of provable sentences. One way to have this happen is for a contradiction to follow from the axioms (1.2.8). Then, by Theorem 1.5.2, every propositional form will have a proof and, in turn, will also be a consequence, but we do not want such a system. Therefore, for all of this to work effectively, it is a requirement that first-order logic is without contradiction. To deal with this concept, we start with a definition.

■ **DEFINITION 7.4.1**

An **S-theory** is a set of S-sentences.

The set of all sentences provable in first-order logic given a set of theory symbols S is an example of an S-theory. The theories that we study should have a familiar property (compare Definition 1.5.1).

■ **DEFINITION 7.4.2**

An S-theory $\mathscr{T}$ is **consistent** [denoted by $\mathrm{Con}(F)$] if $\mathscr{T} \nvdash q \wedge \neg q$ for every S-sentence q. Otherwise, $\mathscr{T}$ is **inconsistent**.

We next generalize Theorem 1.5.2 to first-order logic. Its proof is left to Exercise 2.

■ **THEOREM 7.4.3**

Let $\mathscr{T}$ be an S-theory. The following are equivalent.

- $\mathscr{T}$ is consistent.

- Every finite subset of $\mathscr{T}$ is consistent.

- There is an S-sentence that is not provable from $\mathscr{T}$.

As a consequence of Theorem 7.4.3, a theory $\mathscr{T}$ is inconsistent if either it has a finite subset that is inconsistent or it is able to prove all sentences. This identifies the major weakness of an inconsistent theory. Not only can a contradiction be proved, but any sentence can also be proved. Such a system is certainly worthless.

If a theory is consistent, it can be shown to be a subset of a theory that is the greatest possible consistent theory. We generalize Definition 1.5.3 to name this theory.

■ **DEFINITION 7.4.4**

The S-theory $\mathcal{T}$ is **maximally consistent** if $\mathcal{T}$ is consistent and $\mathrm{Con}(\mathcal{T} \cup \{p\})$ implies $p \in \mathcal{T}$ for all S-sentences p.

The next lemma gives some basic properties of maximally consistent theories.

■ **LEMMA 7.4.5**

Let $\mathcal{T}$ be a maximally consistent S-theory. Let p and q be S-sentences.

- If $\mathcal{T} \vdash p$, then $p \in \mathcal{T}$.

- If $\vdash p$, then $p \in \mathcal{T}$.

- $p \in \mathcal{T}$ or $\neg p \in \mathcal{T}$, but not both.

- If $p \to q, p \in \mathcal{T}$, then $q \in \mathcal{T}$.

PROOF

- Let $\mathcal{T} \vdash p$. Suppose that $\mathcal{T} \cup \{p\} \vdash q \wedge \neg q$ for some S-sentence q. This means that there exists S-sentences $p_0, p_1, \ldots, p_{n-1}$ such that

$$p, p_0, p_1, \ldots, p_{n-1}, q \wedge \neg q$$

  is a proof of $q \wedge \neg q$ from $\mathcal{T} \cup \{p\}$. Since $\mathcal{T}$ proves p, there are S-sentences $q_0, q_1, \ldots, q_{m-1}$ such that

$$q_0, q_1, \ldots, q_{m-1}, p$$

  is a proof of p from $\mathcal{T}$. Therefore,

$$q_0, q_1, \ldots, q_{m-1}, p, p_0, \ldots, p_{n-1}, q \wedge \neg q$$

  is a proof of $q \wedge \neg q$ from $\mathcal{T}$, a contradiction. This implies that $\mathcal{T} \cup \{p\}$ is consistent, so since $\mathcal{T}$ is maximally consistent, $p \in \mathcal{T}$.

- If $\vdash p$, then $\mathcal{T} \vdash p$, so $p \in \mathcal{T}$ by the first property.

- Since $\mathcal{T}$ is consistent, $\mathrm{Con}(\mathcal{T} \cup \{p\})$ or $\mathrm{Con}(\mathcal{T} \cup \{\neg p\})$. Because $\mathcal{T}$ is maximally consistent, $p \in \mathcal{T}$ or $\neg p \in \mathcal{T}$, but not both because $\mathrm{Con}(\mathcal{T})$.

- Let $p \to q$ and p be members of $\mathcal{T}$. By MP, we have that $\mathcal{T} \vdash q$, so again by the first part, $q \in \mathcal{T}$. ■

Given a consistent theory, the construction of the maximally consistent theory that contains it requires the use of a chain.

■ **LEMMA 7.4.6**

If $(\mathscr{C}, \subseteq)$ is a chain of consistent S-theories, $\bigcup \mathscr{C}$ is consistent.

PROOF

Let $\mathscr{C}$ be a chain of consistent sets of S-sentences with respect to $\subseteq$. Suppose $\bigcup \mathscr{C} \vdash q \wedge \neg q$ for some sentence q. Hence, there exists $p_0, p_1, \ldots, p_{n-1} \in \bigcup \mathscr{C}$ such that

$$p_0, p_1, \ldots, p_{n-1} \vdash q \wedge \neg q.$$

This implies that for each $i = 0, 1, \ldots, n-1$, there exists $C_i \in \mathscr{C}$ such that $p_i \in C_i$. Since $\{C_i : i = 0, 1, \ldots, n-1\}$ is a finite chain, arrange them so that C_n is the greatest element with respect to $\subseteq$. This gives

$$\{p_0, p_1, \ldots, p_{n-1}\} \subseteq C_n,$$

which implies that $C_n \vdash q \wedge \neg q$, contradicting the consistency of C_n. ■

The next result proves the existence of a greatest consistent theory by generalizing Theorem 1.5.4 to an arbitrary set of sentences.

■ **THEOREM 7.4.7 [Lindenbaum]**

Every consistent S-theory is a subset of a maximally consistent S-theory.

PROOF

Let $\mathscr{T}$ be a consistent set of S-sentences. Define

$$\mathscr{A} = \{E : \mathscr{T} \subseteq E \text{ and } E \text{ is a consistent set of S-formulas}\}.$$

Note that $\mathscr{A} \neq \varnothing$ since $\mathscr{T} \in \mathscr{A}$. Let $\mathscr{C}$ be a chain in $\mathscr{A}$. Then, $\bigcup \mathscr{C} \in \mathscr{A}$ because

- $\mathscr{T} \subseteq \bigcup \mathscr{C}$.

- $\bigcup \mathscr{C}$ is consistent because each element of the chain $\mathscr{C}$ is consistent by Lemma 7.4.6.

We conclude by Zorn's lemma (5.1.13) that there is a greatest element $M \in \mathscr{A}$ with respect to $\subseteq$. By definition, $\mathscr{T} \subseteq M$ and Con(M). Also, let p be a formula such that Con($M \cup \{p\}$). Then, $M \cup \{p\} \in \mathscr{A}$, but since M is maximal, we conclude that $M \cup \{p\} = M$. Thus, $p \in M$, showing that M is maximally consistent. ■

Soundness

We previously defined a logic to be sound if every theorem is a tautology and complete if every tautology is a theorem (Definition 1.5.5). We now give the corresponding definition for first-order logic (Figure 7.3).

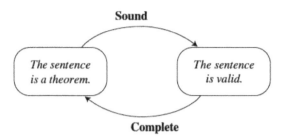

Figure 7.3 Sound and complete logics.

■ DEFINITION 7.4.8

- A logic is **sound** if every theorem is valid.

- A logic is **complete** if every valid sentence is a theorem.

A propositional form p is a tautology if $v(p) = T$ for every valuation v (Definition 1.1.14). This means that any interpretation of p will always yield a true proposition. This, in turn, implies that p will hold in every model. That is,

$$\text{tautologies are valid.}$$

Therefore, it is evident that Definition 7.4.8 is a generalization of Definition 1.5.5 to first-order logic.

As in Section 1.5, we begin with soundness.

■ THEOREM 7.4.9 [Soundness]

Let $\mathcal{T}$ be a set of S-sentences. For any S-sentence p, if $\mathcal{T} \vdash p$, then $\mathcal{T} \vDash p$.

PROOF

Using strong induction, we prove that for all $k \in \mathbb{Z}^+$,

$$\begin{array}{c} \text{if there exists a proof of } p \text{ from } \mathcal{T} \\ \text{consisting of } k \text{ sentences, then } \mathcal{T} \vDash p. \end{array} \qquad (7.20)$$

In the case that $n = 1$, we see that $\vdash p$. This implies that p is an axiom or $p \in \mathcal{T}$ (Definition 1.2.13), so we have that $\mathcal{T} \vDash p$.

Now suppose that $n > 1$ and assume that (7.20) holds for all $k \leq n$. This means that there exists sentences $p_0, p_1, \ldots, p_{n-1}$ provable from $\mathcal{T}$ such that

$$p_0, p_1, \ldots, p_{n-1}, p$$

is a proof of p. Let $\mathfrak{A}$ be any model of $\mathcal{T}$. We have three cases to consider (Theorem 1.4.2).

- $p \in \mathcal{T}$, which implies that $\mathfrak{A} \vDash p$.

- p is derived using MP. This implies there exists S-sentences r and p such that $r, r \to p \in \{p_0, p_1, \ldots, p_{n-1}, p\}$. Since a sentence of a proof is proved from previous sentences of the proof, by induction, $\mathfrak{A} \models r \to p$ and $\mathfrak{A} \models r$. Therefore, $\mathfrak{A} \models p$.

- p is logically equivalent to p_i for some $i = 0, 1, \ldots, n - 1$. This means that $p \leftrightarrow p_i$ is a tautology, so $\mathfrak{A} \models p \leftrightarrow p_i$. By Theorem 7.1.7, we conclude that $\mathfrak{A} \models p$ if and only if $\mathfrak{A} \models p_i$, but by induction, $\mathfrak{A} \models p_i$, so $\mathfrak{A} \models p$. ∎

If $\mathscr{T} = \varnothing$, we can apply Theorem 7.4.9 to conclude the corollary.

■ COROLLARY 7.4.10

First-order logic is sound.

As with Corollary 1.5.11 for propositional logic, we can also prove the next result.

■ COROLLARY 7.4.11

First-order logic is consistent.

Completeness

It is now time to prove the completeness of first-order predicate logic. This result is due to Kurt Gödel (1929) but the proof given here is due to Leon Henkin (1949). As opposed to the soundness theorem (7.4.9), the proof is rather involved. We begin with a definition (compare with EI, Theorem 2.3.14).

■ DEFINITION 7.4.12

Let $\mathscr{T}$ be a set of S-sentences and C a set of constant symbols of S. Define C to be a **witness set** for $\mathscr{T}$ if for all S-formulas $p = p(x)$, there exists $c \in C$ such that

$$\mathscr{T} \vdash \exists x p \to p\frac{c}{x}.$$

For example, let $\mathscr{T} = \{x + 1 = 0, x + (1 + 1) = 0, (x + y) + 1 = 1\}$ be a set of NT-sentences. Extend NT to $S = NT \cup C$, where $C = \{-1, -2\}$. Then, C is a witness set for $\mathscr{T}$ if

$$\mathscr{T} \vdash \exists x(x + 1 = 0) \to (x + 1 = 0)\frac{-1}{x}$$

and

$$\mathscr{T} \vdash \exists x(x + (1 + 1) = 0) \to (x + (1 + 1) = 0)\frac{-2}{x}.$$

Since $(x + y) + 1 = 1$ has two free variables, the witness set does not apply to it.

Recall that for a set of theory symbols S, the notation $L(S)$ refers to the set of all S-formulas (Definition 2.1.13).

■ LEMMA 7.4.13

Let $\mathcal{T}$ be a consistent set of S-sentences. Let C be a set of constant symbols not in S such that $|C| = |L(S)|$. Then, $\mathcal{T}$ can be extended to a consistent set of $(S \cup C)$-sentences that has C as a witness set.

PROOF

Let $\kappa = |L(S)|$ and suppose that $C = \{c_\alpha : \alpha \in \kappa\}$ is a set of new constant symbols. Assume that $c_\alpha \neq c_\beta$ for all $\alpha \in \beta \in \kappa$. Since $|L(S \cup C)| = \kappa$, we identify the formulas of $L(S \cup C)$ with at most one free variable as

$$p_\alpha = p_\alpha(x_\alpha)$$

with $\alpha \in \kappa$. Now define a chain $\mathscr{C} = \{\mathcal{T}_\gamma : \gamma \in \kappa\}$ and a set $\{d_\gamma : \gamma \in \kappa\}$ by transfinite recursion (Corollary 6.1.24) such that for all $\beta \in \kappa$,

- $\mathcal{T}_\beta$ is consistent,

- $\mathcal{T}_\beta$ is a $(S \cup C)$-theory,

- $|\mathcal{T}_{\beta+1} \setminus \mathcal{T}_\beta| = 1$,

- d_β is not among the symbols of the sentences in $\mathcal{T}_\beta$.

Let $\mathcal{T}_0 = \mathcal{T}$ and suppose that $\mathcal{T}_\delta$ and d_δ has been defined for all $\delta \in \alpha$ with the indicated properties.

- Let $\alpha = \beta + 1$. Choose d_β from C such that d_β is not among the symbols of the formulas in $\mathcal{T}_\beta \cup \{p_\beta\}$. This can be done since there are less than κ theory symbols in $\mathcal{T}_\beta \cup \{p_\beta\}$. Now define

$$\mathcal{T}_{\beta+1} = \mathcal{T}_\beta \cup \left\{ \exists x_\beta p_\beta \rightarrow p_\beta \frac{d_\beta}{x_\beta} \right\}. \tag{7.21}$$

In order to obtain a contradiction, suppose that $\mathcal{T}_{\beta+1}$ is inconsistent. Since $\mathcal{T}_\beta$ is consistent, it must be the case that

$$\mathcal{T}_\beta \vdash \neg \left(\exists x_\beta p_\beta \rightarrow p_\beta \frac{d_\beta}{x_\beta} \right).$$

This implies that

$$\mathcal{T}_\beta \vdash \exists x_\beta p_\beta \wedge \neg p_\beta \frac{d_\beta}{x_\beta}.$$

By Com and Simp,

$$\mathcal{T}_\beta \vdash \neg p_\beta \frac{d_\beta}{x_\beta}.$$

Since d_β is not among the symbols of p_β or $\mathcal{T}_\beta$, it is arbitrary, so by UG,

$$\mathcal{T}_\beta \vdash \forall x_\beta \neg p_\beta.$$

That is,

$$\mathscr{T}_\beta \vdash \neg \exists x_\beta p_\beta.$$

Using Simp and Conj, we conclude that

$$\mathscr{T}_\beta \vdash \exists x_\beta p_\beta \wedge \neg \exists x_\beta p_\beta,$$

contradicting the consistency of $\mathscr{T}_\beta$.

- If α is a limit ordinal, define

$$\mathscr{T}_\alpha = \bigcup_{\beta \in \alpha} \mathscr{T}_\beta,$$

which is consistent by Lemma 7.4.6.

We claim that the $(S \cup C)$-theory

$$\bigcup_{\beta \in \kappa} \mathscr{T}_\beta$$

is the desired extension of $\mathscr{T}$. To see this, first note that this union is consistent by Lemma 7.4.6. Also, let $p(x)$ be an $S \cup C$-formula with at most one free variable. This implies that $p = p_\alpha$ and $x = x_\alpha$ for some $\alpha \in \kappa$. Therefore,

$$\exists x_\alpha p_\alpha \rightarrow p_\alpha \frac{d_\alpha}{x_\alpha} \in \mathscr{T}_{\alpha+1},$$

so

$$\bigcup_{\gamma \in \kappa} \mathscr{T}_\gamma \vdash \exists x_\alpha p_\alpha \rightarrow p_\alpha \frac{d_\alpha}{x_\alpha}. \quad \blacksquare$$

Let S be a set of theory symbols and C be a set of constants from S. Let $\mathscr{T}$ be a consistent S-theory. We want to define an S-structure that has a chance of being a model of $\mathscr{T}$. One of the issues that must be overcome is whether any pair of constants should be interpreted as being equal. If so, we want to view those constants as equivalent. To do this, define a relation $\sim$ on C by

$$c_0 \sim c_1 \text{ if and only if } c_0 = c_1 \in \mathscr{T} \tag{7.22}$$

for all $c_0, c_1 \in C$. This defines an equivalence relation under the right conditions.

■ LEMMA 7.4.14

Let C be a set of constants from S. If $\mathscr{T}$ is a maximally consistent S-theory, then $\sim$ is an equivalence relation on C.

PROOF

Let $c_0, c_1, c_2 \in C$.

- Since $\mathcal{T} \cup \{c_0 = c_0\}$ is consistent by **E1** (Axioms 5.1.1), by the maximal consistency of $\mathcal{T}$, we have that $c_0 = c_0 \in \mathcal{T}$.

- Suppose $c_0 = c_1 \in \mathcal{T}$. Then, $\mathcal{T} \vdash c_0 = c_1$, so $\mathcal{T} \vdash c_1 = c_0$ by **E2**. Thus, $c_1 = c_0 \in \mathcal{T}$.

- Let $c_0 = c_1 \in \mathcal{T}$ and $c_1 = c_2 \in \mathcal{T}$. Hence, we have $\mathcal{T} \vdash c_0 = c_1$ and $\mathcal{T} \vdash c_1 = c_2$. Then, $\mathcal{T} \vdash c_0 = c_2$ by **E3**, which yields $c_0 = c_2 \in \mathcal{T}$. ∎

We define the domain of the desired structure to be $C/\sim$, the set of equivalence classes modulo $\sim$ (Definition 4.2.10). Before we define the function $\mathfrak{a}$, we need some actual functions and relations on $C/\sim$. Use the following lemma to define them.

■ **LEMMA 7.4.15**

Let $\mathcal{T}$ be a maximally consistent S-theory and C be a set of constant symbols from S. Let $t_0, t_1, \ldots, t_{n-1}, t'_0, t'_1, \ldots, t'_{n-1}$ be S-terms such that

$$t_0 \sim t'_0, t_1 \sim t'_1, \ldots, t_{n-1} \sim t'_{n-1}.$$

- For every n-ary function symbol f,

$$f(t_0, t_1, \ldots, t_{n-1}) \sim f(t'_0, t'_1, \ldots, t'_{n-1}).$$

- For every n-ary relation symbol R,

$$R(t_0, t_1, \ldots, t_{n-1}) \in \mathcal{T} \Leftrightarrow R(t'_0, t'_1, \ldots, t'_{n-1}) \in \mathcal{T}.$$

PROOF

By (7.22) we have that

$$\mathcal{T} \vdash t_0 = t'_0, \mathcal{T} \vdash t_1 = t'_1, \ldots, \mathcal{T} \vdash t_{n-1} = t'_{n-1}.$$

Let f be an n-ary function symbol. Then,

$$\mathcal{T} \vdash f(t_0, t_1, \ldots, t_{n-1}) = f(t'_0, t'_1, \ldots, t'_{n-1})$$

by **E4** (Axioms 5.1.1), so

$$f(t_0, t_1, \ldots, t_{n-1}) = f(t'_0, t'_1, \ldots, t'_{n-1}) \in \mathcal{T}$$

by Lemma 7.4.5, which implies that $f(t_0, t_1, \ldots, t_{n-1}) \sim f(t'_0, t'_1, \ldots, t'_{n-1})$. Next, let R be an n-ary relation symbol such that $R(t_0, t_1, \ldots, t_{n-1}) \in \mathcal{T}$. This implies that $\mathcal{T} \vdash R(t_0, t_1, \ldots, t_{n-1})$, so we have that $\mathcal{T} \vdash R(t'_0, t'_1, \ldots, t'_{n-1})$ by **E5**. Again, by Lemma 7.4.5, $R(t'_0, t'_1, \ldots, t'_{n-1}) \in \mathcal{T}$. ∎

We now define the functions and relations on the domain $C/\sim$ with $\mathcal{T}$ being maximally consistent.

- Let f be an n-ary function symbol from S. Let $c_0, c_1, \ldots, c_{n-1} \in C$. Define the n-ary function $f_\sim : C/\sim \to C/\sim$ by

$$f_\sim([c_0], [c_1], \ldots, [c_{n-1}]) = [c] \Leftrightarrow f(c_0, c_1, \ldots, c_{n-1}) = c \in \mathcal{T}. \qquad (7.23)$$

Since $f_\sim$ is defined on a set of equivalence classes, we must confirm that $f_\sim$ is well-defined. Let $d_0, d_1, \ldots, d_{n-1} \in C$ such that

$$c_0 \sim d_0, c_1 \sim d_1, \ldots, c_{n-1} \sim d_{n-1}.$$

Then, $[c] = [d]$ because by Lemma 7.4.15,

$$f(c_0, c_1, \ldots, c_{n-1}) \sim f(d_0, d_1, \ldots, d_{n-1}).$$

- Let R be an n-ary relation symbol from S and take $c_0, c_1, \ldots, c_{n-1} \in C$. Define the n-ary relation $R_\sim$ on $C/\sim$ by

$$([c_0], [c_1], \ldots, [c_{n-1}]) \in R_\sim \Leftrightarrow R(c_0, c_1, \ldots, c_{n-1}) \in \mathcal{T}. \qquad (7.24)$$

We must check that the relation holds if different constant symbols are used to represent the classes $[c_0], [c_1], \ldots, [c_{n-1}]$. To do this, let $d_0, d_1, \ldots, d_{n-1} \in C$ such that $c_0 \sim d_0, c_1 \sim d_1, \ldots, c_{n-1} \sim d_{n-1}$. Then, we have that

$$([d_0], [d_1], \ldots, [d_{n-1}]) \in R_\sim$$

since by Lemma 7.4.15,

$$R(d_0, d_1, \ldots, d_{n-1}) \in \mathcal{T}.$$

The previous work makes the interpretation of the function and relation symbols of S obvious. However, what about the constant symbols? To find out, take a constant symbol c of S. If $c \in C$, then c should be interpreted as $[c]$, so suppose that $c \notin C$. Because $\vdash \exists x(x = c)$ and $\mathcal{T}$ is maximally consistent,

$$\exists x(x = c) \in \mathcal{T}.$$

At this point make the further assumption that C is a witness set of $\mathcal{T}$. Then, there exists $d \in C$ such that

$$c = d \in \mathcal{T}.$$

This implies that $[c] = [d]$. Therefore, for every $c \in$ S, there exists $c_\sim \in C$ such that $[c] = [c_\sim]$. We can now define the structure.

■ **DEFINITION 7.4.16**

Let $\mathscr{T}$ be a maximally consistent set of S-sentences with witness set C. Let $\sim$ be the relation (7.22). Define $\mathfrak{A}_{\sim}$ to be the S-structure with domain $C/\sim$ and function $\mathfrak{a}$ such that

- $\mathfrak{a}(c) = [c_{\sim}]$ for all constant symbols $c \in$ S,

- $\mathfrak{a}(R) = R_{\sim}$ for all n-ary relation symbols $R \in$ S,

- $\mathfrak{a}(f) = f_{\sim}$ for all n-ary function symbols $f \in$ S.

Since we are investigating consistent sets of sentences, we do not need to involve a particular interpretation for the structures in the proofs of the following results (Theorem 7.1.31). Using Definition 7.4.16 will suffice.

■ **LEMMA 7.4.17**

Suppose that $\mathscr{T}$ is a maximally consistent set of S-sentences with C as a witness set. Then, $\mathfrak{A}_{\sim} \models t = c$ if and only if $t = c \in \mathscr{T}$, for every constant symbol $c \in$ S and S-term t.

PROOF
Assume that $c \in$ S is a constant symbol and let t be an S-term. Since $\mathscr{T}$ is a set of sentences, t cannot have any free variables.

- Let $t = d$ for some constant symbol d from S. Then,

$$\mathfrak{A}_{\sim} \models d = c \Leftrightarrow [d] = [c] \Leftrightarrow d \sim c \Leftrightarrow d = c \in \mathscr{T}.$$

- Let $t = f(t_0, t_1, \ldots, t_{n-1})$ for some n-ary function symbol f and S-terms $t_0, t_1, \ldots, t_{n-1}$ containing no free variables. Assume that

$$\mathfrak{A}_{\sim} \models f(t_0, t_1, \ldots, t_{n-1}) = c.$$

Since $\mathscr{T}$ is maximally consistent,

$$\exists x(t_i = x) \in \mathscr{T}$$

for $i = 0, 1, \ldots, n - 1$. Since C is a witness set, there exists $c_i \in C$ such that

$$t_i = c_i \in \mathscr{T},$$

which is equivalent to

$$t_i \sim c_i. \tag{7.25}$$

Therefore, by induction,

$$\mathfrak{A}_{\sim} \models t_i = c_i.$$

Hence,

$$\mathfrak{A}_{\sim} \models f(c_0, c_1, \ldots, c_{n-1}) = c.$$

This implies that

$$f_\sim([c_0], [c_1], \dots, [c_{n-1}]) = [c],$$

so by (7.23),

$$f(c_0, c_1, \dots, c_{n-1}) = c \in \mathcal{T}.$$

Therefore, by Lemma 7.4.15 and (7.25),

$$f(t_0, t_1, \dots, t_{n-1}) = c \in \mathcal{T}.$$

The converse is left to Exercise 4. ■

■ LEMMA 7.4.18

If $\mathcal{T}$ is a consistent set of S-sentences with C as a witness set, then $p \in \mathcal{T}$ if and only if $\mathfrak{A}_\sim \vDash p\,[I]$ for every S-sentence p.

PROOF

We can assume that $\mathcal{T}$ is maximally consistent (Exercise 5). Let C be a witness set of $\mathcal{T}$. We prove the theorem by induction on formulas.

- Let $t_0 = t_1 \in \mathcal{T}$ for S-terms t_0 and t_1 with no free variables. Since $\mathcal{T}$ is maximally consistent,

$$\exists x(t_0 = x) \in \mathcal{T}$$

and

$$\exists x(t_1 = x) \in \mathcal{T}.$$

Because C is a witness set for $\mathcal{T}$, there exists $c, d \in C$ such that

$$t_0 = c \in \mathcal{T}$$

and

$$t_1 = d \in \mathcal{T}.$$

This implies that $t_0 \sim c$ and $t_1 \sim d$, so because $t_0 \sim t_1$,

$$[c] = [d].$$

Therefore, $\mathfrak{A}_\sim \vDash c = d$, but because $\mathfrak{A}_\sim \vDash t_0 = c$ and $\mathfrak{A}_\sim \vDash t_1 = d$ (Lemma 7.4.17), $\mathfrak{A}_\sim \vDash t_0 = t_1$. The proof of the converse is Exercise 7(a).

- Let $R(t_0, t_1, \dots, t_{n-1}) \in \mathcal{T}$ for some relation symbol R in S and S-terms $t_0, t_1, \dots, t_{n-1}$ with no free variables. As in the first part of this proof, there exist constants $c_i \in C$ ($i = 0, 1, \dots, n-1$) such that $t_i \sim c_i$. Thus, by Lemma 7.4.15, $R(c_0, c_1, \dots, c_{n-1}) \in \mathcal{T}$, and then by the definition of $R_\sim$, we have that

$$([c_0], [c_1], \dots, [c_{n-1}]) \in R_\sim.$$

Therefore,

$$\mathfrak{A}_\sim \vDash R(c_0, c_1, \dots, c_{n-1}),$$

so $\mathfrak{A}_\sim \vDash R(t_0, t_1 \ldots, t_{n-1}) \, [I]$ as in first part. The proof of the converse is Exercise 7(b).

- Let p be an S-sentence. Then, by induction and the maximal consistency of $\mathcal{T}$,

$$\neg p \in \mathcal{T} \Leftrightarrow p \notin \mathcal{T} \Leftrightarrow \mathfrak{A}_\sim \nvDash p \Leftrightarrow \mathfrak{A}_\sim \vDash \neg p \, [I].$$

- Let p and q be S-sentences. Assume $p \to q \in \mathcal{T}$ and $\mathfrak{A}_\sim \vDash p \, [I]$. By induction, $p \in \mathcal{T}$, so by Lemma 7.4.5 and MP, $q \in \mathcal{T}$. Again by induction, $\mathfrak{A}_\sim \vDash q \, [I]$, so $\mathfrak{A}_\sim \vDash p \to q \, [I]$. See Exercise 7(c) for the proof of the converse.

- Let p be an S-formula with at most one free variable x. By induction and the maximal consistency of $\mathcal{T}$,

$$\mathfrak{A}_\sim \vDash \exists x p \Leftrightarrow p\frac{c}{x} \in \mathcal{T} \text{ for some } c \in C \Leftrightarrow \exists x p \in \mathcal{T}. \blacksquare$$

The hard work of this section leads to the fundamental theorem of model theory.

■ THEOREM 7.4.19 [Henkin]

An S-theory $\mathcal{T}$ is consistent if and only if $\mathcal{T}$ has a model.

PROOF

Let $\mathcal{T}$ be a set of S-sentences. Suppose $\mathcal{T}$ is consistent and let C be a set of constant symbols not found in S such that $|C| = |\mathsf{L}(\mathsf{S})|$. By Lemma 7.4.13, there exists a consistent extension $\overline{\mathcal{T}}$ of $\mathcal{T}$ such that the elements of $\overline{\mathcal{T}}$ are $(\mathsf{S} \cup C)$-sentences and C is a witness set for $\overline{\mathcal{T}}$. By Lemma 7.4.18, there exists a model $\mathfrak{A}$ of $\overline{\mathcal{T}}$. Since the sentences of $\mathcal{T}$ do not contain any of the constants of C, the reduct of $\mathfrak{A}$ to S is a model of $\mathcal{T}$. To see this, let $\mathfrak{B}$ be the indicated reduct and proceed by induction on formulas.

- Let t_0 and t_1 be S-terms such that $(t_0 = t_1) \in \mathcal{T}$. These terms are also $(\mathsf{S} \cup C)$-terms, so since $\mathcal{T} \subseteq \overline{\mathcal{T}}$, we have that $\mathfrak{A} \vDash t_1 = t_2$. Thus, there is an $(\mathsf{S} \cup C)$-interpretation I such that $\mathfrak{A} \vDash t_1 = t_2 \, [I]$. Hence, $I(t_1) = I(t_2)$. Let I' be the restriction of I to S. Since t_1 and t_2 contain no symbols from C, $I'(t_1) = I'(t_2)$, so $\mathfrak{B} \vDash t_1 = t_2 \, [I']$ because $\mathfrak{B}$ is the reduct of $\mathfrak{A}$ to S. By Theorem 7.1.31, we have that $\mathfrak{B} \vDash t_1 = t_2$.

- That $R(t_0, t_1, \ldots, t_{n-1}) \in \mathcal{T}$ implies $\mathfrak{B} \vDash R(t_0, t_1, \ldots, t_{n-1})$ for any n-ary relation symbol $R \in \mathsf{S}$ and S-terms $t_0, t_1, \ldots, t_{n-1}$ is Exercise 8(a).

- Let $\neg p \in \mathcal{T}$. This implies that $\mathfrak{A} \vDash \neg p$. That is, not $\mathfrak{A} \vDash p$. Since $\mathfrak{B}$ is the reduct of $\mathfrak{A}$ to S, not $\mathfrak{B} \vDash p$. In other words, $\mathfrak{B} \vDash \neg p$.

- That $(p \to q) \in \mathcal{T}$ implies $\mathfrak{B} \vDash p \to q$ for S-sentences p and q is Exercise 8(b).

- Suppose that $\exists x p \in \mathscr{T}$, where $p = p(x)$ is an S-formula. Since C is a witness set and $\mathscr{T}$ is maximally consistent, $p\dfrac{c}{x} \in \mathscr{T}$ for some $c \in C$. By induction,

$$\mathfrak{A} \vDash p\frac{c}{x}.$$

Let I be an interpretation of $\mathfrak{A}$. Then, $\mathfrak{A} \vDash p\,[I_x^{a(c)}]$, so we conclude that $\mathfrak{A} \vDash \exists x p$.

To prove the converse, suppose that $\mathscr{T} \vdash q \wedge \neg q$ for some S-sentence q. By the soundness theorem (7.4.9), $\mathscr{T} \vDash q \wedge \neg q$. Hence, if $\mathfrak{A}$ is an S-structure such that $\mathfrak{A} \vDash \mathscr{T}$, then $\mathfrak{A} \vDash q \wedge \neg q$, which implies that $\mathfrak{A} \vDash q$ and not $\mathfrak{A} \vDash q$. Hence, $\mathscr{T}$ does not have a model. ∎

Henkin's theorem (7.4.19) is used to prove the converse to the soundness theorem (7.4.9).

■ COROLLARY 7.4.20 [Completeness]

Let $\mathscr{T}$ be a consistent set of S-sentences. For any S-sentence p, if $\mathscr{T} \vDash p$, then $\mathscr{T} \vdash p$.

PROOF
Suppose that $\mathscr{T} \nvdash p$. This implies that $\mathscr{T} \cup \{\neg p\}$ is consistent, so by Henkin's theorem (7.4.19), there exists an S-structure $\mathfrak{A}$ such that $\mathfrak{A} \vDash \mathscr{T} \cup \{\neg p\}$. Hence, $\mathscr{T} \nvDash p$. ∎

■ COROLLARY 7.4.21

First-order logic is complete.

The next corollary was first proved by Kurt Gödel (1929). It was his doctoral dissertation.

■ COROLLARY 7.4.22 [Gödel's Completeness Theorem]

An S-sentence is a theorem if and only if it is valid.

■ COROLLARY 7.4.23

For all sets of S-sentences $\mathscr{T}$ and S-sentences p, $\mathscr{T} \vdash p$ if and only if $\mathscr{T} \vDash p$.

The use of models to show consistency is now a common technique in mathematical logic. An early example was Hilbert's use of $\mathbb{R} \times \mathbb{R}$ to show that Euclidean geometry is consistent (Hilbert 1899). That is, $\mathbb{R} \times \mathbb{R}$ serves as the domain of a structure that models the postulates of Euclidean geometry. The proof relies on the consistency of the properties of $\mathbb{R} \times \mathbb{R}$, a fact that is left unproved. A later usage is the model used by Gödel that shows that both CH and the axiom of choice (5.1.10) are consistent with **ZF** (Gödel 1940). Both are examples of **relative consistency** proofs. Gödel assumed the consistency of **ZF**, so he had a model for it. From this he defined another model that satisfies CH and, in addition, the axiom of choice. We represent this by writing

$$\text{Con}(\mathbf{ZF}) \rightarrow \text{Con}(\mathbf{ZFC} + \text{CH}).$$

This is a partial answer to one of the ten then unsolved problems that Hilbert presented at the International Congress of Mathematicians at Paris in 1900. An expanded list was published that year in Germany with an English translation released two years later (Hilbert 1902). Hilbert's intention was to outline the important problems that mathematicians should strive to prove in the twentieth century. The first problem involved the continuum hypothesis. Namely, it should be determined whether

> as regards equivalence there are ... only two assemblages of numbers, the countable assemblage and the continuum.

Hilbert's second problem dealt with the axioms of arithmetic. Specifically, he thought that any of the chosen axioms should not be provable from the others. That is, they should be **independent**. More importantly, mathematicians should seek

> [t]o prove that they [the axioms] are not contradictory, that is, that a finite number of logical steps based upon them can never lead to contradictory results.

In this problem we see the notions of finite proof (Definition 1.2.13) and consistency (Definitions 1.5.1 and 7.4.2), which would influence to the development of model theory some 20–30 years after Hilbert's talk and lead to some interesting results.

Exercises

1. Is $\varnothing$ consistent? Explain.

2. Prove Theorem 7.4.3.

3. Prove that the group axioms (7.1.14) and the ring axioms (Axioms 7.1.36) are consistent.

4. From the proof of Lemma 7.4.17, prove that if $f(t_0, t_1, \ldots, t_{n-1}) = c \in \mathscr{T}$, then $\mathfrak{A}_{\sim} \vDash f(t_0, t_1, \ldots, t_{n-1}) = c$ for all constant symbols c, n-ary function symbols f, and S-terms $t_0, t_1, \ldots, t_{n-1}$.

5. In Lemma 7.4.18, why can $\mathscr{T}$ be assumed to be maximally consistent?

6. Let $\mathfrak{A}$ be an S-structure. Define $\text{Th}(\mathfrak{A})$ to be the set of S-sentences that are satisfied by $\mathfrak{A}$. Prove that $\text{Th}(\mathfrak{A})$ is maximally consistent.

7. Prove the given results from the proof of Lemma 7.4.18.
   (a) If $\mathfrak{A}_{\sim} \vDash t_0 = t_1$, then $t_0 = t_1 \in \mathscr{T}$ for all S-terms t_0 and t_1 with no free variables.
   (b) For all S-terms $t_0, t_1 \ldots, t_{n-1}$ with no free variables, $R(t_0, t_1 \ldots, t_{n-1}) \in \mathscr{T}$ if $\mathfrak{A}_{\sim} \vDash R(t_0, t_1 \ldots, t_{n-1})$.
   (c) $\mathfrak{A}_{\sim} \vDash p \rightarrow q$ implies that $p \rightarrow q \in \mathscr{T}$ for all S-sentences p and q.

8. Show the following from the proof of Henkin's theorem (7.4.19).
   (a) If $R(t_1, \ldots, t_n) \in \mathscr{T}$, then $\mathfrak{B} \vDash R(t_1, \ldots, t_n)$ for any relation symbol $R \in S$ and S-terms $t_1, \ldots, t_n$.

(b) If $p \to q \in \mathcal{T}$, then $\mathfrak{B} \models p \to q$ for all S-sentences p and q.

9. Prove Corollary 7.4.22.

10. Let $\mathcal{T}$ and $\mathcal{U}$ be consistent sets of S-sentences. Prove that if $\mathcal{T} \cup \mathcal{U}$ is inconsistent, then there exists an S-sentence p such that $\mathcal{T} \models p$ but $\mathcal{U} \not\models p$.

11. We say that an S-theory $\mathcal{T}$ is **closed under deductions** if for all S-sentences p, if $\mathcal{T} \vdash p$, then $p \in \mathcal{T}$.
- (a) If a theory is closed under deductions, is the theory maximally consistent?
- (b) Let β be an ordinal and suppose that $\{\mathcal{T}_\alpha : \alpha \in \beta\}$ is a family of S-theories such that for all $\gamma, \delta \in \beta$, if $\gamma \in \delta$, then $\mathcal{T}_\gamma \subseteq \mathcal{T}_\delta$. Prove that if $\mathcal{T}_\alpha$ is consistent for all $\alpha \in \beta$, then $\bigcup_{\alpha \in \beta} \mathcal{T}_\alpha$ is consistent.

12. A theory $\mathcal{T}$ is **finitely axiomatizable** if there exists an S-sentence p such that for all S-sentences q, $\mathcal{T} \models q$ if and only if $p \models q$. Let $\{\mathcal{T}_n : n \in \omega\}$ be a chain of finitely axiomatizable S-theories such that $m \leq n$ implies that $\mathcal{T}_m \subseteq \mathcal{T}_n$. Suppose that for all $n \in \omega$, there exists a model of $\mathcal{T}_n$ that is not a model of $\mathcal{T}_{n+1}$. Prove that $\bigcup_{n \in \omega} \mathcal{T}_n$ is not finitely axiomatizable.

13. Suppose that $\mathcal{T}_1$ and $\mathcal{T}_2$ are S-theories. Let $\mathfrak{A}$ and $\mathfrak{B}$ be S-structures. Prove the following.
- (a) If $\mathcal{T}_1 \subseteq \mathcal{T}_2$, then every model of $\mathcal{T}_2$ is a model of $\mathcal{T}_1$.
- (b) If $\mathfrak{A} \subseteq \mathfrak{B}$, then $\mathrm{Th}(\mathfrak{B}) \subseteq \mathrm{Th}(\mathfrak{A})$.
- (c) If $\mathfrak{A} \models \mathcal{T}_1 \cup \mathcal{T}_2$, then $\mathfrak{A} \models \mathcal{T}_1$ and $\mathfrak{A} \models \mathcal{T}_2$.
- (d) $\mathfrak{A} \models \mathcal{T}_1$ if and only if $\mathcal{T}_1 \subseteq \mathrm{Th}(\mathfrak{A})$.

7.5 MODELS OF DIFFERENT CARDINALITIES

The natural numbers and their basic operations of addition and subtraction can be defined using the axioms of **ZFC** (Section 5.2). It can then be proved that ω has basically the same algebraic properties as **N**. Another approach to studying the natural numbers is to break down the subject to its most basic parts. Think about addition of natural numbers and how it is first explained. Adding 4 to 3 to obtain 7, for example, means starting at 3 and adding 1 four times in sequence:

$$3 + 1 = 4,$$
$$4 + 1 = 5,$$
$$5 + 1 = 6,$$
$$6 + 1 = 7.$$

Adding 1 simply means moving to the next natural number, so to add any two natural numbers, all one needs is to know the numerals and have the ability to count (compare Definition 5.2.15). Although not efficient, this will do. Multiplication, the other operation, is based on addition. Multiplying 4 by 3 means writing 4 down 3 times and then adding:

$$4 + 4 + 4 = 12.$$

This means that multiplication is also based on the ability to count (compare Definition 5.2.18). This ability is represented by the successor function,

$$Sn = n + 1,$$

which is the basis of arithmetic (compare Definition 5.2.1). Thus, to understand the natural numbers and how they work, we simply need a successor function that satisfies the right rules.

Peano Arithmetic

Using the theory symbols AR (Example 2.1.4), we make the following axioms. They are a minor modification of the axioms for the system of arithmetic found in *Arithmetices principia* by Giuseppe Peano (1889).

■ **AXIOMS 7.5.1**

- **P1.** $\forall x(\neg\, Sx = 0)$

- **P2.** $\forall x \forall y(Sx = Sy \rightarrow x = y)$

- **P3.** For every AR-formula p with free variable x,

$$p(0) \wedge \forall x[p(x) \rightarrow p(Sx)] \rightarrow \forall x p(x).$$

Denote $\{P1, P2, P3\}$ by **P**. If we define

$$\begin{aligned} \mathfrak{p}(S) &= {}^+, \\ \mathfrak{p}(0) &= \varnothing, \end{aligned} \tag{7.26}$$

then Theorem 5.2.2, Theorem 5.2.6, and Corollary 5.2.12 imply that $\mathfrak{P} = (\omega, \mathfrak{p})$ is a model of **P**. Because of the existence of a model for these axioms, which was constructed using **ZFC**, we conclude the following (Theorem 7.4.19).

■ **THEOREM 7.5.2**

The consistency of **P** is a consequence of **ZFC**.

Since the Peano axioms are intended to be the assumptions of number theory, AR is often extended to AR′ (Example 2.1.4) and the sentences of Axioms 7.5.1 broadened to include axioms involving $+$, $\cdot$, and $<$. These axioms serve as the foundation of number theory.

■ **AXIOMS 7.5.3 [Peano]**

- **PA1.** $\forall x(\neg\, Sx = 0)$

- **PA2.** $\forall x \forall y(Sx = Sy \rightarrow x = y)$

- **PA3.** $\forall x(x + 0 = x)$

- **PA4.** $\forall x \forall y[x + Sy = S(x + y)]$

- **PA5.** $\forall x(x \cdot 0 = 0)$

- **PA6.** $\forall x \forall y(x \cdot Sy = x \cdot y + x)$

- **PA7.** $\forall x(\neg\, x < 0)$

- **PA8.** $\forall x \forall y[x < Sy \leftrightarrow (x < y \vee x = y)]$

- **PA9.** $\forall x \forall y(x < y \vee x = y \vee y < x)$

- **PA10.** For every AR$'$-formula p with free variable x,

$$p(0) \wedge \forall x[p(x) \rightarrow p(Sx)] \rightarrow \forall x p(x).$$

Let **PA** denote the Peano axioms (7.5.3). We call the set of consequences of the Peano axioms **Peano arithmetic**. To find a model for these axioms, extend the function $\mathfrak{p}$ (7.26) to $\mathfrak{p}'$ so that $\mathfrak{p}'(+)$ is the addition of Definition 5.2.15, $\mathfrak{p}'(\cdot)$ is the multiplication of Definition 5.2.19, and using the order of Definition 5.2.10,

$$\mathfrak{p}'(<) = \{(m, n) \in \omega \times \omega : m \in n\}.$$

That $\mathfrak{p}'(<)$ satisfies **PA7–PA9** is left to Exercise 2. Then, as before, $\mathfrak{P}' = (\omega, \mathfrak{p}')$ is a model of the Peano axioms, which is an expansion of the model $\mathfrak{P}$, and we can again apply Theorem 7.4.19 to obtain a consistency result.

■ THEOREM 7.5.4

The consistency of **PA** is a consequence of **ZFC**.

We call $\mathfrak{P}'$ and any AR$'$-structure isomorphic to it a **standard model** of Peano arithmetic. The the elements of the domain of any standard model are called **standard numbers**. Any model of Peano arithmetic that is not isomorphic to $\mathfrak{P}'$ is called a **nonstandard** model of Peano arithmetic.

Observe that **P** and **PA** are sets of axioms separate from **ZFC**. However, because of the work of Section 5.2, both **P** and **PA** can be viewed as consequences of **ZFC**. Specifically, if **P1** is replaced by

$$\forall x(x^+ \neq \varnothing), \tag{7.27}$$

P2 is replaced by

$$\forall x \forall y(x^+ = y^+ \rightarrow x = y), \tag{7.28}$$

and **P3** is replaced by

$$p(\varnothing) \wedge \forall x[p(x) \rightarrow p(x^+)] \rightarrow \forall x p(x) \tag{7.29}$$

for every ST-formula p with free variable x, then

$$\mathbf{ZFC} \models \{(7.27), (7.28), (7.27)\}.$$

That is, a copy of **P** is found among the consequences of **ZFC**. Also, using Definitions 5.2.15 and 5.2.18, it can be shown that a copy of Peano arithmetic is found among
the consequences of **ZFC**.

We will not repeat all our number theory work in Peano arithmetic, although it is
possible to do so. Instead, we give a sample of some basic results to illustrate that
the two systems are the same. Because of the work in Section 5.2, the addition and
multiplication of Peano arithmetic are commutative, associative, satisfy the distributive
and cancellation laws, and have identities in the standard model. That these properties
hold in every model of Peano arithmetic is a separate issue, yet their proofs are similar
to the work of Section 5.2.

■ **THEOREM 7.5.5**

The following AR′-sentences are theorems of **PA**.

- Associative Laws
$$\forall x \forall y \forall z [x + (y + z) = (x + y) + z]$$
$$\forall x \forall y \forall z [x \cdot (y \cdot z) = (x \cdot y) \cdot z]$$

- Commutative Laws
$$\forall x \forall y (x + y = y + x)$$
$$\forall x \forall y (x \cdot y = y \cdot x)$$

- Additive Identity
$$\forall x (0 + x = x)$$

- Multiplicative Identity
$$\forall x (S0 \cdot x = x)$$

- Distributive Law
$$\forall x \forall y \forall z [x \cdot (y + z) = x \cdot y + x \cdot z]$$

PROOF
We prove the first associative law and the multiplicative identity property, leaving
the others to Exercise 9.

- Define
$$p(x) := \forall z [x + (y + z) = (x + y) + z].$$

By **PA3**, we have $p(0)$ because

$$x + (y + 0) = (x + y) + 0.$$

Next, assume $p(k)$. Since S is a function symbol, **PA4** yields

$$S[x + (y + k)] = S[(x + y) + k],$$
$$x + S(y + k) = (x + y) + Sk,$$
$$x + (y + Sk) = (x + y) + Sk.$$

Therefore, $p(Sk)$, so the associative law for multiplication holds by **PA 10**.

- Assuming that the commutative laws have already been proved, **PA6**, **PA5**, and **PA3** imply that

$$S0 \cdot x = x \cdot S0 = x \cdot 0 + x = 0 + x = x + 0 = x,$$

so $S0 \cdot x = x$ by **E3** (Axioms 5.1.1). ∎

To ease our notation when dealing with numbers other than $1 = S0$, define

$$S^0 x = x \tag{7.30}$$

and for all $n \in \omega \setminus \{0\}$,

$$S^n = \underbrace{SS \dots S}_{n \text{ times}} x. \tag{7.31}$$

We will use this notation in the next example.

■ **EXAMPLE 7.5.6**

In the formula,

$$(3 + 2) + 1 = (1 + 3) + 2, \tag{7.32}$$

two properties from Theorem 7.5.5 are being applied. First, we conclude that $(3 + 2) + 1 = 1 + (3 + 2)$ by the commutative law. The associative law is then used to draw the final conclusion. Notice that (7.32) is the standard interpretation of

$$++SSS0SS0S0 = +S0SSS0 + SS0,$$

which, using (7.31), is equivalent to

$$++S^3 0 S^2 0 S^1 0 = +S^1 0 S^3 0 + S^2 0.$$

Also, the distributive law is applied in the formula

$$2 \cdot x + 3 \cdot x = 5 \cdot x,$$

because

$$2 \cdot x + 3 \cdot x = x \cdot 2 + x \cdot 3 = x \cdot (2 + 3) = x \cdot 5 = 5 \cdot x.$$

Given an equation like

$$5 \cdot x + 1 = 11, \tag{7.33}$$

the standard routine is to solve it like this:

$$5 \cdot x + 1 = 11,$$
$$5 \cdot x = 10,$$
$$x = 2.$$

In the first step, -1 was added to both sides, and in the second, $1/5$ was multiplied by both sides. However, only 0 has an additive inverse in Peano arithmetic, and only 1 has a multiplicative inverse. The way around this problem are cancellation laws.

■ **THEOREM 7.5.7 [Cancellation]**

The following AR$'$-sentences are theorems of **PA**.

- $\forall x \forall y \forall z[(x + z = y + z) \rightarrow x = y]$

- $\forall x \forall y \forall z[(x \cdot z = y \cdot z \wedge \neg z = 0) \rightarrow x = y]$.

PROOF

The proof of the cancellation law for multiplication is Exercise 10. For addition, define

$$p(z) := \forall x \forall y(x + z = y + z \rightarrow x = y).$$

First, notice that if $x + 0 = y + 0$,

$$x = x + 0 = y + 0 = y,$$

where the first and last equality hold by **PA3**. Therefore, $p(0)$. For the induction step, suppose $p(k)$. Then, by **PA4** and **PA2**,

$$x + Sk = y + Sk,$$
$$S(x + k) = S(y + k),$$
$$x + k = y + k,$$
$$x = y,$$

where the last step follows by $p(k)$. Therefore, $\forall z p(z)$ by **PA10**. ■

Therefore, in Peano arithmetic, solve (7.33) like this:

$$5 \cdot x + 1 = 11,$$
$$5 \cdot x + 1 = 10 + 1,$$
$$5 \cdot x = 10,$$
$$5 \cdot x = 5 \cdot 2,$$
$$x = 2.$$

The third and last equations follow by cancellation (Theorem 7.5.7).

Compactness Theorem

Having a model that satisfies **P** or **PA** means that basic arithmetic can be done in this model. We know that this happens in the standard model (assuming **ZFC**), so we now want to know whether there are any nonstandard models of Peano arithmetic. To be successful in our search, we turn to some theorems that follow from Henkin's theorem (7.4.19). The first states that the existence of a model rests on the finite. It was first proved by Gödel for countable first-order theories (1930) and by Anatolij Mal'tsev for arbitrary first-order theories (1936).

■ **THEOREM 7.5.8 [Compactness]**

An S-theory $\mathcal{T}$ has a model if and only if every finite subset of $\mathcal{T}$ has a model.

PROOF

Since sufficiency is clear, let $\mathcal{T}$ have the property that every finite subset has a model. This implies that every finite subset of $\mathcal{T}$ is consistent by Henkin's theorem. Therefore, $\mathcal{T}$ is consistent by Theorem 7.4.3, so $\mathcal{T}$ also has a model by Henkin's Theorem. ■

We apply the compactness theorem to find another model of **PA**. Let c be a constant symbol other than 0. For every natural number n, use (7.30) and (7.31) to define the set of $(AR' \cup \{c\})$-sentences

$$\mathcal{F}_n = PA \cup \{S^i 0 < c : i \in \omega \wedge i \leq n\},$$

and let $\mathfrak{A}_n = (\{0, 1, \ldots, n+1\}, \mathfrak{a})$, where

$$\mathfrak{a} \restriction AR' = \mathfrak{p}'$$

and

$$\mathfrak{a}(c) = n + 1.$$

Observe that

$$\mathfrak{A}_n \vDash \mathcal{F}_n,$$

so $\mathcal{F}_n$ is consistent for all $n \in \omega$, which implies that it has a model by Henkin's Theorem (7.4.19). Therefore,

$$\mathcal{F} = \bigcup_{n \in \omega} \mathcal{F}_n$$

has a model by the compactness theorem (7.5.8). Let $\mathfrak{A}$ be the reduct of this model to AR'. Notice that $\mathfrak{A} \vDash PA$, but because it has an element that is interpreted to be greater than every element of its domain, $\mathfrak{A}$ is a nonstandard model of Peano arithmetic (Theorem 7.3.24).

Löwenheim–Skolem Theorems

Now that we know that a nonstandard model of Peano arithmetic exists under **ZFC**, we want to know if there are others. For this, we need a definition. The **power** of an S-structure $\mathfrak{A}$ is denoted by $|\mathfrak{A}|$ and refers to the cardinality of the domain of $\mathfrak{A}$. This means that a model is countable if and only if its power is countable. We introduce a sequence of theorems related to this. They are due to Skolem (1922), Tarski and Vaught (1957), and Leopold Löwenheim (1915).

■ **THEOREM 7.5.9 [Downward Löwenheim–Skolem]**

Every consistent S-theory has a model with power of at most $|L(S)|$.

PROOF

Recall these facts from the proof of Henkin's theorem (7.4.19).

- C is a set of new constant symbols such that $|C| = |L(S)|$.

- $\mathfrak{A}$ is a $(S \cup C)$-structure.

- $\mathfrak{B}$ is the reduct of $\mathfrak{A}$ to S.

We can assume that $\mathfrak{A}$ has the property that every element of the domain of $\mathfrak{A}$ is the interpretation of a constant of C. Then, since $\mathfrak{A}$ and $\mathfrak{B}$ have the same power,

$$|\mathfrak{B}| \leq |L(S \cup C)| = |L(S)|. \ \blacksquare$$

Since the witness set of a single sentence is finite and there are only finitely many symbols in a sentence, $S \cup C$ can be assumed to be finite in the proof of the downward Löwenheim–Skolem theorem (7.5.9). If this is the case, $|L(S \cup C)| = |L(S)| = \aleph_0$, and we have the next corollary.

■ **COROLLARY 7.5.10 [Löwenheim]**

If an S-sentence has a model, it has a countable model.

Since $L(AR')$ is countable, Theorem 7.5.9 also implies the following.

■ **COROLLARY 7.5.11 [Skolem]**

ZFC implies that there is a countable nonstandard model of Peano arithmetic.

In fact, although we do not prove it here, **ZFC** implies that there are $2^{\aleph_0}$ countable nonstandard models of Peano arithmetic.

The title of Theorem 7.5.9 suggests the existence of the following theorem.

■ **THEOREM 7.5.12 [Upward Löwenheim–Skolem]**

If an S-theory has an infinite model, it has a model of cardinality κ for every $\kappa \geq |L(S)|$.

PROOF

Let $\mathcal{T}$ be an S-theory with an infinite model $\mathfrak{A} = (A, \mathfrak{a})$. Take $\kappa \geq |L(S)|$. Choose a set $C = \{c_\alpha : \alpha \in \kappa\}$ of distinct constant symbols not found in S. Extend $\mathcal{T}$ to the $(S \cup C)$-theory $\overline{\mathcal{T}}$ by defining

$$\overline{\mathcal{T}} = \mathcal{T} \cup \{c_a \neq c_b : \alpha \in \beta \in \kappa\}.$$

Let $\mathcal{S}$ be a finite subset of $\overline{\mathcal{T}}$. This implies that there exists

$$C' = \{c_{a_0}, c_{a_1}, \ldots, c_{a_{n-1}}\} \subseteq C$$

such that the constants of the sentences of $\mathcal{S}$ are among the constants of C'. Expand the S-structure $\mathfrak{A}$ to the $(S \cup \{c_{a_0}, c_{a_1}, \ldots, c_{a_{n-1}}\})$-structure $\mathfrak{A}' = (A, \mathfrak{a}')$, where $a' \restriction S = a$ and

$$a'(c_{a_i}) \in A$$

for all $i = 0, 1, \ldots, n - 1$. Since A is infinite, we further assume that

$$a'(c_{a_0}), a'(c_{a_1}), \ldots, a'(c_{a_{n-1}})$$

are distinct. Therefore, $\mathfrak{A}' \vDash \mathscr{S}$, so by the compactness theorem (7.5.8), there exists a model $\mathfrak{B}$ of $\overline{\mathscr{T}}$, and due to the interpretation of the new constants, the power of $\mathfrak{B}$ is κ. Thus, the reduct of $\mathfrak{B}$ to S is a model of $\mathscr{T}$ and has power κ. ■

The upward Löwenheim–Skolem theorem with **ZFC** implies that the Peano axioms have models of all infinite cardinalities, which are nonstandard by definition.

The von Neumann Hierarchy

We constructed a model of **PA** using **ZFC**. If we can find models of **ZFC**, we would have other models of **PA**, plus prove the consistency of **ZFC**. In order to find models of **ZFC**, we begin by searching for models of individual axioms using a definition due to von Neumann (1929). The objective of the definition is to construct a sequence of sets that have the property that every set is in one of the stages of the sequence. However, since von Neumann's definition used functions instead of sets, it was Zermelo (1930) who gave it its more recognizable form. While von Neumann left his base stage empty, Zermelo allowed the first set of the sequence to contain objects, which are called **urelements**, that were not sets yet were allowed to be elements of sets (Exercise 20). These two approaches are combined in the following definition, which is named after von Neumann.

■ **DEFINITION 7.5.13**

Let $V_0 = \varnothing$. Let α be an ordinal.

- $V_{\alpha+1} = \mathbf{P}(V_\alpha)$.

- $V_\alpha = \bigcup_{\beta < \alpha} V_\beta$ if α is a limit ordinal.

This is called the **von Neumann hierarchy**.

As proved in Exercise 22, V_α is a set for every ordinal α by the empty set, union, power set, and replacement axioms (5.1.2, 5.1.6, 5.1.7, 5.1.9). Thus, every element of V_α is a set. For example,

$$V_1 = \{\varnothing\},$$
$$V_2 = \{\varnothing, \{\varnothing\}\},$$
$$V_3 = \{\varnothing, \{\varnothing\}, \{\{\varnothing\}\}, \{\varnothing, \{\varnothing\}\}\}$$
$$\vdots$$

The sets in V_ω are finite and called the **hereditarily finite sets**. There are countably many of these sets because V_n is countable for each $n \in \omega$ (Theorem 6.3.17), while

$|V_{\omega+1}| = 2^{\aleph_0}$. Observe that the cardinality of V_n for each $n \in \omega$ is finite but grows very quickly.

$$|V_0| = 0,$$
$$|V_1| = 1,$$
$$|V_2| = 2,$$
$$|V_3| = 2^2,$$
$$|V_4| = 2^{2^2},$$
$$|V_5| = 2^{2^{2^2}},$$
$$|V_6| = 2^{2^{2^{2^2}}}$$
$$\vdots$$

Before we can use individual stages of von Neumann's hierarchy, we need to know some of their key properties. For this, we start with some lemmas.

■ **LEMMA 7.5.14**

Let α and β be ordinals.

- V_α is a transitive set.

- If $\alpha \subseteq \beta$, then $V_\alpha \subseteq V_\beta$.

PROOF

Let ζ be a limit ordinal containing α and β. Define

$$A = \{\eta \in \zeta : V_\eta \text{ is transitive}\}.$$

Assume that $\text{seg}(A, \delta) \subseteq A$ for the ordinal $\delta \in \zeta$. Let $B \in V_\delta$.

- V_0 is transitive because $V_0 = \varnothing$.

- Suppose that $\delta = \gamma^+$ for some ordinal γ. By definition, $B \in P(V_\gamma)$, so $B \subseteq V_\gamma$. Take $x \in B$. This implies that $x \in V_\gamma$. Because V_γ is transitive, we have that $x \subseteq V_\gamma$. Hence, $x \in P(V_\gamma) = V_\delta$, so $B \subseteq V_\delta$.

- Let δ be a limit ordinal. Then, there exists $\gamma \in \delta$ such that $B \in V_\gamma$. Since V_γ is transitive by hypothesis, $B \subseteq V_\gamma \subseteq V_\delta$.

We conclude that $\delta \in A$. Thus, by transfinite induction (Theorem 6.1.18), $A = \zeta$ and V_α is transitive.

Now, suppose $\alpha \subseteq \beta$ and take $x \in V_\alpha$. Since

$$V_\alpha \in P(V_\alpha) \subseteq V_\beta,$$

$V_\alpha \in V_\beta$. However, the first part shows that V_β is transitive, so $V_\alpha \subseteq V_\beta$. ■

■ LEMMA 7.5.15

If α is an ordinal, then $\alpha \in V_{\alpha^+}$.

PROOF

Let α be an ordinal and take ζ to be a limit ordinal such that $\alpha \in \zeta$ and proceed by transfinite induction.

- Since $V_0 = \varnothing$, we have that $\varnothing \in V_1$.

- Suppose that $\alpha \in V_{\alpha^+}$. Since V_{α^+} is transitive (Lemma 7.5.14), $\alpha \subset V_{\alpha^+}$. Also, $\{\alpha\} \subset V_{\alpha^+}$. Therefore, $\alpha \cup \{\alpha\} \subset V_{\alpha^+}$, so $\alpha^+ \in V_{\alpha^{++}}$.

- Let α be a limit ordinal and assume that $\beta \in V_{\beta^+}$ for all $\beta \in \alpha$. That is, $\beta \subset V_{\beta^+}$ for all $\beta \in \alpha$ since each V_{β^+} is transitive. Hence,

$$\alpha = \bigcup_{\beta \in \alpha} \beta \subseteq \bigcup_{\beta \in \alpha} V_{\beta^+} = V_\alpha,$$

which implies that $\alpha \in V_{\alpha^+}$. ■

The definition of the von Neumann hierarchy along with the fact that every ordinal belongs to a member of the hierarchy suggests that many, if not all, sets also belong to the hierarchy. For this reason, we define the following.

■ DEFINITION 7.5.16

Let V denote the collection of all sets A such that $A \in V_\alpha$ for some ordinal α.

It is the case that all sets belong to the hierarchy. This is the next theorem. Its proof is aided by the use of two terms and a lemma.

- A set A is **grounded** if there exists an ordinal α such that $A \subseteq V_\alpha$.

- The **transitive closure** of A is

$$\mathrm{TC}(A) = \{u : \forall v (A \in v \wedge v \text{ is transitive} \rightarrow u \in v)\}.$$

As the name implies, $\mathrm{TC}(A)$ is a transitive set (Exercise 24).

The proof of the next lemma is left to Exercise 21.

■ LEMMA 7.5.17

Every element of A is grounded if and only if A is grounded.

■ THEOREM 7.5.18

For every set A, there exists an ordinal α such that $A \in V_\alpha$.

PROOF

Suppose that A is a set such that $A \notin V_\alpha$ for all ordinals α. If $A \subseteq V_\beta$ for some ordinal β, then $A \in V_{\beta+}$, which contradicts the hypothesis. Hence, A is not grounded, which implies that $\{A\}$ is not grounded (Lemma 7.5.17). Define

$$B = \{u \in \mathrm{TC}(\{A\}) : u \text{ is not grounded}\}.$$

Since $B \neq \varnothing$, the regularity axiom (5.1.15) implies that there exists $C \in B$ such that $C \cap B = \varnothing$. Let $x \in C$. Since transitive closures are transitive, $x \in \mathrm{TC}(A)$. However, $x \notin B$, so x is grounded. From this we conclude that C is grounded by Lemma 7.5.17, a contradiction. ∎

Therefore, every set is in **V**, but we know by Corollary 5.1.17 that **V** is not a set in that it cannot be built using **ZFC**. However, we sometimes want to refer to such collections even though they are not sets. For this reason the term **class** was introduced, so we call **V** the class of all sets.

We are now ready to use sets from the von Neumann hierarchy to serve as models for axioms from **ZFC** (Section 5.1).

■ **DEFINITION 7.5.19**

Let α be an ordinal. Define the ST-structure $\mathfrak{B}_\alpha = (V_\alpha, \in)$.

Consider

$$\mathfrak{B}_3 = (\{\varnothing, \{\varnothing\}, \{\{\varnothing\}\}, \{\varnothing, \{\varnothing\}\}\}, \in). \tag{7.34}$$

The elements of V_3 are the sets of the model. These elements are equal exactly when they share the same elements (Definition 3.3.7), so

$$\mathfrak{B}_3 \vDash \text{extensionality axiom}.$$

Because the union of any two elements of V_3 is an element of V_3, such as

$$\varnothing \cup \{\varnothing\} = \{\varnothing\}$$

and

$$\{\{\varnothing\}\} \cup \{\varnothing, \{\varnothing\}\} = \{\varnothing, \{\varnothing\}\},$$

we see that

$$\mathfrak{B}_3 \vDash \text{union axiom}.$$

If we take a ST-formula $p(x)$ and $A \in V_3$, then $\{x : x \in A \wedge p(x)\} \in V_3$. Hence,

$$\mathfrak{B}_3 \vDash \text{subset axioms}.$$

Because V_3 is finite,

$$\mathfrak{B}_3 \vDash \text{axiom of choice},$$

and we have that

$$\mathfrak{B}_3 \vDash \text{axiom of regularity}$$

since, for example,

$$\{\varnothing\} \cap \{\{\varnothing\}\} = \varnothing.$$

These results are particular examples of the next general theorem.

■ THEOREM 7.5.20

If α is an ordinal,

- $\mathfrak{B}_\alpha \models$ extensionality axiom,

- $\mathfrak{B}_\alpha \models$ union axiom,

- $\mathfrak{B}_\alpha \models$ subset axioms,

- $\mathfrak{B}_\alpha \models$ axiom of choice,

- $\mathfrak{B}_\alpha \models$ axiom of regularity.

PROOF

Let α be an ordinal and I be an ST-interpretation of $\mathfrak{B}_\alpha$. We check that the extensionality axiom (5.1.4) holds in the model and leave the other parts of the proof to Exercise 25. Take $A, B \in V_\alpha$. Assume that

$$\mathfrak{B}_\alpha \models \forall u (u \in x \leftrightarrow u \in y) \, [(I_x^A)_y^B].$$

That is,

for all $m \in V_\alpha$, $\mathfrak{B}_\alpha \models u \in x \, [((I_x^A)_y^B)_u^m]$ if and only if $\mathfrak{B}_\alpha \models u \in y \, [((I_x^A)_y^B)_u^m]$.

We want to show that $A = B$, so let $a \in A$. Since V_α is transitive (Lemma 7.5.14), $a \in V_\alpha$, which implies that

$$\mathfrak{B}_\alpha \models u \in x \, [((I_x^A)_y^B)_u^a].$$

Therefore,

$$\mathfrak{B}_\alpha \models u \in y \, [((I_x^A)_y^B)_u^a],$$

which implies that $a \in B$. This proves that $A \subseteq B$. A similar proof shows that $B \subseteq A$. Thus, $A = B$, from which follows that

$$\mathfrak{B}_\alpha \models x = y \, [(I_x^A)_y^B].$$

Therefore,

if $\mathfrak{B}_\alpha \models \forall u (u \in x \leftrightarrow u \in y) \, [(I_x^A)_y^B]$, then $\mathfrak{B}_\alpha \models x = y \, [(I_x^A)_y^B]$.

In other words,

$$\mathfrak{B}_\alpha \models \forall u (u \in x \leftrightarrow u \in y) \to x = y \, [(I_x^A)_y^B].$$

Since A and B were arbitrarily chosen,

$$\mathfrak{B}_\alpha \models \forall x \forall y (\forall u [u \in x \leftrightarrow u \in y] \to x = y). \ ■$$

Again, using the ST-structure $\mathfrak{B}_3$ (7.34), we see that the result of pairing two arbitrarily chosen elements of V_3 into a single set might not be an element of V_3. For example,

$$\{\{\varnothing\}, \{\{\varnothing\}\}\} \notin V_3.$$

The same is true regarding power sets. For example,

$$\mathbf{P}(\{\{\varnothing\}\}) = \{\varnothing, \{\{\varnothing\}\}\} \notin V_3.$$

However, both of these sets are elements of V_ω, which suggests that for the pairing (5.1.5) and power set (5.1.7) axioms to hold in stage V_α of the von Neumann hierarchy, α needs to be a limit ordinal.

■ THEOREM 7.5.21

If α is a limit ordinal,

- $\mathfrak{B}_\alpha \vDash$ pairing axiom,

- $\mathfrak{B}_\alpha \vDash$ power set axiom.

PROOF

Let α be a limit ordinal. Take $a, b \in V_\alpha$. Then, there exists $\beta_1, \beta_2 \in \alpha$ such that $a \in V_{\beta_1}$ and $b \in V_{\beta_2}$. Without loss of generality, we can assume that $\beta_1 \subseteq \beta_2$, which implies that $a \in V_{\beta_2}$ by Lemma 7.5.14. Therefore, $\{a, b\} \subseteq V_{\beta_2}$, so

$$\{a, b\} \in \mathbf{P}(V_{\beta_2}) = V_{\beta_2^+} \subseteq V_\alpha.$$

Hence,

$$\mathfrak{B}_\alpha \vDash \forall u \forall v \, \exists x \, \forall w \, (w \in x \leftrightarrow w = u \vee w = v).$$

The proof of the second part of the theorem is left to Exercise 26. ■

Certainly, the empty set will be an element of V_α provided that α is not empty, so the proof of the next theorem is left to Exercise 27. In addition, V_α needs to contain ω to satisfy the infinity axiom.

■ THEOREM 7.5.22

If α is a nonempty ordinal, $\mathfrak{B}_\alpha \vDash$ empty set axiom.

■ THEOREM 7.5.23

If α is an ordinal such that $\omega \in \alpha$, then $\mathfrak{B}_\alpha \vDash$ infinity axiom.

PROOF

Since Lemma 7.5.15 implies that $\omega \in V_{\omega^+}$, by Lemma 7.5.14, we have that

$$\mathfrak{B}_\alpha \vDash \exists x (\{\,\} \in x \wedge \forall u [u \in x \rightarrow \exists y (y \in x \wedge u \in y \wedge \forall v [v \in u \rightarrow v \in y])]). ■$$

Since $\omega \cdot 2 = \omega + \omega$ is a limit ordinal greater than ω, we conclude the following.

■ THEOREM 7.5.24

$$\mathfrak{B}_{\omega \cdot 2} \vDash \mathsf{Z}.$$

The ordinal $\omega + \omega$ requires one of the replacement axioms (5.1.9) to prove its existence (page 317). This means that the proof of the consistency of Z relies on axioms not in

Z, so let us continue to examine the von Neumann hierarchy for a model of **ZFC**. It will not be $\mathfrak{B}_{\omega \cdot 2}$ because not all replacement axioms are true in $\mathfrak{B}_{\omega \cdot 2}$. This is because $\mathfrak{B}_{\omega \cdot 2}$ does not satisfy Theorem 6.1.19. To use a stage of the von Neumann hierarchy to serve as a model for **ZFC**, we need a strongly inaccessible cardinal (Definition 6.5.11), which is the next theorem. We state it without proof.

■ **THEOREM 7.5.25**

$\mathfrak{B}_\kappa \vDash$ **ZFC** if and only if κ is a strongly inaccessible cardinal.

However, the sentence

$$\textit{there exists a strongly inaccessible cardinal} \tag{7.35}$$

cannot be proved or disproved using **ZFC**. This means that (7.35) is independent of the axioms of set theory. It also means that we are ready for the next definition. Do not confuse it with the notion of a complete logic (Definition 7.4.8).

■ **DEFINITION 7.5.26**

An S-theory $\mathscr{T}$ is **complete** if $\mathscr{T} \vdash p$ or $\mathscr{T} \vdash \neg p$ for all S-sentences p, else $\mathscr{T}$ is **incomplete**.

The definition means that **ZFC** is not complete. It turns out that the underlying issue is that **ZFC** satisfies the Peano axioms (7.5.3). The ability to do basic arithmetic guarantees that there exists a sentence that is independent of **ZFC**. This result generalizes to the incompleteness theorems due to Gödel (1931).

■ **THEOREM 7.5.27 [Gödel's First Incompleteness Theorem]**

If the Peano axioms are provable from a consistent theory, the theory is incomplete.

Since **ZFC** is assumed to be consistent and it can be used to deduce the Peano axioms, we conclude that **ZFC** is incomplete. Now suppose that instead of assuming the consistency of **ZFC**, we try to prove that **ZFC** is consistent using its own axioms. Gödel's next theorem proves that we cannot do this, except under one condition.

■ **THEOREM 7.5.28 [Gödel's Second Incompleteness Theorem]**

If a theory proves the Peano axioms and its own consistency, the theory is inconsistent.

Therefore, if **ZFC** could be used to prove (7.35), **ZFC** would prove that it has a model, which would imply that **ZFC** is consistent by Henkin's theorem (7.4.19). Since we believe that **ZFC** is consistent, we conclude that **ZFC** cannot prove the existence of a strongly inaccessible cardinal. If **ZFC** was extended with an axiom that would allow such a proof, there would be another issue with the extension that would prevent it from proving its own consistency, provided that the new theory was consistent.

The statement of the first incompleteness theorem does not explicitly give a mathematical statement that cannot be proved. It was left to later mathematicians to find some. For example, when combined with Gödel's proof of the relative consistency of CH, the proof of Paul Cohen (1963) of

$$\text{Con}(\textbf{ZFC}) \rightarrow \text{Con}(\textbf{ZFC} + \neg\textbf{CH}),$$

shows that both CH and its negation cannot be proved from **ZFC**. This means that the continuum hypothesis is independent of **ZFC**. In general, to prove that a sentence p is independent of a theory $\mathcal{T}$, do two things.

- Find a model $\mathfrak{A}$ such that $\mathfrak{A} \vDash \mathcal{T} \cup \{p\}$. Then, $\mathcal{T} \nvDash \neg p$ by Definition 7.1.20. This implies that $\mathcal{T} \nvdash \neg p$ by Corollary 7.4.23.

- Find another model $\mathfrak{B}$ such that $\mathfrak{B} \vDash \mathcal{T} \cup \{\neg p\}$. From this conclude that $\mathcal{T} \nvDash p$, which implies $\mathcal{T} \nvdash p$.

Using this strategy, other statements have been discovered to be independent of **ZFC**. Some of these are technical set theoretic statements such as Martin's axiom (Martin and Solovay 1970) or the diamond principle (Jensen 1972). There are independent statements in other branches of mathematics as well. An example from group theory is the independence of the Whitehead problem (Shelah 1974), named after the mathematician John H. C. Whitehead (1950).

Exercises

1. Write an AR'-formula equivalent to the given English sentences.
    (a) x is an even number.
    (b) x is an odd number.
    (c) x is a prime.
    (d) x divides y.

2. Prove that the order of Definition 5.2.10 proves the given AR'-sentences.
    (a) $\forall x(\neg\, x < 0)$
    (b) $\forall x \forall y[x < Sy \leftrightarrow (x < y \lor x = y)]$
    (c) $\forall x \forall y(x < y \lor x = y \lor y < x).$

3. Prove that the given AR'-formulas can be proved from Axioms 7.5.3.
    (a) $\forall x(0 < x \lor x = 0)$
    (b) $\forall x[x = 0 \lor \exists y(x = Sy)]$
    (c) $\forall x \forall y(x \cdot y = 0 \rightarrow x = 0 \lor y = 0)$
    (d) $\forall x \forall y(x < y \leftrightarrow Sx < Su)$
    (e) $\forall x \forall y(x < y \lor x = y \lor y < x)$

4. Prove $\textbf{PA} \vDash (x+2) \cdot (x+3) = (x \cdot x + 5 \cdot x) + 6$, where 2, 3, 5, and 6 are understood to mean the appropriate successors of 0.

5. Find examples of the following if possible.
   (a) An S-theory $\mathcal{T}$ such that every two finite models of $\mathcal{T}$ are isomorphic, but there exists two models of $\mathcal{T}$ that are infinite and not isomorphic.
   (b) An S-theory $\mathcal{T}$ such that every two countable models of $\mathcal{T}$ are isomorphic, but there exists two models of $\mathcal{T}$ that are uncountable and not isomorphic.

6. Let $\mathcal{T}$ be an S-theory and p an S-sentence where $\mathcal{T} \vdash p$. Prove that there exists a finite $\mathcal{U} \subseteq \mathcal{T}$ such that $\mathcal{U} \vdash p$.

7. Let $\mathcal{T}$ be an S-theory and p an S-sentence such that $\mathcal{T} \vDash p$ and $p \vDash \mathcal{T}$. Show that there exists a finite subset $\mathcal{U}$ of $\mathcal{T}$ such that $\mathcal{U} \vDash \mathcal{T}$.

8. Prove that the following are equivalent for an S-theory $\mathcal{T}$.
   - $\mathcal{T}$ is consistent.
   - $\mathcal{T}$ has a model.
   - $\mathcal{T}$ has a countable model.
   - Every finite subset of $\mathcal{T}$ is consistent.
   - Every finite subset of $\mathcal{T}$ has a model.
   - Every finite subset of $\mathcal{T}$ has a countable model.

9. Finish the proof of Theorem 7.5.5.

10. Prove that the cancellation law holds for multiplication in Peano arithmetic. See Exercise 7.5.7.

11. If possible, find an AR'-sentence $p \in Th(\mathfrak{P}')$ that is not a consequence of **PA** (Exercise 7.4.6).

12. Demonstrate that there is no finite model of **P** or **PA**.

13. Prove that there exists a model of **P** that is not isomorphic to $\mathfrak{P}$.

14. Prove that there is a countable nonstandard model of Peano arithmetic.

15. The axioms for an ordered field are the ring axioms (7.1.36) plus the following OF-sentences:
   - $\forall x \forall y (x \otimes y = y \otimes x)$,
   - $\exists x \forall y (x \otimes y = y)$,
   - $\forall x [\neg x = 0 \rightarrow \exists y (x \otimes y = 1)]$,
   - $\forall x \forall y (x < y \lor x = y \lor y < x)$.

Find a model for these axioms.

16. Show that there exists a model of the axioms for an ordered field (Exercise 15) such that ω is a subset of the domain of the model and there exists m in the domain so that $n < m$ for every natural number n. In addition, prove that there are infinitely many such m. What does $1/m$ look like?

17. Let $\mathfrak{R} = (\mathbb{R}, 0, +, \cdot, <)$ be a model of the axioms for an ordered field. Prove that there is a countable model of $Th(\mathfrak{R})$. What is the significance of this model?

18. Expand OF to OF $\cup$ $\{E\}$, where E is a binary function symbol. Find a model of the axioms for an ordered field including the following (OF $\cup$ $\{E\}$)-sentences:
   - $\forall x(xE0 = S0)$
   - $\forall x\forall y(xE(Sy) = (xEy)x$

19. Let $\mathcal{T}$ be a theory such that for every $n \in \omega$, there exists $m \in \omega$ such that $m > n$ and $\mathcal{T}$ has a model of power m. Prove that $\mathcal{T}$ has an infinite model.

20. This hierarchy is due to Zermelo. Let V_0 be a set of atoms. For every ordinal α, define $V_{\alpha+} = V_\alpha \cup P(V_\alpha)$, and if α is a limit ordinal, define $V_\alpha = \bigcup_{\beta<\alpha} V_\beta$. Find V_1 and V_2 assuming the given sets of atoms.
   - (a) $V_0 = \varnothing$
   - (b) $V_0 = \{a\}$
   - (c) $V_0 = \{0, 1, 2, 3\}$

21. Prove Lemma 7.5.17.

22. Prove that V_α is a set for every ordinal α.

23. Using Exercise 6.3.22, show that $|V_{\omega+\alpha}| = \beth_\alpha$ for every ordinal α.

24. Let A be a set. Prove that $TC(A)$ is a transitive set.

25. Prove the remaining parts of Theorem 7.5.20.

26. Let α be a limit ordinal. Prove that $\mathfrak{B}_\alpha \vDash \forall x \exists y \forall u(u \in y \leftrightarrow \forall v[v \in u \rightarrow v \in x])$.

27. Prove that $\mathfrak{B}_\alpha \vDash \exists x \forall y \neg(y \in x)$ if $\alpha \neq \varnothing$.

28. Let $\mathfrak{A}$ be an S-structure. Prove that Th($\mathfrak{A}$) is complete.

29. Let κ be a cardinal and $\{\mathcal{T}_\alpha : \alpha \in \kappa\}$ be a chain of complete S-theories such that for all $\gamma \in \delta \in \kappa$ we have that $\mathcal{T}_\gamma \subseteq \mathcal{T}_\delta$. Show that $\bigcup_{\alpha\in\kappa} \mathcal{T}_\alpha$ is complete.

30. Let $\mathcal{S} \subseteq \mathcal{T}$ be S-theories. Prove or show false: If $\mathcal{T}$ is complete, then $\mathcal{S}$ is complete.

APPENDIX

ALPHABETS

Greek Alphabet

Upper	Lower	Name		Upper	Lower	Name
A	α	*alpha*		N	ν	*nu*
B	β	*beta*		Ξ	ξ	*xi*
Γ	γ	*gamma*		O	o	*omicron*
Δ	δ	*delta*		Π	π	*pi*
E	ϵ	*epsilon*		P	ρ	*rho*
Z	ζ	*zeta*		Σ	σ	*sigma*
H	η	*eta*		T	τ	*tau*
Θ	θ	*theta*		Υ	υ	*upsilon*
I	ι	*iota*		Φ	φ	*phi*
K	κ	*kappa*		X	χ	*chi*
Λ	λ	*lambda*		Ψ	ψ	*psi*
M	μ	*mu*		Ω	ω	*omega*

A First Course in Mathematical Logic and Set Theory, First Edition. Michael L. O'Leary.
© 2016 John Wiley & Sons, Inc. Published 2016 by John Wiley & Sons, Inc.

English Alphabet in the Fraktur Font

Upper	Lower	Name		Upper	Lower	Name
𝔄	𝔞	A		𝔑	𝔫	N
𝔅	𝔟	B		𝔒	𝔬	O
ℭ	𝔠	C		𝔓	𝔭	P
𝔇	𝔡	D		𝔔	𝔮	Q
𝔈	𝔢	E		𝔕	𝔯	R
𝔉	𝔣	F		𝔖	𝔰	S
𝔊	𝔤	G		𝔗	𝔱	T
𝔥	𝔥	H		𝔘	𝔲	U
𝔍	𝔦	I		𝔙	𝔳	V
𝔍	𝔧	J		𝔚	𝔴	W
𝔎	𝔨	K		𝔛	𝔵	X
𝔏	𝔩	L		𝔜	𝔶	Y
𝔐	𝔪	M		𝔷	𝔷	Z

REFERENCES

Aristotle (1984). *The Complete Works of Aristotle*, ed. J. Barnes. Princeton, NJ: Princeton University Press.

Boole, G. (1847). *The Mathematical Analysis of Logic : Being an Essay Towards a Calculus of Deductive Reasoning.* Cambridge, UK: Macmillan, Barclay, & Macmillan.

Boole, G. (1854). *An Investigation of the Laws of Thought, on Which Are Founded the Mathematical Theories of Logic and Probabilities.* London: Walton and Maberly.

Boyer, C. B. and U. C. Merzbach (1991). *A History of Mathematics* (2nd ed.). New York: John Wiley & Sons, Inc.

Burali-Forti, C. (1897). Una questione sui numeri transfiniti. *Rendiconti del Circolo Matematico di Palermo 11*(1), 154–164.

Cantor, G. (1874). Ueber eine Eigenschaft des Inbegriffs aller reellen algebraischen Zahlen. *Journal Fur Die Reine Und Angewandte Mathematik 77*, 258–262.

Cantor, G. (1888). Mitteilungen zur Lehre vom Transfiniten. *Zeitschrift für Philosophie und philosophische Kritik 91*, 81–125.

Cantor, G. (1891). Über eine elementare Frage def Mannigfaltigkeitslehre. *Jahresbericht der Deutschen Mathematiker-Vereinigung 1*, 75–78.

Cantor, G. (1932). *Gesammelte Abhandlungen mathematischen und philosophischen Inhalts.* Berlin: Springer-Verlag.

A First Course in Mathematical Logic and Set Theory, First Edition. Michael L. O'Leary.
© 2016 John Wiley & Sons, Inc. Published 2016 by John Wiley & Sons, Inc.

Chang, C. C. and H. J. Keisler (1990). *Model Theory* (3rd ed.). Studies in Logic and the Foundations of Mathematics. Amsterdam: North Holland.

Church, A. (1956). *Introduction to Mathematical Logic*. Princeton, NJ: Princeton University Press.

Ciesielski, K. (1997). *Set Theory for the Working Mathematician*. London Mathematical Society Student Texts. Cambridge, UK: Cambridge University Press.

Cohen, P. J. (1963). The independence of the continuum hypothesis. *Proceedings of the National Academy of Sciences of the United States of America 50*(6), 1143–1148.

Copi, I. M. (1979). *Symbolic Logic* (5th ed.). New York: Macmillan Publishing.

Dauben, J. W. (1979). *George Cantor: His Mathematics and Philosophy of the Infinite*. Princeton, NJ: Princeton University Press.

De Morgan, A. (1847). *Formal Logic: or, the Calculus of Inference, Necessary and Probable*. London: Taylor and Walton.

Dedekind, R. (1893). *Was sind und was sollen die Zahlen?* Braunschweig: F. Vieweg.

Dedekind, R. (1901). *Essays on the Theory of Numbers*, trans. W. W. Beman. Chicago: Open Court.

Descartes, R. (1985). *The Philosophical Writings of Descartes*, eds. J. Cottingham, R. Stoothoff, and D. Murdoch. Cambridge, UK: Cambridge University Press.

Doets, K. (1996). *Basic Model Theory*. Stanford: CSLI Publications.

Drake, F. R. (1974). *Set Theory: An Introduction to Large Cardinals*, Volume 76 of *Studies in Logic and Foundations of Mathematics*. Amsterdam: North-Holland.

Ebbinghaus, H., J. Flum, and W. Thomas (1984). *Mathematical Logic*. Undergraduate Texts in Mathematics. New York: Springer-Verlag.

Ebbinghaus, H. and V. Peckhaus (2007). *Ernst Zermelo: An Approach to His Life and Work*. Berlin: Springer.

Eklof, P. C. (1976). Whitehead's problem is undecidable. *American Mathematical Monthly 83*, 775–788.

Eklof, P. C. and A. H. Mekler (2002). *Almost Free Modules: Set-theoretic Methods* (revised ed.). Amsterdam: North-Holland.

Enderton, H. B. (1977). *Elements of Set Theory*. San Diego: Academic Press.

Euclid (1925). *The Elements*, ed. T. Heath. Reprint, New York: Dover, 1956.

Ewald, W. B. (2007). *From Kant to Hilbert: A Source Book in the Foundations of Mathematics*. Oxford: Oxford University Press.

Fibonacci and L. E. Sigler (2002). *Fibonacci's Liber Abaci: A Translation into Modern English of Leonardo Pisano's Book of Calculuation*. Sources and Studies in the History of Mathematics and Physical Sciences. New York: Springer.

Fraenkel, A. A. (1922). Zu den Grundlagen der Cantor-Zermeloschen Mengenlehre. *Mathematische Annalen 86*, 230–237.

Fraleigh, J. B. (1999). *A First Course in Abstract Algebra* (6th ed.). Reading, MA: Addison-Wesley.

Frege, G. (1879). *Begriffsschrift, Eine Der Arithmetischen Nachgebildete Formelsprache Des Reinen Denkens*. Halle a/S.

Frege, G. (1884). *Die Grundlagen Der Arithmetik: Eine Logisch Mathematische Untersuchung Über Den Begriff Der Zahl*. Breslau: W. Koebner.

Frege, G. (1893). *Grundgesetze Der Arithmetik : Begriffsschriftlich Abgeleitet*. Jena: H. Pohle.

Gödel, K. (1929). *Über die Vollständigkeit des Logikkalküls*. Ph. D. thesis, University of Vienna.

Gödel, K. (1930). Die Vollständigkeit der Axiome des logischen Functionenkalküls. *Monatshefte für Mathematik und Physik 37*, 349–360.

Gödel, K. (1931). Über formal unentscheidbare Sätze der Principia Mathematica und verwandter Systeme, I. *Monatshefte für Mathematik und Physik 38*, 173–198.

Gödel, K. (1940). *Consistency of the Continuum Hypothesis*. Princeton, NJ: Princeton University Press.

Halmos, P. R. (1960). *Naive Set Theory*. New York: Springer-Verlag.

Henkin, L. (1949). *The Completeness of Formal Systems*. Ph. D. thesis, Princeton University, Princeton, NJ.

Herrlich, H. (2006). *Axiom of Choice*. Lecture Notes in Mathematics. Berlin: Springer.

Hilbert, D. (1899). *Grundlagen der Geometrie*. Leipzig: Teubner.

Hilbert, D. (1902). Mathematical problems, trans. M. F. W. Newson. *Bulletin of the American Mathematical Society 8*, 437–479.

Hodges, W. (1993). *Model Theory*. Encyclopedia of Mathematics and its Applications. Cambridge, U.K.: Cambridge University Press.

Hofstadter, D. R. (1989). *Gödel, Escher, Bach: an Eternal Golden Braid*. Reprint, New York: Vintage Books.

Jech, T. (1973). *The Axiom of Choice*, Volume 75 of *Studies in Logic and the Foundations of Mathematics*. Amsterdam: North-Holland.

Jech, T. (2003). *Set Theory: The Third Millennium Edition*. Springer Monographs in Mathematics. Berlin: Springer.

Jensen, R. B. (1972). The fine structure of the constructible hierarchy. *Annals of Mathematical Logic 4*(3), 229–308.

Kaye, R. (1991). *Models of Peano Arithmetic*, eds. D. S. Angus Macintyre, John Shepherdson. Oxford, UK: Clarendon Press.

Kline, M. (1972). *Mathematical Thought from Ancient to Modern Times*. New York: Oxford University Press.

Kneale, W. and M. Kneale (1964). *The Development of Logic*. Oxford, UK: Clarendon Press.

König, J. (1905). Zum Kontinuum-problem. *Mathematische Annalen 60*(2), 177–180.

König, J. (1906). Sur la théorie des ensembles. *Comptes rendus hebdomadaires des séances de l'Académie des sciences 143*, 110–112.

Kunen, K. (1990). *Set Theory: An Introduction to Independence Proofs*, Volume 102 of *Studies in Logic and the Foundations of Mathematics*. Amsterdam: North-Holland.

Kuratowski, K. (1921). Sur la notion de l'order dans la théorie des ensembles. *Fundamenta Mathematicae 2*, 161–171.

Kuratowski, K. (1922). Sur l'opération a de l'analysis situs. *Fundamenta Mathematicae 3*, 182–199.

Leary, C. C. (2000). *A Friendly Introduction to Mathematical Logic*. Upper Saddle River, NJ: Prentice Hall.

Leibniz, G. W. (1666). *Dissertatio de arte combinatoria, in qua ex arithmeticae fundamentis complicationum ac transpositionum doctrina novis praeceptis extruitur, & usus ambarum per universum scientiarum orbem ostenditur; nova etiam artis meditandi, seu logicae inventionis semina sparguntur*. Lipsiae, apud Joh. Simon Fickium et Joh. Polycarp. Seuboldum, Literis Spörelianis.

Levy, A. (1979). *Basic Set Theory*. Perspectives in Mathematical Logic. Berlin: Springer-Verlag.

Löwenheim, L. (1915). Über Möglichkeiten im Relativkalkül. *Mathematische Annalen 76*(4), 447–470.

Łukasiewicz, J. (1930). *Elementy logiki matematycznej*. Warsaw: s.n.

Łukasiewicz, J. (1951). *Aristotle's Syllogistic From the Standpoint of Modern Formal Logic*. Oxford, UK: Clarendon Press.

Mal'tsev, A. I. (1936). Untersuchungen aus dem Gebiete der mathematischen Logik. *Matematicheskii Sbornik 1*(43), 323–336.

Martin, D. A. and R. M. Solovay (1970). Internal Cohen extensions. *Annals of Mathematical Logic 2*(2), 143–178.

Mirimanoff, D. (1917). Les antinomies de Russel et de Burali-Forti et le problème fondamental de la théorie des ensembles. *L'Enseignement Mathématique 19*, 37–52.

Pascal, B. (1665). *Traité du triangle arithmetique, avec quelques autres petits traitez sur la mesme matrière*. Paris: chez G. Desprez.

Peano, G. (1889). *Arithmetices principia: nova methodo*. Augustae Taurinorum [Torino]: Fratres Bocca.

Reid, C. (1996). *Hilbert*. New York: Copernicus.

Rosen, K. H. (1993). *Elementary Number Theory and Its Applications* (3rd ed.). Reading, MA: Addison-Wesley.

Rubin, H. and J. E. Rubin (1985). *Equivalents of the Axiom of Choice, II*, Volume 116 of *Studies in Logic and the Foundations of Mathematics*. Amsterdam: North-Holland.

Rubin, J. E. (1973). The compactness theorem in mathematical logic. *Mathematics Magazine 46*(5), 261–265.

Shelah, S. (1974). Infinite abelian groups, Whitehead problem and some constructions. *Isreal Journal of Mathematics 18*(3), 243–256.

Skolem, T. (1922). Einige Bemerkungen zur axiomatischen Begründung der Mengenlehre. In *Proceedings of the 5th Scandinavian Mathematicians' Congress in Helsinki*, pp. 217–32.

Suppes, P. (1972). *Axiomatic Set Theory*. New York: Dover Publications.

Tarski, A. (1935). Der wahrheitsbegriff in den formalisierten sprachen. *Studia Philosophica 1*, 261–405.

Tarski, A. (1983). *Logic, Semantics, Metamathematics: Papers from 1923 to 1938*, trans. J. H. Woodger. Indianapolis: Hackett Publishing Company.

Tarski, A. and R. L. Vaught (1957). Arithmetical extensions of relational systems. *Compositio Mathematica 13*, 81–102.

van Dalen, D. (1994). *Logic and Structure* (3rd ed.). Berlin: Springer-Verlag.

van Dalen, D., H. C. Doets, and H. de Swart (1978). *Sets: Naive, Axiomatic and Applied*, Volume 106 of *International Series in Pure and Applied Mathematics*. Oxford, UK: Pergamon Press.

Van Heijenoort, J. (1971). *Frege and Gödel: Two Fundamental Texts in Mathematical Logic*. Cambridge MA: Harvard University Press.

Van Heijenoort, J. (1977). *From Frege to Gödel: A Source Book in Mathematical Logic, 1879–1931*. Source Books in History of Sciences. Cambridge, MA: Harvard University Press.

Venn, J. (1894). *Symbolic Logic* (2nd ed.). London: Macmillan.

von Neumann, J. (1923). Zur Einführung der transfiniten Zahlen. *Acta literarum ac scientiarum Regiae Universitatis Hungaricae Francisco-Josephinae, Sectio scientiarum mathematicarum 1*, 199–208.

von Neumann, J. (1928). Über die Definition durch transfinite Induktion und verwandte Fragen der allgemeinen Mengenlehre. *Mathematische Annalen 99*, 373–391.

von Neumann, J. (1929). Über eine Widerspruchsfreiheitsfrage der axiomatischen Mengenlehre. *Journal für die reine und angewandte Mathematik 160*, 227–241.

Whitehead, A. N. and B. Russell (1910). *Principia Mathematica* (2nd ed.). Cambridge, UK: Cambridge University Press.

Whitehead, J. H. C. (1950). Simple homotopy types. *American Journal of Mathematics 72*(1), 1–57.

Wussing, H. (1984). *The Genesis of the Abstract Group Concept*, ed. H. Grant, trans. A. Shenitzer. Cambridge, MA: MIT Press.

Zermelo, E. (1908). Untersuchungen über die grundlagen der mengenlehre. *Mathematische Annalen 65*, 261–281.

Zermelo, E. (1930). Über Grenzzahlen und Mengenbereiche. Neue Untersuchungen über die Grundlagen der Mengenlehre. *Fundamenta mathematicae 16*, 29–47.

Zorn, M. (1935). A remark on method in transfinite algebra. *Bulletin of the American Mathematical Society 41*, 667–670.

INDEX

Abel, Niels, 342
Abelian group, 342
Absolute value, 116
Abstraction, 122
Addition, inference rule, 25
Addition, matrix, 355
Additive identity, 199, 412
Additive inverse, 199
Aleph, 313
Algebraic, 315
Alphabet, 5
    first-order, 68
    second-order, 73
And, 4
Antecedent, 4
Antichain, 183
Antisymmetric, 177
Arbitrary, 88
Argument form, 21
Assignment, 7
Associative, 34, 139, 199, 412
Assumption, 46
Asymmetric, 177
Atom, 3, 6
Automorphism, 380

Axiom scheme, 228
Axiom(s), 24
    choice, 231, 235
    empty set, 227
    equality, 226
    extensionality, 227
    foundation, 234
    Frege–Łukasiewicz, 24
    group, 340
    paring, 228
    power set, 228
    regularity, 234
    replacement, 230
    ring, 353
    separation, 228
    subset, 229
    union, 228
    Zermelo, 231
Axiomatizable, finitely, 409

Basis case, 258
Bernstein, Felix, 301
Beth, 316
Biconditional, 5
Biconditional proof, 107

435

A First Course in Mathematical Logic and Set Theory, First Edition. Michael L. O'Leary.
© 2016 John Wiley & Sons, Inc. Published 2016 by John Wiley & Sons, Inc.

Printed and bound by CPI Group (UK) Ltd, Croydon, CR0 4YY

16/04/2025

14658532-0005